Perspectives of Mesoscopic Physics

Dedicated to Yoseph Imry's 70th Birthday

Perspectives of Mesoscopic Physics

Dedicated to Yoseph Imry's 70th Birthday

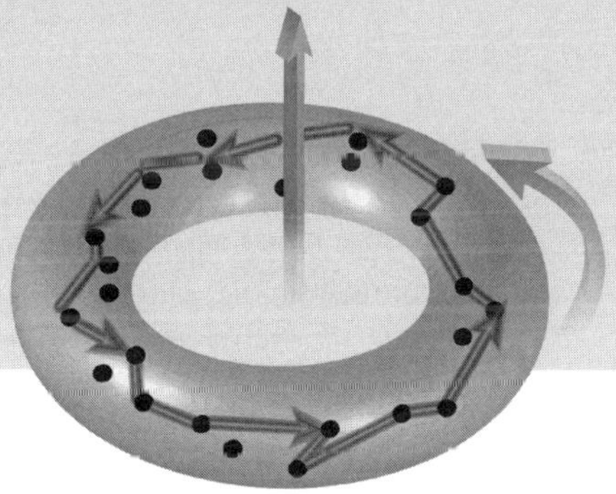

editors

Amnon Aharony
Ora Entin-Wohlman

Ben Gurion University, Israel

World Scientific

NEW JERSEY • LONDON • SINGAPORE • BEIJING • SHANGHAI • HONG KONG • TAIPEI • CHENNAI

Published by

World Scientific Publishing Co. Pte. Ltd.
5 Toh Tuck Link, Singapore 596224
USA office: 27 Warren Street, Suite 401-402, Hackensack, NJ 07601
UK office: 57 Shelton Street, Covent Garden, London WC2H 9HE

British Library Cataloguing-in-Publication Data
A catalogue record for this book is available from the British Library.

PERSPECTIVES OF MESOSCOPIC PHYSICS
Dedicated to Yoseph Imry's 70th Birthday

ISBN-13 978-981-4299-43-5
ISBN-10 981-4299-43-X

Printed by FuIsland Offset Printing (S) Pte Ltd. Singapore

Preface

Professor Yoseph (Joe) Imry was one of the initiators and has been one of the leaders of **mesoscopic physics** for several decades. This book contains articles by leading (theoretical and experimental) scientists working in **nanoscience** and in related fields. Most of the contributions, which contain both reviews of the state of the art and new results, summarize invited talks given at two conferences dedicated to Imry's 70th birthday: "the 101st Statistical Mechanics Conference", held at Rutgers University on May 10–12, 2009, and "Perspectives of Mesoscopic Physics", held at the Weizmann Institute of Science on May 31–June 1, 2009. The book covers a broad range of active research in nanoscience, including topics like quantum interference and the Aharonov–Bohm effect, decoherence and dephasing, quantum noise, electron correlations, the Kondo effect, the quantum Hall effect, nano superconductors, nano magnets, nano electromechanical devices, molecular junctions, ordered and disordered rings, nonequilibrium, glassy behavior and more. We expect this book to be of use to researchers and to graduate students in the broad fields of mesoscopic physics and nanoscience.

We are grateful to all the authors who agreed to contribute to this book, and were very cooperative at the various stages of its growth. We are also grateful to the staff of World Scientific, especially to Ms. Fang Zhang and Mr. Rajesh Babu, who helped with the technicalities of the book's production. The figure on the cover was kindly contributed by Hélène Bouchiat.

Amnon Aharony and Ora Entin-Wohlman
Beer Sheva, January 2010

Contents

About Yoseph Imry

Yoseph Imry was born in 1939 in Tel Aviv. He is married with two daughters and four grandchildren. He obtained his M. Sc. "With Special Distinction" at the Hebrew University, Jerusalem, in 1961 and his Ph. D. in 1966 from the Feinberg Graduate School, Weizmann Institute of Science. His army service was in 1961-2 (First Leutenant, armored corps). During 1962-7 he was a research scientist, Israeli AEC, and in 1967-9 a research associate at Cornell. From 1969 to 1986 he was a Senior Lecturer, Associate Professor and Professor at Tel-Aviv University. From 1986 to 2007 he was a Professor of Physics, Weizmann Inst. of Science, (Max-Planck Professorial Chair, 1988-2009), from 2007 he is a Professor Emeritus, with special research rights. From 2001 to 2004 he was the head of the Goldschleger Center for Nanophysics and from 1995 to 2002, the head of the Albert Einstein Minerva Center for Theoretical Physics, at the Weizmann Institute of Science. He had various visiting appointments at IBM Research, Wissenschaftskolleg Berlin, Brookhaven National Laboratory, the University of California and Yale University. This includes being a Distinguished Science Advisor (1985-1992) at IBM and organizing two programs at ITP (later KITP), UCSB: Mesoscopic Physics in 1991 and Nanophysics in 2001, as well as serving as an advisor on Nanoscience to the president of Ben-Gurion University (2002-7).

Among his awards and honors: he is a Foreign Associate of the *US National Academy of Sciences* (2008), a member of *the Israel Academy of Sciences and Humanities* (2002), an honorary member, *the Indian Academy of Sciences* (2007), titular Member, the *European Academy of Arts, Sciences and Humanities* (Paris, 2003) and a member of *the European Academy of Sciences and Arts* (Salzburg, 1994). He was the incumbent of *the Chaire Internationale de Recherche "Blaise Pascal",* held at the Ecole Normale Superieure, Paris (1999-2001) and the *Lorentz Chair*, University of Leiden (fall, 1996). He is the recipient of the *EMET Prize* (2006), *Israel Prize*, Jerusalem (2001) *Rothschild Prize* (1996), *Weizmann Prize* and an *IBM Research Award* (1987), *Landau Prize* (1983) and the *Humboldt Research Award* (1993).

He served on the editorial board of *Comments on Condensed Matter Physics* (1985-1992), a Divisional Associate Editor for Condensed matter Physics, *Physical Review Letters*, (1996-1999) and the Editorial Council, *Annals of Physics* (NY) (1996-2002). He has participated in about ten academic visiting and advisory committees. He is a member, *Israel Physical Society* and a fellow, *American Physical Society*. He was a member, *the Israel National Nanotechnology Initiative* (2002-6) and is a member, *Council for Peace and Security, an Association of Experts*, (Israel 1997-).

Introduction

I would like to convey my deep gratitude to all the friends and colleagues who contributed to this heart-warming volume. It truly embraces a number of exciting subjects and problems and will offer outstanding reading material for both novices and experts in related fields. Special thanks are due to my long-time friends Amnon Aharony and Ora Entin-Wohlman for the job of editing the book so well.

One of the benefits of being so long in the field is the interactions with a large number of interesting and pleasant people. Almost all of them have become personal and even family friends. Starting with my own students and postdocs, let me first mention (in chronological order) Benny Gavish, Uzy Efron, Koby Achiam, Ora Entin-Wohlman, Moshe Schwartz, Uri Sivan, Claudio Hartsztein, Ady Stern, Avi Shalgi, Amir Yacoby, Niv Shmuely, Nathan Argaman, Uri Gavish, Moshe Schechter and, most recently, Ariel Amir and Hamutal Bary-Soroker. Then there have been those students who were not mine but with whom I interacted strongly, such as Amnon Aharony, Zvi Ovadyahu, Aharon Kapitulnik and Dror Orgad. Among the "grown-ups", I'll mention Leon Gunther, David Bergman, Yehiel Disatnik, Mark Nelkin, Michael Fisher, Alan Evans, David Mukamel, Douglas Scalapino, Pedro Montano, Dan Hone, Eva Andrei, Natan Andrei, Myron Strongin, Daniel Prober, Paul Marcus, Erling Pytte, Geoffrey Grinstein, Alex Malozemoff, Bernard Barbara, Chang Tsuei, Richard Webb, Frank Stern, Alan Fowler, Norman Shiren, Norton Lang, Markus Büttiker, Edouard Brézin, Joel Lebowitz, Isaac Goldhirsch, Zeev Schuss, Hidetoshi Fukuyama, Akira Tonomura, Jean-Louis Pichard, Hélène Bouchiat, Gilles Montambaux, Myriam Sarachik, David Khmel'nitskii, David Thouless, Carlo Beenakker, Hans Mooij, Eytan Domany, Uzy Smilansky, Adam Schwimmer, Zeev Vager, Yigal

Talmi, Zvi Lipkin, Moty Heiblum, Israel Bar-Joseph, Eli Zeldov, Danny Shahar, Alexander Finkel'stein, Hans Weidenmüller, Peter Wölfle, Gerd Schön, Christoph Bruder, Jan on Delft, Sebastien Balibar, James Langer, Walter Kohn, Peter Silvestrov, Oriol Bohigas, Serge Haroche, Claude Cohen-Tanoudji, Philip Stamp, Shahal Ilany, Yuval Oreg, Ehud Altman, Kathryn Moller, Michael Aizenman and especially the late Solly Cohen, Israel Pelah, Yosef Dotan, Shang-keng Ma, Ted Schultz, Rolf Landauer, Albert Schmid, Bernhard Mühlschlegel, Arkadi Aronov and Yehoshua Levinson, all of whom I remember with great affection. Courses and seminars by Julio Racah, Felix Bloch, Avraham Many and Solly Cohen at the Hebrew University as well as many pleasant and instructive interactions with Israel Pelah, shaped me as a Physicist.

Yoseph Imry — Review of Research, 1961–2010

I'll start by discussing my present interests, then review older work. Looking at the material, one can see two trends: coming back to the same problem, sometimes more than once, with a new wrinkle; and often basing the work on very simple common-sense observations. The latter were usually not very well received by the community, until confirmed by more complicated calculations, or by experiment.

1. My present research interests are mostly within the Physics of mesoscopic and disordered systems.

Mesoscopic physics is a relatively new branch of condensed-matter physics, of which I have been one of the originators. It is concerned with how the behavior of systems whose size is in between micro- and macroscopic, crosses-over between these two regimes. These systems can be handled and addressed by more-or-less usual macroscopic methods, but their behavior may still be dominated by the unusual effects of quantum mechanics. Among the interesting new effects found in mesoscopics, there are: the quantization of various conductances, with the universal conductance unit e^2/h (with e being the electron's charge and h Planck's quantum constant), Aharonov–Bohm (AB) oscillations in the conductance of mesoscopic rings with flux period h/e and "persistent currents" flowing without dissipation in such systems in the normal (non-superconducting) resistive state. A nontrivial issue is to explain the magnitude of these currents, on which interesting advances have very recently been made with Ora Entin-Wohlman and our joint student Hamutal Bary-Soroker. This may have to do with fluctuations of mesoscopic superconductivity, which is another exciting subject. The subject of fluctuations and noise, including the case of systems out of equilibrium, is another direction of my research. I hope to be able to relate some of these results with other recent findings in out-of-equilibrium Physics.

Mesoscopic Physics is very much the forerunner of what is called now "Nanophysics". The latter is the small size end of the former. A particular example is the problem of so-called "molecular electronics" having to do with the transport of electrons through a molecular bridge between two conducting electrons. Results on the excitation of molecular vibrations by these currents have been and are being obtained with Amnon Aharony and Ora Entin-Wohlman. This problem and several other ones are related with the generic problem of a small system coupled with, or being "observed" by a much larger one, *e.g.* a reservoir. Relaxation, decoherence and "which path" detection are some of the relevant issues which have reappeared repeatedly in my research.

My research in mesoscopics is strongly coupled with experimental work at the Braun Submicron Center (http://www.weizmann.ac.il/smc/), which, together with Moty Heiblum, I was instrumental in founding (see below, and http://www.weizmann.ac.il/smc/braun_history.html).

Disordered systems encompass a large variety of real-life examples and my interests have recently been centered on disordered many-electron systems. Of great interest to me is the possible spontaneous formation of inhomogeneous granular-type structures, due to even very weak disorder. This may follow in analogy with a similar formation of domains in, for example, magnetic systems with which I was involved in the past with Shang-keng Ma. Such structures seem to be relevant for many superconducting systems. After settling the issue of superconductivity in a nano-grain, The next issue is the disordered and weakly coupled granular superconductor. In more heavily disordered systems such as electron glasses, I am currently interested in the slow relaxation and conductance phenomena. Together with Yuval Oreg and our joint student Ariel Amir, we developed recently a model for "electronic glasses" which boasts an interesting, apparently rather universal, spectrum of relaxation rates, λ, going approximately like $1/\lambda$ at small λ. This new universality class of random matrices yields a natural explanation for the ubiquitous logarithmic relaxation of physical quantities in such glasses and for some of the "aging" phenomena, observed in many of them. We believe that this is the beginning of a powerful paradigm for the understanding of such phenomena in glasses. Recently, we have provided an exact solution for the spectrum of an interesting "exponential matrices" model, which does go like $1/\lambda$ at small λ, with small but interesting corrections. This model may well capture much of the relevant physics of quite a few glassy systems.

2. My research interests over the years have spanned a number of topics, mostly within the realms of condensed-matter and statistical physics.

My (experimental) MSc thesis, performed under the guidance of the late Professor Solly Cohen and with the help of the late Gideon Gilat, then a senior PhD student, used the angular correlations of two successively emitted γ-rays to study the dynamics of the intermediate nuclear state in the solid. The instrument I built to perform the measurements automatically, has worked well for at least 20 years thereafter. A huge hurdle was that it needed two large expensive (by the standards of that time) electrolytic capacitors, and with some effort we got only one. I therefore invented and built a simple "bootstrap" circuit that switched the same capacitor between its two tasks—something that the experts did not expect to work, but it did. It is still one of the achievements I'm most proud of. My (mostly theoretical) PhD work, with the guidance of Israel Pelah, dealt with proton dynamics in hydrogen-bonded systems, including some ferroelectrics, as studied by inelastic neutron scattering and infrared spectroscopy. The tunneling of the proton between two minima along the hydrogen bond played an important role there. From that work followed the one (during the postdoctoral period and later) on phase transitions. Of special interest to me were the questions of finite, but long, range order in low-dimensional (low-D) systems and quasi long range order (decaying as a power-law) in 2D x-y symmetry systems, partly in collaboration with Leon Gunther. We also considered the response of a small conducting ring to an Aharonov–Bohm (AB) flux. It was known that if the ring is normal, nothing would happen in the infinite-system limit. We found however that a current would flow for a finite ring at low temperatures, proportional to the inverse of the ring's radius for ideal noninteracting particles. Moreover, for a ring made of superconducting material quite above the nominal T_c, there should still a finite equilibrium current even when the resistance of the wire making the ring was finite. This novel ability of a small system with respect to the infinite one was in fact an important precursor of mesoscopic Physics, but it was not regarded as having any interest by most colleagues at that time. My interest in fluctuations and scaling as function of system's size started then. At the same time, Alan Evans and I studied the pairing theory of the Bose superfluid, as an attempt to provide a, still needed, microscopic theory of the latter.

After joining Tel Aviv University, the research on what happens to phase transitions in finite systems was continued with David Bergman, examining how the "transition" sharpens with increasing size and resulting in a physically motivated first theory of finite size scaling, leading also to insights on the usual scaling laws. Later, came an approximate but surprisingly good picture of "interdimensional" scaling. In addition to studying the renormalization group method and applying it to several problems of critical phenomena, my interest in small superconductors began, much with Bergman, Gunther, Deutscher and Alexander. The model of Josephson-coupled granular system was first introduced then, later to be followed up with Douglas Scalapino and Leon Gunther and, still later, with Myron Strongin. The last paper by Strongin and myself on this subject, which explains interesting universalities observed in high T_c superconductors, is in preparation now.

During a sabbatical at the University of California, interesting results on fluctuations in quasi 1D systems were produced with Douglas Scalapino, along with more work on granular superconductors and, with Phil Pincus, on weakly coupled chains (one of the interesting results of which has been the great sensitivity of these systems to disorder, in agreement with experiment). At that time my interest in disorder started, resulting in general work on phase transitions in those systems with Amnon Aharony and Shang-keng Ma and on the random-field model with Ma, including the mechanism for the destruction of the ordered state, the number of whose physical applications is still increasing. Later, the three of us contributed to the treatment of dimensional reduction in this model. In 1976, Aharony and I had the idea of constructing an analogy between the spin glass physics and Anderson localization. The idea was fruitful, but we did not understand the relevant symmetries for this problem. This was taken up by Wegner, who did the scaling theory of localization correctly, which, together with weak localization, led to the celebrated and very useful "gang of 4" scaling theory (Abrahams, Anderson, Licciardello and Ramakrishnan). In 1979 I have followed up on the latter using ideas due to Thouless, and produced simple consequences on the Anderson transition. A collaboration with Zvi Ovadyahu based on his experiments followed, resulting in confirmation of most salient features, and of weak localization effects and novel observations of the Altshuler-Aronov interaction effects with their dimensional crossover. The Anderson transition should also result in a divergence of the static and $q = 0$ dielectric constant, and in the whole low-frequency and large wavelength dielectric function, with interesting behavior in the critical regime of frequency and wavenumber, due to the scale-dependent diffusion. This very fundamental paper, with Bergman and Gefen was not easy to get published, it finally appeared as a "brief report" in Phys Rev. Interestingly, an adaptation of some of the ideas of that paper to percolating clusters by Gefen, Aharony and Alexander had a much warmer reception by the community. Bergman and I published a few years earlier, results on the divergence of the dielectric function in the classical, percolation transition, but we believe that the quantum case is much deeper.

In parallel with the above, I became interested in the nonlinear aspects of Josephson Physics. Chaos, with Braiman, and solitons, with Ben Jacob, were discovered in those systems and a theory of such effects in suitable junctions and including AC and DC squids was produced. Another unrelated noteworthy development was the proof with Moshe Schwartz that the supersolid phase does **not** exist in perfect lattices (in spite of speculations by several prominent scientists), but a finite density of defects is necessary for it. Our speculation on Bose condensation of vacancies was in line with the former ideas of Andreev and Lifshitz. This problem came to the forefront in recent years, due to novel experiments. It appears that these experiments are fully consistent with our ~30 year old statements, even when unknown to the experimentalists…

Is the transition to the spin-glass phase a real, sharp transition? It occurred to me during a sabbatical at IBM Yorktown heights that since the usual susceptibility has only a cusp at the transition, a better way to look into the problem would be by observing the **nonlinear** susceptibility which should diverge at the transition, if any. Experiments by Alex Malozemoff and Bernard Barbara confirmed the predicted divergence and the full scaling with temperature and magnetic field. Our joint paper was the first such evidence for the spin-glass transition in a real system.

Around that time, the technology of small scale fabrication was developed enough so that electronic devices on the micron and submicron scale could be produced and observed. Thus, the time was ripe for the application of small-size thinking to these problems, *i.e.* for mesoscopic Physics. The motivation for that increased strongly due to the discovery of the Quantum Hall effect around that time. The relationship with the IBM lab, which was at the forefront in these technologies, and especially the collaboration and deep friendship with the late and esteemed friend and colleague Rolf Landauer, and with Markus Büttiker, were extremely important. The first contribution was the full understanding, with them, how to show that the aforementioned finite equilibrium currents in a small resistive ring do not decay (hence termed "persistent currents"). However, it was even more interesting and relevant to consider transport properties. Landauer had created, years before, an innovative interesting presentation of the conductance of a 1D system in terms of the transmission through it. But, it was thought that its applicability was confined to 1D. It was obvious to me that the Landauer picture gave an appropriate tool to study mesoscopic transport phenomena and I pushed that idea on several fronts. An interesting first question was what would be the resistance of a normal ring as a function of an AB flux through it. The original Landauer formula can still be used if the ring is connected to the measurement system with 1D leads. The calculation was done with Gefen and Azbel, resulting in periodic oscillations of the ring's resistance as a function of the flux with the fundamental period ***h/e*** (the single-electron flux quantum). Not too many people believed in this, until the experimental proof, by Webb, Washburn, Umbach and Laibovitz at IBM and by Chadrasekhar, Rooks, Wind and Prober at Yale came about. Two years before our work, a related problem was considered in a beautiful paper by Altshuler, Aronov and Spivak (AAS), using perturbation theory in the disorder. This too gave a periodic AB conductance oscillation, but the fundamental period was found to be ***h/2e*** (the two-electron flux quantum)—the first harmonic of the ***h/e*** period! This prediction was very quickly verified experimentally, on ingeniously prepared long ~micron diameter cylinders, by Sharvin and Sharvin. Is the fundamental flux period ***h/e*** or ***h/2e?*** Fortunately, experimentalists continued the quest for the ***h/e*** period. In the beginning, though, there was a persistent aperiodic component that was **reproducible** for the same mesoscopic sample. Fortunately, Richard Webb insisted and proved experimentally that that was a real effect. I was convinced that it was due to the "non-AB" portion of the magnetic field penetrating the real material and sensitive to the specific defect arrangement of each sample (and therefore would vanish by averaging over an ensemble of many mesoscopic samples prepared with similar macroscopic properties but differing by the detailed defect arrangement—which may be termed "the impurity ensemble") and this was confirmed with numerical simulations by Stone. This started the subject of "mesoscopic fluctuations" which turned out to have been a very fundamental mesoscopic phenomenon. In 1984, I heard from Gefen that a way to understand the ***h/e*** oscillation could be to interpret it as a mesoscopic fluctuation, so it will occur in any sample, albeit with a sample-specific phase (*i.e.* paramagnetic or diamagnetic at small flux) and magnitude. Thus, the ***h/e*** component being a mesoscopic fluctuation, should average out for the impurity ensemble, but some of the ***h/2e*** component would survive. All this was confirmed by Stone's simulations in our joint work, and later by Murat, Gefen and myself. The reason that the AAS calculation produced the ***h***/2e fundamental frequency was that the perturbative calculation used the impurity-ensemble averaged propagators (as appropriate for a very long cylinder which may be thought of as comprised of many connected rings!).

In 1985 came the theoretical discovery by Altshuler, and then by Lee and Stone, of the fundamental property of the universal magnitude of the conductance fluctuations (at ***T=0***), irrespective of *e.g.* disorder strength (as long as it was not too strong), system's size and material. Using the two-terminal multichannel Landauer formula (see below) and a presentation due to Jean-Louis Pichard, I was able to demonstrate the relationship of the above universality with the random-matrix description of the system's transfer matrix. This started a large chapter of research by numerous people. The semiclassical picture due to Berry was used by Argaman, Smilansky and myself to obtain the spectral correlations for diffusing electrons, and later found to be a useful tool for other properties too. The nontrivial generalization of the Landauer formula to many channels was done with Landauer, Büttiker, my student Pinhas and myself. I then understood a little later the distinction between the two-terminal and four-terminal situations—which even Landauer objected to in the beginning, but did accept later. According to this, I found that even an ideal system should have a finite two-terminal conductance (due to its contact resistances with the outside world, given in fact by the number of conduction channels in units of the quantum conductance unit e^2/h. Fortunately, this strange prediction was soon confirmed by experiments in both the Delft and Cambridge groups. The generalization of the above to any number of terminals, done later by Büttiker, enabled the understanding of a number of further mesoscopic effects that were seen experimentally.

This started the golden age of mesoscopic Physics, which also went serious in including sophisticated electron interaction and correlation effects, as well as those coupling the system to a reservoir (see the mention of "decoherence" below).

It would be ideal to combine the theoretical expertise created in the burgeoning field of mesoscopics with the appropriate level experimental research, which necessitates some advanced technology. However, starting with the missing capacitor story of my youth and observing the lamentable situation of experimental condensed matter Physics in Israel, this looked like a very nontrivial task. Negotiations with several institutions took place in 1985, during my sabbatical at Yale (where I taught Solid State theory, helped build their group in this field, and learned from Daniel Prober about the requirements for experimental research in the field). It looked like the only institution in Israel that perhaps had the ability and perhaps the will to do this is the Weizmann Institute, which I joined in 1986 and where I spend a large fraction of my effort for the first few years in convincing my colleagues and the management to go about this. Fortunately, Moty Heiblum was looking for a way to return to Israel, and accepted to come, build and head what became the Braun Submicron Center, which succeeded even more than my expectations, and the rest is history! In parallel, the theory group (comprised in the beginning by Shimon Levit and Yuval Gefen, besides myself) was also strengthened with prominent younger Israelis: Ady Stern and Yuval Oreg and, very recently, Ehud Altman. In addition we had the great benefit of having three outstanding additions from the ex-Soviet Union high-level theorists: Alexander Finkel'stein, and the late Yehoshua Levinson and Arkadi Aronov who also became dear friends. They added a new dimension to our theory research and left their clear marks on the group. Unfortunately, Arkadi was with us only three years and Yehoshua passed away two years ago. In addition to Israel Bar-Joseph who joined the experimental group from the very beginning, and Yacoby who joined for a number of years, Eli Zeldov, whom I convinced to join us, is beautifully active in superconductivity and the younger generations are presented by Danny Shahar and Shahal Ilani.

I was very fortunate to have at that time a few of my best graduate students ever (where they all became good friends and colleagues later).

After a first interesting paper on mesoscopic thermoelectric transport by Uri Sivan (on which I became active again very recently), work on the magneto-transport in the localized phase was done by him with Ora Entin-Wohlman and Claudio Hartsztein, motivated again to an extent by experimental results due to Zvi Ovadyahu. Uri also did the spectral correlations and ac transport in the localized phase. Upon finishing, he was a postdoc with Moty Heiblum and became a leading experimentalist, and a Professor and founding head of the RBNI Nanoscience Center at the Technion. Ady Stern dealt with the decoherence due to the coupling with the environment. This was thoroughly studied with the help by Yakir Aharonov, and resulted not only in a rather full understanding of the problem including the equivalence of the effects of environment noise **on** the system and the trace left **by** the system, but also in a useful formula for the dephasing rate for weak coupling, which is consistent with the special more complicated earlier important formulation by Altshuler, Aronov and Khmelnitskii. Doron Cohen and I proved from this formulation rather generally that for weak coupling the dephasing rate must vanish in the ***T=0*** limit. Since extra dephasing seems to be observed at low, but finite, temperatures in many real conductors, it must be due to an extra mechanism such as a measurement current being not small enough, magnetic impurities or two-level systems (that are ubiquitous in disordered systems). The last one was suggested and worked out with Fukuyama and Schwab. Ovadyahu demonstrated experimentally that reducing the current into the real linear transport regime often eliminates the extra dephasing, although we found that the two-level systems are relevant as well. In the noble metals it was demonstrated by the Saclay-Michigan group that magnetic impurities are the culprit. The pair-breaking due to those was relevant for the explanation of the magnitude of the persistent current mentioned above. Later, Sivan, Aronov and myself showed in agreement with experiment, that in a quantum dot the inelastic electron-electron scattering may cause the spectrum to become effectively continuous above a characteristic energy termed the Thouless energy of the dot. There are some interesting subtleties here, which have been considered by others and are not fully settled yet. Amir Yacoby did his MSc work with me clarifying the conditions for getting a quantized conductance, obtaining the leading corrections to it and understanding (with Norton Lang), the focusing of electrons emitted by a source). This was a rather high-level theoretical endeavor and with a little more work, it would become an excellent theoretical PhD. However, Amir felt that his real vocation is experiment, started an experimental Phd from scratch with Moty Heiblum and, like Uri Sivan, became a leading experimentalist and a Professor at Weizmann. More recently, he accepted such a position at Harvard. In the late part of his PhD work, Amir tried to measure the transmission phase of a quantum dot embedded in an AB interferometer, by looking at the phase of the AB flux oscillations discussed above. It became immediately clear that this can not be done with a closed interferometer (conserving electrons), where Onsager symmetry locks the phase of the AB oscillations. Therefore, the interferometer has to lose electrons to another lead. These losses must be strong enough, but not too strong in order not to perturb the phase shift. With Aharony, Entin-Wohlman, Levinson, Schiller and, later with Professor Bert Halperin, we worked out the precise details of how and to what extent the interferometer should be opened to effect a good measurement. We also found that the transmission phase can be determined by fitting the detailed measurements to the theory for a closed interferometer (but not via a naïve shift of the AB phase). The AB interferometer is also a nice system for studies of dephasing, and controlled decoherence by "which path" detection was thoroughly studied there. The measurements of the transmission phase by the Heiblum group discovered several unexpected phenomena. A notable one was the "phase jump"—a sharp decrease of the transmission phase by π near the middle of the Coulomb blockade valley between consecutive transmission resonances. Many explanations have been advanced to this phenomenon, and it is becoming clear that the crux of the matter is a switching of the occupation of a broad and a narrow level, suggested by Peter Silvestrov and worked out in our joint papers, including the effect of spin. This switching is due to a combination of the interaction and the coupling to the leads. Earlier, it was suggested by Weidenmuüler et al that a similar occupation switching might be more simply due to a peculiar shape of the dot. It seems, though, that the considered dots do not have the necessary shape for this.

It is interesting that when the single electrons are localized by disorder, two electrons can propagate together provided their combined energy is high enough (but still much below the mobility edge), I interpreted it in terms of the Thouless' picture for localization and the increase with energy of the two-electron density of states (DOS). Interestingly, my association with Silvestrov started from his criticism on this work! Later, Aharony, Entin Wohlman and I provided some exact results for related problem.

In the later 90's I also dealt with two further major topics: mesoscopic superconductivity, with my student Moshe Schechter and with Levinson and quantum noise also with Levinson and with our joint student Uri Gavish. Moshe found novel results in nano-superconductivity, the most surprising one being that the superconducting correlation energy of a nanograin can be relatively much larger than that of the bulk. Together with Jan von Delft of Munich, we identified two pairing energy scales for such systems. These results are crying for a follow up, and I do hope to get to that soon. With Uri and Levinson we understood the Physics of noise correlators for both equilibrium and shot-nose which gives insights on their detectability. Interesting limits on the latter as obtained with an amplifier in the quantum realm were later confirmed and generalized, in collaboration with Bernie Yurke of Bell Labs. An important generalization of the fluctuation-dissipation theorem to non equilibrium steady states (*e.g.* nonlinear differential conductance) was found as well.

Two recent problems I have been active on are the interpretation of the Casimir force as the balance between radiation pressures, and the physical understanding of the quantization of the ac conductance (discovered by Büttiker and confirmed experimentally by Glattli's group) in terms of "delayed currents"—a picture due to Zohar Ringel, a joint MSc student of Entin-Wohlman and myself.

Yoseph Imry
Rehovot, January 2010

Chapter 1

The Kondo Screening Cloud: What It Is and How to Observe It

Ian Affleck

Department of Physics and Astronomy, University of British Columbia, Vancouver, BC, Canada V6T 1Z1

The Kondo effect involves the formation of a spin singlet by a magnetic impurity and conduction electrons. It is characterized by a low temperature scale, the Kondo temperature, T_K, and an associated long length scale, $\xi_K \equiv \hbar v_F/(k_B T_K)$ where v_F is the Fermi velocity. This Kondo length is often estimated theoretically to be in the range of .1 to 1 microns but such a long characteristic length scale has never been observed experimentally. In this review, I will examine how ξ_K appears as a crossover scale when one probes either the dependence of physical quantities an distance from the impurity or when the impurity is embedded in a finite size structure and discuss possible experiments that might finally observe this elusive length scale.

1.1. Introduction

The Kondo model gives a simplified description of the interaction of a single magnetic impurity with the conduction electrons in a host, with Hamiltonian:

$$H - \mu N = \sum_{\vec{k}\sigma} \Psi^\dagger_{\vec{k}\sigma} \Psi_{\vec{k}\sigma} \epsilon_k + J \vec{S}_{\text{imp}} \cdot \vec{S}_{\text{el}}(r = 0). \tag{1.1}$$

Here $\vec{S}_{\text{imp}}$ is the impurity spin operator and $\vec{S}_{\text{el}}$ is the electron spin density at position $\vec{r}$:

$$\vec{S}_{\text{el}}(\vec{r}) \equiv \Psi^\dagger_\alpha \frac{\vec{\sigma}_{\alpha\beta}}{2} \Psi_\beta(\vec{r}) \tag{1.2}$$

where $\Psi_\alpha(\vec{r})$ annihilates a conduction electron of spin α at $\vec{r}$ and repeated spin indices are summed over. We will focus on the case $S_{\text{imp}} = 1/2$ although much of the discussion here carries over to higher spin. The dispersion relation will often be assumed to be that of free electrons:

$$\epsilon_k = \frac{k^2}{2m} - \epsilon_F, \tag{1.3}$$

although that is not crucial. (For a general review of the Kondo model, see Ref. [1].) Essentially the same model can be used to describe a quantum dot in the Coulomb blockade regime in a spin 1/2 state, interacting with one or more metallic leads (which can be formed by a 2 dimensional electron gas or by quantum wires in a semi-conductor inversion layer).

The dimensionless parameter which measures the strength of the Kondo interaction is $J\nu \equiv \lambda_0$ where ν is the density of states at the Fermi energy. For the free electron dispersion in D dimensions, this is:

$$\nu_D = k_F^{D-1}/(c_D v_F) \tag{1.4}$$

where $c_3 = 2\pi^2$, $c_2 = 2\pi$ and $c_1 = \pi$. Typically $J\nu \ll 1$ suggesting that perturbation theory could be useful. However, perturbation theory encouters infrared divergences at low energy scales. In general, an nth order term in perturbation theory for some physical quantity characterized by an energy E is proportional to $[J\nu \ln(\mathcal{D}/E)]^n$, where $\mathcal{D}$ is an ultra-violet scale of order the bandwidth or Fermi energy. It is found that the leading logarithmic divergences can be summed by expressing perturbation theory in terms of the renormalized coupling constant at scale E:

$$\lambda(E) \approx \lambda_0 + \lambda_0^2 \ln(\mathcal{D}/E) + \dots \tag{1.5}$$

Note that $\lambda(E)$ increases in magnitude, as E is lowered, assuming it is initially positive (antiferromagnetic). $\lambda(E)$ obeys the renormalization group equation:

$$\frac{d\lambda}{d\ln E} = \beta(\lambda) = -\lambda(E)^2 + (1/2)\lambda(E)^3 + \dots \tag{1.6}$$

where $\beta(\lambda)$ is the β-function. Keeping only the quadratic term in this RG equation, we find:

$$\lambda(E) \approx \frac{\lambda_0}{1 - \lambda_0 \ln(\mathcal{D}/E)} \tag{1.7}$$

Note that λ_0, the bare coupling, is the value of the renormalized coupling at the ultraviolet scale, $\mathcal{D}$. As the energy scale is lowered, λ becomes of O(1) at the Kondo energy scale:

$$k_B T_K = \mathcal{D} e^{-1/\lambda_0}. \tag{1.8}$$

This renormalization group equation, (1.6) is derived by integrating out high energy Fourier modes, reducing the effective bandwidth from $O(D)$, in energy units, to $O(E)$. Once the bandwidth becomes narrow (as happens at low energies) the width in wave-vector is related to the width in energy by $E = v_F\hbar|k - k_F|$ where v_F is the Fermi velocity. Thus we may equally well describe $\lambda(E)$ as the effective coupling at wave-vector scale $E/(\hbar v_F)$ as at energy E. From this perspective, $k_B T_K/(\hbar v_F)$ is the characteristic wave-vector scale at which the effective Kondo coupling becomes large. Thus it is natural to introduce the Kondo length scale:

$$\xi_K \equiv \hbar v_F/(k_B T_K) \tag{1.9}$$

(We henceforth set k_B and $\hbar$ to one.) It seems appropriate to think of the effective Kondo coupling as growing at large distances, becoming large at length scales of order ξ_K. Equivalently, we might expect physical quantities depending on a length scale r to be scaling functions of the ratio r/ξ_K rather than depending on r and ξ_K separately.

There is an interesting analogy here with Quantum Chromodynamics (QCD) the theory of the strong interactions in high energy physics. The renormalization group equation for the effective QCD coupling, g_{QCD}, (describing the interactions of gluons with themselves and with quarks) is the same as Eq. (1.6) at quadratic order. It also gets large as the energy scale is lowered, and small as it is raised. In this case, one switches back and forth from energy units to momentum units using the velocity of light, c. The characteristic energy scale, Λ_{QCD} where g_{QCD} becomes $O(1)$ is of order 1 GeV, the mass of the proton, and the corresponding length scale is or order the Compton wavelength of the proton, 1 Fermi or 10^{-15} m. If the quarks inside a proton are probed with high energy (short wavelength) photons they appear nearly free. On the other hand, they exhibit confinement (with an interaction growing linearly with separation) at long distances. In high energy physics it is commonplace to go back and forth freely between energy and length units using c (and $\hbar$). However, there are well understood pitfalls in taking this picture of the effective coupling getting weak at high energies and strong at low energies, too literally. (These pitfalls occur both in the energy and distance picture, and do not seem to be particularly related to the distance viewpoint.) In this review article, we will be concerned with the validity of the corresponding picture for the Kondo model.

The δ-function form of the Kondo interaction implies that only the s-wave harmonic interacts with the impurity so that the model becomes fundamentally 1-dimensional. After linearizing the dispersion relation about the Fermi surface, the low energy effective Hamiltonian becomes:

$$H = \frac{iv_F}{2\pi}\int_0^\infty dr \left[\psi_L^\dagger \frac{d}{dr}\psi_L - \psi_R^\dagger \frac{d}{dr}\psi_R\right] + v_F\lambda\vec{S}_{\text{imp}}\cdot\vec{S}_{\text{el}}(0). \qquad (1.10)$$

Here r is the radial coordinate and $\psi_{L/R}$ represent incoming and outgoing waves, with the boundary condition:

$$\psi_L(0) = \psi_R(0). \qquad (1.11)$$

These are defined in terms of the s-wave part of the 3D electron annihilation operator, $\Psi(\vec{r})$ by:

$$\Psi(\vec{r}) = \frac{1}{\sqrt{2}\pi r}\left[e^{-ik_F r}\psi_L(r) - e^{ik_F r}\psi_R(r)\right] + \ldots \qquad (1.12)$$

where the ... represents higher spherical harmonics. Note that we have normalized the fermion fields as in Ref. [2] so that:

$$\{\psi_L^\dagger(r), \psi_L(r')\} = 2\pi\delta(r-r'). \qquad (1.13)$$

In the limit of zero Kondo interactions, $\lambda_0 = 0$, this describes a relativistic Dirac fermion with the Fermi velocity playing the role of the velocity of light. In such a model it is natural to go back and forth between energy and length units using v_F. Note that it is crucial to this estimate of ξ_K that only one channel (the s-wave) couples to the Kondo impurity, allowing a mapping into a 1D model. While this happens in a variety of circumstances, including the case of a quantum dot coupled to 2DEG's, there are also important cases where it can fail, which will be discussed later.

An intuitive picture of this Kondo length scale can be obtained from considering the low energy strong coupling behavior of the model. This is most easily understood from a 1 dimensional tight-binding version of the model with Hamiltonian:

$$H = -t\sum_{j=0}^{\infty}(c_j^\dagger c_{j+1} + h.c.) + JS_{\text{imp}}\cdot\vec{S}_{\text{el}}(0). \qquad (1.14)$$

The strong coupling limit is easily understood. When $J \gg t > 0$, one electron gets trapped at the origin to form a singlet with the impurity spin. Note that this "Kondo screening" actually corresponds to the formation of an entangled state between the impurity spin and one conduction electron. The other electrons can do whatever they want except that they cannot enter or leave the origin since this would break up the Kondo singlet, costing an energy of $O(J)$. They effectively feel an infinite repulsion from the origin, corresponding to a $\pi/2$ phase shift. This is simply a boundary condition on otherwise free electrons.

While the strong coupling limit is trivial, we are actually interested in the case of weak bare coupling, at low energies and long distances, where the effective coupling becomes strong. It is known from various approximate and exact calculations that the low energy physics (at $E \ll T_K$) is described by a local spin singlet state and non-interacting electrons with a $\pi/2$ phase shift. On the other hand, the physics at intermediate energy scales, of $O(T_K)$ is complicated. To form a spin singlet with an S = 1/2 impurity, only one electron is needed and an intuitive picture is that one electron is "removed from the Fermi sea" for this purpose. However, unlike the simple case of large bare coupling, it would be quite wrong to think that this electron is localized at the origin. The natural length scale over which we may think of this electron's wave-function being non-negligible is ξ_K. Such a naive picture must be used with caution. At best, it is valid only at long distances and low energies. If we probe the screened impurity with a long wavelength probe, this picture may apply. At shorter distances, it certainly breaks down. While this is only an intuitive picture, it nonetheless seems reasonable that ξ_K will appear as a characteristic scale in any distance dependent physical property of a Kondo system. The nature of the crossover at ξ_K is the subject of this review.

Nozières' local Fermi liquid theory is well-known to provide a powerful way of studying the behaviour of distance-independent quantities at low $T \ll T_K$. It turns out to also be useful for studying distance-dependent quantities at large distances $r \gg \xi_K$. In this approach, the screened impurity spin is eliminated from the effective Hamiltonian which, in lowest approximation, just consists of non-interacting electrons with a $\pi/2$ phase shift. Then, interactions at the origin are added for these phase shifted electrons, which originate from virtual excitations of the screened impurity complex. We may simply list all possible interactions at the origin allowed by symmetry. Assuming exact particle-hole symmetry, there is only one important leading irrelevant interaction:

$$H_{\text{int}} = -\frac{v_F^2}{6T_K}\vec{J}_L^2(0). \tag{1.15}$$

Here the current operator is defined as:

$$\vec{J}_L(r) \equiv \psi_L^\dagger(r)\frac{\vec{\sigma}}{2}\psi_L(r). \tag{1.16}$$

H_{int} contains a dimension 2 operator, so its coupling constant must have dimensions of inverse energy. On general grounds, we expect that this coupling constant will be of order the crossover scale, $1/T_K$. The factor of 1/6 in Eq. (1.15) is just a matter of convenience. With this normalization, a calculation of the impurity susceptibility, to first order in this interaction gives:

$$\chi(T) \to \frac{1}{4T_K}. \tag{1.17}$$

[This factor of 1/4 in $\chi(T)$ is conventional since in the high $T \gg T_K$ limit we obtain the free impurity result, $\chi(T) \to 1/(4T)$.] We emphasize that various other definitions of T_K are in common use, differing by factors of O(1).

Perhaps surprisingly, distance dependent aspects of Kondo physics are much less well-studied than energy dependent ones. At a theoretical level this may be a consequence of the fact that it is very difficult to get distance dependence from two of the most powerful methods of studying the Kondo effect: the Numerical Renormalization Group (NRG) method, introduced by Wilson,[3] and the Bethe ansatz (BA) solution, discovered by Andrei[4] and Weigmann.[5] The difficulty with NRG lies in the "logarithmic discretization" or the fact that an effective 1D tight binding model is introduced in which the hopping parameter decays exponentially with distance from the impurity. This is, in fact, only an approximation to the full problem and one which seems to fail to keep track of length-dependence properly. However, more recently a way around this difficulty is being exploited[6,7] which involves a more complicated and numerically costly version of NRG. The difficulty with BA seems to be that while exact wave-functions are calculated, they are given in such a complicated form that it is very challenging to calculate any Green's functions with them. Thus much of our understanding of length dependence comes from perturbation theory and analytic RG arguments, various mean field theories, exact diagonalization of short systems and Density Matrix Renormalization Group (DMRG) numerical results which are also restricted to fairly short systems (up to 32 sites). The reasons why this exponentially large scale, $\xi_K = v_F/T_K$, has never been seen experimentally are probably that the associated crossover effects at this length scale are rather weak and subtle and that the basic Kondo model may not be adequate to describe the experimental systems used. Some inprovements to the model that might be neccessary are: including charge fluctuations (as in the Anderson model) taking into account a finite density of magnetic impurities, taking into account a finite density of non-magnetic impurities and taking into account electron-electron interactions even away from the impurity location. Thus searches for effects at this length scale represent both an experimental challenge to find sufficiently ideal systems and also an opportunity to study the limits of validity of the basic model.

We will be concerned with two types of length dependence. The first type, analysed in Sec. 1.2, involves a single impurity in an infinite host, described by the Hamiltonian of Eq. (1.1). We consider various observables as a function of distance from the impurity. The first one (Sec. 1.2.1) is the Knight shift, measurable in nuclear magnetic resonance (NMR) experiments. This is simply the magnetic polarization of the electrons as a function of distance from the impurity, in linear response to a magnetic field (applied to both the impurity and the conduction electrons). The second (Sec. 1.2.2) is the charge density (Friedel) oscillations, as a function of distance from the impurity. These could be observed by scanning tunneling microscopy (STM) for a magnetic impurity on a metallic surface. The third (Sec. 1.2.3) is the ground state equal time correlation function of the impurity spin and electron spin density at distance r from the impurity. While probably not experimentally observable, this has a rather direct interpretation in terms of measuring the spatial probability distribution for the electron forming the spin singlet with the impurity. The second type of length dependence, analysed in Sec. 1.3, occurs in mesocopic samples containing a single localized S $= 1/2$ where the size of some part of the device is comparable to ξ_K. Here we consider four different situations. Two of them involve a S $= 1/2$ quantum dot coupled to a finite length one-dimensional quantum wire. In the first case, (Sec. 1.3.1) this wire is closed into a ring and we consider the persistent current through it as a function of the ring length and magnetic flux. In the second case, (Sec. 1.3.2) the ring is straight with the quantum dot coupled to one end and the other end open. In Sec. 1.3.3 we discuss an S $= 1/2$ quantum dot embedded in the centre of a finite length quantum wire which is tunnel-coupled to leads. In Sec. 1.3.4, we contrast these strictly 1D models with the case of a magnetic impurity inside a three dimensional magnetic sample containing non-magnetic disorder: the Kondo box model.[8] Given certain simplifying assumptions about the disorder, the characteristic Kondo length scale is much smaller than v_F/T_K. It turns out that this is not just a consequence of the 3 dimensionality but also of the disorder. In an ideal 3D Kondo box ξ_K would again be the characteristic length, as we discuss. Electron-electron interactions away from the impurity are ignored in all sections except 1.3.2, a quantum dot at the end of a finite quantum wire, where they are included using Luttinger liquid techniques. Section 1.4 contains conclusions.

1.2. Impurity in an Infinite Host

In this section, we consider the Hamiltonian of Eq. (1.1) for a single impurity interacting with electrons in an infinite volume. This is the traditional model for magnetic impurities in a metal. Note, however, that we ignore the presence of other impurities. A simplifying feature of this situation is that the mapping onto a single channel 1D system is exact. As we will see in Sec. 1.3.4, this is not generally the case when we put a 3D system into a finite box. Reflections off the boundary of the "box" greatly complicate the model.

An immediate question that arises is: how dilute must the magnetic impurities be to justify a single impurity approximation? Given the screening cloud picture, one might come to the pessimistic conclusion that the average separation of impurities should be much larger than ξ_K:

$$n_{\rm imp}^{-1/3} \gg \xi_K ?? \tag{1.18}$$

(Here $n_{\rm imp}$ is the impurity density.) Such low densities are rarely, perhaps never, acheived in experiments on metals with dilute concentrations of magnetic impurities. Since typical ratios of ξ_K to lattice constant can be in the thousands this would require densities $n_{\rm imp} \ll 10^{-9}$ per unit cell. The reason[9] that the condition is much weaker than this is basically that screening cloud wave-functions around different impurities become nearly orthogonal even when their centres are much closer together than their size. To see this consider two spherically symmetric wave-functions, centred around the points $\pm\vec{R}/2$ but otherwise identical, only containing Fourier modes within a narrow band of wave-vectors of width ξ_K^{-1} around k_F. The overlap is:

$$O(R) = \int d^3r \psi^*(|\vec{r}+\vec{R}/2|)\psi(|\vec{r}-\vec{R}/2|). \tag{1.19}$$

In terms of the Fourier transform of the screening cloud wave-function, this becomes:

$$O(R) = \int \frac{d^3k}{(2\pi)^3} e^{i\vec{k}\cdot\vec{R}} |\psi(k)|^2 = \int_0^\infty dk k^2 |\psi(k)|^2 \frac{\sin(kR)}{2\pi^2 kR}. \tag{1.20}$$

We expect the screening cloud wave-function to decay on the length scale ξ_K but to oscillate at the Fermi wave-vector. It should be built out of wave-vectors within ξ_K^{-1} of k_F. Thus we may assume:

$$|\psi(k)|^2 \approx (\xi_K/k_F^2)f[(k-k_F)\xi_K] \tag{1.21}$$

where the scaling function $f(y)$ obeys the normalization conditon:

$$\int dyf(y) = 2\pi^2 \tag{1.22}$$

in order that $O(0)=1$. Thus,

$$O(R) = (1/2\pi^2)\int dyf(y)\sin[k_FR + (R/\xi_K)y]/k_FR. \tag{1.23}$$

For $R \ll \xi_K$, this reduces to:

$$O(R) = \sin(k_FR)/k_FR \tag{1.24}$$

independent of the details of the wave-function. Thus screening clouds centred on different impurities have negigible overlap provided that they are separated by a distance $R \gg k_F^{-1}$. A simple estimate of the condition for validity of the single impurity Kondo model is provided by the Nozìeres exhaustion principle. There must be enough available conduction electron states, within ξ_K^{-1} of the Fermi wave-vector, to form linearly independent screening wave-functions around each impurity. This gives the conditions on the average impurity separation:

$$n_{imp}^{-1/3} \gg \xi_K^{1/3}k_F^{-2/3} \quad (3D) \tag{1.25}$$

$$n_{imp}^{-1/2} \gg \xi_K^{1/2}k_F^{-1/2} \quad (2D) \tag{1.26}$$

$$n_{imp}^{-1} \gg \xi_K \quad (1D). \tag{1.27}$$

1.2.1. *Knight Shift*

The Knight shift is proportional to the r-dependent polarization of the electron spin density in linear response to an applied magnetic field. In NMR experiments, the total effective magnetic field felt by a nucleus is measured, from the nuclear resonance frequency. In addition to the applied magnetic field there is an additional contribution arising from the hyperfine interaction between the nuclear spin and the surrounding electron spins. Assuming for simplicity that this hyperfine interaction is very short range, the Knight-shift is just proportional to the local spin density at the location of the nucleus. For sufficiently weak applied fields, this is $\propto \chi(r)$, the local susceptibility. In the absence, of the magnetic impurity, this is simply the

r-independent Pauli susceptibility, χ_0. The Kondo interaction leads to an additional r-dependent term, which vanishes far from the impurity. Let

$$\vec{S}_{\rm el} \equiv \int d^3r S_{\rm el}(\vec{r}), \tag{1.28}$$

be the total electron spin operator. The total conserved spin operator is:

$$\vec{S}_{\rm tot} \equiv \vec{S}_{\rm imp} + \vec{S}_{\rm el}. \tag{1.29}$$

The r-dependent magnetic susceptibility is:

$$\chi(r) \equiv \langle S^z_{\rm el}(\vec{r}) S^z_{\rm tot}\rangle / T. \tag{1.30}$$

[We ignore the possible difference of g-factors for impurity and conduction electrons for simplicity; this is discussed in Ref. [10,11].] $\chi(r)$ contains a constant term, present at zero Kondo coupling, which is simply the usual Pauli susceptibility, $\nu/2$. Since only the s-wave harmonic couples to the impurity spin, the other terms come entirely from the s-wave component of the electron field. It is convenient to take advantange of the boundary condition of Eq. (1.11), to make an "unfolding transformation", regarding $\psi_R(r)$ as the continuation of $\psi_L(r)$ to the negative r-axis:

$$\psi_R(r) = \psi_L(-r). \tag{1.31}$$

The s-wave part of the spin density operator may then be written:

$$\vec{S}^{\rm s-wave}_{\rm el}(\vec{r}) \approx \frac{1}{8\pi r^2}[\vec{J}_L(r) + \vec{J}_L(-r) + e^{2ik_F r}\psi^\dagger_L(r)\frac{\vec{\sigma}}{2}\psi_L(-r) + e^{-2ik_F r}\psi^\dagger_L(-r)\frac{\vec{\sigma}}{2}\psi_L(r)] \tag{1.32}$$

We expect this the continuum limit approximation to be valid at sufficiently long distances, $r \gg k_F^{-1}$. Thus, we see that the local susceptibility takes the form[13] at $r \gg k_F^{-1}$:

$$\chi(r) - \nu/2 \to \frac{1}{8\pi^2 r^2 v_F}[\chi_{\rm un}(r) + \chi_{2k_F}(r) 2\cos 2k_F r] \tag{1.33}$$

where the uniform and $2k_F$ parts are slowly varying functions of r. These can be written in the 1D field theory as:

$$\chi_{\rm un}(r) \equiv v_F \int_0^\beta d\tau \langle [\psi^\dagger_L(r,\tau)\frac{\sigma^z}{2}\psi_L(r,\tau) + \psi^\dagger_L(-r,\tau)\frac{\sigma^z}{2}\psi_L(-r,\tau)] S^z_{\rm tot}(0)\rangle$$

$$\chi_{2k_F}(r) \equiv -v_F \int_0^\beta d\tau \langle \psi^\dagger_L(r,\tau)\frac{\sigma^z}{2}\psi_L(-r,\tau) S^z_{\rm tot}(0)\rangle. \tag{1.34}$$

Here $\vec{S}_{\rm tot}$ is defined as in Eq. (1.29) but we may restrict the electron part to its s-wave harmonic only, since the other harmonics give zero contribution to the Green's functions of Eq. (1.34):

$$\vec{S}_{\rm el} = \frac{1}{2\pi}\int_{-\infty}^{\infty} dr \psi_L^\dagger(r)\frac{\vec{\sigma}}{2}\psi_L(r) + \ldots . \tag{1.35}$$

It can be proved,[10,11] to all orders in perturbation theory, that

$$\chi_{\rm un}(r) = 0. \tag{1.36}$$

In the 1D field theory, $\chi_{2k_F}(r)$ obeys a simple renormalization group equation with zero anomalous dimension which expresses how physical quantities change under a change of ultraviolet cut off and of bare coupling:

$$\left[\mathcal{D}\frac{\partial}{\partial\mathcal{D}} + \beta(\lambda)\frac{\partial}{\partial\lambda}\right]\chi_{2k_F}(T,\lambda,\mathcal{D},rT/v_F) = 0. \tag{1.37}$$

This simple form follows because only operators defined at $r = 0$ have anomalous dimensions in a boundary field theory. Thus there is no anomalous dimension for $\psi_L^\dagger(r)\psi_L(-r)$. Furthermore, $\vec{S}_{\rm tot}$ has zero anomalous dimension because it is conserved, commuting with the full Hamiltonian. This RG equation implies that $\chi_{2k_F}(T,\lambda_0,D,rT/v_F)$ which is, a priori a dimensionless function of 3 dimensionless variables, λ_0, (the bare coupling), $\mathcal{D}/T$ and rT/v_F, can be written as a function of the renormalized coupling at scale T, $\lambda(T)$ and rT/v_F only. i.e. both the bare coupling, λ_0 and the ultraviolet cut-off, D can be eliminated in favour of a single variable, $\lambda(T)$. This is a basic consequence of renormalizability. An equivalent statement is that we may express χ_{2k_F} as a function of the ratio rT/v_F and $\lambda(r)$, the effective coupling at energy scale v_F/r. This follows from the RG equation, Eq. (1.6) which implies:

$$F[\lambda(r)] - F[\lambda(T)] = \ln(v_F/rT) \tag{1.38}$$

where $F(\lambda)$ is the indefinite integral of $1/\beta(\lambda)$:

$$F(\lambda) \equiv \int^{\lambda} \frac{1}{\beta(\lambda')}d\lambda'. \tag{1.39}$$

Equation (1.38) can be solved to express $\lambda(T)$ in terms of $\lambda(r)$ and a function of rT/v_F.

Results on $\chi_{2k_F}(r)$ were presented[10–12] to third order in pertubation theory in the Kondo coupling, λ_0, at finite T. In the $T \to 0$ limit these become:

$$\chi_{2k_F} = \frac{\pi v_F}{8rT}[\lambda_0 + \lambda_0^2\ln(\tilde{\Lambda}r) + \lambda_0^3\ln^2(\tilde{\Lambda}r) + .5\lambda_0^3\ln(\tilde{\Lambda}r) - \lambda_0^3\ln(\mathcal{D}/T) + \text{constant}\cdot\lambda_0^3]. \tag{1.40}$$

Here $\tilde{\Lambda} \equiv 4\pi D/v_F$. The presence of the $\lambda_0^3\ln(D/T)$ term is very important. If this term were absent, we could safely take $T \to 0$ without obtaining any infrared divergence in perturbation theory, at sufficiently small r. In that case, it would be convenient to express χ_{2k_F} in terms of $\lambda(r)$ only. At small $\lambda(r)$ perturbation theory would apparently be valid. However, the presence of this term implies that this doesn't work. We can write instead:

$$\chi_{2k_F} \approx \pi\frac{v_F}{8rT}[\lambda(r) + \text{constant}\cdot\lambda^3(\text{r})][1 - \lambda(\text{T})]. \tag{1.41}$$

An important lesson from this expression is that the behaviour of the local susceptibity does not become perturbative, at low T, even at small r. This contradicts the naive expectation from the analogy with QCD, discussed in Sec. 1.1. Nonetheless, it does suggest that there may be some crossover, or change in behaviour as we go from the region $r \ll \xi_K$ where $\lambda(r) \ll 1$ to $r \gg \xi_K$ where $\lambda(r)$ is large. A conjecture was made in Ref. [11] for the behaviour at all $r \ll v_F/T$:

$$\chi_{2k_F}(r) \to \frac{\pi v_F}{r}f(r/\xi_K)\chi(T), \tag{1.42}$$

where $\chi(T)$ is the total, r-independent, impurity susceptility; i.e. the linear response of the total magnetization to an applied field, with the bulk Pauli term subtracted off. Here f is some scaling function, depending on r/ξ_K. This conjecture is consistent with both the perturbative results given above, Fermi liquid results at long distances and low T, $r \gg \xi_K$, $T \ll T_K$, and also with results on the k-channel Kondo model in the large k limit. $\chi(T)$ is well-known from a variety of methods, including BA. At $T \gg T_K$ it has the pertubative behaviour:

$$\chi(T) \to \frac{1-\lambda(T)}{4T} \approx \frac{1}{4T}\left[1 - \frac{1}{\ln(T/T_K)}\right]. \tag{1.43}$$

At low $T \ll T_K$, $\chi(T) \to 1/(4T_K)$. For intermediate T it is quite well approximated by $\chi(T) \approx .17/[(T + .65T_K)]$. We only know the form of the other scaling function, $f(r/\xi_K)$ asymptotically, at short and long distances. At short distances, $r \ll \xi_K$ it is perturbative:

$$f(r) \to \lambda(r) \approx \frac{1}{\ln(\xi_K/r)}. \tag{1.44}$$

In the opposite limit, $r \gg \xi_K$ we may also determine[13] $f(r)$ from Fermi liquid theory. In the limit of low $T \ll T_K$ and large $r \gg \xi_K$ first order perturbation theory in the Fermi liquid interaction of Eq. (1.15) gives the proposed form, Eq. (1.42) with:

$$f(r/\xi_K) \to 4r/(\pi\xi_K) + 1/3 + O(\xi_K/r). \tag{1.45}$$

At intermediate length scales, of $O(\xi_K)$, the function $f(r/\xi_K)$ must somehow crossover between $1/\ln(r/\xi_K)$ and $4r/(\pi\xi_K)$. Determining the function in this region requires numerical methods.

At $T \gg T_K$, we found that the infrared divergences of perturbation theory are cut-off and weak coupling behaviour ensues, at any distance. There is no crossover at $r \approx \xi_K$ in this case. Our result to third order in λ is:

$$\chi_{2k_F}(r) \approx (3\pi^2/4)\lambda(T)^2[1-\lambda(T)]e^{-2\pi rT/v_F}. \tag{1.46}$$

Our proposed scaling form for the local susceptibility when $r \ll v_F/T$:

$$\chi(r) - \nu/2 \to \frac{\cos 2k_F r}{4\pi r^3} f(r/\xi_K)\chi(T) \tag{1.47}$$

makes it clear that NMR measurements of the Knight shift would be a very difficult way of detecting the Kondo screening cloud. Typically an NMR signal is only picked up for nuclei within a few times k_F^{-1} of an impurity spin. Over this range, the r-dependence of Eq. (1.47) is dominated by the first factor, $\cos 2k_F r/r^3$. The slowly varying factor in the envelope of the oscillations, $f(r/\xi_K)$, could not be measured unless signals could be picked up out to distances of $O(\xi_K)$. At such large distances, the Knight shift is very small due to the factor of $1/r^3$. In fact, this form is qualitatively consistent with existing NMR experiments which looked for the Kondo screening cloud.[14] An important further complication in the experiments is that a given nucleus will feel a contribution to its Knight shift from many magnetic impurities at various distances. These contributions are expected to be simply additive, provided that the impurities are sufficiently diulte; see discussion at the beginning of Sec. 1.2. Nonetheless, it would be extremely difficult to disentangle these various contributions and require extremely accurate measurements of the Knight shift to determined the tiny contributions from distance impurities.

An important lesson here is that the Kondo screening cloud is difficult to detect because it is so large. To see it, one must look for a crossover at large distances but at these distances the effects are weak precisely because the distances are so large. The Kondo screening "cloud" is perhaps better described as a "faint fog".[15] It is worth noting that the situation is much better for a sample of reduced dimensionality. Our formula for the local susceptibility, Eq. (1.33), carries over directly to a 2D or 1D sample, with the factor of $1/r^2$ simply becoming proportional to $1/r^{D-1}$. In the 1D case, $\chi(r)/\cos(2k_F r)$ crosses over from $1/[r\ln(\xi_K/r)]$ for $r \ll \xi_K$ to a constant at $r \gg \xi_K$, actually growing with distance. However, the diluteness condition is also modified in 1D to

$$n_{\rm imp}^{-1} \gg \xi_K \tag{1.48}$$

as discussed at the beginning of Sec. 1.2.

1.2.2. *Density Oscillations*

While the Kondo effect is associated with the spin degrees of freedom of the conduction electrons, it also has an interesting effect on the charge density in the vicinity of the impurity.[7,16] This is perhaps a bit surprising, in light of spin-charge separation in 1D. It turns out that density oscillations only occur when particle-hole (p-h) symmetry is broken. In particular, none occur in a half-filled tight-binding model, which has exact particle-hole symmetry. Importantly, the Kondo interaction itself preserves p-h symmetry.

These density oscillations turn out to be quite a bit simpler to analyse than the Knight shift, $\chi(r)$, discussed in Sec. 1.2.1. This is because, at long distances, $r \gg k_F^{-1}$, they can be expressed as a half Fourier transform of a well-studied scaling function of ω/T_K, the $\mathcal{T}$-matrix, $\mathcal{T}(\omega)$. Due to the δ-function nature of the Kondo interaction, the exact retarded electron Green's function in the Kondo model can be written:

$$\begin{aligned} -i\int_0^\infty dt e^{i\omega t}\langle\{\psi(\vec{r},t),\psi^\dagger(\vec{r}\,',0)\}\rangle \\ = G(\vec{r},\vec{r}\,',\omega) = G_0(\vec{r}-\vec{r}\,',\omega) + G_0(\vec{r},\omega)\mathcal{T}(\omega)G_0(-\vec{r}\,',\omega). \end{aligned} \tag{1.49}$$

Here G_0 is the Green's function for the non-interacting case:

$$G_0(r,\omega) = \int \frac{d^3k}{(2\pi)^3} e^{i\vec{k}\cdot\vec{r}} \frac{1}{\omega - \epsilon_k + i\eta} \tag{1.50}$$

The $\mathcal{T}$-matrix is used to determined the resistivity and thus plays an important role in the theory of the Kondo model. Although it is not accessible by BA, NRG techniques have been developed to calculate it, in addition to perturbative and Fermi liquid results and other approximations. The density is obtained from the Green's function in the standard way:

$$\rho(r) = -\frac{2}{\pi} \int_{-\infty}^{0} d\omega \mathrm{ImG}(\tilde{r}, \tilde{r}, \omega). \tag{1.51}$$

Note that $G(\vec{r}, \vec{r}, \omega)$ itself has trivial r-dependence, determined entirely by $G_0(r)$. Only the frequency-dependence reflects the Kondo physics. However, once we integrate over ω in Eq. (1.51), this ω-dependence introduces some interesting r-dependence. In the limit $r \gg k_F^{-1}$, we may use the asymptotic expression for $G_0(r)^2$. In D-dimensions, for the free-electron dispersion relation this is:

$$G_0(r,\omega)^2 \to -\frac{1}{v_F^2} \left(\frac{-ik_F}{2\pi r} \right)^{D-1} \exp(2ik_F r + 2i\omega r/v_F). \tag{1.52}$$

This asymptotic form of G_0 can be calculated for other dispersion relations and has a similar form. Thus we see that, at $r \gg k_F^{-1}$ the density oscillations induced by a Kondo impurity are given by the half-Fourier transform of the $\mathcal{T}$-matrix. Furthermore, for such large r, we expect this half Fourier transform to be dominated by frequencies $\leq v_F/r \ll v_F k_F \propto D$. In this frequency range, the $\mathcal{T}$-matrix is a universal scaling function of ω/T_K.

It is important in calculating the density oscillations to take into account additional potential scattering, arising from the magnetic impurity, in addition to the Kondo interaction. This was ignored in Sec. 1.2.1 since it is not very important for the Knight shift. This corresponds to an additional term in the Kondo Hamiltonian, Eq. (1.1) of the form:

$$\delta H = V \sum_{\sigma} \psi_\sigma^\dagger(0) \psi_\sigma(0). \tag{1.53}$$

We refer to the s-wave phase shift at the Fermi surface due to the potential scattering as δ_P. The $\mathcal{T}$-matrix, at $\omega \ll D$ can be expressed in terms of δ_P, a universal function, $t_K(\omega/T_K)$ and the density of states in D-dimensions, ν_D:

$$\mathcal{T}(\omega) = \left\{ e^{2i\delta_P} [t_K(\omega/T_K) + i] - i \right\} / (2\pi\nu_D). \tag{1.54}$$

Note that, in the particle-hole symmetric case, $\delta_P = 0$, $\mathcal{T} = t_K/(2\pi\nu_D)$. By rescaling the frequency integration variable, and using $v_F/T_K = \xi_K$, we see that the half-Fourier transform occuring in $\rho(r)$ may be expressed in terms of a universal scaling function, $F(r/\xi_K)$:

$$\int_{-\infty}^{0} d\omega e^{2i\omega r/v_F} t_K(\omega/T_K) \equiv [v_F/(2r)][F(rT_K/v_F) - 1]. \tag{1.55}$$

Particle-hole symmetry of the model with no potential scattering (from which t_K can be determined) implies that $t_K^*(\omega/T_K) = -t_K(-\omega/T_K)$. Furthermore, since t_K is related to the retarded Green's function it is analytic in the upper half plane. These two conditions imply that $F(r/\xi_K)$ is real. On the other hand, note that it depends on both real and imaginary parts of t_K. This makes it challenging to determine the density oscillations numerically, since while reasonably accurate results may exist for Im t_K, this is not the case for the real part. From Eqs. (1.51), (1.54) and (1.55) we obtain the formula for the density oscillations at long distances from the impurity, $r \gg k_F^{-1}$:

$$\rho(r) - \rho_0 \to \frac{C_D}{r^D}[\cos(2k_F r - \pi D/2 + 2\delta_P)F(r/\xi_K) - \cos(2k_F r - \pi D/2)] \tag{1.56}$$

where the constant is $C_3 = 1/(4\pi^2)$, $C_2 = 1/(2\pi^2)$ and $C_1 = 1/(2\pi)$. The formula applies in D = 1, 2 or 3, with a spherically symmetric dispersion relation. It also applies to the 1D tight-binding model in which case r is restricted to integer values. Note that in this case, there is exact p-h symmetry at half-filling with no potential scattering, corresponding to $k_F = \pi/2$ and $\delta_P = 0$. In this case, the right hand side of Eq. (1.56) vanishes identically as expected. p-h symmetry breaking is necessary to get non-zero density oscillations and the potential scattering phase shift, δ_P plays an important role in them.

The universal Kondo $\mathcal{T}$-matrix, $t_K(\omega/T_K)$, can be calculated at higher frequencies, $\omega \gg T_K$ using perturbation theory in the Kondo interaction. It can be calculated at low frequencies using Fermi liquid theory. We find that these expansions can be simply inserted into Eq. (1.55) to obtain expansions for $F(r/\xi_K)$ and hence the density oscillations at short distances, $r \ll \xi_K$ and long distances, $r \gg \xi_K$. The weak coupling result is

$$t_K(\omega) = -(3i\pi^2/8)[\lambda_0^2 + 2\lambda_0^3 \ln(\mathcal{D}/\omega) + \ldots]. \tag{1.57}$$

This may be written:

$$t_K(\omega/T_K) \to -(3\pi^2 i/8)\lambda^2(|\omega|) \approx -3\pi^2 i/[8\ln^2(|\omega|/T_K)]. \tag{1.58}$$

The $\mathcal{T}$-matrix vanishes slowly at high frequencies. Substituting this weak coupling expansion into the integral defining F, Eq. (1.55) and integrating term by term gives:

$$F = 1 - (3\pi^2/8)[\lambda_0^2 + 2\lambda_0^3 \ln(\tilde{\Lambda} r) + \ldots] \tag{1.59}$$

where $\tilde{\Lambda}$ is of order v_F/D. We recognize the weak couplig expansion of the effective coupling at scale r:

$$F(r/\xi_K) \to 1 - (3\pi^2/8)\lambda^2(r) \approx 1 - 3\pi^2/[8\ln^2(\xi_K/r)]. \tag{1.60}$$

Note that in the weak coupling limit $F \to 1$ and then Eq. (1.56) reduces to the standard formula for the Friedel oscillations induced by a non-magnetic impurity which produces a phase shift at the Fermi surface of δ_P (in the s-wave channel only).

Our results indicate that weak coupling behaviour occurs for the density oscillations at short distances $r \ll \xi_K$ even at $T = 0$. This is quite different than what we found for the Knight shift in Sec. 1.2.1. In that case we found, in Eq. (1.40), that the infrared divergences of perturbation theory were not completely cut off at short distances. However our conjecture, Eq. (1.42), implied that the non-perturbative aspect simply led, at low $T \ll v_F/r$, to a factor of $\chi(T)$ in $\chi(r)$ which was otherwise linear in $\lambda(r)$, Eq. (1.44). The density oscillations therefore show simpler behaviour at short distances.

At low frequencies, $\omega \ll T_K$, the Kondo $\mathcal{T}$-matrix can be calculated from Fermi liquid theory:

$$t_K \to -i[2 + i\pi\omega/(2T_K) - 3\pi^2\omega^2/(16T_K^2) + \ldots]. \tag{1.61}$$

Substituting into Eq. (1.55), gives:

$$F(r/\xi_K) \to -1 + \pi\xi_K/(4r) - 3\pi^2\xi_K^2/(32r^2) + \ldots \tag{1.62}$$

At large distances, $r \gg \xi_K$, $F \to -1$. From Eq. (1.56), we see that this is equivalent to shifting $\delta_P \to \delta_P + \pi/2$ while setting $F = 1$, the non-interacting value. In this limit the Kondo scattering simply changes the phase shift by $\pi/2$, the familiar result. Thus, while we obtain the standard Friedel oscillation formula at both short and long distances, there is a non-trivial and universal crossover, determined by $F(r/\xi_K)$, at intermediate distances.

The density oscillations were calculated using an improved NRG method in Ref. [7]. In this approach Wannier states are introduced both at the impurity location and at the point of interest, r, fixing the problem with spatial resolution in the usual NRG approach. Results are shown in Fig. 1.1. The oscillations can be parametrized, at $r \gg k_F^{-1}$, by the form of Eq. (1.56). The function $F(r/\xi_K)$ fits the asymptotic predictions at large and small arguments, crossing over between 1 and -1 as expected. In fact, it is fairly well fit, throughout the crossover by the simple "one spinon approximation"[16,17]

$$t_K(\omega/T_K) \approx -2i/[1 - i\pi/(2T_K)] \tag{1.63}$$

giving an approximate formula for F:

$$F(u) \approx 1 + (8u/\pi)e^{4u/\pi}\mathrm{Ei}(-2u/\pi), \tag{1.64}$$

where $E_i(x)$ is the exponential-integral function.

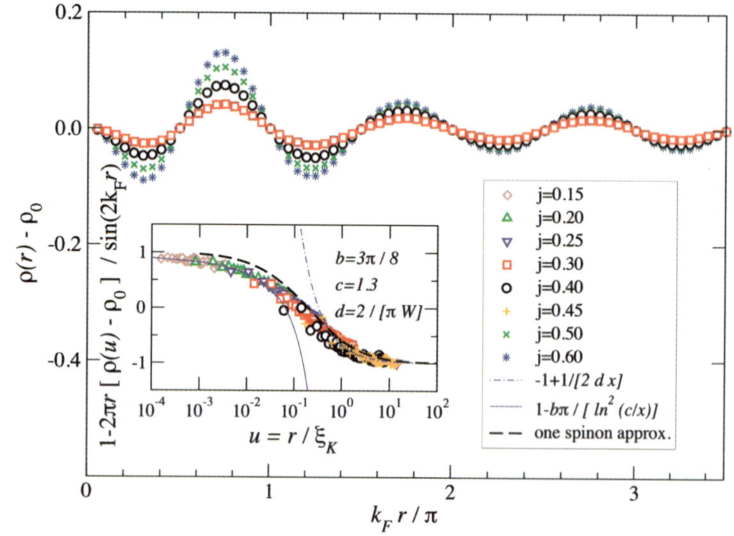

Fig. 1.1. NRG results on charge oscillations around a Kondo impurity coupled to 1D conduction electrons with particle-hole symmetry from Ref. [7]. Note that the oscillations vanish at $k_F r/\pi \in \mathbb{N}$. As shown in the inset, the properly rescaled envelope function of the oscillations (extracted as $\rho - \rho_0$ at the local maxima) for different Kondo couplings nicely collapse into one universal curve except for the points where $r \sim k_F^{-1}$. In the inset we show the analytical results for the asymptotics as well: Note the good agreement between the analytical results and the numerics.

Experimental measurement of the Kondo screening cloud via density oscillations would again be extremely challenging. The $1/r^D$ factor in Eq. (1.56) implies that the oscillations would be extremely small at the length scale ξ_K. (Note that the number of oscillations that would need to be measured is of order $\xi_K k_F \gg 1$.) As for the Knight shift, the situation is improved in lower dimensions. The Kondo effect is apparently observed for magnetic impurities on metallic surfaces using Scanning Tunelling Microscopy (STM). In cases, where the Kondo interaction is predominantly with surface conduction electron states, Eq. (1.56) with $D = 2$ should apply. The STM tunelling rate for tip energy E is usually assumed to be proportional Eq. (1.51) but with the lower limit of integration replaced by E. This quantity could also be calculated from knowledge of t_K, but only when $E \gg T_K$ do we recover a scaling function of r/ξ_K only.

1.2.3. *Impurity spin correlation function*

Another quantity which gives a very direct picture of the Kondo screening cloud is the equal time ground state correlator of the impurity spin with the spin density at a distance r:

$$K(\vec{r}, T) \equiv \langle 0|S^z_{\text{el}}(\vec{r}, 0)S^z_{\text{imp}}(0)|0\rangle\rangle. \tag{1.65}$$

Although this quantity was also discussed at finite T in Ref. [10,11], we restrict our discussion here to the $T = 0$ case. Unlike for the local susceptibility, no divergences are encountered in perturbation theory by taking $T \to 0$, at least to the (third) order studied. $K(r)$ can easily be seen to obey an exact sum rule. If we consider a finite system with an even number of electrons in total (including the impurity spin as an electron) then the ground state is a spin singlet:

$$S_{\text{tot}}(0)^z|0\rangle = 0. \tag{1.66}$$

Thus we have:

$$0 = \langle 0|S^z_{\text{imp}}(0)S_{\text{tot}}(0)^z|0\rangle = 1/4 + \int d^3r K(\vec{r}). \tag{1.67}$$

A heuristic picture of $K(r)$ is obtained by writing:

$$S^z_{\text{el}}(\vec{r}) \approx s^z_{\text{el}}\rho(\vec{r}) \tag{1.68}$$

Here $\rho(\vec{r})$ is the probability of the screening electron being at the location $\vec{r}$ and s^z_{el} is a S=1/2 operator, representing the spin of the screening electon, with $\langle S^z_{\text{imp}}s^z_{\text{el}}\rangle = -1/4$. $\rho(\vec{r}) = |\phi(\vec{r})|^2$ where $\phi(\vec{r})$ is the screening cloud wave-function. Then we obtain $K(r) \approx -\rho(r)/4$ and the sum rule of Eq. (1.67) is simply the normalization condition on the screening electron's wave-function. As for the local susceptibility, discussed in Sec. 1.2.1, at $r \gg k_F^{-1}$ we may decompose K into uniform and $2k_F$ parts:

$$K(r) \to \frac{1}{8\pi^2 v_F r^2}[K_{\text{un}}(r) + 2K_{2k_F}(r)] \tag{1.69}$$

where $K_{\rm un}$ and K_{2k_F} can be calculated in the 1D field theory. Perturbation theory in the Kondo coupling, valid at distances $r \ll \xi_K$, gives:[10,11]

$$K_{\rm un} \approx -\frac{\pi v_F \lambda(r)^2(1+\lambda_0/2)}{2r}$$
$$K_{2k_F} \approx \frac{\pi v_F \lambda(r)(1+\lambda_0/2)}{8r} \tag{1.70}$$

Several differences are to be noted with the local susceptibility. First of all, we have obtained a finite result at $T=0$, as noted above. Secondly the uniform part is now non-zero. Thirdly, the result cannot be expressed in terms of the renormalized coupling at scale r only, but also involves a factor containing the bare coupling, λ_0. This can be understood[10,11] from the fact that the impurity spin operator has a non-zero anomalous dimension:

$$\gamma_{\rm imp} \approx \lambda^2/2 + \ldots \tag{1.71}$$

However, this is not very important since this factor goes to 1 in the scaling limit of weak bare coupling. On the other hand, at $r \gg \xi_K$ we may calculate[10,11,18] $K(r)$ using Fermi liquid theory giving:

$$K_{2k_F} \to -(1/2)K_{\rm un} \to \text{const} \times \frac{\rm \xi_K v_F}{\rm r^2}. \tag{1.72}$$

Our heuristic interpretation then suggests a slow power-law decay for the the screening cloud electron probability density:

$$\rho(r) \approx \text{const} \times \frac{\rm \sin^2 k_F r \xi_K}{\rm r^4} \quad (\rm r \gg \xi_K). \tag{1.73}$$

On the other hand, at shorter distances, $r \ll \xi_K$, $K(r)$ is not negative definite and the heuristic picture of $K(r)$ is not valid.

1.3. Spin in a Mesoscopic Device

Given the extreme difficulty of observing the Kondo screening cloud from length-dependent measurements in a macroscopic sample, discussed in Sec. 1.2, together with the fact that mesoscopic devices routinely contain components with dimensions of order 1 micron, it is natural to consider whether the best way of finally observing the Kondo screening cloud experimentally might be via size-dependent effects in such a device. We are interested in devices containing a quantum dot in the Coulomb blockade regime with a gate voltage tuned so that the number of electrons in the dot is odd, with a $\rm S = 1/2$ ground state. For simplicity, we will also assume that the tunelling from dot to leads is sufficiently weak that only virtual charge fluctuations of the dot need be considered so that the Kondo model is appropriate, rather than the Anderson model. Alternatively, we might consider one or more magnetic impurity atoms in a mesoscopic sample.

An important point is that, associated with a finite sample size, we have a finite spacing between energy levels. If the sample is 1 dimensional, then the level spacing will generally be $\Delta \approx v_F/L$ where L is the sample size. Thus the condition $v_F/T_K \approx L$ is equivalent to $\Delta \approx T_K$. Thus, it sometimes argued that observing size dependence in this situation does not really show the existence of a Kondo screening cloud but "merely" that T_K is the characteristic energy scale for the Kondo model. However, since as discussed in the previous sections, the Kondo screening cloud concept is really just the inevitable consequence of being able to convert an energy scale to a length scale using a factor of velocity, size effects can provide realizations of it. We may equally well say that the behaviour changes when the finite size gap becomes larger than T_K or else when the Kondo screening cloud no longer fits inside the sample. If the sample should instead be regarded as 2 or 3 dimensional the finite size level spacing may be $\ll v_F/L$ depending on which energy levels are important and other details, discussed in Sec. 1.3.4.

The Kondo model is used both to describe magnetic impurities in metals and also gated semi-conductor heterostructure quantum dots. The heterostructure (such as GaAs/AlGaAs) provides a fairly clean 2D electron gas (2DEG) buried some distance (often around 100 Angstroms) below the surface of the semi-conductor wafer. Gate voltages are applied to the surface to define point contacts and quantum dots. A quantum dot refers to an island of electrons, with the electron number (typically around 100 or less) controlled in unit steps by a gate voltage. When this number is odd, the quantum dot usually has an S = 1/2 ground state. The quantum dot can be connected by narrow point contacts to the left and right sides of the 2DEG. If the point contacts are close to being pinched off they only permit one channel of electrons to pass through, giving a $2e^2/h$ conductance. In this case, a simplified model of the system involves effectively 1D leads. The quantum dot has a charging energy, U, as well as a gate voltage, such that the energy as a function of electron number, N, is $U(N - N_0)^2$ for a value of the parameter N_0 controlled by the gate. It is often convenient, theoretically, to use a 1D tight-binding model; universality of Kondo physics implies that such details are not important. The corresponding Hamiltonian is:

$$H = H_0 + H_d \tag{1.74}$$

where

$$H_0 = -\left[t\sum_{j=-\infty}^{-2} c^\dagger_{j\alpha}c_{j+1,\alpha} - t\sum_{j=1}^{\infty} c^\dagger_{j\alpha}c_{j+1,\alpha} - t'c^\dagger_{-1\alpha}c_{0\alpha} - t'c^\dagger_{0\alpha}c_{1\alpha} + h.c.\right] \tag{1.75}$$

and

$$H_d = \epsilon_0 n_0 + U n_{0\uparrow}n_{0\downarrow}. \tag{1.76}$$

Here $n_{0\alpha}$ is the occupation number at site 0 for spin α, with $n_0 \equiv n_{0\uparrow} + n_{0\downarrow}$. t' represents the tunelling amplitudes, through the left and right point contacts, to the leads. (We take these to be equal for simplicity but the generalization is straightforward.)

If the tunnelling amplitudes are sufficiently weak compared to U, then the quantum dot makes only virtual charge fluctuations and the system is in the Coulomb blockade regime. The conductance through the dot then usually begins to decrease as the temperature is lowered, if N_0 is close to an integer. In the case where N_0 is near an *odd* integer, so that the quantum dot behaves as an S $= 1/2$ impurity, we may make a Shrieffer-Wolff transformation to the corresponding Kondo model with Hamitonian:

$$H = H_0 + H_K \tag{1.77}$$

where now:

$$H_0 = -\left[t\sum_{j=-\infty}^{-2} c^\dagger_{j\alpha}c_{j+1,\alpha} - t\sum_{j=1}^{\infty} c^\dagger_{j\alpha}c_{j+1,\alpha}\right] \tag{1.78}$$

$$H_K = J(c_{-1} + c_1)^\dagger \frac{\sigma}{2}(c_1 + c_{-1}) \cdot \vec{S} \tag{1.79}$$

where $\vec{S}$ is the spin operator for the site 0:

$$S \equiv c_0^\dagger \frac{\vec{\sigma}}{2} c_0 \tag{1.80}$$

and the Kondo coupling is:

$$J = 2t'^2\left[\frac{1}{-\epsilon_0} + \frac{1}{U + \epsilon_0}\right]. \tag{1.81}$$

In general, a potential scattering term is also generated, of the same order of magnitude as J. To simplify the discussion, we focus here on the particle-hole symmetric case, $\epsilon_0 = -U/2$, where the potential scattering vanishes. We also assume that the system is at half-filling so that the particle-hole symmetry is exact. It is then easy to calculate the zero temperature conductance for this Kondo model in either limit, $J \ll t$ or $J \gg t$. The second limit is not physical for the underlying Anderson model of Eq. (1.74) but it is convenient to consider nonetheless as a low energy fixed point Hamiltonian. When $J = 0$ the the Kondo Hamiltonian contains no terms linking left and right sides of the dot, so the conductance vanishes. When $J \gg t$, one electron gets trapped in the symmetric orbital, with annihilation operator:

$$c_s \equiv (c_1 + c_{-1})/\sqrt{2} \tag{1.82}$$

and forms a singlet with the impurity spin. At first sight, one might think that this would block the transport through the impurity. However, this is not so due to transport through the antisymmetric orbital with annihilation operator:

$$c_a \equiv (c_1 - c_{-1})/\sqrt{2}. \tag{1.83}$$

Electrons entering from sites -2 or 2 can hop into this antisymmetric orbital without breaking the Kondo singlet. If instead they hop into the symmetric orbital there is a large energy cost of $O(J)$. At large J/t we may obtain a low energy effective Hamiltonian from H_0 by projecting $c_{\pm 1}$ onto c_a:

$$Pc_{\pm 1}P = \pm c_a/\sqrt{2}. \tag{1.84}$$

This gives:

$$H_{\text{low}} = -t\left[\sum_{j=-\infty}^{3} c^\dagger_{j\alpha}c_{j+1,\alpha} - \sum_{j=4}^{\infty} c^\dagger_{j\alpha}c_{j+1,\alpha} + \frac{1}{\sqrt{2}}c^\dagger_{-2\alpha}c_{a\alpha} - \frac{1}{\sqrt{2}}c^\dagger_{2\alpha}c_{a\alpha} + h.c.\right] \tag{1.85}$$

The tranmission probability for this non-interacting electron problem is easily calculated:[20]

$$T(k) = \sin^2 k. \tag{1.86}$$

In particular, at half-filling, $k_F = \pi/2$, the transmission probability at the Fermi energy is one (corresponding to a resonance). The conductance for this non-interacting model is given by the Landauer formula:

$$G = \frac{2e^2}{h}T(\epsilon_F) = \frac{2e^2}{h}. \tag{1.87}$$

Calculations show that, for a small bare Kondo coupling, the conductance grows with decreasing T, saturating at the perfect tranmission value, $2e^2/h$, for $T \ll T_K$. These results can be obtained by expressing the conductance as a frequency integral of the imaginary part of the $\mathcal{T}$-matrix, introduced in Sec. 1.2.2, at finite temperature, times the derivative of the Fermi function. Analytic formulas can be obtained for $G(T)$ in both the high T and low T limits using perturbation theory and Fermi liquid theory respectively. Calculations at intermediate T are typically based on less controlled approximations. This increase of conductance with decreasing T, due to the Kondo effect, only sets in at low temperatures, following an initial decrease due to the onset of the Coulomb blockade.

Note that this behaviour is the inverse of what happens for a magnetic impurity in a bulk metal where it is the resistivity which grows with lowering T, not the conductance. The impurity behavior is more closely related to that for a side-coupled quantum dot where the conductance is $2e^2/h$ for zero Kondo coupling (high T) and vanishes at low $T \ll T_K$.

1.3.1. *Persistent current in a ring containing a Kondo impurity*

A calculationally simple situation in which to observe finite size effects is to close the leads embedding the quantum dot into a ring. The Hamiltonian is that of Eqs. (1.77-1.79) but the leads are now of finite length with periodic boundary conditions:

$$H_0 = -t \sum_{j=1}^{L-2} (c^\dagger_{j\alpha} c_{j+1,\alpha} + h.c.) \tag{1.88}$$

$$H_K = J(c_{L-1} + c_1)^\dagger \frac{\vec{\sigma}}{2} (c_{L-1} + c_1) \cdot \vec{S}. \tag{1.89}$$

While the conductance is now not measurable, one can instead calculate the persistent current in response to an enclosed magnetic flux, $\Phi = (\hbar c/e)\alpha$. The current, at $T = 0$, is given by the derivative of the ground state energy with respect to the flux:

$$j = -(e/\hbar) dE_0/d\alpha. \tag{1.90}$$

A perturbative calculation gives:[19,20]

$$\begin{aligned} j_e(\alpha) &= \frac{3\pi v_F e}{4L} \{[\sin\tilde{\alpha}[\lambda + \lambda^2 \ln(Lc)] + (1/4 + \ln 2)\lambda^2 \sin 2\tilde{\alpha}\} + O(\lambda^3) \\ j_o(\alpha) &= \frac{3\pi v_F e}{16L} \sin 2\alpha[\lambda^2 + 2\lambda^3 \ln(Lc')] + O(\lambda^4), \end{aligned} \tag{1.91}$$

for N even and odd respectively, where c and c' are constants of O(1) which we have not determined and:

$$\tilde{\alpha} = \alpha \quad (N/2 \text{ even})$$
$$\tilde{\alpha} = \alpha + \pi \quad (N/2 \text{ odd}). \tag{1.92}$$

N is the number of electrons, including the electron on the quantum dot, $j = 0$. At half-filling, N is just the total number of lattice sites, including the origin; $N = L$. The fact that j is $O(\lambda)$ for N even but not for N odd is easily understood. The unperturbed ground state consists of a partially filled Fermi sea and a decoupled impurity spin. For N even, there are an odd number, $N-1$, of electrons in the Fermi sea. The unpaired electron at the Fermi surface forms a spin singlet with the impurity in first order degenerate perturbation theory. On the other hand, for N odd, there are no unpaired electrons in the non-interacting Fermi sea so it is necessary to go to second order in λ. A very important property of these results is that, to the order worked, the persistent current only depends on the effective Kondo coupling at the length scale L: $\lambda(L) = \lambda + \lambda^2 \ln L$. That is, logarithmic divergences are encountered in next to leading order, as is standard for many calculations in the Kondo model, but they only involve $\ln L$, not $\ln T$. The current is finite at $T = 0$, for finite L. This suggests that the finite size of the ring is acting as an infrared cut-off on the growth of the Kondo coupling. Provided that $\lambda \ln L \ll 1$, the higher order corrections are relatively small and perturbation theory appears trustworthy. This is equivalent to the condition $\xi_K \gg L$. We may say that, in this case, the screening cloud doesn't "fit" inside the ring so the Kondo effect (growth of the effective coupling to large values) doesn't occur. We expect that higher order perturbation theory would preserve this property, giving $v_F e/L$ times functions of the renormalized coupling at scale L (and the flux) only. This follows because the current obeys a renormalization group equation with zero anomalous dimension. This in turn follows from the fact that the current is conserved $d\langle j(x)\rangle/dx = 0$ so it can be calculated at a point far from the impurity spin. This scaling form implies, equivalently, that we can write the current as:

$$j_{e/o} = \frac{ev_F}{L} f_{e/o}(L/\xi_K, \tilde{\alpha}) \tag{1.93}$$

for even and odd N respectively, where f_e and f_o are universal scaling functions, depending only on the ratio L/ξ_K and not separately on the bare Kondo coupling λ_0 and cut-off $\mathcal{D}$. In the limit $L \gg \xi_K$, the persistent current can be calculated using the non-interacting effective low energy Hamiltonian of Eq. (1.85), i.e. Fermi liquid theory. This gives:

$$j_e(\alpha) = -\frac{2ev_F}{\pi L}[\tilde{\alpha} - \pi] \quad (N \text{ even}), \tag{1.94}$$

$$j_o(\alpha) = -\frac{ev_F}{\pi L}([\alpha] + [\pi - \pi])] \quad (N \text{ odd}). \tag{1.95}$$

Here

$$[\alpha] \equiv \alpha \ (\mathrm{mod}\ 2\pi), \quad |[\alpha]| \leq \pi. \tag{1.96}$$

These small L and large L limits are plotted in Fig. 1.2. Note that the current has the same sign for all α in both limits. Furthermore it has the same period: 2π for N even and π for N odd. For intermediate lengths, L of order ξ_K, it is necessary to do a numerical calculation. A combination of exact diagonalization and DMRG [21] supports the scaling behavior of Eq. (1.93) giving scaling functions that interpolate smoothly, as a function of ξ_K/L. See Fig. 1.3.

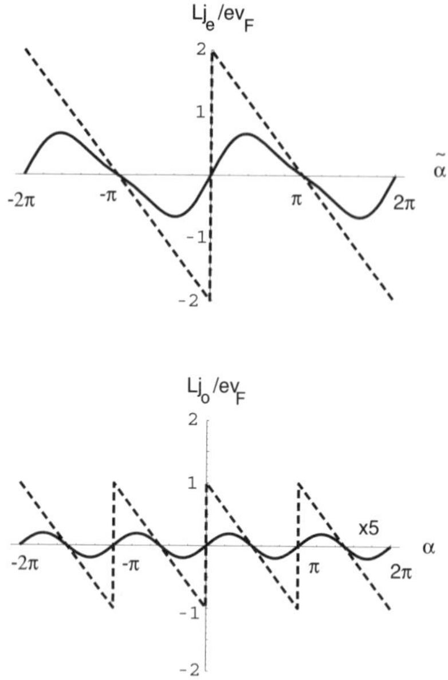

Fig. 1.2. Persistent current vs. flux for an even or odd number of electrons in the weak coupling limit, for $\xi_K/L \approx 50$, from Eq. (1.91) (solid line) and in the strong coupling limit, $\xi_K/L << 1$, from Eq. (1.95) (dashed line). j_o is multiplied $\times 5$ at $\xi_K/L = 50$ for visibility. The solid lines are obtained from Eq. (1.91) using the effective coupling $\lambda(L) \approx 1/\ln(\xi_K/L)$. (From [19].)

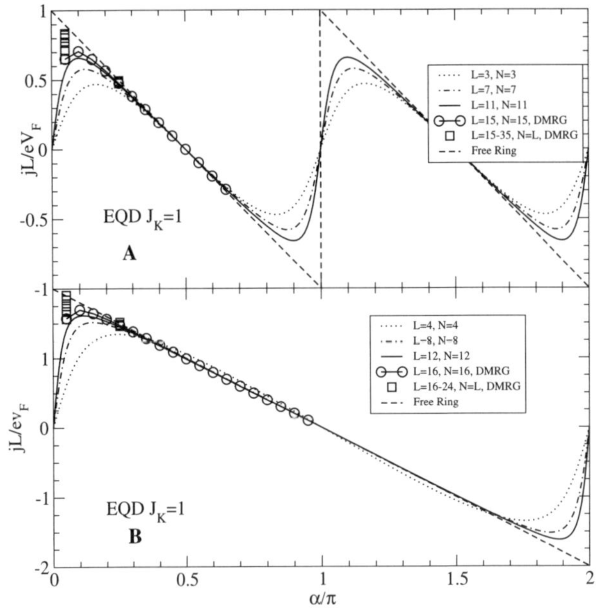

Fig. 1.3. The current at 1/2-filling for $N = 4p - 1$ (A) and $N = 4p$ (B) for the embedded quantum dot. Results[21] are shown for a number of system sizes with $J_K = 1$ as a function of α/π. The results for the small system sizes are obtained using exact diagonalization methods while the larger system sizes (circles) have been obtained using DMRG techniques. From Ref. [21].

1.3.2. *Charge steps in a finite length quantum wire terminated by a quantum dot*

Another mesoscopic system which is quite readily analysed theoretically involves a linear quantum wire of finite length with a quatum dot, in the Kondo regime, at one end. (See Fig. 1.4.) We assume that the system is very weakly tunnel-coupled to an electron reservoir and that a uniform gate voltage, V_G, is applied to the wire, corresponding to a chemical potential $\mu = -eV_G$, in addition to the gates controlling the occupancy of the quantum dot and the tunelling between dot and wire. As μ is varied, the number of electrons in the system, in its ground state will vary in single steps at particular values of μ. This defines a "charge staircase". See Fig. 1.5. The width of the steps, with an even or odd number of electrons in the system, is a sensitive indicator of the strength of the effective Kondo coupling. The steps, at discrete values of μ, could be detected by some

Coulomb blockade technique. Such measurements are likely to be far easier than measuring persistent currents in closed rings. For simplicity, we treat the wire as having only a single channel - i.e. as being an ideal 1D system. A great advantage of this system is that the charge staircase can be determined very accurately from the Bethe ansatz solution [4,5] of the Kondo model. Remarkably, this remains true, to some extent, even when we include short range Coulomb interactions throughout the wire. In this case we take advantage of the more recent Bethe ansatz solution of an S=1/2 Heisenberg antiferromagnet with a weakly coupled spin at one end [22]. We first discuss the case with no Coulomb interactions (apart from those in the quantum dot, leading to the effective Kondo model) and later include bulk Coulomb interactions in the wire.

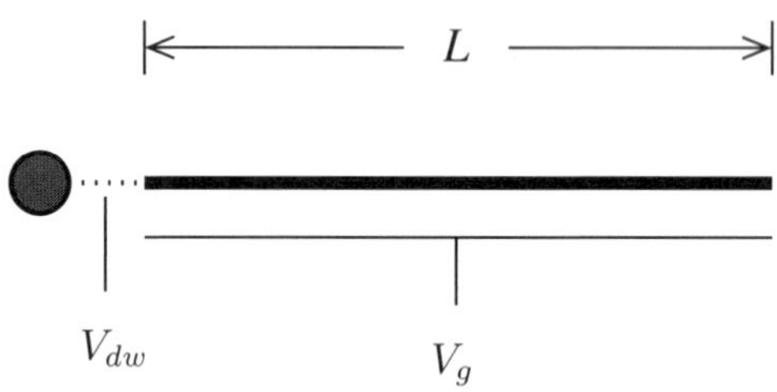

Fig. 1.4. Possible experimental setup. V_{dw} controls the tunneling t' between the small dot (on the left) and the wire and V_g varies the chemical potential in the wire.

1.3.2.1. *Interactions in quantum dot only*

Again it is convenient to consider a tight-binding model. In the Kondo limit this is

$$H = H_0 + H_K \tag{1.97}$$

with

$$H_0 = -t\sum_{j=1}^{L-2}[(c_{j\alpha}^\dagger c_{j+1,\alpha} + h.c.) - \mu c_j^\dagger c_j] \tag{1.98}$$

and

$$H_K = Jc_1^\dagger \frac{\vec{\sigma}}{2} c_1 \cdot \vec{S}. \tag{1.99}$$

Note that this time we have "open" boundary conditions. The chain terminates at site $L-1$. We are interested in how the total electron number in the system, N, changes at $T = 0$ in the grand canonical ensemble, as we vary μ. We define N to include the single electron on the quantum dot. Considering a small range of μ, the steps with even N all have the same width, as do the steps with odd N. We are interested in the ratio R:

$$R \equiv \frac{\delta\mu^o}{\delta\mu^e} \tag{1.100}$$

where $\delta\mu^{e/o}$ is the width of the interval of chemical potential over which N has a fixed even (odd) value.

The limits of weak and strong Kondo coupling are readily analysed.[23] See Fig. 1.5. When $J = 0$, finite width steps only occur for N odd since the energy levels in the quantum wire are doubly occupied and a single electron resides on the quantum dot. In the opposite limit of strong Kondo coupling, finite width steps only occur for N even. One electron is removed from the Fermi sea to screen the spin and the remaining electrons doubly occup free fermion energy levels, appropriately $\pi/2$ phase shifted. Thus $R = \infty$ at zero coupling and $R = 0$ at strong coupling. Again it is possible to calculate the corrections to these limits at leading order in renormalization group improved perturbation theory at weak coupling and Fermi liquid theory at

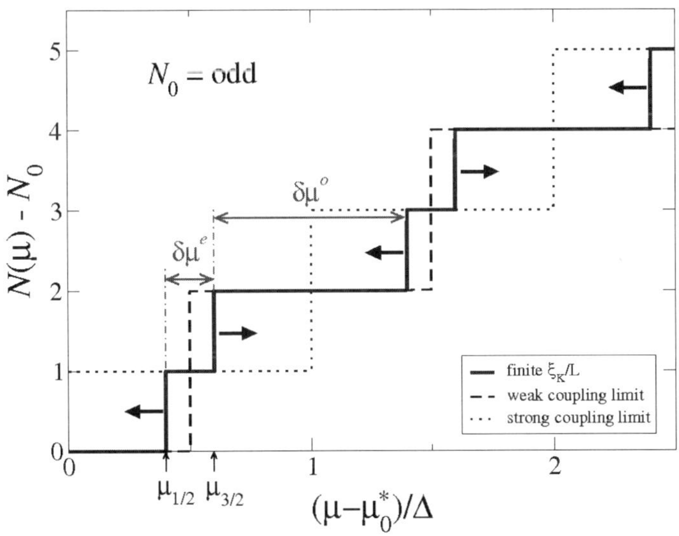

Fig. 1.5. Charge quantization steps for the wire coupled to a small quantum dot. The arrows indicate the direction in which the single steps move as $\lambda(L)$ grows.

strong coupling. At weak coupling we find:

$$\frac{1}{R} \approx (3/2)[\lambda_0 + \lambda_0^2 \ln(L/c)] + \dots \tag{1.101}$$

As for the persistent current, we see that the length of the quantum wire cuts off the growth of the Kondo coupling, at $T = 0$ and we may write:

$$\frac{1}{R} \to (3/2)\lambda(L) \approx \frac{3}{2\ln(\xi_K/L)}, \quad (L \ll \xi_K). \tag{1.102}$$

In the other limit, first order perturbation theory in the Fermi liquid interaction of Eq. (1.15) gives:

$$R \to \frac{\pi\xi_K}{4L}, \quad (L \gg \xi_K). \tag{1.103}$$

Again, we see that R appears to be a scaling function of ξ_K/L only.

For this case, we can check this scaling hypothesis, and determined the function $R(L/\xi_K)$ to high precision for all L/ξ_K by using the Bethe ansatz solution.[4,5] The result is shown in Fig. 1.6. Excellent agreement is obtained with the asymptotic weak ($L \ll \xi_K$) and strong ($\xi_K \ll L$) coupling limits. Weak particle-hole symmetry breaking has a negligible effect on these results, since it simply shifts the entire charge staircase by a constant.[23]

1.3.2.2. *Including Coulomb interactions in the quantum wire*

So far, we have ignored the effect of bulk Coulomb interactions, only considering the Coulomb interaction on the quantum dot, in the Anderson impurity model, which leads to the Kondo interaction. This neglect may be justified in many cases, based on Landau Fermi liquid theory and the fact that the Kondo interaction just involves low energy electron states. However, it is certainly not justified when the Kondo screening takes place in a 1D quantum wire. In this case, Fermi liquid theory breaks down, being replaced by Tomonaga-Luttinger Liquid (TLL) theory. In 1D, Coulomb interactions lead to major changes in the interaction with an impurity. It turns out however, that these changes are minimized when the magnetic impurity is at the end of a quantum wire, so that the conclusions of the previous sub-sub-section will not be modified too drastically.

We modify the model of Eq. (1.98) and (1.99) by adding screened bulk Coulomb interactions. For example, we could consider the Hubbard model with only on-site repulsion:

$$H_c = U\sum_j \hat{n}_{j\uparrow}\hat{n}_{j\downarrow} \tag{1.104}$$

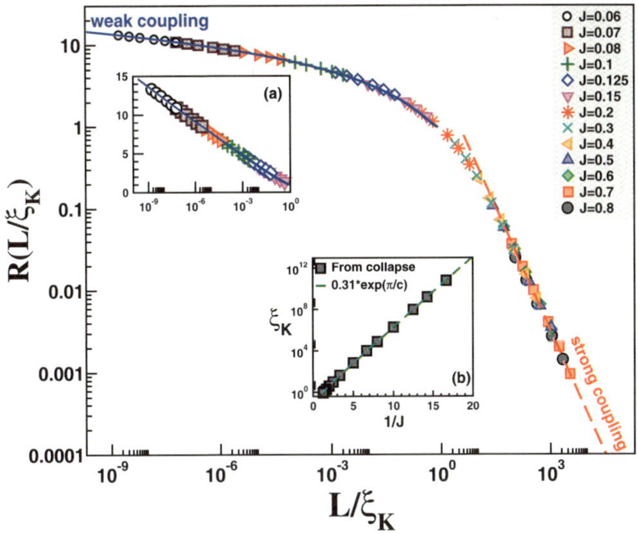

Fig. 1.6. Universal ratio of even and odd charge steps, $R = \delta\mu^{o}/\delta\mu^{e}$, as a scaling function of ξ_K/L from [23]. Bethe ansatz results[23] obtained for various systems sizes (N_0 = 51, 101, 201, 501, 1001, 2001) and 13 different values of the Kondo exchange J, indicated by different symbols. For each value of J, the system lengths L have been rescaled $L \to L/\xi_K(J)$ in order to obtain the best collapse of the data using the strong coupling curve Eq. (1.103) (dashed red line) as a support for the rest of the collapse. The weak coupling regime for $L \ll \xi_K$, enlarged in the inset (a), is described by the weak coupling expansion Eq. (1.101) with constant $\simeq 0.33$ (continuous blue curve). Inset (b): The Kondo length scale, extracted from the universal data collapse of the main panel (black squares), is described by theexpected exponential dependence on Kondo coupling (dashed green line).

where $\hat{n}_{j\alpha}$ is the number operator for electrons of spin α at site j. We expect our results to apply to more realistic models, not based on tight-binding approximations and with longer range screened Coulomb interactions. We begin by writing the 1D electron annihilation operators in terms of left and right movers, similar to what we did for the s-wave projected 3D theory in Eq. (1.12).

$$c_j \approx e^{ik_F j}\psi_R(j) + e^{-ik_F j}\psi_L(j). \tag{1.105}$$

Here $\psi_{R/L}$ vary slowly on the lattice scale, containing Fourier modes of the c_j's in narrow bands around $\pm k_F$, appropriate for the low energy effective theory. The Coulomb interactions reduces, in this low energy approximation, to four different terms which can be conveniently written in terms of charge and spin densities (or currents) for left and right movers:

$$J_{R/L} \equiv \sum_{\alpha} \psi^{\dagger}_{R/L\alpha} \psi_{R/L\alpha}, \quad \vec{J}_{R/L} \equiv \sum_{\alpha\beta} \psi^{\dagger}_{R/L\alpha} \frac{\vec{\sigma}_{\alpha\beta}}{2} \psi_{R/L\beta}. \tag{1.106}$$

Three of the Coulomb interaction terms are proportional to $(J_L^2 + J_r^2)$, $J_L J_R$ and $(\vec{J}_L^2 + \vec{J}_R^2)$. To make further progress, we bosonize the fermions, and introduce separate bosons for the charge and spin degrees of freedom. Upon doing this, the first two interaction terms only involve the charge boson. They change the velocity, v_c for charge excitations and also rescale the boson field by the Luttinger parameter for the charge sector, K_c. The third term changes the velocity, v_s for spin excitations. Importantly, all of these terms leave a non-interacting model of two massless bosons, for charge and spin. The fourth Coulomb interaction (exchange) term is written explicitly as:

$$H_{ex} = -2\pi g_0 v_s \int dx \vec{J}_L \cdot \vec{J}_R, \tag{1.107}$$

where the coupling constant, g_0, is proportional to the strength of the Coulomb interactions. This can be written entirely in terms of the spin boson operators but corresponds to a non-trivial interaction. The exchange interaction, g_0 is marginally irrelevant, its effects getting weaker at lower energy, so that asymptotically, free boson behaviour is obtained. Note that this is the opposite to the behaviour of the Kondo coupling which gets stronger at lower energies. We assume the system is not at half-filling so that we can ignore Umklapp interactions.

An important consideration for the charge steps in the wire-dot system we are considering is the boundary conditions on the low energy fields, $\psi_{R/L}$ at $j = 1$, the end of the chain next to the quantum dot. In the limit, $J = 0$ these are simply:

$$\psi_L(0) + \psi_R(0) = 0. \tag{1.108}$$

These can be seen, for example, by adding a "phantom site" at $j = 0$ and requiring that $c_0 = 0$. Since the low energy fields $\psi_{R/L}$ vary slowly over one lattice spacing, the Kondo interaction can be written in terms of $\vec{J}_R(0)$ only:

$$c_1 \approx \psi_R(0) e^{ik_F} + \psi_L(0) e^{-ik_F} \approx 2i \sin k_F \psi_R(0) \tag{1.109}$$

$$c_1^{\dagger} \frac{\vec{\sigma}}{2} c_1 \approx 4 \sin^2 k_F \vec{J}_R(0). \tag{1.110}$$

Since $\vec{J}_R$ can be expressed in terms of the spin boson only, the Kondo interaction only involves the spin degrees of freedom. This is *not* the case if the impurity interacts with the quantum wire far from its end. In that case, we must use the bosonization expression:

$$\psi_L^\dagger(j)\frac{\vec{\sigma}}{2}\psi_R(j) \approx e^{2ik_F j}\exp[i\sqrt{2\pi K_c}\phi_c]\vec{n} \tag{1.111}$$

where $\vec{n}$ can be expressed in terms of the spin boson. Thus, in this situation the Kondo interaction mixes the charge and spin sectors leading to quite different physics. Here, we have deliberately considered the case of the end-coupled spin in order to preserve as much as the physics of the usual Kondo model as possible.

The charge staircase is determined by the ground state energy, as a function of number of particles. In the case where we set to zero both the Kondo interaction, λ_0 and the bulk marginal coupling, g_0, this is:

$$E_0 - \mu N = \frac{\pi v_c}{4K_c L}(N - N_0)^2 + \frac{\pi v_s}{L}s^2 \tag{1.112}$$

where s is the spin quantum number of the ground state: 0 for N odd and 1/2 for N even. (Recall that we include the electron on the decoupled quantum dot in our definition of N. Here s is the spin of the electrons in the wire, not including the quantum dot.) Including both exchange and Kondo interactions, it is convenient to write:

$$E_0 - \mu N = \frac{\pi v_c}{4K_c L}(N - N_0)^2 + \frac{\pi v_S}{L}f_s(\lambda_0, g_0, L) \tag{1.113}$$

where $s = 0$ or $1/2$ for N odd or even and f_s are some functions to be determined. Then defining:

$$f \equiv f_{1/2} - f_0 \tag{1.114}$$

it is easily seen that the ratio of step widths is:

$$\tilde{R} = \frac{1 + uf}{1 - uf} \tag{1.115}$$

where

$$u \equiv v_s K_c / v_c. \tag{1.116}$$

Coulomb interactions made $v_c < v_s \approx v_F$ and also reduce K_c from its non-interacting value of 1. Both these effects reduce u from its non-interacting value of 1 thus bringing $\tilde{R}$ closer to 1. For cleaved edge overgrowth quantum wires $u \approx .5$ and for carbon nanotubes $u \approx .1$.

In lowest order perturbation theory,

$$f \approx 1 - \frac{3}{2}g(L) - 3\lambda(L), \quad (L \ll \xi_K) \tag{1.117}$$

Here we have replaced both the exchange and Kondo couplings by their renormalized values at scale L. The exchange interaction lowers the ground state energy when the wire (excluding the quantum dot) has an $s = 1/2$ ground state, corresponding to N even. The effective exchange coupling is:

$$g(L) \approx 1/\ln(L/L_1) \tag{1.118}$$

where L_1 is a short length scale, of order k_F^{-1} or smaller. The exchange interactions actually modify the β-function for the Kondo coupling:

$$\frac{d\lambda}{d\ln L} = \lambda^2 + g\lambda + \dots \tag{1.119}$$

(On the other hand, the Kondo interaction doesn't affect the renormalization of $g(L)$ since the Kondo interaction only occurs at the end of the wire and the exchange interaction occurs throughout the wire.) Note that the exchange interaction causes the Kondo coupling to grow faster as the length scale is increased and consequently reduces ξ_K for a given bare Kondo coupling:

$$\xi_K \propto \exp[\sqrt{a/\lambda_0 + b}] \tag{1.120}$$

for some constants a and b depending on g_0. For long lengths, $L \gg \xi_K$, we may again use Fermi liquid theory. Note that the effective exchange coupling, $g(L)$ continues to get smaller as we increase L so we may continue to treat it perturbatively. When $\xi_K \ll L$ however, one electron is removed from the Fermi sea to screen the spin of the quantum dot so that the remaining nearly free fermions system has one fewer electron. This switches the $s = 0$ and $s = 1/2$ ground states and changes the sign of the term in f linear in $g(L)$:

$$f \approx -1 + \frac{3}{2}g(L) + \frac{\pi\xi_K}{2L}, \quad (\xi_K \ll L). \tag{1.121}$$

In general we may write:

$$f = f(\xi_K/L, g(L)). \tag{1.122}$$

The dependence on $g(L)$ destroys the universal scaling with L/ξ_K but in practice this effect can be small since $g(L)$ tends to be small and slowly varying.

Since the Kondo effect for a Luttinger liquid, with the impurity spin end-coupled, only involves the spin degrees of freedom, it is essentially unchanged if the charge degrees of freedom are gapped, as would occur in the Hubbard model at half-filling. In fact, we may take the large U limit at half-filling to obtain the S=1/2 Heisenberg antiferromagnetic chain with one weakly coupled impurity spin at the end. The only important effect of this large U limit is that the bare exchange coupling, g_0 becomes O(1). This impurity model has been solved via Bethe ansatz methods[22] and this allows an exact calculation[23] of the function $f(L/\xi_K, g(L))$ for a particular (large) g_0. The results are shown, for $L = 1000$ and a range of Kondo couplings, in Fig. 1.7. The expected asymptotic limits, $\pm(1 - 3g(1000)/2)$ are obtained for $L/\xi_K \ll 1$ and $\gg 1$. Note that, even for this large bare exchange coupling, the deviations from the model with no bulk Coulomb interactions is quite small. For conducting quantum wires, we expect g_0 to be considerably smaller and the deviations to be even less. Thus the main effect of the Coulomb interactions is to reduced $\tilde{R}$ due to the parameter u in Eq. (1.115). If R is the step width ratio in the non-interacting model, we may write it in the interacting case as approximately:

$$\tilde{R}(\xi_K/L) = \frac{(1-u) + (1+u)R(\xi_K/L)}{(1+u) + (1-u)R(\xi_K/L)}. \tag{1.123}$$

This varies from $\tilde{R}(\xi_K/L) = R(\xi_K/L)$ at $u = 1$ (no Coulomb interactions) to $\tilde{R}(\xi_K/L) = 1$ at $u = 0$ (strong Coulomb interactions). Varying ξ_K/L at fixed u, $\tilde{R}$ goes from $(1+u)/(1-u)$ at $L \ll \xi_K$ to $(1-u)/(1+u)$ at $\xi_K \gg L$. Coulomb interactions weaken the even-odd alternation in the step widths but only trivially modify the dependence on ξ_K/L.

1.3.3. *Transport through a finite wire containing a quantum dot*

The previous examples have involved closed systems, making the measurement of the Kondo effect challenging. Most experiments observing the Kondo effect in quantum dots involve transport measurements; the systems are "open", connected to external leads. A natural way of observing finite size effects in a transport measurement is to embed a quantum dot, in the Kondo regime, in a finite length quantum wire which is weakly connected to external leads by tunnel junctions,[24,25] as illustrated in Fig. 1.8. Assuming for simplicity, perfect parity symmetry, the Hamiltonian is:

$$H = H_L + H_W + H_{LW} + H_K \tag{1.124}$$

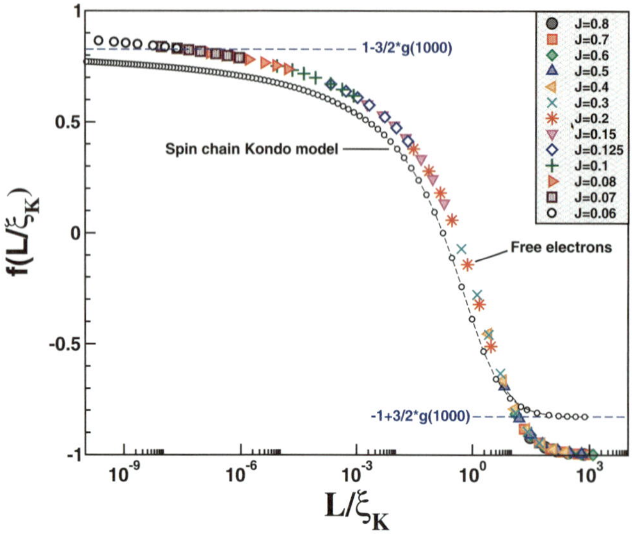

Fig. 1.7. Effect of the marginally irrelevant bulk operator on the scaling function $f(\xi_K/L, g(L))$ calculated from the Bethe ansatz from [23]. The symbols correspond to the noninteracting case, $g(L) = 0$, as in Fig. 1.6. The black circles are obtained[23] with the Bethe ansatz solution of the spin chain Kondo model for $L = 1000$ ($g(1000) \simeq 0.115$).

where L labels the leads, W labels the quantum wire and K the Kondo interaction. Here:

$$H_L = -t \left[\sum_{j=-\infty}^{-L-2} + \sum_{j=L+1}^{\infty} \right] (c_j^\dagger c_{j+1} + h.c.) \tag{1.125}$$

$$H_W = -t \left[\sum_{j=-L}^{-2} + \sum_{j=1}^{L-1} \right] (c_j^\dagger c_{j+1} + h.c.) + \epsilon_W \left[\sum_{j=-L}^{-1} + \sum_{j=1}^{L} \right] c_j^\dagger c_j \tag{1.126}$$

$$H_{LW} = -t_{LW} (c_{-L-1}^\dagger c_{-L} + c_L^\dagger c_{L+1} + h.c.) \tag{1.127}$$

$$H_K = J(c_{-1}^\dagger + c_1^\dagger) \frac{\vec{\sigma}}{2} (c_{-1} + c_1) \cdot \vec{S}. \tag{1.128}$$

Here spin indices are suppressed and we have ignored a potential scattering term, as well as Coulomb interactions in the wire, for simplicity. Note that we have included a chemical potential term, ϵ_W for the wire. When $t_{LW} = 0$ the wire + dot system is closed and we expect that the length, L, of the wire will act as an infrared cut-off on the growth of the Kondo coupling, leading to scaling in ξ_K/L as discussed in the previous two subsections. What happens when the wire is connected to the external leads? When $\xi_K \ll L$, the Kondo screening cloud fits inside the leads and, at low $T \ll T_K$, we therefore expect perfect transmission through the quantum dot. The conductance is then simply that of a finite length non-interacting wire, W, connected by tunelling amplitude t_{WL} to the external leads. The conductance is then easily calculated by the Landauer formula which expresses it in terms of the transmission probability for an electron at the Fermi energy to pass through the wire. In this case the Kondo temperature, T_K is unaffected by the finite wire length, L. On the other hand, when $L \ll \xi_K$ the Kondo screening cloud must leak into the external leads. In fact, we might then think of the quantum wire plus dot as acting like a large quantum dot. (But note that, for simplicity, we haven't included an associated charging energy in the Hamiltonian.)

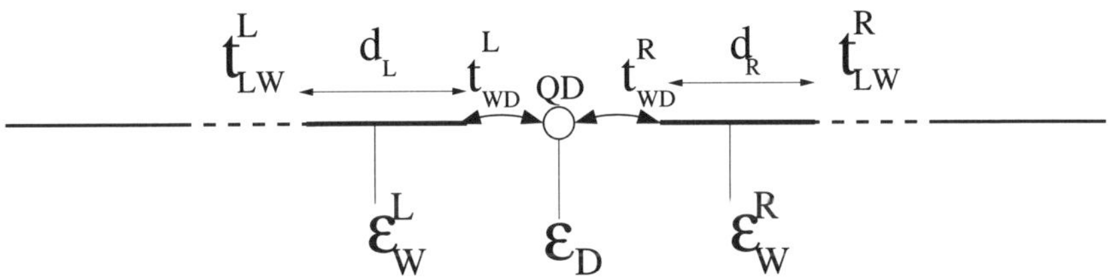

Fig. 1.8. Quantum dot embedded in finite length quantum wire. ϵ_D and $\epsilon_W^{L/R}$ control respectively the dot and wires gate voltage. Here we only consider the parity symmetric case whre all "L" and "R" quantities are equal.

Ignoring the Kondo interactions, the weak tunnelling between wire and leads, t_{LW}, leads to a broadening of the discrete spectrum of the wire. That is to say, the local density of states, ocurring in the Kondo interaction, has peaks at the energies given by the discrete spectrum of the wire, with spacing $\Delta = \pi v_F/L$ and broadenned by an amount $\delta \propto t_{LW}^2$. We assume that $\delta \ll \Delta$. When $L \ll \xi_K$, it was found that the conductance of the device depends very sensitively, at low temperatures, on the precise value of ϵ_W, which shifts the broadened energy levels of the wire, relative to the Fermi energy, moving them on or off resonance. The renormalization of the Kondo coupling may be calculated straightforwardly in second order, using this density of states. Off resonance, there is little renormalization of the Kondo coupling down to exceedingly low energy scales. On resonance, there is again little renormalization until the effective band width is reduced to δ, the width of the resonance peak. At that stage the large density of states leads to a rapid growth of the effective Kondo coupling. In this situation it is convenient to define an effective Kondo temperature, T_K which depends on the length, L, of the wire and also on the gate voltage applied to the wire, ϵ_W. The non-resonant effective Kondo temperature, $T_K^{\mathrm{NR}} \approx T_K$ when $T_K \gg \Delta$, i.e. when the screening cloud fits inside the wire. However, T^{NR} drops rapidly to zero as T_K is reduced below Δ. On the other hand, while the on-resonance Kondo temperature, T_K^{R}, also tracks T_K when $T \gg \Delta$, as T_K is reduced further, T_K^{R} levels off at a value of order δ, the width of the resonance controlled by t_{LW}, as illustrated in Fig. 1.9. Correspondly, the on-resonance conductance through the device, when $\delta \ll T_K \ll \Delta$, stays very small down to temperatures of order δ before shooting up towards the unitary limit, quite unlike the behaviour in the case $\Delta \ll T_K$, when the conductance approaches the unitary limit as the temperature is lowered below T_K. This is illustrated in Fig. 1.10. The dependence of the low $T \ll \Delta$ conductance on the gate voltage, ϵ_W, applied to the wire, is also interesting. When the screening cloud doesn't fit inside the wire, $L \ll \xi_K$, there are resonant peaks with spacing in ϵ_W given by the level spacing of the decoupled wires of length L, Δ. On the other hand, when the screening cloud fits inside the wire, $\xi_K \ll L$, the resonant peaks in low T conductance are separated by $\Delta/2$, the level spacing for the doubled wire of length $2L$.

1.3.4. *The Kondo Box Model*

In 1999, Thimm, Kroha and van Delft introduced[8] a model of a Kondo spin in a finite size system, which they referred to as "the Kondo box" model. It behaves quite differently than the models reviewed so far in this article, and here we focus on this difference. The experimental application that these authors discussed, a single magnetic impurity in a disordered ultra-small metallic grain, was rather different than those that we have discussed so far. They adopted a simplified Anderson impurity model in which:

(1) All energy levels of the unperturbed system (at $J_K = 0$) were assumed to be equally spaced in energy near the Fermi energy.
(2) All unperturbed states with energies near the Fermi energy were assumed to have equal hybridization with the impurity.

The corresponding Anderson model is:

$$H = \sum_{i\alpha} \epsilon_i c_{i\alpha}^\dagger c_{i\alpha} + \epsilon_0 \sum_\alpha d_\alpha^\dagger d_\alpha + U \hat{n}_{d\uparrow} n_{d\downarrow} + v \sum_{i,\alpha} (c_{i\alpha}^\dagger d_\alpha + h.c). \quad (1.129)$$

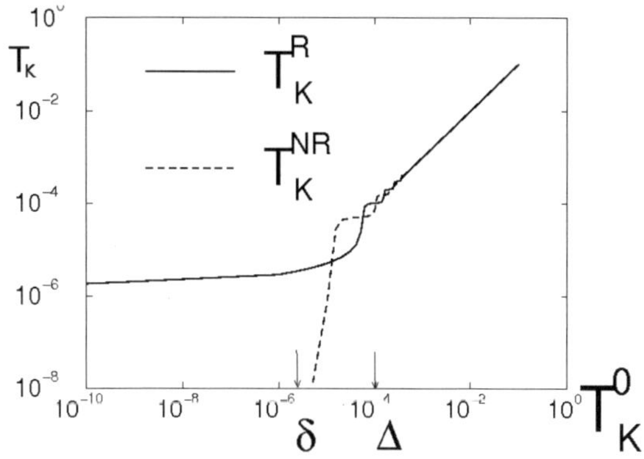

Fig. 1.9. The curves[25] represent $T_K = f(T_K^0)$ in a log-log scale keeping the same values for Δ_n, δ_n for ϵ_W on resonance (plain curve which becomes almost flat at low $T_K^0 \ll \delta_n$), and ϵ_W off resonance (dashed curve which drops sharply at low T_K^0). Both curves coincide at $T_K^0 > \Delta_n$.

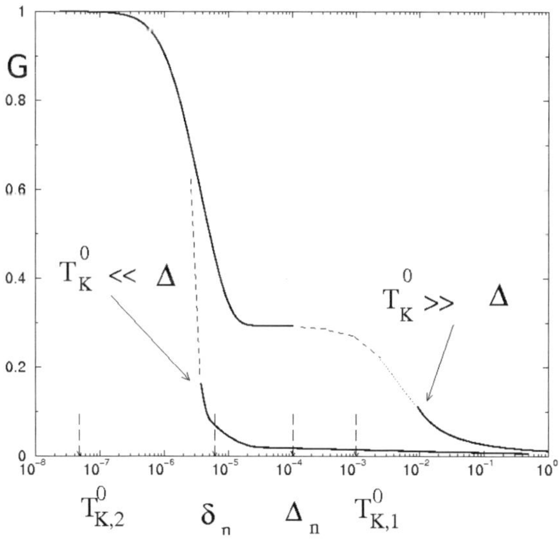

Fig. 1.10. On-resonance conductance[25] as a function of temperature in a symmetric device (assuming ϵ_W is on resonance) for both cases $\Delta_n \ll T_{K,1}^0$ (right curve) and $\Delta_n \gg \delta_n \gg T_{K,2}^0$ (left curve), from [25]. The curves in plain style correspond to the perturbative calculations plus the Fermi liquid result for the first case only. We have schematically interpolated these curves (dotted lines) where neither the perturbative nor the Fermi liquid theory applies.

where the ϵ_i are equally spaced. Taking the large U limit, and ignoring potential scattering, the corresponding Kondo model is:

$$H = \sum_{i\alpha} \epsilon_i c_{i\alpha}^\dagger c_{i\alpha} + J \sum_{i,j} c_i^\dagger \frac{\vec{\sigma}}{2} c_j \cdot \vec{S} \tag{1.130}$$

where $\vec{S}$ is the spin of the magnetic impurity, $\vec{S} = d^\dagger \vec{\sigma} d/2$ and $J = v^2/U$. The (uniform) energy level spacing, near the Fermi energy was estimated as:

$$\Delta = \frac{\pi^2}{mk_F \mathcal{V}} \tag{1.131}$$

where $\mathcal{V}$ is the volume of the grain. Note that in the 1D models considered in this section, the level spacing is proportional to $1/L$, not $1/\mathcal{V}$. The behaviour of the Kondo box model follows immediately from the discussion of the 1D models earlier in this section, if we make the corresponding replacement for Δ. As usual, finite size effects become important when $T_k \approx \Delta$, the bulk Kondo temperature is of order the level spacing. This implies[8] that grains of order 10 nm. in linear dimension are required to see finite size effects, considerably smaller than the estimate of 1 micron in the 1D case. The behaviour of this model seems to contradict the naive Kondo screening cloud picture since there are essentially no finite size effects even when the linear dimension of the grain is much smaller than v_F/T_K.

In fact, the behaviour of this model depends strongly on the assumptions 1 and 2, above, which certainly don't always apply. To make this point it is instructive, as a thought experiment, to consider another model of an ultra-small grain which is far less realistic than even the above one. Imagine that

(1) The grain is completely clean with no disorder
(2) The Kondo interaction is short range and spherically symmetric, the standard assumptions in the Kondo model
(3) The grain is perfectly spherical with the impurity atom exactly at the centre

In this case, we may expand in spherical harmonics and project onto the s-wave, as usual. The s-wave model is essentially the one studied in Sec. 1.2.1, with the impurity spin at one end of a finite length 1D wire. Now finite size effects set in when the length L is of order $v_F/T_K = \xi_K$ as discussed above. The essential difference between the two models (or two sets of assumptions) is that in the second case most of the states of the metal have zero (or negligible) coupling to the impurity, whereas in the first case, all states are assumed to have equal coupling. In the second model only the s-wave states couple, a negligible fraction of the total set of states. This happens because the higher harmonic wave-functions all vanish at the origin, where the impurity is assumed to be located. Note also that the Kondo temperature will be much lower for this spherical model since the relevant density of states, appearing in the dimensionless Kondo coupling, scales with the volume in the first case but only as the length in the second.

The behaviour of the second model is very strongly dependent on the assumptions, as can be seen by relaxing only the third of them. Suppose that we allow the surface of the grain to have a non-spherical shape. Then the usual wave-functions, involving products of spherical harmonics and spherical Bessel functions, are no longer eigenstates of the non-interacting model. If we insist on using this basis, then the impurity only interacts with the s-wave as usual. However, the reflections of the spherical harmonic wave-functions off the walls of the grain mix these wave-functions together. Actually, analogous phenomena occur in the 1D models considered earlier in this section if we take more realistic models of the quantum wires with several channels,[20,25] corresponding to transverse wave-functions. Consider, for example, the model in Sec. 1.3.2 of a quantum dot at the end of a finite length quantum wire, which has several channels. The quantum dot hybridizes with some linear combination of channels in the wire. We can always adopt a basis such that one basis state is the one hybridizing with the quantum dot. However, these states are mixed by scattering off the other end of the wire (away from the quantum dot). A simple, intuitive picture, is that the screening cloud wave-function is now a linear superposition of the various channels, or the screening cloud folds back upon itself many times inside the wire, due to the reflections off the ends. If the wire has N channels, its effective length could be as long as LN. An important point is that embedding a Kondo spin in a finite system typically destroys one of the key simplifying features of the Kondo model — that the impurity spin only couples to one channel of unperturbed eigenstates (eg. the s-wave). The only case where we can be sure this assumption is still valid is when the finite system is an ideal single channel 1D system.

Depending on various assumptions about the grain shape, magnetic impurity location and disorder, one could imagine the Kondo temperature at which finite size effects set in ranging anywhere from order $\pi v_F/L$ to order $\pi^2/mk_F\mathcal{V}$. In all cases, we expect that the condition for finite size effects to be important is that T_K be of order the finite size gap, suitably weighted by the hybridization of the corresponding wave-functions with the impurity. Only in single-channel situations does the simple Kondo screening cloud picture apply.

Note that this complication makes observation of the screening cloud relatively easier in bulk situations where there are no boundaries to mix the hybridizing wave-function with the other channels. However, even there, it seems likely that scattering off other Kondo impurities might have a similar effect.

1.4. Conclusions

In the standard Kondo model, describing a single magnetic impurity in an infinite host, the length scale $\xi_K = v_F/T_K$ appears as a characteristic crossover length scale for many (perhaps essentially all) physical observables. Some intuition about the nature of the crossover is provided by imagining that, in the ground state, the impurity spin forms a spin singlet with a single conduction electron, whose wave-function is constructed out of states within energy T_K of the Fermi surface and hence is spread out over a distance of order ξ_K. One must be careful to only use this picture at low $T \ll T_K$ and even then it sould be used with caution. Naively, we might expect to see weak coupling behaviour, even at $T = 0$, at distances $r \ll \xi_K$ and strong coupling behaviour, outside the screening cloud, at $r \gg \xi_K$. While this seems to apply to the density oscillations and static impurity spin correlations, the first of these expectations turns out be be wrong, or at least to need some modifications, in the case of the Knight shift. The crossover effects at ξ_K appear to be very hard to observe experimentally, precisely because they occur at such long distances. In a D-dimensional sample, they are suppressed by a factor of $1/r^D$ (or $1/r^{D-1}$) making their observation more favourable in systems of reduced dimensionality.

In a mesoscopic single-channel 1D systems with a component of finite length, L, there is again typically a cross-over in physical quantities when ξ_K is of order L. In all cases, weak coupling behaviour occurs when $L \ll \xi_K$ so that the screening cloud doesn't "fit" inside the device and strong coupling behaviour occurs when $\xi_K \ll L$ so that finite size effects become unimportant. Except in ideal circumstances, mesoscopic systems do not have only a single channel. While this is unimportant for the Kondo effect in bulk, it can have important consequences for the observation of the Kondo screening cloud in a mesoscopic device. For an N-channel system, the effective length "seen" by the Kondo screening cloud may be as large as LN.

Throughout most of this paper we have ignored Coulomb interactions between the conduction electrons, only taking them into account at the magnetic impurity or quantum dot. While this may be justified at low T by Landau's Fermi liquid theory, for D = 2 or 3, it would be necessary for the inelastic scattering length to exceed ξ_K for our analyses to apply. In the D = 1 case, such interactions are important, even at $T = 0$. We analysed them only in a particularly simple case, the quantum dot end-coupled to a single channel quantum wire. While they somewhat reduced effects associated with the Kondo screening cloud, we concluded that the effects are still present and qualitatively unchanged.

Another effect which can interfere with observation of the screening cloud is non-magnetic disorder. We only touched on this briefly in Sec. 1.3.4, is our review of the "Kondo box" model. In addition a finite density of magnetic impurities further complicates the situation.

Throughout essentially all of this review, we have considered only the Kondo model, not the more realistic Anderson impurity model. We expect similar crossover effects to also occur in that case but with another characteristic length scale entering, where the crossover from the free orbital to (unstable) local moment fixed point occurs. The Kondo crossover can only be observed if this other crossover occurs at a significantly shorter length than ξ_K.

After many years of theoretical and theoretical investigations, the Kondo screening cloud, as predicted by the basic Kondo model, remains undetected experimentally. This is likely due to a combination of experimental difficulties, and limitations of the basic Kondo model. It is to be hoped that further experimental and theoretical progress will eventually bring this long search to a happy conclusion.

Acknowledgements

I would like to thank all my collaborators in this work. In chronological order: Erik Sørensen, Victor Barzykin, Pascal Simon, Rodrigo Pereria, Nicolas Laflorencie, Bert Halperin, Hubert Saleur and Lazlo Borda. I have also benefited from discussions on this topic with Jan von Delft, Junwu Gan, Nikolai Prokof'ev and Fred Zawadowski. This research was supported in part by NSERC and CIfAR.

References

1. A.C. Hewson, *The Kondo problem to heavy fermions*, [Cambridge University Press, Cambridge (1993)].
2. I. Affleck and A.W.W. Ludwig, Nucl. Phys. **B360**, 641 (1991).
3. K.G. Wilson, Rev. Mod. Phys. **47**, 773 (1975).
4. N. Andrei, Phys. Rev. Lett. **45**, 379 (1980); N. Andrei, K. Furuya and J.H. Lowenstein, Rev. Mod. Phys. **55**, 331 (1983).
5. P.B. Weigmann, JETP Lett. **31**, 364 (1980).
6. L. Borda, Phys. Rev. **B75**, 041307 (2007).
7. I. Affleck, L. Borda and H. Saleur, Phys. Rev. **B77**, 180404 (2008).
8. W.B. Thimm, J. Kroha and J. von Delft, Phys. Rev. Lett. **82**, 2143 (2001).
9. V. Barzkyin and I. Affleck, Phys. Rev. **B61**, 6170 (2000).
10. V. Barzykin and I. Affleck, Phys. Rev. Lett. **76**, 4959 (1996).
11. V. Barzykin and I. Affleck, Phys. Rev. **B57**. 432 (1998).
12. J. Gan, J. Phys.:Cond. Mat. **6**, 4547 (1994).
13. E. Sørensen and I. Affleck, Phys. Rev. **B53**, 9153 (1995).
14. J.P. Boyce and C.P. Slichter, Phys. Rev. Lett. **32**, 61 (1974); Phys. Rev. **B13**, 379 (1976).
15. G. Bergman, Phys. Rev. **B77**, 104401 (2008).
16. F. Mezei and G. Grüner, Phys. Rev. Lett. **29**, 1465 (1972).
17. F. Lesage and H. Saleur, J. Phys. **A30**, L457 (1997).
18. H. Ishii, J. Low Temp. Phys **32**, 457 (1978).
19. I. Affleck and P. Simon, Phys. Rev. Lett. **86**, 2854 (2001).
20. P. Simon and I. Affleck, Phys. Rev. **B64**, 085308, 2001.
21. E.S. Sørensen and I. Affleck, Phys. Rev. Lett. **94**, 086601 (2005).
22. H. Frahm and A.A. Zvyagin, J. Phys. **9**, 9939 (1997).
23. R.G. Pereira, N. Laflorencie, I. Affleck and B.I. Halperin, Phys. Rev. **B77**, 125327 (2008).
24. P.S. Cornaglia and C.A. Balseiro, Phys. Rev. **B66**, 115303 (2002); ibid 174404 (2002); Phys. Rev. Lett. **90**, 216801 (2003).
25. P. Simon and I. Affleck, Phys. Rev. Lett. **89**, 206602 (2002); Phys. Rev. **B68**, 115304 (2003).

Chapter 2

Plasmonics, Nanoplasmonics, and SPASER's

David J. Bergman
Raymond and Beverly Sackler School of Physics & Astronomy,
Faculty of Exact Sciences, Tel Aviv University,
IL-69978 Tel Aviv, Israel
Email: bergman@post.tau.ac.il
Tel: +972-3-640-8543, Fax: 972-3-642-9306

Plasma oscillations in a fluid of itinerant electrons or holes have attracted renewed attention following the emergence of composite media where nanometer-sized metal particles are embedded in a dielectric host, or nanometer-sized holes are punched in a thin metal film. This has led to the discovery of some new physical phenomena, e.g., enhanced light transmission (ELT) by very small holes, (i.e., smaller than the wavelength) and a nanometric device, known as SPASER, that can amplify or generate coherent light whose wavelength is much larger than the size of the device. Some of these new phenomena are described and discussed.

2.1. Surface Plasmons and Surface Plasmon Resonances

A plasma oscillation is a collective excitation wherein a charged fluid moves periodically in space with regard to an oppositely charged background which is either fixed, as in the case of electron gas in a solid metal, or has a very different inertia or mass, as in the case of a gaseous plasma of itinerant ions and electrons. The restoring force in such an oscillation is the Coulomb interaction between the opposite charges, but the Coulomb interactions between pairs of like-charged particles are also very important. Because these two-particle interactions are long ranged, the microscopic treatment of the dynamics, including interaction with an external electromagnetic (EM) field, is quite complicated. If we are interested in EM wavelengths that are much greater than the mean inter-particle distance, and in photon energies that are much less than the Fermi energy of an electron gas, then the EM response of such a system can be characterized by a frequency dependent complex electric permittivity $\epsilon(\omega)$. The characteristic frequency of plasma oscillations in a three-dimensional electron gas is called the plasma frequency ω_p, and it is given by

$$\omega_p^2 = \frac{4\pi n e^2}{m},$$

where n is the density of itinerant charge carriers and m is their mass. A simple approximate expression for $\epsilon(\omega)$ of a metal is

$$\epsilon(\omega) = \epsilon_0 - \frac{\omega_p^2}{\omega^2}\frac{\omega^2\tau^2}{1+\omega^2\tau^2} + \frac{4\pi i n e^2 \tau}{1+\omega^2\tau^2}.$$

Here ϵ_0 is the approximately frequency independent electric permittivity due to other constituents of the system, e.g., the strongly bound positive ions in the metal, and τ is the relaxation time that characterizes the interactions of the itinerant charge carriers with each other and with the rest of the system. If $\omega_p\tau > 1$, then when ω increases through the value $\omega_p/\sqrt{\epsilon_0}$, $\mathrm{Re}\,\epsilon(\omega)$ passes continuously through zero from negative to positive values. At that point the metal ceases to be opaque and reflecting and begins to enable EM radiation to propagate through it.

Somewhat less well known is the fact that when $\mathrm{Re}\,\epsilon(\omega) < 0$, a wave can propagate along the flat interface between that metal and any conventional dielectric medium — see Fig. 2.1. That wave decreases exponentially with increasing distance from the interface in both the bulk metal and the bulk dielectric, over a distance that depends on the wavelength. Such a wave is called a surface plasma wave or surface plasmon. In the dielectric it consists mostly of an electric field **E**, but in the metal it is a strongly coupled combination of an oscillating **E**-field and an oscillating electron density. Also notable is the fact that the magnetic field is much weaker than the electric field and the direction of **E** in the interface plane is the same as that of the wave vector, i.e., the wave has a strong longitudinal component of **E**. If the interface is the x, y-plane and the wave propagates along y with wave vector k_y, and the metal is at $z > 0$ and the wave number squared there is $k_1^2 = \epsilon_1\omega^2/c^2 < 0$, while the dielectric is at $z < 0$ and the wave number squared there is $k_2^2 = \epsilon_1\omega^2/c^2 > 0$, then

$$\mathbf{E} = \left(0, A_y e^{ik_y y - |k_z z|}, A_z e^{ik_y y - |k_z z|}\right).$$

The coefficient A_z and the imaginary wave vector k_z have different values for $z > 0$ and $z < 0$:

$$k_{iz}^2 = k_i^2 - k_y^2 = \epsilon_i \frac{\omega^2}{c^2} - k_y^2 < 0.$$

This expression will always be negative in the metal, since $\epsilon_1 < 0$. However, for this expression to be negative in the dielectric, we must have $k_y^2 > \epsilon_2 \omega^2 / c^2 \equiv k_2^2 > 0$. This means that the wavelength of the surface plasma wave $\lambda \equiv 2\pi / k_y$ will always be shorter than that of an EM plane wave with the same frequency in the bulk dielectric, and that there is no lower limit on λ of the surface plasmon.

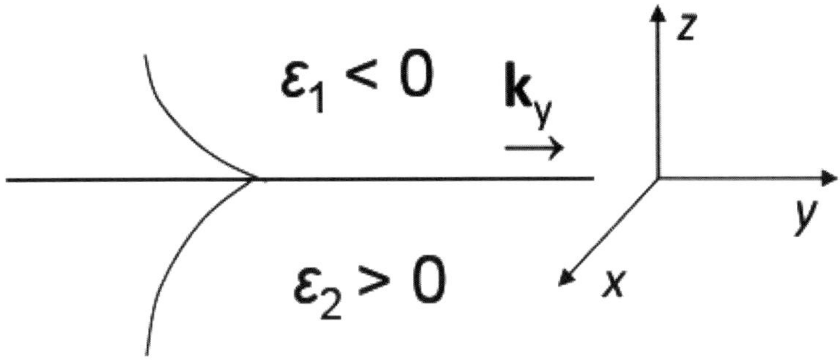

Fig. 2.1. Schematic drawing of a surface plasma wave propagating along a flat interface between a medium where $\epsilon_1 < 0$ (e.g., a metal) and a medium where $\epsilon_2 > 0$ (e.g., a conventional dielectric).

If instead of a planar interface between large volumes of metal and dielectric we now consider a small metal particle, we can expect that a surface plasma wave which propagates along such a surface will interfere with itself and lead to a discrete sequence of standing waves or surface plasmon resonances. Moreover, if $1/k_y$, $1/k_{iz}$, and $1/k_i$ are all much larger than the linear size of the metal particle, then those resonances will be eigenstates of Maxwell's equations in the quasistatic limit, where only the electric field is significant. That is why these resonances have also been called "electrostatic resonances" or "quasistatic resonances". These eigenstates were first discussed in Refs. 1,2, where they were introduced as a tool for calculating the electrical response of an isolated inclusion and of a composite medium made of many such inclusions. Somewhat later similar eigenstates of the full Maxwell equations were considered as a tool for calculating the EM scattering by a collection of macroscopic particles.[3]

An important property of these resonances is that their spatial size is independent of the wavelength of any bulk EM wave and depends only on the geometry of the microstructure. Thus, in the case of an isolated inclusion without any sharp geometric features the size of all the eigenfunctions is essentially that of the inclusion. If the inclusion has sharp features, then some of the eigenfunctions will be spatially localized near those features[4] — see Fig. 2.2. Finally, in the case of a cluster of inclusions with some inter-inclusion gaps that are very small, some of the eigenfunctions will be strongly localized in those gaps. Examples of these situations are shown in Fig. 2.3. In these examples it is clear that the spatial size of the exhibited resonances is not only smaller than any EM characteristic length, like wavelength or skin depth, but is also smaller than the particle sizes.

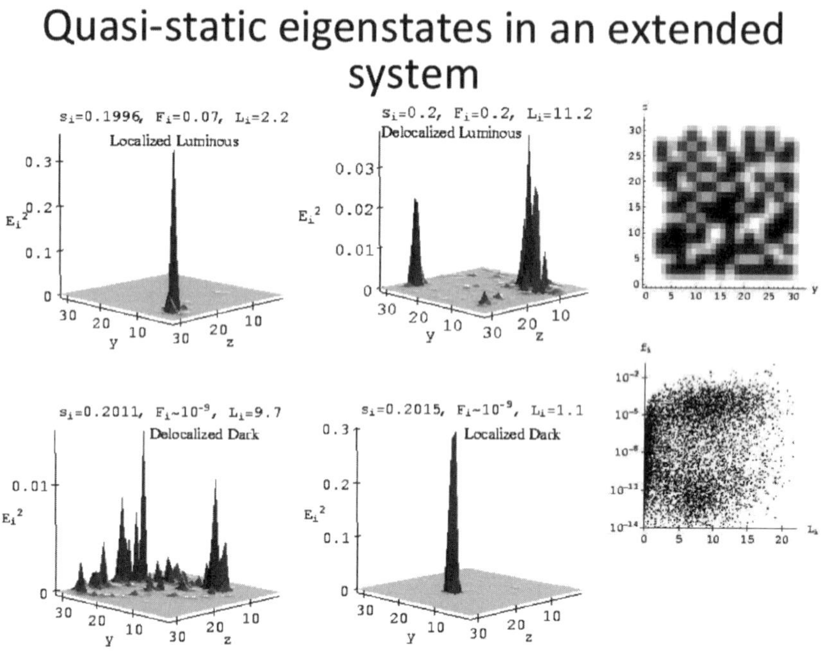

Fig. 2.2. Four of the many quasistatic eigenstates of a large disordered array of inclusions in a thin film shown on the right side. These were calculated by numerical solution of the appropriate boundary value problem based on the appropriate partial differential equation. Note how some of these eigenstates are localized in the close vicinity of a sharp geometic feature of the microstructure (after Ref. 4).

A simple example of such a resonance is the electrostatic dipole resonance of a spherical inclusion embedded in an otherwise uniform host,

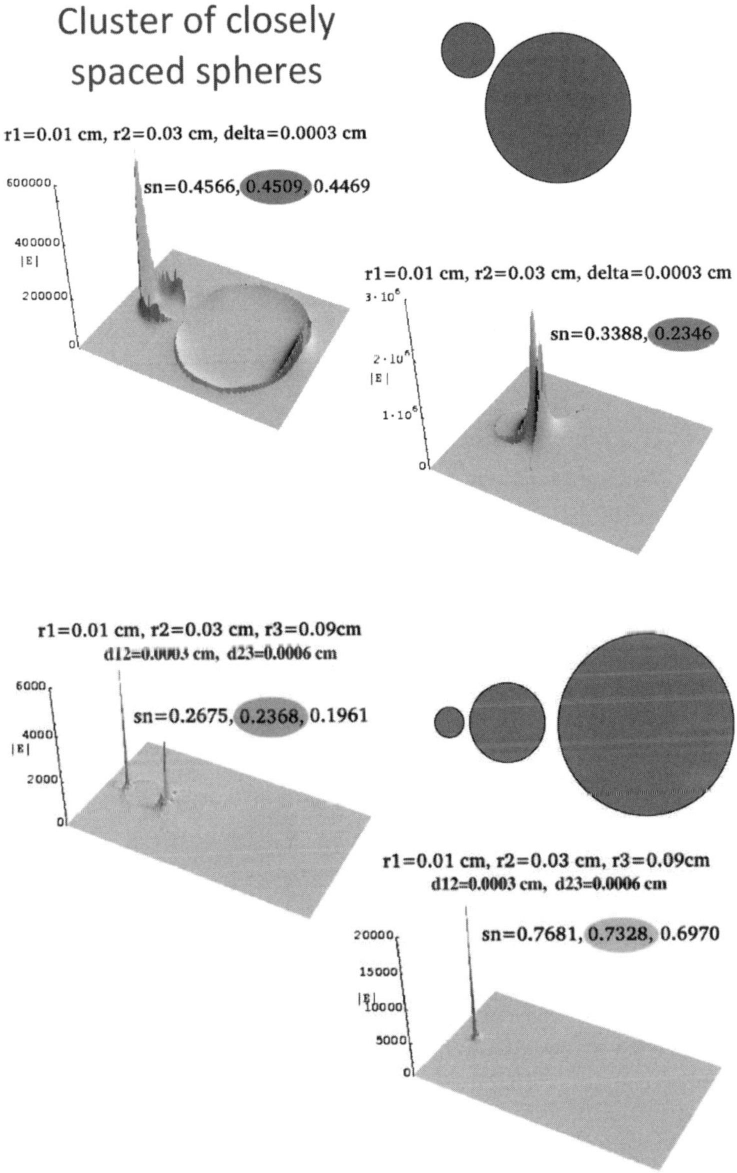

Fig. 2.3. Some quasistatic EM eigenstates of a two-sphere cluster and a three-sphere cluster which exhibit a strongly localized character in the inter-sphere gaps. The highlighted numbers denote eigenvalues [i.e., the special values of $\epsilon_2/(\epsilon_2 - \epsilon_1)$] that are well separated from any neighboring eigenvalues. That is important if one wishes to excite just one resonance.

which appears when the electric permittivity ratio is

$$\frac{\text{inclusion } \epsilon}{\text{host } \epsilon} \equiv \frac{\epsilon_1}{\epsilon_2} = -2.$$

This can be seen by recalling that when a uniform external electric field $\mathbf{E}_0$ is applied to that system, an electric dipole $\mathbf{p}$ is induced in the particle, given by

$$\mathbf{p} \propto a^3 \frac{\epsilon_1 - \epsilon_2}{\epsilon_1 + 2\epsilon_2} \mathbf{E}_0,$$

where a is the sphere radius — see Fig. 2.4. Actually, the sphere has an infinite sequence of order-l multipole resonances that are characterized by the following eigenvalues[1]

$$\frac{\epsilon_1}{\epsilon_2} = -\frac{l+1}{l}.$$

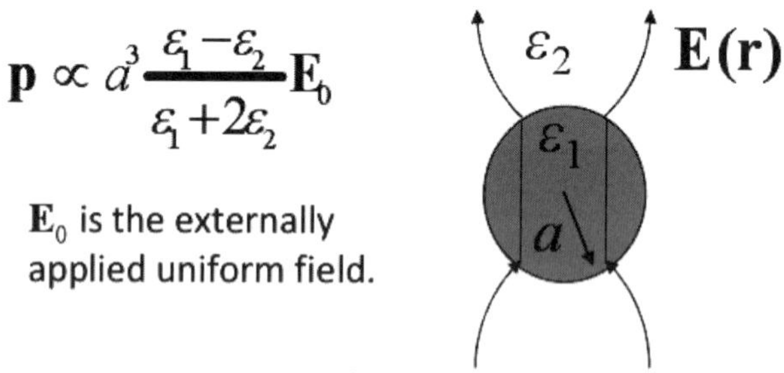

Fig. 2.4. The distorted electric field $\mathbf{E}(\mathbf{r})$ which results when a uniform external field $\mathbf{E}_0$ is applied to a single spherical inclusion. The induced electric dipole moment is $\mathbf{p}$.

Examples of some eigenstates in a tapered wedge of a thin film appear in Fig. 2.5.[5] Some of these eigenstates stretch out over the entire volume of the wedge, but one of them is strongly confined to the close vicinity of its apex. Examples of some strongly localized eigenstates in a linear cluster of two and three spheres with decreasing radii and separations appear in Fig. 2.3.

An important consequence of the existence of nano-localized quasistatic EM resonances is the possibility to concentrate or focus the energy of an incident EM wave in a sub-wavelength volume. This has led to the idea of a nanolens.[6]

Closely related to the eigenstates exhibited in Fig. 2.5 is the possibility of sending a surface plasma wave down a tapering conical surface toward its

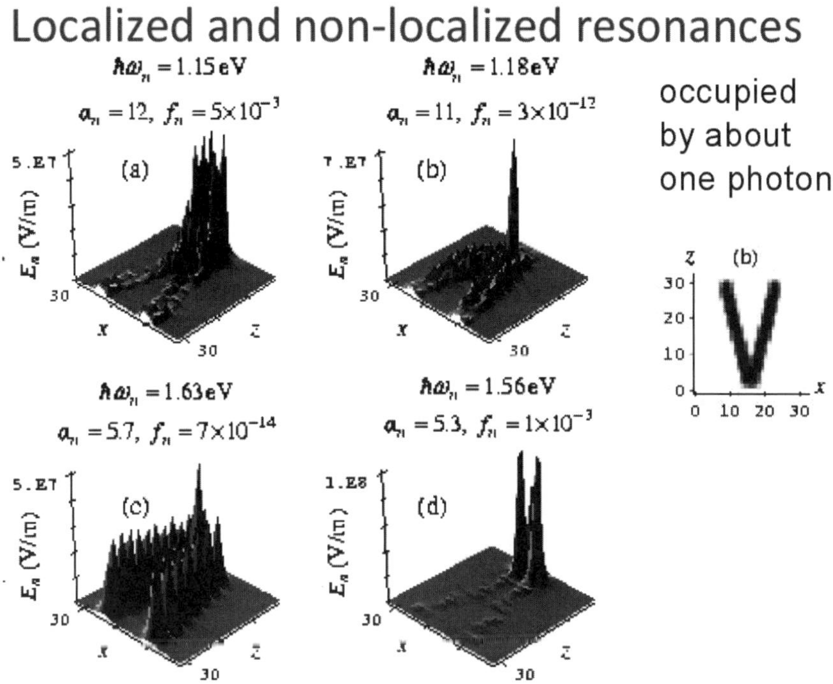

Fig. 2.5. Four of the many quasistatic eigenstates of a V-shaped narrow wedge, shown on the right side. These were calculated by numerical solution of the appropriate boundary value problem based on the appropriate partial differential equation. Note how some of these eigenstates are localized in the close vicinity of the apex (after Ref. 5).

apex, thereby also achieving a nano-focusing effect. A detailed theoretical treatment of this process shows that the wavelength of the surface plasma wave decreases adiabatically and the energy density increases adiabatically as the apex is approached.[7] When the wave reaches the apex, there is almost no reflection: A very hot spot appears there which enables the observation of a variety of strong-field, nonlinear phenomena, e.g., enhanced Raman scattering by appropriate molecules which are situated there.[8,9]

2.2. Small Quasistatic Resonances Applied to Nanolens and Nano-SPASER

The microgeometry shown in Fig. 2.5 was recently implemented in a pioneering experiment performed by Ewold Verhagen, Kobus Kuipers, and Albert Polman.[10,11] In Fig. 2.6 are shown measurements of of the EM intensity (i.e., the energy density) as measured by counting the number of visible-light photons ($\lambda = 550$ nm and 660 nm) emitted from levels of Er^{3+} ions which were excited in a sequence of transitions brought about by a strong basic incident field with an infrared wavelength $\lambda_{\text{inc}} = 1490$ nm. It is clear from these figures that a hot spot is generated at the apex whose diameter is 60 nm.

Another possible application of a quasistatic resonance with a sub-wavelength spatial size is for accumulating enough energy that the local **E**-field is strong enough to bring about induced emission of additional quanta into that resonance. These quanta are not photons, since the oscillating field is not just an **E**-field but a combination of **E**-field and a collective oscillatory motion of the electron fluid in the metal. Since the number of these quanta reflect the energy of an appropriate surface plasma resonance, they should rightly be called "surface plasmons", or just "plasmons". A source for emitting these plasmons into the surface plasma resonance can be a collection of excited atoms or molecules or quantum dots that have a pair of states with the right energy gap and are spatially located where the **E**-field of the resonance is strong.

If the gain of such a stimulated emission process is large enough to overcome the losses in the system, then the number of coherent plasmons occupying the resonance will build up to some saturated value that is significantly greater than 1. Such a device is known as a SPASER (acronym for "surface plasmon amplification by stimulated emission of radiation"). This is, admittedly, somewhat of a misnomer, since what is emitted into the resonance is not pure EM radiation, i.e., photons, but quanta of the surface plasma resonance, namely plasmons.

This device was first proposed in Ref. 5, where the conditions for implementing it in practice were discussed and calculated. Recently such a device was demonstrated in an experiment by Noginov et al.[12] In Fig. 2.7, which is copied from that reference, one can see that the 530 nm spectral line emitted by individual particles, in the form of a 14 nm diameter gold sphere coated by a 44 nm diameter dye-doped silica shell, narrows significantly when the excitation of the dye molecules is increased above a certain threshold. In that regime, the emission intensity increases linearly with the intensity of excitation.

An interesting further extension of some of the ideas discussed above is to exploit a nanolens with an ultra-small quasistatic resonance as the basis for an ultra-small SPASER.[13]

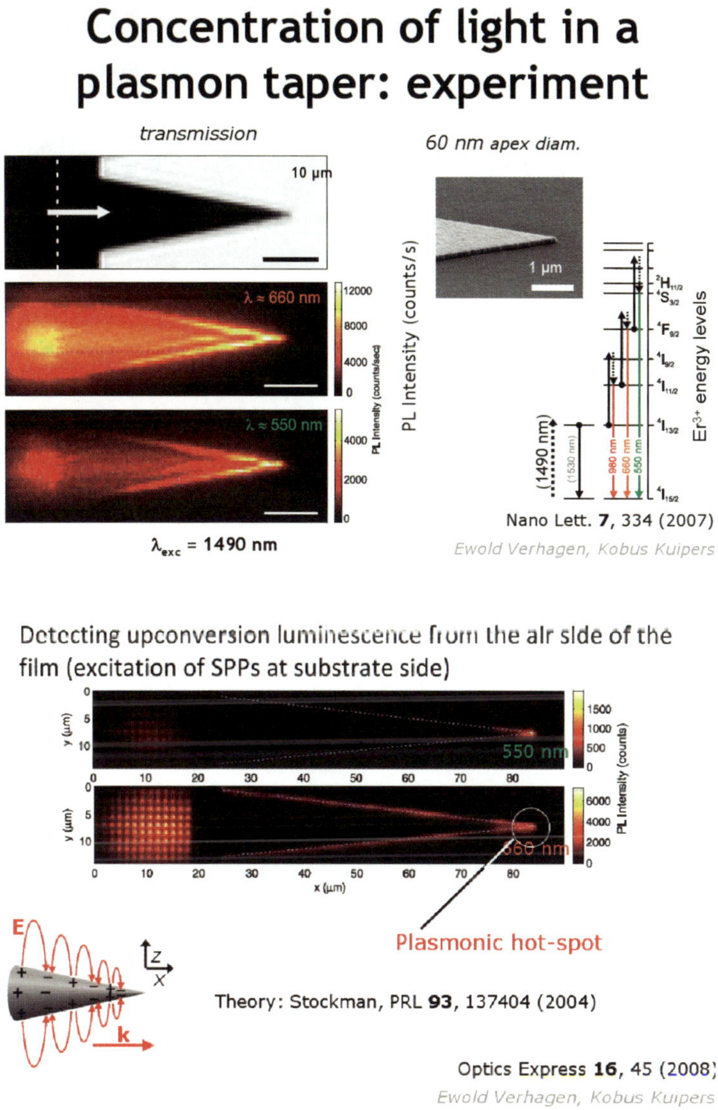

Fig. 2.6. Experiments where a thin gold film tapered wedge was illuminated by a coherent infrared pulse at the wide edge. The wedge was formed on a sapphire substrate, in which Er^{3+} ions were embedded near the sapphire/gold interface. These allowed a much higher frequency surface plasma wave to be excited due to a sequence of transitions between excited states of that ion. Finally, some of those higher frequency waves traveled down the wedge, along its surface, and formed a hot spot at the apex (after Refs. 10,11.)

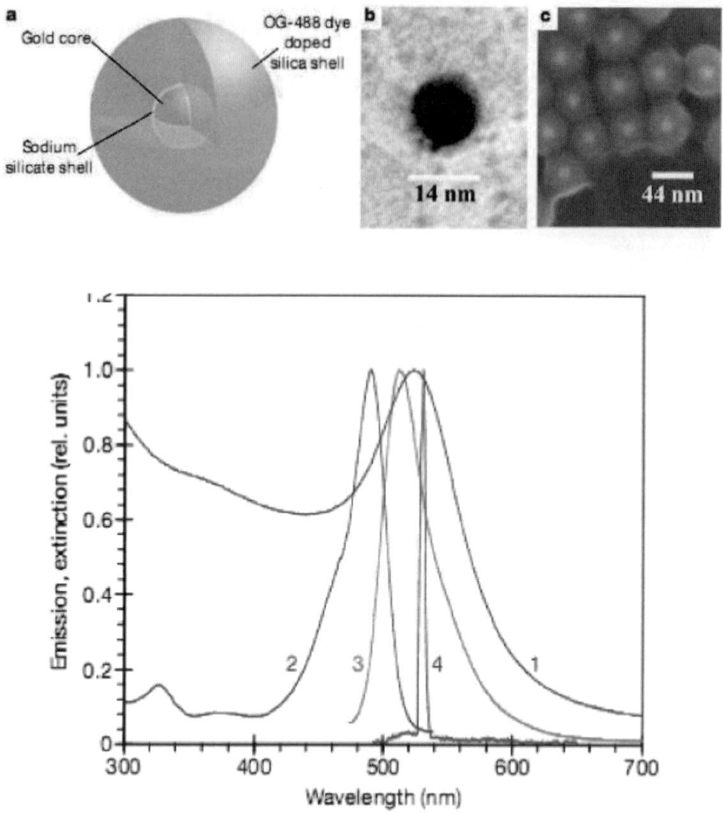

Fig. 2.7. Top: (a) Schematic picture of a gold nanosphere (diameter 14 nm) coated by a thin sodium silicate shell and a thicker, dye-doped, silica shell, of total diameter 44 nm. (b) TEM image of the gold core. (c) SEM image of a cluster of the coated nanoparticles. Bottom: Spectroscopic results under EM radiation which excites the dye molecules in a collection of suspended, separated nanoparticles: Normalized extinction (1), excitation of the dye molecules (2), spontaneous emission (3), and stimulated emission (4). Stimulated emission is signaled by a significant narrowing of the emission spectrum, and appears as soon as the incident intensity exceeds a critical threshold value (after Ref. 12).

2.3. Summary and Conclusions

I described some of the early history of nano-plasmonics and how the surface plasmon resonances have become a useful tool in the discussion of the EM response of metal nano-structures.

The theoretical prediction of nanolens and SPASER were two of the results based on those resonances. The subsequent recent experimental realization of such devices shows that those resonances can be a useful tool in the theoretical and numerical study of nanometric optics and plasmonics.

Other applications of that tool which were not discussed here include coherent control of femtosecond optical pulses in a nano-structure[14,15] and the synthesis of nano-structures with a macroscopic response that is both electric and magnetic. This means that the macroscopic electric permittivity ϵ_e as well as the macroscopic magnetic permeability μ_e are both significantly different from 1. That includes the possibility that they are both negative, which would mean that the material is a "left-handed EM medium".[16,17]

Acknowledgments

Partial support for this research was provided by grants from the U.S.-Israel Binational Science Foundation, the Russia-Israel Science Foundation, and the Israel Science Foundation.

References

1. D. J. Bergman, The dielectric constant of a two-component granular composite—a practical scheme for calculating the pole spectrum, *Phys. Rev. B* **19**, 2359–2368 (1979).
2. D. J. Bergman, The dielectric constant of a simple - cubic array of identical spheres, *J. Phys. C* **12**, 4947–4960 (1979).
3. D. J. Bergman and D. Stroud, Theory of resonances in the electromagnetic scattering by macroscopic bodies, *Phys. Rev. B* **22**, 3527–3539 (1980).
4. M. I. Stockman, S. V. Faleev, and D. J. Bergman, Localization versus Delocalization of Surface Plasmons in Nanosystems: Can One State Have Both Characteristics?, *Phys. Rev. Letters* **87**, 167401-1–167401-4 (2001).
5. D. J. Bergman and M. I. Stockman, Surface plasmon amplification by stimulated emission of radiation: Quantum generation of coherent surface plasmons in Nanosystems, *Phys. Rev. Letters* **90**, 027402-1–027402-4 (2003).
6. K. Li, M. I. Stockman, and D. J. Bergman, Self-similar chain of metal nanospheres as an efficient nanolens, *Phys. Rev. Letters* **91**, 227402-1–227402-4 (2003).
7. M. I. Stockman, Nanofocusing of optical energy in tapered plasmonic waveguides, *Phys. Rev. Letters* **93**, 137404 (2004).
8. A. Hartschuh, E. J. Sanchez, X. S. Xie, and L. Novotny, *Phys. Rev. Letters* **90**, 095503 (2003).
9. J. Kneipp, X. Li, M. Sherwood, U. Panne, H. Kneipp, M. I. Stockman, and K. Kneipp, Gold nanolenses generated by laser ablation—efficient enhancing structure for surface enhanced Raman scattering analytics and sensing, *Anal. Chem.* **80**, 4247–4251 (2008).

10. E. Verhagen, L. Kuipers, and A. Polman, Enhanced nonlinear optical effects with a tapered plasmonic waveguide, *Nano Lett.* **7**, 334–337 (2007).
11. E. Verhagen, A. Polman, and L.(Kobus) Kuipers, Nanofocusing in laterally tapered plasmonic waveguides, *Optics Express* **16**, 45–52 (2008).
12. M. A. Noginov, G. Zhu, A. M. Belgrave, R. Bakker, V. M. Shalaev, E. E. Narimanov, S. Stout, E. Herz, T. Suteewong, and U. Wiesner, Demonstration of a spaser-based nanolaser, *Nature* **460**, 1110–1113, (2009).
13. K. Li, X. Li, M. I. Stockman, and D. J. Bergman, Surface plasmon amplification by stimulated emission in nanolenses, *Phys. Rev. B* **71**, 115409-1–115409-5 (2005).
14. M. I. Stockman, S. V. Faleev, and D. J. Bergman, Coherent control of femtosecond energy localization on nanoscale, *Phys. Rev. Letters* **88**, 067402-1–067402-4 (2002).
15. M. I. Stockman, M. F. Kling, U. Kleinberg, and F. Krausz, Attosecond nanoplasmonic-field microscope, *Nature Photonics* **1**, 539–544 (2007).
16. D. R. Smith, W. Padilla, D. C. Vier, S. C. Nemat-Nasser, and S. Schultz, A composite medium with simultaneously negative permeability and permittivity, *Phys. Rev. Lett.* **84**, 4184–4187 (2000).
17. S. O'Brien and J. B. Pendry, Photonic band-gap effects and magnetic activity in dielectric composites, *J. Phys.: Condens. Matter* **14**, 4035–4044 (2002).

Chapter 3

Entanglement, Bunching, and Counting Statistics

Christoph Bruder[1] and Wolfgang Belzig[2]

[1]*Department of Physics, University of Basel, CH-4056 Basel, Switzerland*

[2]*Department of Physics, University of Konstanz, D-78457 Konstanz, Germany*

This article briefly reviews the creation of entanglement in mesoscopic devices, in particular, devices based on a superconducting injector. We then discuss whether and how current correlation measurements can be used to confirm the existence of entanglement.

3.1. Introduction

In contrast to classical systems, quantum-mechanical systems can exist in coherent superpositions of different states, and this quantum coherence gives rise to a variety of peculiar electronic transport properties.[1] A particularly interesting form of quantum superposition that appears in systems consisting of several subsystems was called "Verschränkung" or *entanglement* by Schrödinger in 1935. Entanglement is the basis of many of the fascinating properties of quantum mechanics. Some of the most dramatic manifestations of entangled quantum states are quantified in Bell inequalities, whose violation can serve as a probe of entanglement. In particular, maximally entangled states violate these Bell inequalities maximally. Consequently, they cannot be described by local hidden-variable theories and are decidedly non-classical. Entangled states are an important resource for applications like quantum cryptography, which give the possibility of secure communication, quantum teleportation, and quantum computing, that enables an exponential increase in speed for the solutions of certain problems like e.g. the factorization of large integers (Shor algorithm).

To discuss entanglement, we need to consider a system consisting of several subsystems. Pure entangled states can be characterized by their inability to be 'factorized' in a product of factors living in different of these subsystems. The prototype entangled state of two electrons is the spin singlet state

$$|\psi\rangle = \frac{1}{\sqrt{2}}(|\uparrow\downarrow\rangle - |\downarrow\uparrow\rangle) \tag{3.1}$$

that clearly cannot be expressed as a product of factors referring to one spin. A variety of sources of such entangled spin singlet states has been discussed in the literature (see Ref. 2 for a recent review). One attractive and conceptually easy idea is to use two semiconductor quantum dots[3] containing one electron each. The presence of a weak tunnel coupling in combination with the Coulomb repulsion between two electrons sitting on the same dot leads to an effective antiferromagnetic Heisenberg coupling of the two electron spins that can be used to generate a singlet state like in Eq. (3.1).

The ground state of a conventional even-parity BCS superconductor can be written as

$$|BCS\rangle = \prod_k (u_k + v_k c^\dagger_{k\uparrow} c^\dagger_{-k\downarrow})|0\rangle \tag{3.2}$$

where u_k, v_k are even in k and $c^\dagger_{k\uparrow}$ creates an electron with wave vector k and spin up. It is easy to see that the terms with wave vectors k and $-k$ in the product lead to the combination $c^\dagger_{k\uparrow}c^\dagger_{-k\downarrow} - c^\dagger_{k\downarrow}c^\dagger_{-k\uparrow}$, i.e., a singlet or entangled pair. In other words, the ground state of a superconductor consists of entangled electron pairs, the so-called Cooper pairs. It is therefore tempting to use a superconductor as a source of entangled particles, but this requires extracting a pair and separating the two members of the Cooper pair, e.g., by using two quantum dots.[4–6] Related electron entangler devices with an SC injector attached to normal-metal leads have been studied by a number of groups.[7–13] For proposal of solid-state entanglers in non-interacting structures, see.[14–17]

3.2. Detection and Verification of Entangler Output

The theoretical proposals mentioned before have triggered a number of experimental attempts to realize an electron entangler based on a superconducting injector. Before discussing some of these experiments, we would like to study the question how to probe and verify entanglement once it has been created.

In principle, once a Cooper pair has been extracted from the superconductor and separated in a beam splitter, correlated measurements of the spins of the two electrons along different axes can be used to verify its singlet character. The probabilities $C_{\vec{a},\vec{b}}$ of a correlated detection of two spins along directions $\vec{a}$ and $\vec{b}$ in two different output ports can be combined to form the so-called Bell parameter

$$B = C_{\vec{a},\vec{b}} - C_{\vec{a},\vec{b}'} + C_{\vec{a}',\vec{b}} + C_{\vec{a}',\vec{b}'}\,, \tag{3.3}$$

which obeys the Bell inequality $|B| \leq 2$ for classical correlations. It can be easily shown that the Bell parameter for the quantum-mechanically entangled singlet state shown in Eq. (3.1) can violate the Bell inequality. Hence, the quantum-mechanical correlation is stronger than any classical correlation can be. In an electric circuit, one could relate the spin correlator to a spin-resolved cross-correlation measurement.[7–9]

However, such cross-correlation measurements are difficult in practice. First, it is usually more difficult to measure the cross noise, which is typically an order of magnitude smaller than the auto-correlation signal. Second, one has to realize spin measurements along two different non-collinear directions in a solid-state environment in a controllable manner. This is possible using differently shaped ferromagnetic islands, whose magnetization direction can be controlled by an external magnetic field. However, the integration of normal, superconducting and ferromagnetic elements in such a complex circuit has not yet been achieved.

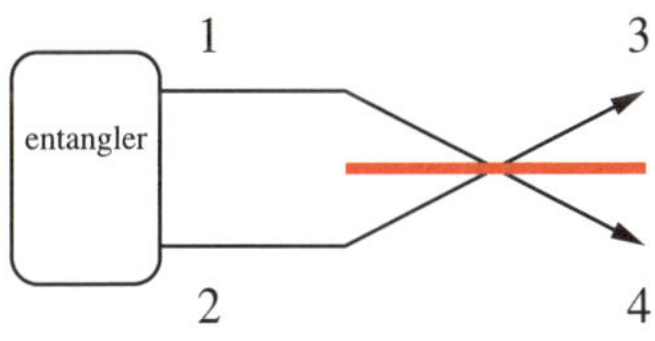

Fig. 3.1. An entangled spin-singlet pair injected in leads 1, 2 is recombined at a 50:50 beam splitter (red bar). The symmetry of the orbital wave function leads to bunching and doubling of the shot-noise power. Adapted from Ref. 18.

An interesting proposal to detect entanglement has been put forward in Refs. 18–23 that describes a method for detecting pairwise entanglement between mobile electrons in two mesoscopic wires based on the measurement of the current noise in one of the wires. The idea is to recombine the two output lines of the entangler in a 50:50 beam splitter, see Fig. 3.1. Since the orbital wave function of a singlet pair is symmetric (like the wave function of two bosons), there will be bunching effects in the output of the beam splitter, leading to an enhancement of the low-frequency noise by a factor 2 in the output lines.[18] The proposal has the big advantage, that only the noise has to be measured, which is experimentally simpler than the cross-correlations. The difficulty here lies in the necessity of preparing a suitable beam of entangled pairs, which might however be overcome by a pulsed generation of the pairs.

The use of entanglement witnesses to detect entanglement in a transport experiment has been proposed in Ref. 24,25.

In the following section, we would like to investigate the correlations in a continuous current generated by a superconducting beam splitter.

3.3. Current Correlations and Counting Statistics

The current in mesoscopic structures fluctuates in time. Due to the coherent nature of the transport, these fluctuations provide information on the quantum physics of the underlying transport mechanism.

The current noise, i. e., the second moment of the full counting statistics, is of particular interest. It can be used as a diagnostic tool to probe the nature and the quantum statistics of the charge carriers.[26]

It is also possible to measure *nonlocal* current-current correlations, the so-called cross-correlations. Negative cross-correlation originate from the Pauli exclusion principle, which leads to a noiseless stream of incoming particles at zero temperature.[27] The electrons are scattered at a beam splitter one by one. An electron leaving in one lead cannot leave in the other, therefore the time-dependent current fluctuations in the two leads are anticorrelated. This is a consequence of the fermionic correlation of the incoming beam. A natural question arising is what happens if the incoming particles are correlated differently. This can be achieved by using a superconducting terminal, where particles are injected by Andreev reflection, i. e., as correlated electron-hole pairs. This can eventually lead to positive cross-correlations, as was shown for different geometries.[28–31]

In the following we address the correlations in beam splitters from the point of view of full counting statistics.[32] In this approach one calculates not only the average current and the current noise, but the full distribution of electrons transferred in a certain time interval. For a general three-terminal setup we denote the probability that N_1 particle leave through terminal 1 and N_2 through terminal 2 by $P(N_1, N_2)$. Here we have used that the particle count in the third terminal is fixed by particle conservation and we have suppressed the dependence on the counting time t_0. A quantity, that is directly accessible by a Green's function method,[33] is the *cumulant-generating function* (CGF) $S(\chi_1, \chi_2)$, defined by

$$e^{-S(\chi_1,\chi_2)} = \sum_{N_1,N_2} P(N_1, N_2) e^{i\chi_1 N_1 + i\chi_2 N_2} . \tag{3.4}$$

Knowing the CGF is therefore equivalent to knowing the full counting statistics. All cumulants can be obtained as derivatives with respect to the counting fields, e. g. the cross-correlations between terminal 1 and 2 is given by $S_{12} = (\partial^2/\partial\chi_1\partial\chi_2)\, S(\chi_1, \chi_2)|_{\chi_1=\chi_2=0}$.

In the following, we will study a three-terminal beam splitter, in which one terminal acts as a source of particles and two terminals act as drains. This can usually be achieved by applying the same voltage between the source and the terminals 1 and 2. In particular we will address a superconducting or a normal-metal source, whereas the drains are always normal metals. Specific results are obtained for a setup with three tunnel junctions, in which all terminals are connected to a central node by tunnel junctions with conductances g_S and $g_1 = g_2 = g_N/2$, see Fig. 3.2(a). Other types of three-terminal devices with one superconducting and two normal-metal leads have been studied in the literature.[34]

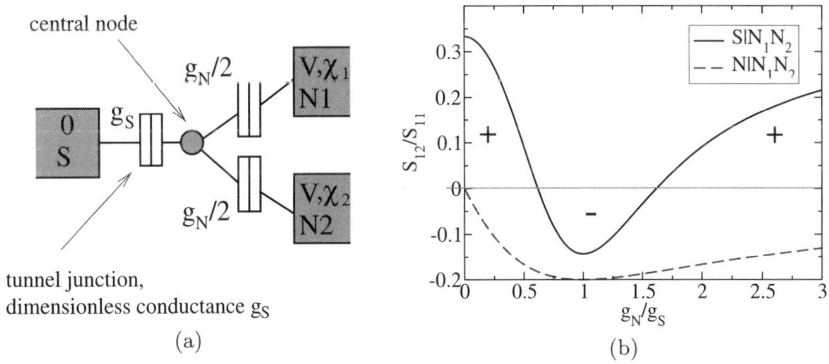

Fig. 3.2. (a) Three-terminal beam splitter. One superconducting or normal-metal terminal (S or N) and two normal terminals (N_1 and N_2) are tunnel-coupled by conductances g_S and $g_1 = g_2 = g_N/2$ to a common central node. A current is passed from S/N into the two normal leads that are kept at the same voltage. Ideal passive charge counters are indicated by the counting fields χ_1 and χ_2. (b) Solid line: Current cross-correlations for a superconducting beam splitter $\mathrm{S|N_1N_2}$. Dashed line: current cross-correlations for a normal beam splitter $\mathrm{N|N_1N_2}$. For $g_N/g_S < (3-\sqrt{5})/2$, the cross-correlations are positive in the superconducting case. In the normal case, the cross-correlations are always negative.

Introducing $p = g_S g_N/(g_S^2 + g_N^2)$ it was shown[30] that the CGF has the form

$$S(\chi_1, \chi_2) = -\frac{M^S}{\sqrt{2}}\sqrt{1 + \sqrt{1 + p^2(e^{i\chi_1} + e^{i\chi_2})^2 - 4p^2}}\,, \tag{3.5}$$

where $M^S = Vt_0\sqrt{g_S^2 + g_N^2}/e$. The inner argument contains counting factors for the different possible processes. A term $\exp(i(\chi_k + \chi_l) - 1)$ corresponds to an event in which two charges leave the superconducting terminal and one charge is counted in terminal k and one charge in terminal l. The prefactors are related to the corresponding probabilities. We note that $S(\chi_1 + \pi, \chi_2 + \pi) = S(\chi_1, \chi_2)$, and consequently the total number of electrons transferred is always even. This is a result of the Andreev reflection process, which allows only transfer in pairs.

The corresponding CGF for the case in which the superconductor is replaced by a normal metal is

$$S^N(\chi_1, \chi_2) = -M^N\sqrt{1 + p^N(e^{i\chi_1} + e^{i\chi_2}) - 2p^N}\,, \tag{3.6}$$

where $p^N = 2g_S g_N/(g_S + g_N)^2$ and $M^N = Vt_0(g_S + g_N)/2e$. Here even and odd numbers of particles can be transferred, since $S^N(\chi_1, \chi_2) \neq S^N(\chi_1 + \pi, \chi_2 + \pi)$.

We will now calculate the cross-correlations of the currents in the two normal leads. These are defined as $S_{ij} = 2\int dt \langle \delta I_i(t)\delta I_j(0)\rangle$, where $I_{i(j)}(t)$ are the time-dependent currents in terminal $i(j)$, and can be calculated from the CGF using $S_{ij} = (2e^2/t_0)(\partial^2/\partial\chi_i\partial\chi_j)\, S(\chi_1, \chi_2)|_{\chi_1=\chi_2=0}$. In the low-energy limit ($T = 0$, $V \ll \Delta$), we obtain:

$$S_{12} = \frac{G|eV|}{2}\left(1 - 5\frac{g_S^2 g_N^2}{(g_S^2 + g_N^2)^2}\right), \quad S_{11} = \frac{G|eV|}{2}\left(3 - 5\frac{g_S^2 g_N^2}{(g_S^2 + g_N^2)^2}\right),$$

here, $G = g_S^2 g_N^2/(g_S^2 + g_N^2)^{3/2}$.

For completeness we also cite the results for the case, in which the superconductor is in its normal state:

$$S_{12}^N = -\frac{G^N|eV|}{2}\frac{g_S g_N}{(g_S + g_N)^2}, \quad S_{11}^N = \frac{G^N|eV|}{2}\left(2 - 3\frac{g_S g_N}{(g_S + g_N)^2}\right),$$

here, $G^N = g_S g_N/(g_S + g_N)$.

In Fig. 3.2(b) we show the cross-correlations, emphasizing the difference between the superconducting state and the normal state. Interestingly, the transport properties are invariant under inversion of the conductance ratio g_N/g_S. In the following we will thus distinguish the case of weak (or strong) proximity effect determined by $g_N \not\approx g_S$ and the case of optimal proximity effect defined by $g_N \approx g_S$. Below we will use the terminology weak for both the weak and the strong proximity regimes equally.

Figure 3.2(b) shows that it is possible to obtain positive cross-correlations in the superconducting beam splitter, whereas they are manifestly negative for the normal beam splitter. The cross-correlations are positive in the limit of weak proximity. Thus, it should be able to derive and interpret them from simple arguments. Indeed, from a calculation of the full counting statistics,[30] it follows that in the weak proximity limit the statistics consists of *independent* events of pair tunneling. The possible events are pair tunneling into terminal N_1 or N_2 and correlated tunneling into both terminals. These events occur with equal probabilities, however, in the limit of weak proximity these events are *uncorrelated.* Thus, two-particle tunneling events into the same lead do *not* contribute to cross-correlations. In contrast, tunneling of two particles into different leads is *automatically* positively correlated. Thus, it follows quite generally, that the cross-correlations are positive.

A beam splitter with one superconducting and two ferromagnetic normal leads coupled to a normal-metal island was also investigated[35] and shown to exhibit positive cross-correlations even in the presence of spin relaxation. Finally, we would like to mention that positive cross-correlations in normal-metal structures have also been predicted in a number of circumstances;[36–40] the effect of inelastic scattering modeled by a voltage probe[36] has been experimentally confirmed.[41]

3.4. Quantitative Evaluation of the Beam Splitter Statistics

To get explicit expressions for the distribution functions $P(N_1, N_2)$ described by Eqs. (3.5,3.6), we will now prove that the CGF (3.5) and (3.6) imply that the counting statistics of a symmetric beam splitter has in general the form

$$P(N_1, N_2) = P_0(N_1 + N_2) \binom{N_1 + N_2}{N_1} \left(\frac{1}{2}\right)^{N_1+N_2}, \qquad (3.7)$$

here, $P_0(Q)$ is the probability that Q electrons are transferred in total.

To this end we show that a probability distribution (3.7) results in a CGF, which has the form

$$S(\chi_1, \chi_2) = S_0 \left(\frac{e^{i\chi_1} + e^{i\chi_2}}{2}\right). \qquad (3.8)$$

The CGF $S_0(e^{i\chi})$ is the cumulant-generating function corresponding to the total charge transfer probability $P_0(Q)$ and is defined by

$$e^{-S_0(e^{i\chi})} = \sum_Q P_0(Q) e^{i\chi Q} . \tag{3.9}$$

The full CGF of the distribution (3.7) is given by

$$e^{-S(\chi_1,\chi_2)} = \sum_{N_1,N_2} e^{i\chi_1 N_1 + i\chi_2 N_2} P_0(N_1+N_2) \binom{N_1+N_2}{N_1} \left(\frac{1}{2}\right)^{N_1+N_2} . \tag{3.10}$$

Replacing the sum over N_1 and N_2 by a sum over $Q = N_1 + N_2$ and N_1, the sum over N_1 can be evaluated and we obtain

$$\sum_Q \left(\frac{1}{2}\right)^Q \left(e^{i\chi_1} + e^{i\chi_2}\right)^Q P_0(Q) , \tag{3.11}$$

which is precisely the definition (3.9) of S_0 with $\exp(i\chi)$ replaced by $(\exp(i\chi_1) + \exp(i\chi_2))/2$.

The CGFs for the total charge transfer statistics for the superconducting splitter is

$$S_0^S(e^{i\chi}) = -\frac{M^S}{\sqrt{2}} \sqrt{1 + \sqrt{1 + p^2(e^{i2\chi} - 1)}} , \tag{3.12}$$

and

$$S_0^N(e^{i\chi}) = -M^N \sqrt{1 + \frac{p^N}{2}(e^{i\chi} - 1)} \tag{3.13}$$

for the normal case. It is interesting to observe that the parity effect in the number of total transferred charges is only evident in the π-periodicity of S_0, whereas the splitting among the two drains of the total number of charges is the same for the normal and the superconducting case.

The total probability distribution $P_0(Q)$ can be found in the limiting case $g_N/g_S \ll 1$ or vice versa. It vanishes for odd values of Q; for even values it turns out to be a Poisson distribution describing an uncorrelated transfer of $Q/2$ Cooper pairs,

$$P_0^S(Q) = e^{-\frac{\bar{Q}}{2}} \left(\frac{\bar{Q}}{2}\right)^{\frac{Q}{2}} / \left(\frac{Q}{2}\right)! \tag{3.14}$$

Here we have defined the average number of transferred electrons $\bar{Q} = t_0 V G/e$.

Similarly, we obtain in the normal case in the same limit the total charge transfer probability distribution

$$P_0^N(Q) = e^{-\bar{N}_1 - \bar{N}_2} \frac{(\bar{N}_1 + \bar{N}_2)^Q}{Q!}, \tag{3.15}$$

for even and odd Q. Here we have abbreviated the average number transferred into terminal i by $\bar{N}_i$. In contrast to the superconducting case the counting statistics factorizes

$$P^N(N_1, N_2) = e^{-\bar{N}_1} \frac{\bar{N}_1^{N_1}}{N_1!} e^{-\bar{N}_2} \frac{\bar{N}_2^{N_2}}{N_2!}. \tag{3.16}$$

Thus, the distribution in the normal case is the product of two Poisson distributions of charge transfers into the two terminals. In the superconducting case such a factorization is not possible.

3.5. Recent Experiments

The theoretical proposals for superconducting entanglers discussed in Section 3.1 and the works on cross-correlations in Section 3.3 have triggered a number of transport experiments on superconductor-normal beam splitters.[42,43]

In 2009, three Cooper-pair splitter experiments were performed.[44–46] In the Basel experiment,[44] a superconductor injects Cooper pairs into an InAs semiconductor quantum wire split into two quantum dots created by gate electrodes. In the second experiment,[45] the semiconductor quantum wire is replaced by a carbon nanotube, which is again split into two quantum dots by gate electrodes. Both experiments show a non-local dependence of the current through one quantum dot on the gate voltage applied to the other dot. This is an indirect proof for the splitting and spatial separation of Cooper pairs.

In the third experiment,[46] the experimentalists managed to observe the cross-correlations of currents in two normal-metal copper wires directly connected to a superconductor, without an intermediate quantum dot. In this circuit, elastic co-tunneling from one normal lead to the other yields a negative contribution to the cross-correlations that may mask the positive value discussed in Section 3.3. The experimental observation of positive cross-correlations in a certain parameter range in the Evanston experiment[46] is therefore a strong hint for the splitting of Cooper pairs.

However, all of these experiments are not yet at the point to probe entanglement. An entanglement probe requires spin-resolved current cross-correlation measurements, like discussed e.g. in Refs. 47–49. In these papers, the results for beam splitters with two normal leads are generalized to ferromagnetic contacts, which can be either collinear or non-collinear. The study of full counting statistics allows for a full identification of the probabilities of elementary events. These are Cooper-pair transfers, and the dependence of the angles of the magnetization directions of the detectors are compatible with the singlet nature of the Cooper pairs. If the detectors are highly polarized, these results can be used to violate a Bell inequality.

3.6. Conclusion

We have discussed entangled states and some aspects of their generation in solid-state entanglers, in particular, sources of spin-entangled electron pairs based on a superconducting injector. After describing various proposals of how to detect and verify the output of an entangler, we have concentrated on current correlations and the full counting statistics of superconducting beam splitters and given explicit expressions for the cumulant-generating function and the distribution functions.

For a multi-terminal structure with one superconducting and two normal-metal arms coupled by tunnel junctions to a central normal-metal island, we have shown that positive cross-correlations of the currents in the two arms are a generic feature. Remarkably, in the region in which the proximity effect has the strongest impact on the conductance, the cross-correlations are negative. This can be explained by a Pauli exclusion principle for Cooper pairs in a transport process. If superconductivity in the input lead of the beam splitter is suppressed, the cross-correlations are negative.

Three recent beam-splitter experiments indicate that it is indeed possible to separate Cooper pairs. However, positive cross-correlations alone are not sufficient to prove the existence of entanglement. To confirm that these structures are indeed sources of entangled pairs, measurements of spin-resolved current correlations are needed.

Acknowledgments

This work was financially supported by the Swiss NSF and the NCCR Nanoscience.

References

1. Y. Imry, *Introduction to Mesoscopic Physics, 2nd ed.* (Oxford University Press, Oxford, 2002).
2. G. Burkard, J. Phys.: Condens. Matter **19**, 233202 (2007).
3. D. Loss and D.P. DiVincenzo, Phys. Rev. A **57**, 120 (1998).
4. M.-S. Choi, C. Bruder, and D. Loss, Phys. Rev. B **62**, 13569 (2000).
5. P. Recher, E.V. Sukhorukov, and D. Loss, Phys. Rev. B **63**, 165314 (2001).
6. P. Recher and D. Loss, Phys. Rev. B **65**, 165327 (2002).
7. G.B. Lesovik, T. Martin, and G. Blatter, Eur. Phys. J. B **24**, 287 (2001).
8. N.M. Chtchelkatchev, G. Blatter, G.B. Lesovik, and T. Martin, Phys. Rev. B **66**, 161320(R) (2002).
9. P. Samuelsson, E. V. Sukhorukov, and M. Büttiker, Phys. Rev. Lett. **91**, 157002 (2003).
10. T. Martin, A. Crepieux, and N.M. Chtchelkatchev, in *Quantum Noise in Mesoscopic Physics*, edited by Yu.V. Nazarov (Kluwer, Amsterdam, 2003).
11. V. Bouchiat, N. Chtchelkatchev, D. Feinberg, G.B. Lesovik, T. Martin, and J. Torres, Nanotechnology **14**, 77 (2003).
12. E. Prada and F. Sols, Eur. Phys. J. B **40**, 379 (2004).
13. O. Sauret, D. Feinberg, and T. Martin, Phys. Rev. B **70**, 245313 (2004).
14. C.W.J. Beenakker, C. Emary, M. Kindermann, and J.L. van Velsen, Phys. Rev. Lett. **91**, 147901 (2003).
15. A.V. Lebedev, G. Blatter, C.W.J. Beenakker, and G.B. Lesovik, Phys. Rev. B **69**, 235312 (2004).
16. P. Samuelsson, E.V. Sukhorukov, and M. Büttiker, Phys. Rev. Lett. **92**, 026805 (2004).
17. A.V. Lebedev, G.B. Lesovik, and G. Blatter, Phys. Rev. B **71**, 045306 (2005).
18. G. Burkard, D. Loss, and E.V. Sukhorukov, Phys. Rev. B **61**, R16303 (2000).
19. J.C. Egues, P. Recher, D.S. Saraga, V.N. Golovach, G. Burkard, E.V. Sukhorukov, and D. Loss, in *Quantum Noise in Mesoscopic Physics*, edited by Yu.V. Nazarov (Kluwer, Amsterdam, 2003).
20. F. Taddei and R. Fazio, Phys. Rev. B **65**, 075317 (2002).
21. L. Faoro, F. Taddei, and R. Fazio, Phys. Rev. B **69**, 125326 (2004).
22. P. Samuelsson, E.V. Sukhorukov, and M. Büttiker, Phys. Rev. B **70**, 115330 (2004).
23. J.C. Egues, G. Burkard, D. Saraga, J. Schliemann, and D. Loss, Phys. Rev. B **72**, 235326 (2005).
24. G. Burkard and D. Loss, Phys. Rev. Lett. **91**, 087903 (2003).
25. L. Faoro and F. Taddei, Phys. Rev. B **75**, 165327 (2007).
26. Ya. M. Blanter and M. Büttiker, Phys. Rep. **336**, 1 (2000).
27. M. Büttiker, Phys. Rev. B **46**, 12485 (1992).
28. T. Martin, Phys. Lett. A **220**, 137 (1996); M.P. Anantram and S. Datta, Phys. Rev. B **53**, 16390 (1996).
29. J. Torres and T. Martin, Eur. Phys. J. B **12**, 319 (1999).
30. J. Börlin, W. Belzig, and C. Bruder, Phys. Rev. Lett. **88**, 197001 (2002).
31. P. Samuelsson and M. Büttiker, Phys. Rev. Lett. **89**, 046601 (2002).

32. L. S. Levitov, H. W. Lee, and G. B. Lesovik, J. Math. Phys. **37**, 4845 (1996); B. A. Muzykantskii and D. E. Khmelnitskii, Phys. Rev. B **50**, 3982 (1994).
33. Yu. V. Nazarov, Ann. Phys. (Leipzig) **8**, SI-193 (1999); W. Belzig and Yu. V. Nazarov, Phys. Rev. Lett. **87**, 197006 (2001); W. Belzig and Yu. V. Nazarov, Phys. Rev. Lett. **87**, 067006 (2001).
34. M. Schechter, Y. Imry, and Y. Levinson, Phys. Rev. B **64**, 22451 (2001).
35. D. Sanchez, R. Lopez, P. Samuelsson, and M. Büttiker, Phys. Rev. B **68**, 214501 (2003).
36. C. Texier and M. Büttiker, Phys. Rev. B **62**, 7454 (2000).
37. F. Taddei and R. Fazio, Phys. Rev. B **65**, 134522 (2002).
38. A. Cottet, W. Belzig, and C. Bruder, Phys. Rev. Lett. **92**, 206801 (2004).
39. A. Cottet, W. Belzig, and C. Bruder, Phys. Rev. B **70**, 115315 (2004).
40. V. Rychkov and M. Büttiker, Phys. Rev. Lett. **96**, 166806 (2006).
41. S. Oberholzer, E. Bieri, C. Schönenberger, M. Giovannini, and J. Faist, Phys. Rev. Lett. **96**, 046804 (2006).
42. D. Beckmann, H.B. Weber, and H. von Löhneysen, Phys. Rev. Lett. **93**, 197003 (2004); S. Russo, M. Kroug, T.M. Klapwijk, and A.F. Morpurgo, Phys. Rev. Lett. **95**, 027002 (2005); P. Cadden-Zimansky and V. Chandrasekhar, Phys. Rev. Lett. **97**, 237003 (2006); A. Kleine, A. Baumgartner, J. Trbovic, and C. Schönenberger, EPL **87**, 27011 (2009).
43. B.R. Choi, A.E. Hansen, T. Kontos, C. Hoffmann, S. Oberholzer, W. Belzig, C. Schönenberger, T. Akazaki, and H. Takayanagi, Phys. Rev. B **72**, 024501 (2005).
44. L. Hofstetter, S. Csonka, J. Nygard, and C. Schönenberger, Nature **461**, 960 (2009).
45. L.G. Herrmann, F. Portier, P. Roche, A. Levy Yeyati, T. Kontos, and C. Strunk, Phys. Rev. Lett. **104**, 026801 (2010).
46. J. Wei and V. Chandrasekhar, arXiv:0910.5558
47. A. Di Lorenzo and Yu.V. Nazarov, Phys. Rev. Lett. **94**, 210601 (2005).
48. J.P. Morten, D. Huertas-Hernando, W. Belzig, and A. Brataas, EPL **81**, 40002 (2008).
49. J.P. Morten, D. Huertas-Hernando, W. Belzig, and A. Brataas, Phys. Rev. B **78**, 224515 (2008).

Chapter 4

Transport Currents and Persistent Currents in Solids: Orbital Magnetism and Hall Effect of Dirac Electrons

Hidetoshi Fukuyama[1], Yuki Fuseya[2] and Akito Kobayashi[3]

[1]*Department of Applied Physics and Research Institute for Science and Technology, Tokyo University of Science, Kagurazaka, Shinjuku-ku, Tokyo 162-860, Japan*
E-mail: fukuyama@rs.kagu.tus.ac.jp

[2]*Division of Materials Physics, Department of Materials Engineering Science, Graduate School of Engineering Science, Osaka University, Toyonaka, Osaka 560-8531*

[3]*Institute for Advanced Research, Nagoya University, Furo-cho, Chikusa-ku, Nagoya 464-8602*

Features of electronic currents in solids are truly diverse depending on circumstances, e.g. non-equilibrium transport currents leading to dissipation and persistent currents flowing in equilibrium. Differences between these currents may be clear in many cases, while there are some where they are not. Results of theoretical studies on the latter cases will be introduced briefly focusing on the inter-band effects of magnetic fields in orbital magnetisms and Hall effects of Dirac electrons.

4.1. Introduction

Orbital magnetism is due to currents flowing in equilibrium state, i.e. persistent currents, caused by external magnetic field. It is clear that such currents flow even in insulators as is evidenced by the existence of atomic diamagnetism. On the other hand, the celebrated Landau diamagnetism is for conduction electrons. The relationship between these in solids, where energy spectra of electrons are described by Bloch bands leading to metallic or insulating states depending on the location of the

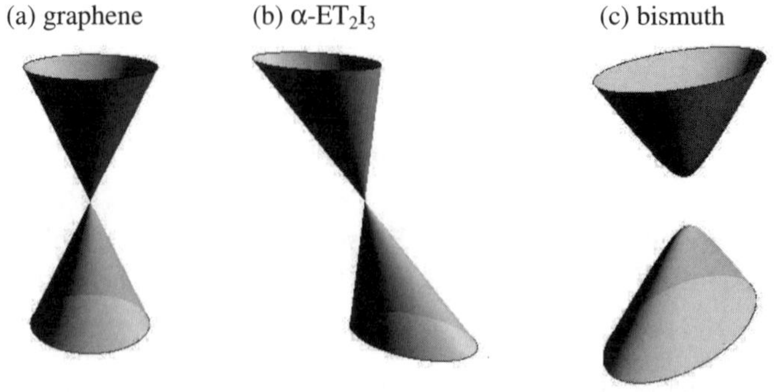

Fig. 4.1. Energy spectra of electrons for (a) graphene, (b) α-ET_2I_3 and (c) bismuth.

Fermi energy, has not been explored in full details. At the same time, the Hall effect in weak magnetic field, which reflects the changes of electronic transport current caused by the field, is non-equilibrium properties in the presence of finite electric field. The relationship between these two phenomena, orbital magnetism and Hall effect, has not been explored so far. In this short note, some of recent results of studies on this problem are reported.

4.2. Inter-band Effects of Magnetic Field on Dirac Electrons

The simplest and most transparent way to take account of magnetic field for electrons in solids, i.e. Bloch electrons, is to introduce the Peierls phase in the transfer integral. The orbital susceptibility for Bloch electrons based on this approximation is given by the Landau-Peierls (LP) formula[1]. However in this approximation the fact that the vector potential representing magnetic field has finite matrix elements between Bloch bands in any circumstances is totally ignored. Actually LP formula does not reflect the effects of weak periodic potential even in its second order[2]. More dramatically theoretical results based on LP formula are in complete disagreement with the experimental observation of large diamagnetism in semimetals, such as graphite and bismuth. It has been clarified that the inter-band effects of magnetic field play crucial roles for the understanding of large diamagnetism in both of these semimetals[3,4]. Natural question then is on the possible consequences of such inter-band effects of magnetic field on the Hall effect, which has been studied recently[5-8]. Actual systems of interest are two-dimensional graphenes and molecular solids, α ET_2I_3, and bulk crystal of bismuth. It is to be noted that these are described as Dirac electrons[9-11]. The former two, graphenes and α ET_2I_3, are described by 2x2 Weyl equation for massless Dirac electrons (Dirac cones, whose energy and wave number at the tips may be called "crossing energy" and "crossing point") similarly but with distinct differences between them: there is a finite tilting of the cones in α-ET_2I_3 since their crossing points are located at off-symmetry points in the Brillouin zone[12]. On the other hand the proper model for bismuth, where spin-orbit interactions are very strong, is 4x4 Dirac equations with spatial anisotropy of velocity. The energy spectra of electrons described by these Dirac equations are shown in Fig. 4.1(a) for graphenes, in Fig. 4.1(b) for α-ET_2I_3 and in Fig. 4.1(c) for bismuth, respectively.

In order to study such subtle inter-band effects of magnetic field on orbital susceptibility and Hall effect on firm ground, the exact formulas for orbital susceptibility, χ, and Hall conductivity, σ_{xy}, are employed which are derived by use of the Luttinger-Kohn representation[13] suited to identify the gauge-invariance associated with vector potential[14,15]. They are given as follows.

$$\chi = \frac{e^2}{c^2} T \sum_n \sum_k \mathrm{Tr} G\gamma_x G\gamma_y G\gamma_x G\gamma_y , \tag{4.1}$$

$$\sigma_{\mu\nu} = \frac{1}{i\omega} K_{\mu\nu}^{\alpha}(\boldsymbol{q},\omega) A_{q\alpha}, \tag{4.2}$$

$$\begin{aligned} K_{\mu\nu}^{\alpha} = &-\frac{e^3}{2mc}\left(q_\mu \delta_{\nu\alpha} - q_\nu \delta_{\mu\alpha}\right) T \sum_n \sum_k \mathrm{Tr}\left[G_- \gamma_\mu G \gamma_\mu G - G_- \gamma_\mu G_- \gamma_\mu G\right] \\ &-\frac{e^3}{2c}\left(q_\mu \delta_{\nu\alpha} - q_\nu \delta_{\mu\alpha}\right) T \sum_n \sum_k \mathrm{Tr}\left[G_- \gamma_\nu G_- \gamma_\mu G \gamma_\mu G \gamma_\nu - G_- \gamma_\mu G_- \gamma_\mu G \gamma_\nu G \gamma_\nu \right. \\ &+\gamma_\mu G \gamma_\nu G_- \gamma_\mu G_- \gamma_\nu G_- - \gamma_\mu G \gamma_\nu G_- \gamma_\nu G_- \gamma_\mu G_- \\ &\left.+G_- \gamma_\mu G \gamma_\mu G \gamma_\nu G \gamma_\nu - G_- \gamma_\mu G \gamma_\nu G \gamma_\mu G \gamma_\nu\right]. \end{aligned} \tag{4.3}$$

where γ is the velocity matrix $\gamma_\mu = \partial H / \partial k_\mu$, the Green function is given by $G = \left[i\tilde{\varepsilon}_n - H + \mu\right]^{-1}$ ($G_- = \left[i\tilde{\varepsilon}_n - i\omega_m - H + \mu\right]^{-1}$) with $\tilde{\varepsilon}_n = \varepsilon_n + \Gamma \mathrm{sgn}(\varepsilon_n)$ ($\varepsilon_n = (2n+1)\pi T$), μ and Γ being the chemical potential and the spectrum broadening, respectively, and the vector potential with finite Fourier component $\boldsymbol{q}$, $\boldsymbol{A}_q$, is introduced to represent even spatially uniform magnetic field B to make gauge invariance visible by taking the limit of $q \to 0$, i.e. $i\boldsymbol{q} \times \boldsymbol{A}_q = B$.

4.3. Graphenes and α-ET_2I_3

The general model applicable to both graphenes and molecular solids α-ET_2I_3 is given as follows[10]

$$H = \sum_{\mu=0,1,2,3} \boldsymbol{k} \cdot \boldsymbol{V}_\mu \sigma_\mu , \tag{4.4}$$

where $\sigma_{1,2,3}$ are the Pauli matrices and σ_0 is the identity matrix. The momentum $\boldsymbol{k}$ is measured from the crossing point. This has been deduced from the effective Hamiltonian describing the motion of electrons around the crossing point, and then it is rigorous in the vicinity of the crossing point. The band energy dispersion is given as follows

$$E = \boldsymbol{k} \cdot \boldsymbol{V}_0 \pm \sqrt{\sum_{\mu=x,y,z} \left(\boldsymbol{k} \cdot \boldsymbol{V}_\mu\right)^2} . \tag{4.5}$$

In graphenes, the vector $\boldsymbol{V}_0 = \boldsymbol{0}$ and the velocity of cones is isotropic, *i. e.* $\boldsymbol{V}_x = (\mathrm{v}, 0)$, $\boldsymbol{V}_y = (0, \mathrm{v})$, and $\boldsymbol{V}_z = \boldsymbol{0}$. In α-ET_2I_3. the velocity strongly depends on the direction of the motion. The vectors $\boldsymbol{V}_x$, $\boldsymbol{V}_y$, $\boldsymbol{V}_z$ result in the anisotropy of cones. The vector $\boldsymbol{V}_0$ tilts the axis of the cones, which makes difference between the velocities in a direction and in the opposite direction. Based on the band calculation, we can take $\boldsymbol{V}_0 = (\mathrm{v}_0, 0)$, $\boldsymbol{V}_x = (\mathrm{v}_x, 0)$, $\boldsymbol{V}_y = 0$, $\boldsymbol{V}_z = (0, \mathrm{v}_z)$ with $\mathrm{v}_0 = 0.8\mathrm{x}10^5$m/s and $\mathrm{v}_\mathrm{x} = \mathrm{v}_\mathrm{z} = 1.0\mathrm{x}10^5$m/s for analytical calculations in the following. In this case the highest velocity is about 10 times larger than the lowest one in the opposite direction.

The results of calculations of σ_{xy} and χ based on Eqs. (4.1)–(4.4) are shown in Figs. 4.2(a) and (b) for α-ET_2I_3 together with those for graphenes as a special case as a function of X= μ / Γ.

The orbital diamagnetism in Fig. 4.2(b) has large values around the crossing energy and strong μ-dependences in the region of $|X| < 1$. The orbital diamagnetism in α-ET_2I_3 exhibits the same X-dependence as that of graphene, but is enhanced by tilting.

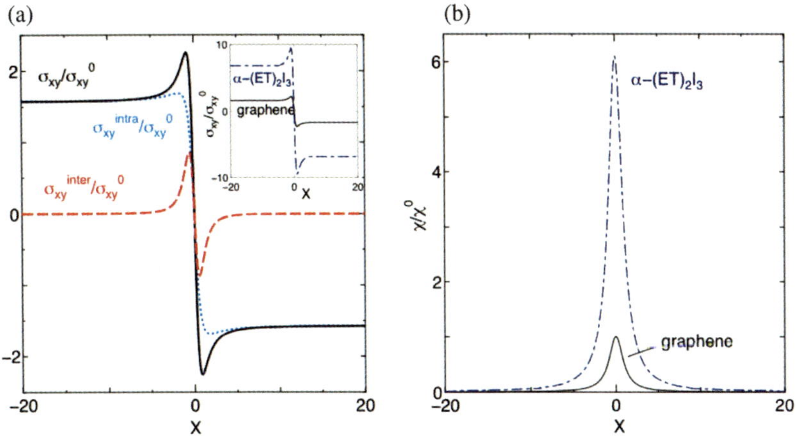

Fig. 4.2. Dependences on chemical potential, μ ,(scaled by the damping energy Γ, i.e. $X = \mu / \Gamma$) of Hall conductivity (a) and orbital susceptibility (b) in the absence of tilting (graphenes) and for α-ET_2I_3 with the choices of v_0 = 0.8x10^5m/s and v = 1.0x10^5m/s deduced from band calculations.

In Fig. 4.2(a), the Hall conductivity σ_{xy} (the solid line) is given by the sum of $\sigma_{xy}{}^{inter}$ (the dashed red line) and $\sigma_{xy}{}^{intra}$ (the dotted blue line), which are defined as contributions from all the states below the Fermi energy and those at the Fermi energy, respectively. These two terms increase with increasing tilting, although they exhibit different tilting-dependences with each other. It is seen that the contributions to $\sigma_{xy}{}^{inter}$ are confined in energy region where the orbital susceptibility takes large values.

An important fact Fig. 4.2(a) indicates is that the Hall coefficient, R_H is vanishing and changing signs at the crossing energy. This implies that the convention to deduce the effective carrier density, n, by $n = (ecR_H)^{-1}$ is totally invalid, since $n = 0$ is expected if the Fermi energy is located at the crossing energy. Another interesting feature expected for α-ET_2I_3, whose energy spectra relative to the crossing energy is not symmetric, i.e. without "electron-hole symmetry", is that the chemical potential is expected to depend on temperature and can pass though the crossing energy at low temperature in the presence of finite but very small amount (even of the order of ppm) of doped carriers. If this happens, the Hall coefficient can undergo very sharp change of sign as a function of chemical potential as shown in Fig. 4.3(c). Such is actually been observed experimentally[15].

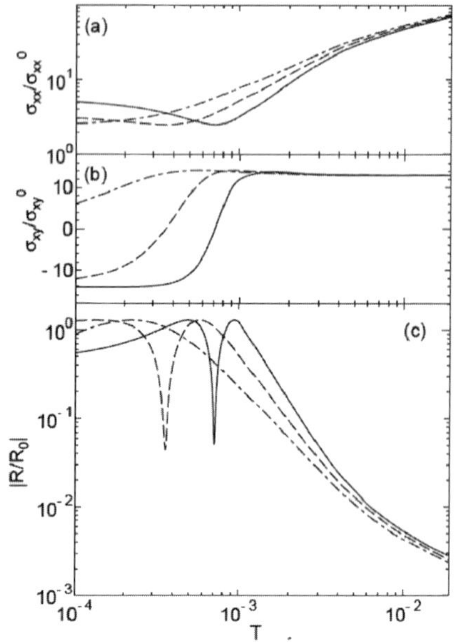

Fig. 4.3. Temperature dependence of (a) σ_{xx}, (b) σ_{xy}, and (c) the Hall coefficient, where the solid, dashed, and dot-dashed lines are calculated for finite doping expressed as the chemical potential off the crossing energy at absolute zero $\Delta\mu = 2.0\text{x}10^{-4}$, $1.0\text{x}10^{-4}$, and 0, respectively.

4.4. Bismuth

The effective model for Bi is given by[11]

$$H = \frac{E_{\mathrm{G}}}{2}\beta + i\boldsymbol{k}\cdot\left[\sum_{\mu=1}^{3}\boldsymbol{W}(\mu)\beta\alpha_{\mu}\right]$$
$$= \begin{pmatrix} E_{\mathrm{G}}/2 & 0 & \boldsymbol{k}\cdot\langle 1|\boldsymbol{v}|3\rangle & \boldsymbol{k}\cdot\langle 1|\boldsymbol{v}|4\rangle \\ 0 & E_{\mathrm{G}}/2 & \boldsymbol{k}\cdot\langle 2|\boldsymbol{v}|3\rangle & \boldsymbol{k}\cdot\langle 2|\boldsymbol{v}|4\rangle \\ \boldsymbol{k}\cdot\langle 3|\boldsymbol{v}|1\rangle & \boldsymbol{k}\cdot\langle 3|\boldsymbol{v}|2\rangle & -E_{\mathrm{G}}/2 & 0 \\ \boldsymbol{k}\cdot\langle 4|\boldsymbol{v}|1\rangle & \boldsymbol{k}\cdot\langle 4|\boldsymbol{v}|2\rangle & 0 & -E_{\mathrm{G}}/2 \end{pmatrix}, \tag{4.6}$$

where E_{G} is the band gap, and β, α_{μ} are the 4x4 matrices that appear in the Dirac theory. The quantities $\langle i|\boldsymbol{v}|j\rangle$ are the matrix elements of the velocity operator, and the indexes i, j denote the four band-edge wave function (2 for bands and 2 for spins). We have measured energies from the center of the band gap. It is to be noted that this model naturally leads to small effective mass and corresponding large g-factor of spin Zeeman splitting of states near the band gaps[11,17]. In order to see the essence of inter-band contributions this model is simplified by assuming the isotropic velocity v as follows:

$$H = \frac{E_G}{2} + iv\sum_{\mu} k_\mu \beta \alpha_\mu . \tag{4.7}$$

The Hall conductivity, σ_{xy}, and the orbital susceptibility, χ, are calculated on the basis of the exact formulas as in the previous section. The final expressions are

$$\sigma_{xx} = -\frac{e^2}{\pi^3 v}\int_{-\infty}^{\infty} d\varepsilon \int_0^{\infty} dX \left[F_1(\varepsilon, X) - F_2(\varepsilon, X)\right] f'(\varepsilon - \mu), \tag{4.8}$$

$$F_1(\varepsilon, X) = \frac{X^2\left(\varepsilon^2 + \Gamma^2 - X^2/3 - E_G^2/4\right)}{\left\{(\varepsilon + i\Gamma)^2 - X^2 - E_G^2/4\right\}\left\{(\varepsilon - i\Gamma)^2 - X^2 - E_G^2/4\right\}},$$

$$F_2(\varepsilon, X) = \frac{X^2\left\{(\varepsilon + i\Gamma)^2 - X^2/3 - E_G^2/4\right\}}{2\left\{(\varepsilon + i\Gamma)^2 - X^2 - E_G^2/4\right\}^2} + \text{c.c.}, \tag{4.9}$$

$$\sigma_{xy} = \frac{e^3 vB}{12\pi^2 c}\int_{-\infty}^{\infty} d\varepsilon \left[F_3(\varepsilon) f(\varepsilon - \mu) + F_4(\varepsilon) f'(\varepsilon - \mu)\right] \operatorname{sgn}(\varepsilon), \tag{4.10}$$

$$F_3(\varepsilon) = \frac{\varepsilon + i\Gamma}{\left[(\varepsilon + i\Gamma)^2 - E_G^2/4\right]^{3/2}} + \text{c.c.},$$

$$F_4(\varepsilon) = \frac{\left(E_G^2/4 - \varepsilon^2\right)^2 - 2\Gamma^4 - \Gamma^2 E_G^2/4 + 2i\Gamma^3\varepsilon - i\Gamma\mu\left(E_G^2/4 - \varepsilon^2\right)}{2\Gamma^2\varepsilon^2\sqrt{\varepsilon^2 - \Gamma^2 - E_G^2/4 + 2i\Gamma\varepsilon}} + \text{c.c.}, \tag{4.11}$$

$$\chi = \frac{4}{15}\frac{e^2 v}{c^2\pi^2}\int_{-\infty}^{\infty} d\varepsilon \left[\frac{1}{\sqrt{(\varepsilon + i\Gamma)^2 - E_G^2/4}} + \text{c.c.}\right] f(\varepsilon - \mu)\operatorname{sgn}(\varepsilon). \tag{4.12}$$

The Hall coefficient is defined as $R_H = \sigma_{xy}/B\sigma_{xx}^2$.

The results of calculations of σ_{xx}, σ_{xy}, R_H, $\sigma_{xy}^{\text{inter}}$ and χ for this model as a function of the chemical potential, μ, are shown in Fig. 4.4(a)–(e). Here, $\sigma_{xx0} = e^2/\pi^2 v$, $\sigma_{xy0} = e^3 v/12\pi^2 c$, $\chi_0 = 4e^2 v/15c^2\pi^2$.

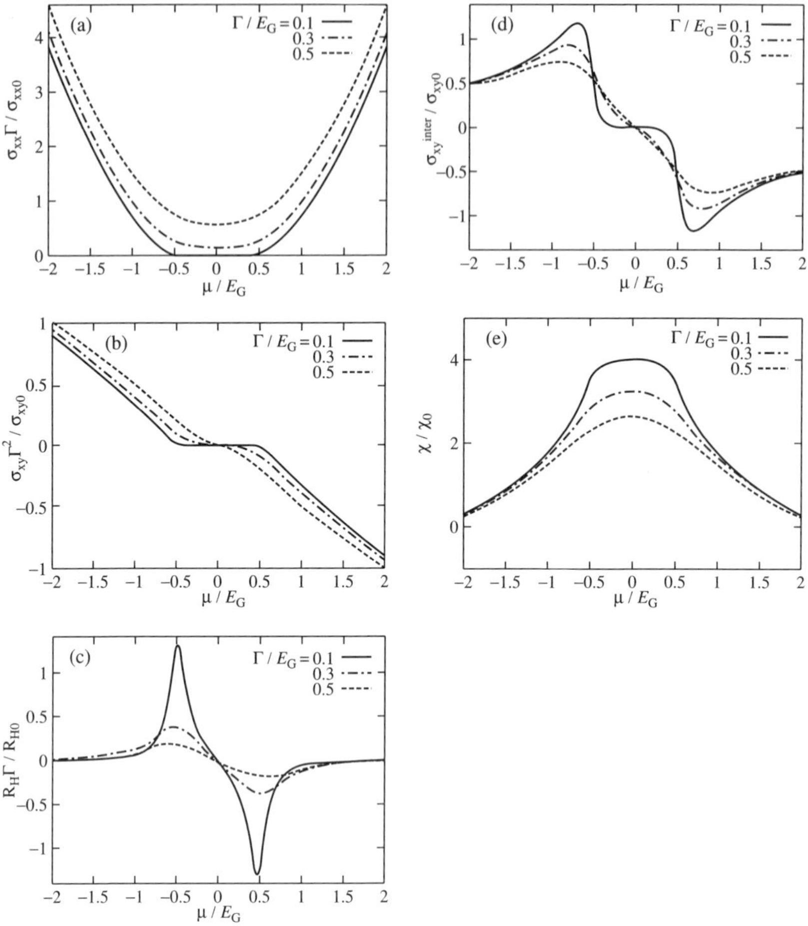

Fig. 4.4. The results of calculations for (a) the conductivity σ_{xx}, (b) the Hall conductivity σ_{xy}, (c) the Hall coefficient, (d) inter-band contribution $\sigma_{xy}{}^{\text{inter}}$, (e) the orbital susceptibility χ, as a function of chemical potential μ.

The inter-band contributions are obtained by subtracting the intra-band contribution, $\sigma_{xy}^{\text{intra}}$, from the exact value σ_{xy} : $\sigma_{xy}^{\text{inter}} = \sigma_{xy} - \sigma_{xy}^{\text{intra}}$. The intra-band contribution can be calculated within the Bloch band picture as[14]

$$\sigma_{xy}^{\text{intra}} = -\frac{4}{3}\frac{e^3 B}{\pi c}\sum_{\boldsymbol{k}}\int_{-\infty}^{\infty} dx\left[\left(\partial_x \xi\right)^2 \partial_y^2 \xi - \partial_x \xi \partial_y \xi \partial_{xy} \xi\right], \qquad (4.13)$$

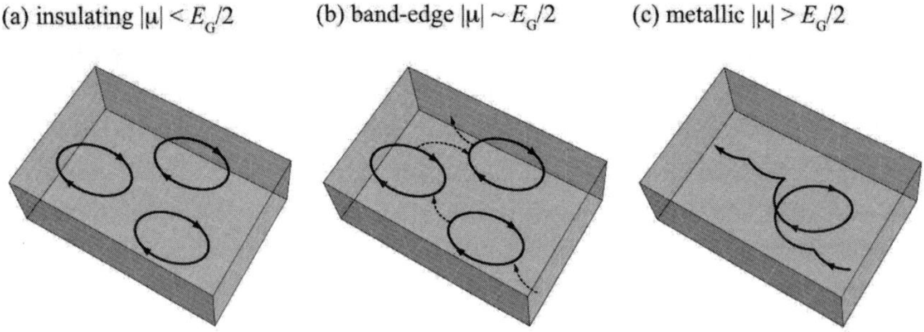

Fig. 4.5. Schematic motion of electrons in a solid for (a) the insulating ($|\mu| < E_G/2$), (b) the band-edge ($|\mu| \sim E_G/2$), and (c) the metallic region ($|\mu| > E_G/2$).

where $\xi = \sqrt{v^2k^2 + E_G^2/4}$ and $\partial_\mu \xi = \partial\xi/\partial k_\mu$. Then we have the final expression for the present model:

$$\sigma_{xy}^{\mathrm{intra}} = -\frac{e^3 vB}{6\pi^3 c}\sum_{n=\pm}\int_{-\infty}^{\infty} d\varepsilon \int_0^{\infty} dX \frac{nX^4}{\left[E_n(X)\right]^3}\frac{4\Gamma^3}{3\left[\{\varepsilon - E_n(X)\}^2 + \Gamma^2\right]^3} f'(\varepsilon-\mu), \quad (4.14)$$

where $E_\pm(X) = \pm\sqrt{X^2 + E_G^2/4}$.

It is seen that $\sigma_{xy}^{\mathrm{inter}}$ exist near the band edges (Fig. 4.4(d)), where χ (Fig. 4.4(e)) takes large values except for insulating region. This strongly suggests that the nature of $\sigma_{xy}^{\mathrm{inter}}$ is closely related to that of orbital current. The physical picture of this correlation can be understood as follows as schematically shown in Fig. 4.5. In the insulating region, the current flows locally under a magnetic field, generating the diamagnetism (Fig. 4.5(a)). This current is non-dissipative, so that it does not contribute to Hall currents. In the band-edge region, on the other hand, the local orbital currents hybridize with the conduction electrons, which will lead to the mixing between diamagnetic currents and Hall currents. This will be the origin of the $\sigma_{xy}^{\mathrm{inter}}$ in the band-gap region.

In the clean limit, i.e. $\Gamma \to 0$, it is analytically seen that $\sigma_{xy}^{\mathrm{inter}}$ is vanishingly small for $-0.5 < \mu/E_G < 0.5$ (band gap region). In this energy region orbital susceptibility shows maximum diamagnetism as seen in Fig. 4.4(e). This implies that orbital currents leading to susceptibility are not contributing to Hall effect, and they are totally independent if the system is clean. Once disorder is introduced, e.g. by impurities, however, current flow will get intermixed and clear separation of these will be no longer clear because of finite σ_{xy} in the band gap region. To identify characteristic features of current flows in such cases, careful studies are needed on the impurity states which have been seen to anomalously large g-factor[18] because of strong spin-orbit interactions. This coupling between orbital currents and spins is also an interesting subject in the context of spin-Hall effect[19], which deserves detailed studies.

Acknowledgments

One of authors (HF) is very thankful for Joe Imry for stimulating and enlightening discussions at various stages of researches on quantum transport properties in solids.

References

1. R. Peierls, *Z. Phys.*, **80**, 763 (1933).
2. K. Yamaji and R. Kubo, *J. Phys. Soc. Jpn.*, **25**, 330 (1958).
3. J. W. McClure, *Phys. Rev.,* **104**, 666 (1956); *ibid.*, **119**, 606 (1960).
4. H. Fukuyama and R. Kubo, *J. Phys. Soc. Jpn.*, **28**, 570 (1970).
5. H. Fukuyama, Ann. Phys.(Leipzig), **15** (2006) 520.
6. H. Fukuyama, *J. Phys. Soc. Jpn.*, **76**, 043711 (2007).
7. A. Kobayashi, Y. Suzumura and H. Fukuyama, *J. Phys. Soc. Jpn.*, **77**, 064718 (2008).
8. Y. Fuseya, M. Ogata and H. Fukuyama, *Phys. Rev. Lett.*, **102**, 066601 (2009).
9. P. R. Wallace: *Phys. Rev.* **71** 622 (1947).
10. A. Kobayashi, S. Katayama, Y. Suzumura and H. Fukuyama, *J. Phys. Soc. Jpn.*, **76**, 034711 (2007).
11. P. A. Wolff, *J. Phys. Chem. Solids*, **25**, 1057 (1964).
12. S. Katayama, A. Kobayashi and Y. Suzumura, J. Phys. Soc. Jpn. **75** 054705 (2006).
13. J. M. Luttinger and W. Kohn, *Phys. Rev.,* **97**, 869 (1955).
14. H. Fukuyama, H. Ebisawa and Y. Wada, *Prog. Theor. Phys.*, **42**, 494 (1969); H. Fukuyama, *ibid.*, **42**, 1284 (1970).
15. H. Fukuyama, *Prog. Theor. Phys.*, **45**, 704 (1971).
16. N. Tajima, S. Sugawara, M. Tamura, R. Kato, Y. Nishino, and K. Kajita, private communication.
17. M. H. Cohen and E. I. Blount, *Phil. Mag.*, **5**, 115 (1960).
18. K. Sakai, C. Ishii and H. Fukuyama, *J. Phys. Soc. Jpn.*, **50**, 3590 (1981).
19. S. Murakami, N. Nagaosa and S.-C. Zheng, *Science*, **301**, 1348 (2003).

Chapter 5

Bloch and Josephson Oscillations in a Ring of an Ideal Bose Gas

Leon Gunther

Department of Physics and Astronomy,
Tufts University,
Medford, Massachusetts 02155

We show that an Ideal Bose gas that is contained within a very thin ring exhibits phenomena analogous to the Bloch and Josephson oscillations of a charged Ideal Fermi gas in a thin ring. If the walls of the ring are constrained to have an angular velocity ω, the angular momentum has an anomalous component that is periodic in ω, with a period equal to the quantum of angular velocity $\omega_0 = \hbar/mR^2$. If a constant applied torque is applied to the walls, there will be component of the angular momentum of the gas that is periodic in time, with a "Josephson frequency" given by $f_J = \tau/N\hbar$. Finally, we show that the oscillations are an automatic feature of the quantum regime of any ring of an ensemble of identical particles, even with particle interactions.

5.1. Introduction

This paper is based upon the research that Joe and I carried out around 1970. Motivated by the papers of Byers and Yang[1] and Bloch,[2] we demonstrated that the Bloch oscillations associated with a superconductor were present in a thin ring of an Ideal Fermi gas, albeit with a flux quantum determined by a single electron charge (thus, hc/e), in contrast with $hc/2e$ for the BCS superconductor. We also showed that there were analogous oscillations in the behavior of a rotating ring of an ideal Bose gas. None of our results were published. However, the results for the Ideal Fermi gas ultimately were and still are being developed much further for a real metal in the normal state (with important contributions made by Joe and his coworkers), including elastic scattering of electrons off impurities, electron scattering off magnetic impurities, electron-electron scattering, and inelastic electron-phonon scattering. See Joe's monograph[3] for more details as well as the very recent paper by Joe and collaborators.[4] The essential predictions for the normal metal ring were ultimately observed in a number of beautiful experiments.[5] On the other hand, the work on the Bose gas was left in the bins of old notes. The purpose of this paper is to revive this problem and to honor my relationship with Joe.

One of the key steps in the formulation of this problem is, for me at least, based upon a talk by Gordon Baym presented at the 1967 Summer School held at St. Andrews, Scotland on "Mathematical Methods in Solid State and Superfluid Theory".[6] Baym pointed out that if a fluid is placed in a rotating cylinder, the energies in the partition function must be those with respect to the rotating frame of reference. We will later see how this requirement leads to Bloch and Josephson oscillations in a ring of an Ideal Bose gas.

By coincidence, it was at this Summer School that I was fortunate to meet Joe. We had a marriage of like minds immediately. Joe was on his way to the US for a leave of absence, fresh from his work on demonstrating that while a one-dimensional Ising model with short range interactions does not exhibit a phase transition in the thermodynamic limit (that is, as the number of spins N goes to infinity), a finite chain of spins could exhibit long range order that is not significantly weaker than that obtained in the thermodynamic limit in a system with long range interactions. I had come to other corresponding conclusions about finite systems during my post-doc at Orsay that year. There I had heard Bernard Jancoviçi[7] give a talk on novel behavior of the susceptibility of a finite 2D harmonic lattice in spite of the accepted result that a 2D harmonic lattice was unstable. His analysis revealed that the long wavelength divergence of fluctuations that destroyed long range order were cut off in a finite system. That same year, I had also heard Vladimir Tkachenko[8] give a talk about his results demonstrating that a rotating lattice of vortices in a superfluid was unstable due to the diverging fluctuations in the vibrations of the vortices. I was led to note that the divergence vanished in a finite lattice, due to a cutoff in the long-wavelength fluctuations that was proportional to log-N.

Our meeting at St. Andrews led to many years of wonderful times together as friends as well as in research that reflected our shared view that theory must always strive to honestly reflect experimental conditions. I believe that this philosophy was the foundation for Joe's incredible ability to translate theory into a language that experimentalists could not only understand but also use to produce wonderful experimental results. Joe would often come up with great suggestions for new areas of research. However, for me it was even more delightful that whenever I had an idea, Joe was there to analyze it and lead us with an explosion of further developments, ways of understanding the phenomena and simplifying the analysis.

I had the pleasure of working with Joe and his colleagues on research that revealed that the study of finite systems — which are what we actually study in the laboratory — could not only make some non-existence theorems based upon infinite systems of not too great importance practically, but also teach us much about systems in the thermodynamic limit and reveal novel behavior in real, finite systems. We take all this for granted since the rise of nanoscopic physics in the '70s.

5.2. Comparison of a Ring of an Ideal Fermi Gas with a Ring of an Ideal Bose Gas

5.2.1. *Energy of the Quantum States*

Our ring has a radius R and cross-sectional dimensions that are so small that the first excited state with respect to either dimension has an energy much larger than the thermal energy kT. As a result, the ring is effectively one-dimensional. The rotational momentum is quantized in the inertial lab frame. Thus we have

$$p = n\frac{\hbar}{R} \tag{5.1}$$

where n is any integer (including zero).

For both the Ideal Fermi gas, as well as the Ideal Bose gas (in the rotating frame of reference of a rotating ring), the energy of the states (simply kinetic energy) that is relevant in the partition function can be expressed as

$$E_n = \frac{p_n^2}{2m} = \frac{\hbar^2(n-\phi)^2}{2mR^2} \tag{5.2}$$

where m is the particle mass and n is any integer from $-\infty$ to $+\infty$. Note that for the Fermi case, this energy is in the lab frame, while for the Bose case, this energy is in the rotating frame.

In the case of the **Fermi** gas, the charged particles are in the presence of an external magnetic field parallel to the axis of the ring. The parameter ϕ is the ratio

$$\phi = \frac{\Phi}{\Phi_0} \tag{5.3}$$

where Φ is the total magnetic flux through the ring and Φ_0 is the flux quantum given by

$$\Phi_0 = \frac{hc}{e} \tag{5.4}$$

The total flux is a sum of the flux due to the external field and the flux produced by an electric current of the charged particles in the ring.

In the case of the **Bose** gas, the situation is a bit more complex. We have a ring of gas that is bounded by a wall that is rotating at a fixed angular velocity ω. The rotational momentum is quantized in the inertial lab frame, and is thus given by equation (5.1). There is a corresponding quantum of velocity v_0 and quantum angular velocity ω_0

$$v_0 = \frac{\hbar}{mR} \qquad \omega_0 = \frac{\hbar}{mR^2} \tag{5.5}$$

We assume that the system has come to thermodynamic equilibrium. As we mentioned above, the energy in the partition function is then the energy in the rotating frame. The rotational velocity in the rotating frame is given by

$$v_{rot} = v_{lab} - \omega R \tag{5.6}$$

Thus the unit of quantum velocity in the rotational frame is

$$v_n = n\frac{\hbar}{mR} - \omega R \tag{5.7}$$

$$\tag{5.8}$$

Thus we obtain for the Bose gas equation (5.2), with

$$\phi = \frac{\omega}{\omega_0} \tag{5.9}$$

Note that the energy of a state can be written as

$$E_n = \frac{\hbar\omega_0}{2}(n-\phi)^2 \tag{5.10}$$

Thus, $e_0 \equiv \hbar\omega_0/2$ is the characteristic energy of the system.

5.2.2. *Electric Current in the Fermi Gas*

In the case of the Fermi gas, it is the electric current that exhibits Bloch oscillations as a function of the flux. It is given by

$$I = \frac{e}{2\pi R}\sum_{n=-\infty}^{\infty} v_n F_n \tag{5.11}$$

$$= \frac{e}{2\pi R}\sum_{n=-\infty}^{\infty} \frac{\hbar}{mR}(n-\phi)F_n = \frac{e\hbar}{2\pi mR^2}\sum_{n=-\infty}^{\infty}(n-\phi)F_n \tag{5.12}$$

where F_n is the Fermi function

$$F_n = \frac{1}{\exp[(E_n-\mu)/kT]+1} \tag{5.13}$$

Here μ is the chemical potential.

From equation (5.12) we see that the current is a periodic function of the magnetic flux, with a period equal to the flux quantum $\Phi_0 = hc/e$.

5.2.3. *Angular Momentum in the Bose Gas*

For the Bose gas we are interested in how the angular momentum L depends upon the fixed angular velocity. In the lab frame it is given by

$$L = \hbar \sum_{n=-\infty}^{\infty} n f_n \tag{5.14}$$

where f_n is the Bose function

$$f_n = \frac{1}{\exp[(E_n-\mu)/kT]-1} \tag{5.15}$$

Here μ is the chemical potential.

We can separate the total angular momentum into two parts:

$$L = L_{class} + L_{anom} \tag{5.16}$$

The first part is what we obtain in the classical limit:

$$L_{class} = \mathcal{I}\omega = N\hbar\phi = \hbar\sum_{n=-\infty}^{\infty}\phi f_n \tag{5.17}$$

where $\mathcal{I} = NmR^2$ is the total moment of inertia.

The second part is the anomalous part — the angular momentum in the rotating frame — which would normally vanish since then the walls would carry the entire gas along with it.

$$L_{anom} = \hbar \sum_{n=-\infty}^{\infty} (n - \phi) f_n \tag{5.18}$$

From equation (5.18) we see that the anomalous angular momentum is a periodic function of the applied angular velocity, with a period equal to the quantum of angular velocity, $\omega_0 = \hbar/mR^2$. Thus, the anomalous angular momentum corresponds to the above Bloch oscillations of the electric current.

5.2.4. *Crossover Temperatures for Oscillations*

It is clear that the discreteness of the energy levels is responsible for the observability of Bloch oscillations. As we raise the temperature, the discreteness becomes less relevant. Generally, the energy level spacing is given by

$$\Delta E_n = \left[(n+1)^2 - n^2\right] e_0 = (2n+1)e_0 \tag{5.19}$$

In the **classical regime**, the average energy per particle is $\sim kT$. The crossover temperature is determined by setting the characteristic energy level spacing equal to kT. The corresponding quantum number is $n = \sqrt{kT/e_0}$. Thus, the relevant energy level spacing is $2e_0\sqrt{kT/e_0} = 2\sqrt{kTe_0}$. And finally, we obtain the crossover temperature T_C from the equation $\sqrt{kTe_0} = kT$:

$$T_C \sim e_0/k \tag{5.20}$$

For **Fermions**, the characteristic energy is the Fermi energy ϵ_F, which corresponds to a quantum number $n \sim N/2$. Thus the relevant energy level spacing is $\Delta E_{N/2} \sim Ne_0$. Again, the crossover temperature T_F is determined by $\Delta E = kT$, so that it is given by

$$T_F \sim Ne_0/k \tag{5.21}$$

In fact, it can be shown that the persistent current is given by[9]

$$I = N \sum_{p=1}^{\infty} \frac{(-1)^{Np}}{2\pi p} \frac{2\pi^2 pkT/Ne_0}{\sinh[2\pi^2 pkT/Ne_0]} \tag{5.22}$$

Hence, the actual crossover temperature is better represented by

$$T_F = Ne_0/(2\pi^2 k) \tag{5.23}$$

For **Bosons**, the situation is entirely different. We know that bosons are attracted into the same state. The 3D Bose gas undergoes a Bose-Einstein condensation. While there is no condensation in 1D or 2D, there is nevertheless a relatively high occupation of the states with low energies — with quantum number n of order unity — as a result of the minus sign in the denominator of the distribution function.

We can learn a lot from an analysis of the situation at essentially absolute zero. In this case, all particles are in the ground state. Let $\alpha(\phi) \equiv -\mu(\phi)/kT$

Then we must have

$$f_0 = \frac{1}{\exp[e_0\phi^2/kT + \alpha] - 1} = N \tag{5.24}$$

For simplicity, we take $\phi = 0$. Then,

$$\alpha = \ln(1 + 1/N) \sim 1/N \tag{5.25}$$

For low temperatures, we still expect $\alpha \sim 1/N$. We want $f_{n\neq 0}$ to be of order unity for n on the order of unity but negligibly small for n not of order unity.

Since

$$f_n = \frac{1}{\exp[e_0 n^2/kT + \alpha] - 1} \sim \frac{1}{\exp[e_0 n^2/kT + 1/N] - 1} \tag{5.26}$$

we must have $kT < Ne_0$. Thus, the crossover temperature T_B for our Bose gas is the same as it is for the Fermi gas, namely

$$T_B \sim Ne_0/k \tag{5.27}$$

It is more straightforward to show that for the low energy states to dominate, we must have

$$f_n = \frac{1}{\exp[(e_0(n-\phi)^2 - \mu)/kT] - 1} \cong \frac{1}{(e_0(n-\phi)^2 - \mu)/kT} \tag{5.28}$$

for the dominant states.

We will define a characteristic temperature

$$T_B \equiv Ne_0/2\pi^2 k \tag{5.29}$$

The chemical potential is determined by the N, T and ϕ through the equation

$$N = \sum_{n=-\infty}^{\infty} f_n \cong \sum_{n=-\infty}^{\infty} \frac{1}{(e_0(n-\phi)^2-\mu)/kT} \tag{5.30}$$

We will use the approximate expression for N from now on. Let us introduce the reduced **reduced chemical potential** $m \equiv \mu/e_0$ and the **reduced temperature** $t \equiv T/T_B$. Then it is straightforward to show that

$$N\frac{e_0}{kT} \equiv \frac{2\pi^2}{t} = \frac{1}{(\phi^2-m)}$$
$$+\frac{\Psi\left(\sqrt{m}-\phi+1\right)-\Psi\left(-\sqrt{m}-\phi+1\right)+\Psi\left(\sqrt{m}+\phi+1\right)-\Psi\left(-\sqrt{m}+\phi+1\right)}{2\sqrt{m}}$$

Here $\Psi(z)$ is the diGamma function.[12] From this expression we see that the reduced chemical potential is expressible entirely in terms of the reduced angular velocity and the reduced temperature, thus confirming our choice of crossover temperature $T_B \equiv Ne_0/(2\pi^2 k)$.

The **reduced anomalous angular momentum**, $\ell \equiv L_{anom}/N\hbar$ can be similarly expressed:

$$\ell = -\sqrt{m} + \frac{t}{2\pi^2}\left(\Psi\left(\sqrt{m}+\phi+1\right) - \Psi\left(-\sqrt{m}-\phi+1\right) - \frac{1}{\sqrt{m}+\phi}\right) \tag{5.31}$$

5.3. Bloch Oscillations - Numerical Results

Below we will summarize the numerical results we obtained using MAPLE. The procedure is to find the chemical potential given the number of particles N using the equation

$$N = \sum_{n=-\infty}^{\infty} \frac{1}{\exp[(e_0(n-\phi)^2-\mu)/kT]-1} \tag{5.32}$$

In Fig. 5.1 we see plots of the chemical potential as a function of the reduced angular velocity — $\phi = \omega/\omega_0$ — for two temperatures: $T_B/100$ (solid curve) and $10T_B$ (dashed curve). The chemical potential is expressed in units of e_0. Note that the chemical potential oscillates with great amplitude in conjunction with the Bose tendency to keep particles in the same state. This strong variation is in great contrast with the situation for Fermions, for which there is a much smaller, though sometimes important, variation that needs to be taken into account.[10,11]

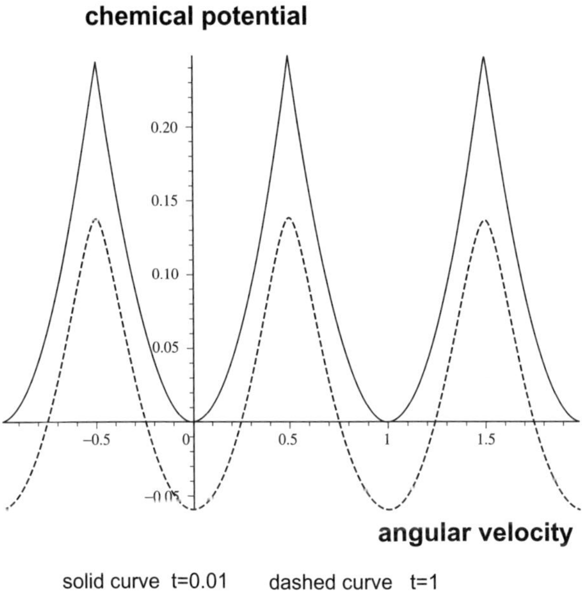

Fig. 5.1. Reduced Chemical Potential vs Reduced Angular Velocity

In Fig. 5.2 we plot the total angular momentum in units of $N\hbar$ as a function of the reduced angular velocity ϕ, for the temperature $T_B/100$ (solid curve) and for the classical regime(dash-dot curve). In Fig. 5.3 we plot the anomalous angular momentum in units of $N\hbar$ as a function of the reduced angular velocity ϕ, for two temperatures: $T_B/100$ (solid curve) and $10T_B$ (dashed curve).

5.4. Josephson Oscillations

As shown by Bloch,[2] Josephson oscillations in a conductor can be explained in terms of the Bloch oscillations of the current in the presence of a time dependent magnetic flux. Generally,

$$V = \frac{1}{c}\frac{\partial \Phi}{\partial t} \tag{5.33}$$

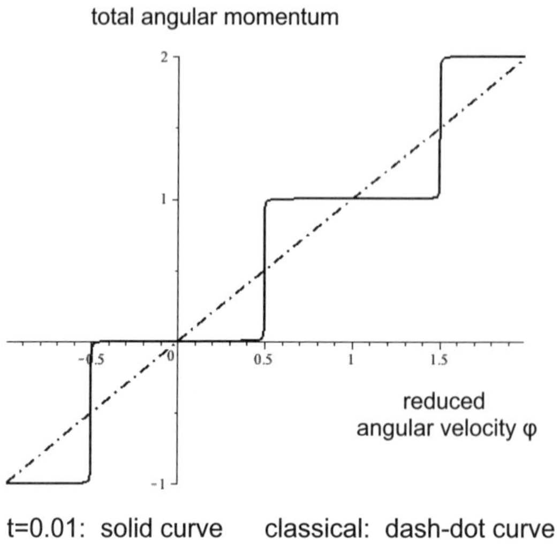

Fig. 5.2. Total Angular Momentum vs. Angular Velocity

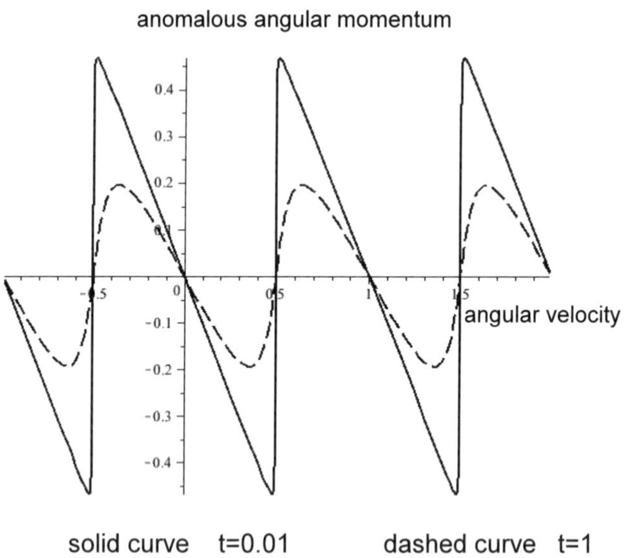

Fig. 5.3. Anomalous Angular Momentum vs. Angular Velocity

If the voltage is constant, the flux increases linearly in time, so that

$$\phi = \frac{cVt}{\Phi_0} = \frac{eV}{h}t \tag{5.34}$$

Then, if the electrons remain in quasi-thermodynamic equilibrium, the current will oscillate with the Josephson frequency, given by

$$f_J = \frac{eV}{\hbar} \tag{5.35}$$

Now let us turn to the ring of an Ideal Bose gas. indexring of ideal Bose gas The corresponding experimental condition is to have a constant torque τ applied to the ring wall. We must remember that the resulting angular velocity of the wall is not the angular velocity of the gas. However, the angular velocity of the wall determines the state of state of the gas assuming, as above, that quasi-static thermodynamic equilibrium is maintained. We have with a constant torque

$$L(\phi) = \tau t \tag{5.36}$$

Thus,

$$\phi(t) = L^{-1}(\tau t) \tag{5.37}$$

where L^{-1} is the inverse function of L. To obtain the corresponding Josephson frequency we use the fact that $L_{anom}(\phi)$ is a periodic function of ϕ:

$$L_{anom}(\phi + 1) = L_{anom}(\phi)) \tag{5.38}$$

Since

$$L_{anom}(\phi) = L(\phi) - L_{class}(\phi) = \tau t - N\hbar\phi \tag{5.39}$$

we easily find that the Josephson frequency is

$$f_J = \frac{\tau}{N\hbar} \tag{5.40}$$

A plot of the angular velocity vs. time is shown in Fig. 5.4 for a temperature $T_B/100$. The time axis is in units of the Josephson period and the angular velocity is in units of the quantum of angular velocity. The dash-dot curve is the classical result, for which the entire gas moves with the wall, so that the angular velocity is linear in time.

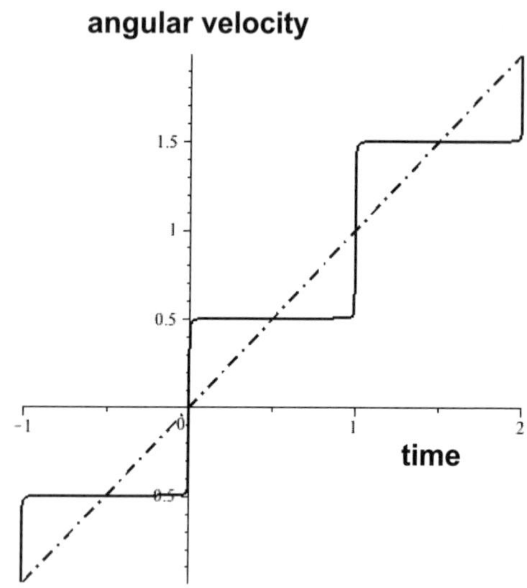

Fig. 5.4. Angular Velocity vs. Time

Let us try to make sense of the graph. We will assume absolute zero, for which the jumps in the graph are discontinuous. First, we note that from zero to one time unit, $t_J = 1/f_J$, the wall has one-half the quantum of angular velocity — that is $\omega_0/2$ — whereas a single particle has an angular velocity that must an integral number of quanta. Furthermore, in the course of this time interval, the total change in angular momentum is $\Delta L = \tau t_J = N\hbar$, which corresponds to all of the particles having a single quantum of angular velocity. Therefore, during this time interval, the particles are continuously making a transition from the $n = 0$ state to the $n = 1$ state. At the end of the time interval, the entire gas is moving twice as fast as the walls! The subsequent behavior is then obvious and will not be discussed here.

5.5. Proof of Bloch & Josephson Oscillations in a Ring of Interacting Particles

THEOREM: *In the quantum regime, a ring of identical* ***interacting particles****, whether they be Bosons or Fermions, has an anomalous component of the angular momentum in response to a fixed angular velocity of the wall — a component that is periodic in the angular velocity, with a period* ω_0.

The proof makes use of the technique used by Byers and Yang[1] in proving that in the presence of a uniform magnetic field, magnetic flux is quantized in a hollow superconducting cylinder.[13]

Lemma 1:
The energy eigenvalues in the rotating frame are periodic functions of the angular velocity, with a period equal to the quantum of angular velocity.

Proof:
The coordinates of the particles can be taken to be the angles $\{\theta_j\}$, with $j = 1...N$. The Schroedinger equation in the lab frame is given by

$$\widehat{H}\Psi = E\Psi \tag{5.41}$$

where the Hamiltonian operator $\widehat{H}$ in the inertial lab frame is given by

$$\widehat{H} = -\frac{\hbar^2}{2mR^2}\sum_{j=1}^{N}\frac{\partial^2}{\partial\theta_j^2} + V(\{\theta_j\}) \tag{5.42}$$

In the rotating frame of reference, the Schroedinger equation is given by

$$\widehat{H}_{rot}\Psi_{rot} = E_{rot}\Psi_{rot} \tag{5.43}$$

where the Hamiltonian operator $\widehat{H}_{rot}$ is given by

$$\widehat{H}_{rot} = \frac{\hbar^2}{2mR^2}\sum_{j=1}^{N}(-i\frac{\partial}{\partial\theta_j} - \phi)^2 + V(\{\theta_j\}) \tag{5.44}$$

Note that the energy eigenvalue is a function of ϕ: $E_{rot} = E_{rot}(\phi)$. Also, the boundary condition on the wave function is

$$\Psi_{rot}(\theta_i + 2\pi) = \Psi_{rot}(\theta_i) \tag{5.45}$$

for any θ_i.

Now let us define the function Ψ'_{rot}:

$$\Psi_{rot} = e^{i\phi \sum_j \theta_j} \Psi'_{rot} \tag{5.46}$$

This new function satisfies the equation (note that the Hamiltonian corresponds to the inertial frame and is independent of ϕ):

$$\widehat{H}\Psi'_{rot} = E_{rot}(\phi)\Psi'_{rot} \tag{5.47}$$

where the energy eigenvalue depends upon ϕ because the boundary condition on Ψ'_{rot} is

$$\Psi'_{rot}(\theta_i + 2\pi) = e^{-2\pi i\phi}\Psi'_{rot}(\theta_i) \tag{5.48}$$

for a given θ_i, with all other θ_j kept fixed.

We then note that if ϕ changes by unity $\phi \to \phi+1$, the boundary condition doesn't change. Since the boundary condition determines the specific solution to the Schroedinger equation and hence the eigenvalue, we see that the eigenvalues, with subscript label s, satisfy $E_{rot,s}(\phi+1) = E_{rot,s}(\phi)$. **QED**

Lemma 2:
The partition function in the rotating frame is a periodic function of the angular velocity, with a period equal to the quantum of angular velocity.

Proof:
This result follows automatically from the fact that the partition function in the rotating frame is given by

$$Z_{rot} = \sum_s e^{-E_{rot,s}/kT} \tag{5.49}$$

Lemma 3:
The angular momentum in the rotating frame, $L_{rot} \equiv L_{anom}$ is given by

$$L_{anom} = kT\frac{\partial \ln Z_{rot}}{\partial \omega} = -\frac{\partial F_{rot}}{\partial \omega} \tag{5.50}$$

where F_{rot} is the free energy in the rotating frame.[14]

Proof:
The partition function in the rotating frame can be expressed as

$$Z_{rot} = Tr\ e^{-\widehat{H}_{rot}/kT} \tag{5.51}$$

while the Hamiltonian in the rotating frame can be expressed as

$$\widehat{H}_{rot} = \frac{\widehat{L}^2_{rot}}{2I} + V(\{\theta_j\}) \tag{5.52}$$

where

$$\widehat{L}_{rot} = \widehat{L} - I\omega = \hbar \sum_j (-i\frac{\partial}{\partial \theta_j}) - I\omega \tag{5.53}$$

It easily follows that

$$L_{anom} = \frac{1}{Z_{rot}} Tr\ e^{-\widehat{H}_{rot}/kT} \widehat{L}_{rot} = kT \frac{\partial \ln Z_{rot}}{\partial \omega} \tag{5.54}$$

The original theorem follows trivially since the periodicity of the partition function is passed on to the angular momentum. In addition, since the energy eigenvalues can be shown to be *even* functions of the angular velocity, the angular momentum is an *odd* function of the angular velocity. And finally, we should note that while we have demonstrated the periodicity of the angular velocity, there is no guarantee that its amplitude is non-vanishing.

5.6 Summary and Discussion

We have shown that a ring of an Ideal Bose gas exhibits both Bloch and Josephson oscillations at a low temperature. Let us consider a concrete numerical example to assess the feasibility (albeit at present remote) of observing these oscillations. We will assume a spin-polarized condensate of $N = 10,000$ atoms of hydrogen as a Boson with low mass – $1.7 \times 10^{-24} gm$ and a ring radius of $0.1 mm$. We obtain the following:

$$\omega_0 = 5.9 \text{ rad/sec}, \quad e_0 = 2.9 \times 10^{-27} \text{ ergs}, \quad T_B = 30\ nK.$$

It is difficult to imagine how one could observe this anomalous behavior with current experimental techniques. While there has been great progress in confinement of BE condensates, the anomalies presented here require the presence of a confining toroidal wall that can be controllably rotated.

Josephson oscillations of an electric current in a ring are the natural expansion of the long-known oscillations (usually referred to as "Bloch oscillations") in the velocity of electrons in a periodic lattice that are driven by an electric field $\mathcal{E}$. Here, as long as electrons remain in one band (Zener tunneling being absent) the electrons move from one end of a band to a zone edge, where they are reflected to the opposite edge. The frequency is given by $f = e\mathcal{E}a/h$, where a is the lattice spacing. In the case of ring, the lattice spacing of a periodic lattice is replaced by the circumference of the ring. The potential V in the Josephson frequency $f_J = eV/h$ is related to the electric field through $\mathcal{E}a = V$. Similarly, the anomalous behavior presented in this paper has its counterpart in motion of atoms in a periodic lattice. In fact, T. Salger et al[15] have recently presented evidence of Bloch oscillations of a BE condensate Rb atoms moving in a periodic optical lattice. Here, the optical lattice is accelerated so as to produce an effective external force.

References

1. N. Byers and C.N. Yang, Phys. Rev. Lett. **7**, 46 (1961).
2. F. Bloch, Phys. Rev. Lett. **21**, 1241 (1968).
3. Y. Imry, *Introduction to Mesoscopic Physics* (Oxford University Press, Oxford, GB, 2002), 2nd ed.
4. H. Bary-Soroker, O. Entin-Wohlman, and Y. Imry, Phys. Rev. Lett. **101**, 057001 (2008).
5. See, for example, L. P. Lévy, et al, Phys. Rev. Lett. **64**, 2074 (1990).
6. See the Proceedings of the 1967 Summer School held at St. Andrews, Scotland on "Mathematical Methods in Solid State and Superfluid Theory" (Oliver and Boyd, Edinburgh, 1967).
7. B. Jancoviçi , Phys. Rev. Lett. **19**, 20 (1967)
8. V. K. Tkachenko, Sov. Phys. (JETP) **23**, 1049 (1966).
9. In deriving this result, it is assumed that the chemical potential is independent of the flux. This will be so for the Ideal Fermi gas when the temperature is much less than the Fermi temperature μ/k. Note that the crossover temperature is much less than the Fermi temperature by a factor of about N.
10. B. L. Altshuler, Y. Gefen, and Y. Imry, Phys. Rev. Lett. **62**, 66 (1991).
11. It is interesting that in the quest for a solution to equation (5.32) using MAPLE, there are multiple solutions corresponding to the various 'bands'. Thus, one must restrict the search to solutions with the lowest energy.
12. I. S. Gradshteyn and I.M. Ryzhik, Table of Integrals, Series, and Products, (Academic Press, USA, 1966)
13. It is easy to show that the current is a periodic function of the flux, though Byers and Yang don't seem to state so explicitly.
14. This result is analogous to the expression for the electric current in a ring: $I = -\partial F/\partial\Phi$.
15. T. Salger et al, Phys. Rev. A **79**, 011605(R) (2009).

Chapter 6

Exploring Decoherence in Cavity QED

Serge Haroche, Igor Dotsenko, Sébastien Gleyzes, Michel Brune, and Jean-Michel Raimond

Laboratoire Kastler Brossel de l'Ecole Normale Supérieure, 24 rue Lhomond 75231 Paris Cedex 05 France
&
Collège de France, 11 Place Marcelin Berthelot, 75005, Paris France

We review here the cavity QED studies performed at ENS to explore environment-induced decoherence on photonic Schrödinger cat states prepared by dispersive methods. These experiments, which were at an early stage at the time Jo Imry was visiting our group in Paris in the 1990's, have been steadily improved since, with current decoherence times three orders of magnitude larger than in the first demonstration. Future developments of these studies which will implement feedback correcting procedures and investigate non-locality effects are also briefly discussed.

6.1. Introduction. Early Decoherence Experiment in Cavity QED

One of us (SH) has met Jo Imry for the first time on the occasion of a Nobel Symposium in Sweden in 1997. Soon after, he visited us in Paris as a Blaise Pascal Professor. At that time, we were working at ENS on improving our first decoherence experiment which involved photons trapped in a high-Q cavity. We soon realized through discussions with Jo that, with his colleagues at the Weizmann institute, he was interested in the same kind of physics, occurring in mesoscopic systems [1]. In the ENS quantum optics set-up [2] as well as in the Weizmann mesoscopic circuit device [3], similar environment-induced decoherence phenomena were at work. The formalisms used to describe them were however quite different and we learned a lot from the confrontation of our disparate view points. We could appreciate on several occasions since then the depth of Jo's thinking and his wide knowledge in physics. As a token of our admiration for Jo and his career, it is a pleasure for us to contribute to the book published on the occasion of his seventieth birthday.

Our goal in this chapter is to briefly describe the history of the decoherence studies in Cavity QED at ENS. Decoherence, as it is understood according to the standard interpretation of quantum mechanics, is the loss of quantum coherence of "large" system state superpositions, due to their coupling to their environment [4]. By "large", it is meant that one can define a "distance" between the states involved in the superposition and that this distance corresponds to many quanta of a relevant observable. The rate of decoherence is generally shown to increase linearly with this distance, becoming exceedingly large for truly macroscopic systems. This explains why such systems are described by statistical mixtures of states obeying classical logic (a cat is either dead or alive) and not by the quantum logic of state superpositions (Schrödinger's cat is coherently suspended between life and death). In order to catch decoherence in action, it is thus necessary to study systems at the border between the quantum and the classical, small enough so that they can be prepared before decoherence has washed out all the state quantum features, and large enough so that one can define and measure a size-dependent decoherence rate. Studying decoherence in the laboratory thus involves to master delicate experimental techniques involving the manipulation of systems with large numbers of quanta interacting with a well defined and controlled environment.

The ENS experiments in this field, which we started in the mid 1990's, were performed on photonic Schrödinger cat states trapped in a microwave superconducting cavity [5]. A sketch of the experimental set-up (see Fig. 6.1) shows the main ingredients of the experiment. A coherent state of a microwave field (at 51 GHz) is first injected in the high-Q cavity C by the classical source S. This field interacts with a circular Rydberg atom. This atom is prepared in an initial quantum state $|g\rangle$ in box B. Before entering C, it is then brought into a coherent superposition of $|g\rangle$ with an adjacent Rydberg states $|e\rangle$ in the auxiliary

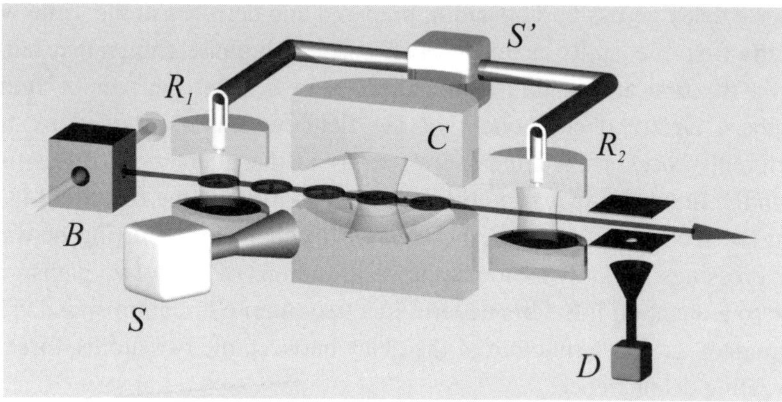

Fig. 6.1. Sketch of the Cavity QED set-up for the study of decoherence.

cavity R_1 coupled to the source S'. The circular Rydberg atom and the field in C are slightly out of resonance. As a result, their interaction cannot lead to energy exchange (absorption or emission of photons), but only to mutual frequency shifts. Due to these transient shifts, the $|g\rangle$ and $|e\rangle$ parts of the atom's state impart to the cavity field two opposite phase shifts at once, thus generating an entangled atom-field state in which the two relevant atomic states are correlated to two field components with different phases.

Detecting directly the atom's state after it leaves the cavity would project the atom-field system into one or the other of its two components, thus destroying the superposition. To keep the quantum ambiguity, a second auxiliary cavity R_2, placed downstream, admixes again, before atomic measurement in detector D, the two states $|e\rangle$ and $|g\rangle$. This second pulse plays the role of a quantum eraser, making it impossible to know, upon atomic detection, whether the atom crossed the cavity C in state $|e\rangle$ or $|g\rangle$. As a result, the field collapses after atomic detection in a linear superposition of two fields with different phases, a quantum optics version of the famous cat imagined by Schrödinger.

In order to probe finally the quantum coherence of this superposition, a second atom, playing the role of a probe or "quantum mouse", is sent after a delay across C. This atom, prepared and detected in the same way as the first one, splits again into two parts each phase component left in C by the first atom. As a result, there is in C, after the pair of atoms, a phase unshifted component of the field, which is obtained by two different processes. It either corresponds to the interaction of the field with the first atom in $|g\rangle$ and the second in $|e\rangle$, or with the first atom in $|e\rangle$ and the second in $|g\rangle$. In the absence of decoherence occurring between the crossings of the two atoms, the contributions of these two paths give rise to a quantum interference term in a two-atom correlation signal. This term decreases as a function of the delay between the two atoms, directly revealing decoherence.

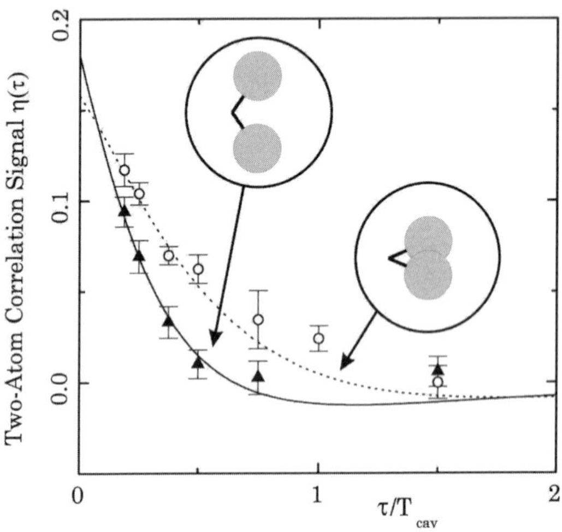

Fig. 6.2. Experimental observation of decoherence in the 1996 ENS experiment: the two atom correlation signal $\eta(\tau)$ is recorded for two values of the angle separating the Schrödinger cat state components (represented in phase space in insets). Points are experimental and curves are theoretical fits (from [2]).

In our 1996 experiment, we have observed this quantum coherence decay (see Fig. 6.2) and we have shown that its rate is proportional to the distance between the two field states measured in dimensionless units in phase space [2]. This experiment revealed various aspects of decoherence, its relationship with the concepts of complementarity, entanglement, "which path" information and quantum eraser effects. It was also a direct illustration of a central aspect of a quantum measurement. The phase of the field could indeed be seen as a "meter" pointing along two different directions and thus measuring the energy of the atom. The decoherence of the photonic cat state thus corresponds to the loss of coherence between the meter states in a measurement, an essential feature of measurement theory.

In this early experiment, the cavity damping time was 160 microseconds and the decoherence time of a cat state containing a few photons was in the 10 microsecond range, barely enough to send the probe atom through the set-up before the field has become classical. Ten years later, we have revisited these studies with a photon trap able to keep the field up to 130 milliseconds. This was achieved by a much better control of the cavity mirrors geometry and superconducting coatings [6]. This three order of magnitude improvement in the field lifetime allows us now to probe the decaying cat state with tens of atoms crossing *C* before a single photon is lost, with field containing larger photon numbers than in the 1996 experiment. We can now not only record two-atom correlation signals, but use the information provided by large atom numbers to reconstruct fully the state of the quantum field in the cavity [7]. Movies of these states evolution reveals directly and completely the decoherence phenomenon [7]. These studies also open the way to new experiments in which decoherence will be counteracted by active feedback [8,9]. Generalization of these experiments to non-local fields trapped in two cavities is also under consideration [10]. We describe these recent developments and the perspectives they open for the exploration of the quantum-classical boundary in the next sections of this Chapter. We restrict our review here to photonic Schrödinger cat states prepared by dispersive methods involving the interaction of the field with non-resonant atoms. We have also performed various studies on similar Schrödinger cat states prepared by resonant methods. These experiments are closely related to the study of the Rabi oscillation collapse and revival phenomenon [11,12]. For a more complete analysis of the physics of Schrödinger cat states of light trapped in a cavity, see Chapter 7 of the book "Exploring the quantum: atoms, cavities and photons" [13].

6.2. Quantum Non-demolition Detection of Photons and Photonic Schrödinger Cats

The first study performed with our very high-Q photon box was a quantum non-demolition (QND) measurement of light at the single photon level [14]. This experiment used essentially the same set-up as the Schrödinger cat's one (see Fig. 6.1). The generation of superposition of field states with different phases is in fact a natural intermediate step of the QND process leading to the generation of photon number states [15]. Let us recall briefly here how this QND measurement is performed, which will give us a different way to look at the photonic cat states and shed a new light on the way they are prepared.

Our QND scheme is based on the detection of the dispersive light-shifts induced by the microwave trapped in the cavity C on the transition frequency between two energy states of probe atoms interacting one by one with the field [16]. Measuring this shift amounts to a measurement of the photon number. Since the probe atoms and the field are non-resonant, the procedure does not modify the field intensity, which ensures its non-demolition character. The field, a microwave around 51 GHz, is confined between two highly reflecting superconducting mirrors forming the Fabry-Perot cavity C (distance between mirrors is 2.7 cm). This resonator traps microwaves for times exceeding a tenth of a second [6]. Atoms prepared in box B into the circular Rydberg state $|e\rangle$ (state of maximum angular momentum with principal quantum number 51) cross the cavity one by one and their transition frequency towards the final level $|g\rangle$ (circular state with principal quantum number 50) is monitored by Ramsey interferometry, according to the standard procedure used in atomic clocks. The Ramsey interferometer is formed by the two auxiliary cavities R_1 and R_2 whose role in the generation and analysis of Schrödinger cat states of light was already discussed in Sec. 6.1.

The superposition $|e\rangle + |g\rangle$ of the two Rydberg states $|e\rangle$ and $|g\rangle$ is prepared by a classical microwave $\pi/2$ pulse applied in R_1. This superposition, corresponding to a non-zero expectation value of the atomic dipole, evolves freely as the atom drifts across the cavity and accumulates a phase shift which, to lowest order in the atom-field coupling, is a linear function of the photon number in *C*. A second $\pi/2$ Ramsey pulse is then applied in R_2, with an adjustable phase ϕ with respect to the pulse applied in R_1. This second pulse transforms the superpositions $|e\rangle \pm e^{i\phi}|g\rangle$ back into $|e\rangle$ or $|g\rangle$. The atomic energy is finally measured by detector *D*, using a state-selective field ionization procedure. The combination of the R_2 pulse with the atom's energy measurement in state $|e\rangle$ or $|g\rangle$ thus amounts to detecting the atomic dipole components corresponding to phases ϕ or $\phi+\pi$ respectively. Due to the dependence of the atomic dipole phase-shift upon the field intensity, the probability for finding an atom in $|e\rangle$ or in $|g\rangle$ is an oscillating function of the photon number.

This atomic clock device has the capability to detect single photons because of the huge sensitivity of the circular Rydberg atoms to microwaves [5,13]. The atom-field coupling is defined by the vacuum Rabi frequency $\Omega/2\pi$ which is, in units of $h/2\pi$, the product of the Rydberg atom dipole matrix element by the electric field associated to one photon in *C*. Its value in our set-up is $\Omega/2\pi = 50$ kHz. To lowest order in Ω, the phase-shift per photon is $\Phi_0 = \Omega^2/2\delta$ independent of the photon number for $n < (\delta/\Omega)^2$. For small detunings δ, analytically calculable non-linear contributions to the phase shift become significant. Choosing a value $\delta/2\pi = 70$ kHz, barely larger than $\Omega/2\pi$, we achieve a phase shift of π per photon around $n = 1$, for atoms crossing the cavity at v = 250 m/s (24 μs crossing time). This phase shift can be adjusted to any smaller value by merely increasing δ.

A single atom is in principle enough to determine the field's energy if the photon number can take only the values 0 or 1 [14]. For such small fields, we adjust the phase shift per photon to the value π and choose the phase of the Ramsey interferometer so that, ideally, the atoms are detected in $|e\rangle$ or $|g\rangle$ when the field contains 1 or 0 photon, respectively. When the experiment is performed with a cavity cooled at 0.8 K, a temperature for which the mean thermal photon number is 0.05, we observe telegraphic sequences of atoms switching at random times from being detected predominantly in $|e\rangle$ or in $|g\rangle$ (see Fig. 6.3). This gives a

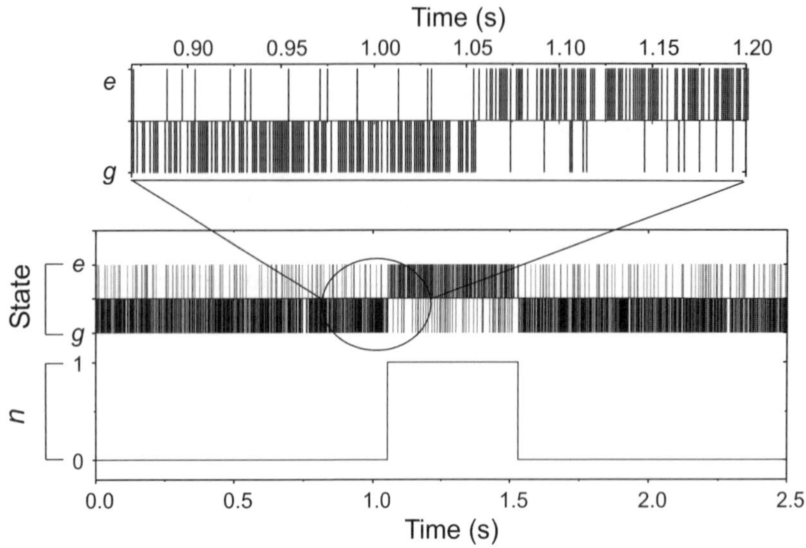

Fig. 6.3. Repetitive QND measurement of a small thermal field containing 0 or 1 photon in C. Lower and upper bars in each trace: atoms detected in $|g\rangle$ and $|e\rangle$, respectively (adapted from [14]).

direct evidence of single photons appearing and disappearing in the cavity due to the thermal fluctuations in its mirrors.

For larger fields, information is extracted progressively by sending sequences of atoms, each contributing partially to the knowledge of the photon number [17]. We assume at the start of the measurement a minimal knowledge, corresponding to a flat photon number distribution $P^{(0)}(n) = 1/(N+1)$, where N is the maximum number of photons expected in the field. The phase shift per photon is adjusted to the value $2\pi/(N+1)$. In practice, we perform experiments with $N = 7$, the test field being a small coherent field produced by a classical microwave source and injected in the cavity by diffraction on the mirrors' edges. It is expected to have a Poisson distribution of its photon number with a mean of the order of 3. The atomic dipole can then point along $N+1 = 8$ different directions corresponding to the photon numbers 0, 1, 2, … , 7. By detecting the dipole of the first atom along the direction ϕ, we exclude the photon number for which the dipole would point along the direction

$\phi + \pi$. Other photon numbers are not ruled out, but their probability is multiplied by a cosine function of the photon number, and the inferred photon distribution becomes an oscillating function of n, noted $P^{(1)}(n)$. As a result of this multiplication, the photon numbers for which the dipole is aligned in a direction close to ϕ become more likely than those for which the dipole is aligned in a direction close to $\phi + \pi$. This simple Bayesian argument is applied progressively to a sequence of atoms crossing C, each atom contributing by a multiplying cosine factor to the previously inferred photon probability distribution. The inferred distribution after the kth detected atom, $P^{(k)}(n)$, makes more and more n-values unlikely. The inferred $P^{(k)}(n)$ distribution eventually converges for $k \sim 100$ towards a delta-like peak corresponding to a single photon number. The measuring time T_m is of the order of 20 ms, much shorter than the cavity damping time (T_c = 130 ms). Resuming the same measurement on another realization of the same initial field generally yields randomly another photon number. The histogram of the photon numbers obtained from a large number of independent measurements reproduces the ensemble statistical distribution $\Pi(n)$ of the photon number in the initial coherent field. We verify that it is, as expected, given by a Poisson law.

When the measurement is pursued on a single field realization after the convergence has occurred, the same photon number is, for some time at least, obtained again. This reproducibility is an essential feature of an ideal QND measurement. Eventually, however, the perturbing effect of field relaxation cannot be neglected. It brings the photon number down, step by step, towards vacuum. Figure 6.4 shows field trajectories representing the photon number expectation value versus time, for six measuring sequences performed on independent realizations of the same initial coherent field. After converging towards a specific value (5 or 7 for the upper traces, 4 for the four lower traces), the photon number remains at first steady on a "plateau", then evolves in each realization along a staircase-like trajectory. These staircases may exhibit an occasional upward kick due to the creation of a thermal photon (last frame in lower traces). All trajectories clearly exhibit the quantum jumps of the field. The times of these jumps are random. Their statistics has been reconstructed from an analysis of a large number of realizations and has been found in good agreement with the predictions of the quantum electrodynamics master equation describing field relaxation [17].

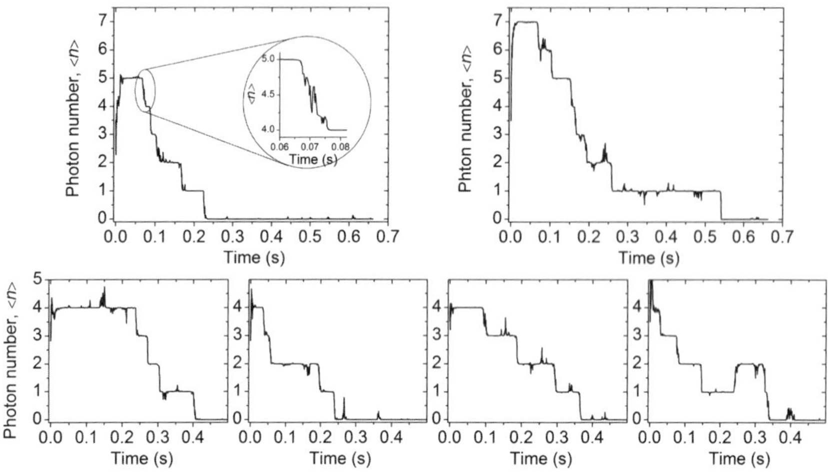

Fig. 6.4. Evolution of the inferred photon number expectation value $\langle n \rangle$ observed on a sample of six field realizations. The measurement converges to $n = 5$ and $n = 7$ in the upper traces, to $n = 4$ in the lower traces and the field subsequently relaxes down step by step towards vacuum (adapted from [17]).

Let us now make the connection between the QND photon counting and the Schrödinger cat state experiment described in Sec. 6.1. Along a single measuring QND sequence, the field, initially in a coherent state, evolves into a Fock state. The phase of the field is strongly affected in the process, which is a direct consequence of quantum complementarity or, equivalently, of the Heisenberg uncertainty relations between time and energy (or phase and photon number). The initial coherent state has a well-defined phase, which requires fuzziness on the photon number. As the photon number gets progressively pinned down to a definite value, the phase gets more and more blurred. This blurring occurs by successive phase doublings, since an atom interacts with C in a superposition of two states shifting the phase of the field in opposite directions. Each atom produces two field components out of each field component left by the preceding atom, doubling at each stage the number of components in the field state superposition until complete phase randomization is obtained. Of particular interest are the states obtained after detecting the first atom, which are the Schrödinger cat states described in Sec. 6.1. These states can be viewed as resulting from the field collapse induced by the partial QND measurement provided by the first probe atom crossing C.

A particularly simple situation occurs if the phase shift per photon for the first atom is set to $\Phi_0 = \pi$, this value remaining approximately constant in the range of n-values across the whole photon number distribution in the field. The atomic dipole phase then assumes (modulo 2π) one value, ϕ, for all even photon number, and the value $\phi + \pi$ for all odd photon numbers. Detecting with the Ramsey interferometer phase ϕ the final state of the atom thus amounts to measuring the photon number parity ("even" if the atom is found in $|e\rangle$, "odd" if in $|g\rangle$). This measurement projects the field into an eigenstate of this observable, i.e. into a superposition of even or odd Fock states. A coherent state $|\alpha\rangle$ of complex amplitude α is a Poissonian superposition of Fock states with mean photon number $n_{\mathrm{m}} = |\alpha|^2$, spanning a range of ~ $|\alpha|$ photon numbers around n_{m}. This state can obviously be expanded as $|\alpha\rangle = [(|\alpha\rangle + |-\alpha\rangle) + (|\alpha\rangle - |-\alpha\rangle)]/2$, a superposition of an even $(|\alpha\rangle + |-\alpha\rangle)$ and an odd $(|\alpha\rangle - |-\alpha\rangle)$ photon number state. Note that the photon number parity measurement, which ideally requires only a single probe atom, is enough to count the photon number if it does not exceeds the value 1. It is precisely what was done in the QND experiment described in [14].

The photon number parity measurement by the first atom interacting with the field thus projects an initial coherent state into one of these two states which are quantum superpositions of coherent states with opposite classical phase. One prepares randomly either one of these two states, whose photon number parity is known only after the atom has been detected in $|e\rangle$ or in $|g\rangle$. These superposition states are indeed the Schrödinger cat states introduced in Sec. 6.1, in the particular case where the phase-shifts induced on a coherent field by an atom in level $|e\rangle$ or $|g\rangle$ are respectively $\pm\ \pi/2$. We thus get an alternative interpretation of the preparation procedure of these states. Whereas we described it above as resulting from the interaction with an atom following two interfering paths in a Ramsey interferometer, it appears here as the collapse of a coherent field undergoing a simple measurement of its photon number parity. The connection between QND measurement and Schrödinger cat physics is discussed in details in [13].

6.3. Full Reconstruction of Schrödinger Cat States and Movies of their Decoherence

The QND procedure provides not only a method to prepare Schrödinger cat states of light, it can also be adapted to reconstruct fully their quantum state in a time-resolved way, yielding a direct and complete view of the decoherence phenomenon [7]. By repeating the cat state preparation many times and separating the events leading to even and odd photon numbers, we can prepare two ensembles of realizations of even and odd Schrödinger cat states (which we will call the "signal"). Let us now show how subsequent probe atoms (crossing C after the first QND atom which has prepared the cat) are used to reconstruct the even and odd cat state "signals". If we just kept measuring the photon number on these two ensembles of even and odd cat states, we would build histograms which would tend towards the probability distribution $\Pi(n)$ of the photon number in these states. This amounts to reconstructing the diagonal matrix elements ρ_{nn} of the signal field density operator in the Fock state basis. This measurement provides though no information about the off-diagonal elements $\rho_{nn'}$ ($n \neq n'$) which describe the phase coherence of the signal fields.

This missing information can be obtained by mixing the signal with "reference" coherent fields of variable complex amplitude α, then measuring by our QND method the photon number distribution of the resulting field for a large sample of different α values. The method bears strong similarities with the homodyne procedure used in quantum optics to reconstruct states of optical light beams [19]. The mixing is experimentally achieved by pulse-injecting in the cavity, via scattering on the mirrors' edges, coherent fields with calibrated amplitudes which interfere classically with the initial field. This coherent-field addition is a translation in phase-space described by the Glauber unitary translation operator $D(\alpha) = \exp(\alpha a^\dagger - \alpha^* a)$, where a and $a^\dagger$ are the photon annihilation and creation operators. The translated field writes $\rho^{(\alpha)} = D(\alpha)\, \rho\, D(-\alpha)$ and the QND measurement of its photon number distribution yields the diagonal elements

$$\rho^{(\alpha)}_{nn} = \Sigma_{n'n''}\, D_{nn'}(\alpha)\, \rho_{n'n''}\, D_{n''n}(-\alpha), \qquad (6.1)$$

which admix in a linear combination all the signal field off-diagonal matrix elements $\rho_{n'n''}$. Determining the $\rho^{(\alpha)}_{nn}$ distributions for a large enough sample of different α values yields a set of equations constraining the $\rho_{n'n''}$ which we seek to reconstruct. Solving this set of equation is achieved by the method of maximum entropy [20].

The complete experimental procedure requires a double sampling. For each α value of the reference field, ~ 100 realizations of the signal field are prepared and the photon number distribution $\rho^{(\alpha)}_{nn}$ is obtained from the QND measurement performed with sequences of atoms crossing C in each of these realizations. Between realizations, a large flux of resonant atoms prepared in state $|g\rangle$ are sent across C to wipe out the remaining field and reinitialize the vacuum in the cavity before the next copy of the signal field is prepared. Since we want to take a snapshot of the field's state in a time short compared to its decoherence time, we usually cannot perform a complete QND measurement in each sequence, which would take about 20 ms. Collecting the data from ~ 8 atoms requires a time window of ~ 4 ms and leads to an inferred photon number distribution $P^{(8)}(n)$ which has not yet collapsed to a single photon number value. Yet, the average of such partially converged distributions over the ensemble of field realizations yields the statistical photon number distribution $\Pi^{(\alpha)}(n) = \rho^{(\alpha)}_{nn}$. Once $\rho^{(\alpha)}_{nn}$ has been determined, the procedure is resumed with another complex amplitude, and so on. About 600 sampling α values are required to constrain all the relevant matrix elements $\rho_{n'n''}$ for signal fields expanding on Fock states with $n < 10$. A full reconstruction requires about 60000 field realizations, with information collected from 0.5 million atoms, within a time of the order of 12 hours.

The QND nature of our measurement procedure is here essential. It is because the measurement does not change the ensemble statistical properties of the photon number distribution that we can collect information in each sequence from an arbitrary number of atoms, as long as data acquisition is performed in a time short compared to the field evolution time. Similar field reconstructions have been recently achieved in Circuit QED experiments, using a field-demolishing procedure to extract the photon number information [21]. There, one elementary measurement only could be performed on each field realization. The data acquisition time was still manageable because the qubit-field oscillator interaction time and the field damping time were much smaller than in our experiment. In our set-up, it takes about 400 microsecond to couple one atom qubit to the field and the reconstruction would have been impossible if the sequence had to be reinitialized after each atom.

Once the density operator has been reconstructed, we compute the field's Wigner function through a simple mathematical formula (see Appendix of [13] for a review of the main properties of Wigner functions of a harmonic oscillator). This function, which is a real distribution in the two-dimensional phase space of the quantum field oscillator, contains the same complete information about the system as the density matrix, but its form gives a clearer insight into the field's physical features. Classical fields such as coherent states or thermal fields have Gaussian Wigner functions. Non-classical fields such as Fock states or Schrödinger cats are expected to have non-Gaussian Wigner functions presenting oscillations and taking negative values at some points in phase space. These features are the signature of quantum interference phenomena. They are also very fragile and prone to be rapidly destroyed by environment-induced decoherence.

To test the reconstruction procedure, we have first applied it to a coherent field with an average photon number $n_{\mathrm{m}} = 2.5$, see Fig. 6.5(a). We obtained, as expected, a Gaussian peak in phase space, which is centered at a point whose coordinates yield the amplitude and the phase of the field. The width of this Gaussian describes the minimal conjugated quantum uncertainties of these quantities. The fidelity of this reconstruction is very high (F = 98%). We have then used the same method to reconstruct the Wigner functions of the even and odd cat states prepared by the interaction of a coherent field with a first QND measuring atom. The obtained Schrödinger cat's Wigner functions (Figs. 6.5(b) and (c)) exhibit two Gaussian peaks associated to the two field components, with an oscillating feature taking negative values in between, accounting for the quantum coherence of the states. The amplitude D^2 = 12. If we translated one of the components towards

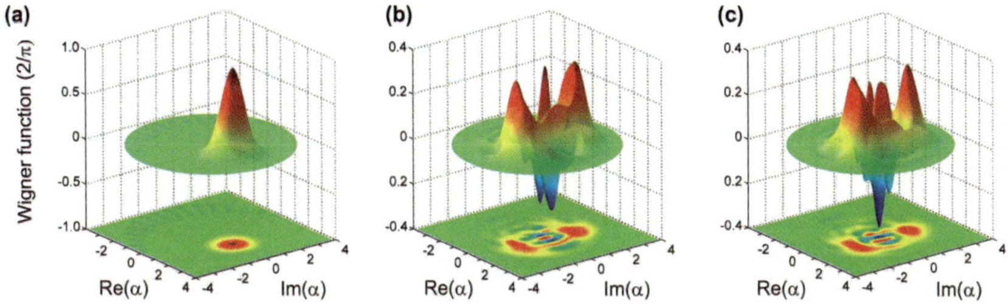

Fig. 6.5. Reconstructed Wigner functions. (a) Coherent state with n_m = 2.5 photons. (b) and (c) Even and odd parity Schrödinger cat state with n_m = 3.5. The fidelities F (overlap between the expected states and the reconstructed ones) are 98%, 64% and 61%, respectively (adapted from [7])

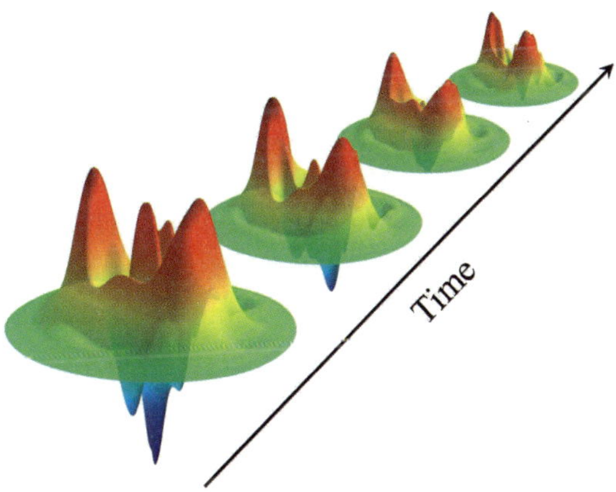

Fig. 6.6. Wigner functions reconstructed after four successive delays (t = 1.3, 4.3, 15.8 and 22.9 ms) exhibiting clearly the evolution of the Schrödinger cat state into a statistical mixture.

vacuum, the other would be a coherent state with $n_m = 12$. In other words, these Schrödinger cat states have the same "size" as a superposition of the vacuum with a 12 photon coherent state. The only difference between the even and odd cat states is the sign of the interference pattern between the two Gaussian peaks. Note that similar Wigner functions with a somewhat smaller distance between their components have been obtained with optical light beams [22]. More closely related to our work are the cat states recently reconstructed in Circuit QED experiments [21].

A time-resolved state-reconstruction procedure has allowed us to realize movies of the Schrödinger cat's evolution, revealing the decoherence process in action [7]. The straightforward way to obtain these movies would be to prepare identical cat state and reconstruct their Wigner functions after letting them evolve in the cavity for a variable time delay. The movie would be finally built by arranging these snapshots in sequences separated by the elementary time interval required for the reconstruction (4 ms). The problem with this method is that it would require several hours for each frame, making the total shooting time prohibitively long.

We have overcome this difficulty by using a simple trick. To obtain at once all the data corresponding to a given value of α, we translate the field in phase space immediately after the cat's state preparation by that amount, then detect successive sequences of atoms divided into elementary 4 ms time windows, accumulating data over an overall 50 ms time interval. This provides a direct record of the evolution of the translated state rather than that of the state itself. The two dynamics are closely related, however. Translating by α an initial field state before letting it evolve due to relaxation during time t is equivalent to letting it evolve during that time, then translating it by $\alpha \cdot \exp(-t/2T_c)$ [7]. We thus analyze the data obtained at time t as if they corresponded to a translation rescaled by the factor $\exp(-t/2T_c)$. This is more efficient than letting the field evolve before translating it because we exploit all the data of a long sequence instead of only recording a short time window for each delay. We have experimentally checked the equivalence between the two methods by comparing the results for one time delay and have verified that the reconstructed Schrödinger cat states are, within the effect of noise, undistinguishable.

Figure 6.6 shows four snapshots of a Schrödinger cat state at increasing times, clearly revealing decoherence. Over the 23 ms of between the first and the last frame, the classical Gaussian peaks have hardly evolved, while the interference feature between them has all but vanished. The initial state has been turned into a statistical mixture of two coherent states. The decay of the quantum coherence, measured as the amplitude of the Wigner function central feature, is plotted in Fig. 6.7. It obeys to an exponential law with a time constant in remarkable agreement with the theory of environment-induced decoherence [4]. The decoherence time T_d is found close to $T_c/2D^2$, with a small correction accounting for residual thermal effects [23]. By comparing the decoherence times of two cat states with different mean photon numbers, we have verified, in agreement with theory, that decoherence occurs at a rate proportional to the distance in phase space between the cat state's classical components.

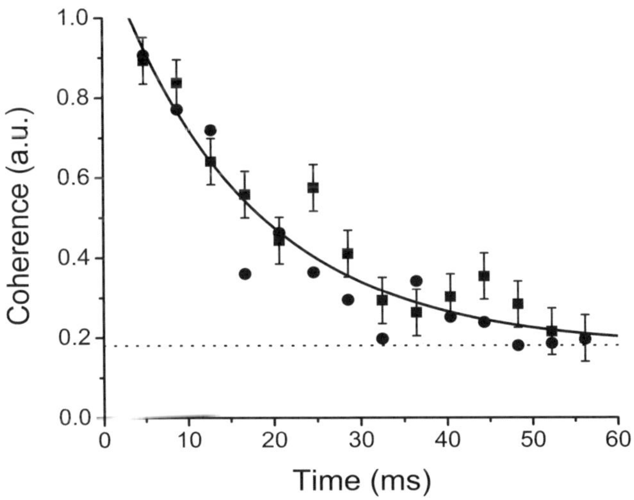

Fig. 6.7. Decay of the amplitude of the interference term in the even (squares) and odd (circles) Schrödinger cat Wigner functions (n_m = 3.5). The points are experimental and the line is a common exponential fit which includes an offset (dotted line) accounting for residual noise (from [7]).

Note that the 1996 two-atom correlation decoherence signal (see Fig. 6.2) corresponded to the measurement of the field Wigner function at the origin of phase space, while we now can obtain this function at all points. The decoherence time was then three orders of magnitude smaller than in the present study.

6.4. Perspectives: Implementing Quantum Feedback and Studying Mesoscopic Non-locality

We have so far monitored passively the decoherence of a photonic Schrödinger cat. In the future, we plan to go one step further and to use feedback methods in order to counteract the effects of decoherence and to maintain a state superposition "alive" in the cavity for a time longer than the natural decoherence time. The principle of the method is simple and within reach of the present status of our experiment [9]. It involves the Ramsey set-up described above and two different sets of atoms sent one by one across the cavity. One set consists in non-resonant probe atoms sent in C at regular intervals, short compared to the decoherence time. The other set is made of resonant source atoms able to emit a single photon in the cavity while undergoing a π-Rabi pulse in the cavity field. The probe atoms are used to check the cat state parity according to the QND procedure described above. As soon as a parity jump is observed by detecting a change of quantum state in one of the probe atoms, a resonant source atom, prepared in the upper state $|e\rangle$ of the Rydberg transition is sent across C. The Ramsey fields are switched-off for a brief time interval and the source atom is made resonant with the field by using the Stark effect induced by applying a small electric field across the mirrors. The interaction time of this source atom is set so that it undergoes a π-Rabi pulse and delivers, with near unit probability a single photon in the cavity. This atom thus adds a quantum of energy in the field and restores, with a good approximation, the parity of the photon number to the value it had before the jump. Numerical simulations show that the resulting field is, after this correction, nearly identical to the Schrödinger cat state before the jump had occurred. Adding a photon by such a process is a unitary operation which cannot exactly undo the effect of the non-unitary quantum jumps. If we repeat the process to track the evolution of the field as it keeps being fed-back by the atoms correcting the effects of successive jumps, the small errors produced at each step cumulate and lead to a slow diffusion of the field quantum phase. Numerical simulations show however that the field state can be preserved in a quantum superposition for a much longer time than the free evolving field in the cavity. The quantum interference terms should now vanish at the same rate as the field energy and not much faster as it is the case under uncontrolled decoherence. By performing this experiment we will demonstrate the feasibility of an error correcting scheme based on the direct QND monitoring of field relaxation.

In another class of future experiments, we plan to extend these Schrödinger cat studies to fields stored in two cavities and entangled by their common interaction with Rydberg atoms interacting in turn with each cavity field [10]. The resulting entangled field would be Schrödinger cat states of a new kind, involving mesoscopic ensembles of atoms suspended between two different locations. The Wigner function of these fields could be reconstructed in a four-dimension phase space by a method similar to the one described above. The reconstruction will involve the detection of probe atoms performing QND measurements, whose cavity crossings will be preceded by field translation operations. A combination of Wigner function values evaluated at four points in this phase space should form a quantity violating a generalized Bell inequality. Checking this inequality experimentally will provide novel tests of non-locality. The Bell violating quantity should evolve in time and cross the quantum-classical boundary after a finite delay following the non-local entangled state preparation. Measuring this delay would indicate how non-locality depends on the "size" of the entangled parts of a physical system suspended between two separated spatial positions.

Acknowledgments

This work was supported by the Agence Nationale pour la Recherche (ANR), by the Japan Science and Technology Agency (JST) and by the European Union under the Integrated Projects SCALA and CONQUEST.

References

1. Jo Imry, *"Introduction to mesoscopic physics"*, Oxford University Press (1997).
2. M. Brune, E. Hagley, J. Dreyer, X. Maitre, A. Maali, C. Wunderlich, J.-M. Raimond and S. Haroche, *Phys. Rev. Lett.* **77**, 4887 (1996).
3. Y. Imry, *"Mesoscopic Physics and the Fundamentals of Quantum Mechanics"*, Physica Scripta (Proceedings of the Nobel Symposium) **T76**, 171 (1998).
4. W. H. Zurek, *Rev. Mod. Phys* **75,** 715 (2003).
5. J.-M. Raimond, M. Brune and S. Haroche, *Rev. Mod. Phys.* **73**, 565 (2001).
6. S. Kuhr, S. Gleyzes, C. Guerlin, J. Bernu, U. B. Hoff, S. Deléglise, S. Osnaghi, M. Brune and J.-M. Raimond, *Appl. Phys. Lett.* **90,** 164101 (2007).
7. S. Deléglise, I. Dotsenko, C. Sayrin, J. Bernu, M. Brune, J.-M. Raimond and S. Haroche, *Nature (London)* **455**, 510 (2008).
8. I. Dotsenko, M. Mirrahimi, M. Brune, S. Haroche, J.-M. Raimond and P. Rouchon, *Phys. Rev.* A **80**, 013805 (2009).
9. S. Zippilli, D. Vitali, P. Tombesi and J.-M. Raimond, *Phys. Rev.* A **67**, 052101 (2003).
10. P. Milman, A. Auffeves, F. Yamagushi, M. Brune, J.-M. Raimond and S. Haroche, *Eur. Phys. J. D.* **32**, 233 (2005).
11. A. Auffeves, P. Maioli, T. Meunier, S. Gleyzes, G. Nogues, M. Brune, J.-M. Raimond and S. Haroche, *Phys. Rev. Lett.* **91**, 230405 (2003).
12. T. Meunier, S. Gleyzes, P. Maioli, A. Auffeves, G. Nogues, M. Brune, J.-M. Raimond and S. Haroche, *Phys. Rev. Lett.* **94**, 010401 (2005).
13. S. Haroche and J.-M. Raimond, *"Exploring the quantum: atoms, cavities and photons"*, Oxford University Press (2006).
14. S. Gleyzes, S. Kuhr, C. Guerlin, J. Bernu, S. Deléglise, U. B. Hoff, M. Brune, J.-M. Raimond and S. Haroche, *Nature (London)* **446**, 297 (2007).
15. M. Brune, S. Haroche, J.-M. Raimond, L. Davidovich and N. Zagury, *Phys. Rev.* A **45**, 5193 (1992).
16. M. Brune, S. Haroche, V. Lefèvre, J.-M. Raimond and N. Zagury, *Phys. Rev. Lett.* **65**, 976 (1990).
17. C. Guerlin, J. Bernu, S. Deléglise, C. Sayrin, S. Gleyzes, S. Kuhr, M. Brune, J.-M. Raimond and S. Haroche, *Nature (London)* **448**, 889 (2007).
18. M. Brune, J. Bernu, C. Guerlin, S. Deléglise, C. Sayrin, S. Gleyzes, S. Kuhr, I. Dotsenko, J.-M. Raimond and S. Haroche, *Phys. Rev. Lett.* **101**, 240402 (2008).
19. A. I. Lvovsky and M. G. Raymer, *Rev. Mod. Phys.* **81**, 299-332 (2009).
20. V. Buzek and G. Drobny, *J. Mod. Opt.* **47**, 2823-39 (2000).
21. M. Hofheinz, H. Wang, M. Ansmann, R. C. Bialczak, E. Lucero, M. Neeley, A. D. O'Connell, D. Sank, J. Wenner, J. M. Martinis and A. N. Cleland, *Nature (London)* **459**, 546 (2009).
22. A. Ourjoumtsev, H. Jeong, R. Tualle-Brouri and P. Grangier, *Nature (London)* **448**, 784 (2007).
23. M. S. Kim and V. Buzek, *Phys. Rev.* A **46**, 4239 (1992).

Chapter 7

Fractional Charge Determination via Quantum Shot Noise Measurements

M. Heiblum

Braun Center for Submicron Research, Department of Condensed Matter Physics,
Weizmann Institute of Science, Rehovot, 76100, Israel

Charge excitations in a two dimensional electron gas, under a quantizing magnetic field and in the fractional quantum Hall effect regime, flow in one dimensional-like strips along the edges of the sample. These excitations (quasiparticles) may be independent or condense into an interacting chiral Luttinger liquid. Adding a backscattering potential, which reflects a forward propagating quasiparticle to a backward propagating one, partitions the stream of quasiparticles and induces quantum shot noise. The noise is proportional to quasiparticles charge and may be affected by their mutual interactions. The dependence of the determined charge on the temperature, excitation energy, and partitioning will be describes for a few fractional states, revealing in some cases a *universal* behavior.

7.1. Introduction

Shot noise refers to time-dependent fluctuations in an electrical current, being a direct consequence of the *particle*-like nature of the electrons and the stochastic nature of their injection. Shot noise generally provides information on the charge and the mutual correlations of the particles — not necessarily given by the time-averaged current. The average current, according to Landauer formula, merely probes the average transmission coefficient of the electrons through the conductor. In a truly classical stochastic (Poisson-like) electron emission processes the time average, or the ensemble average, of the squared current fluctuations, is $\left\langle i^2 \right\rangle_{\Delta f} = S_i(f) \cdot \Delta f = 2eI_t \cdot \Delta f$, where $S_i(f)$ is the generally *white* spectral density, I_t the average transmitted current, and Δf the bandwidth of the measurement. In mesoscopic conductors, however, where electrons scatter only elastically, shot noise is suppressed. The reason for this suppression is fundamental, resulting from the fermionic nature of the electrons [2]. In contrast, in classical conductors, where electrons scatter many times and thermalize, shot noise, in the sense of being proportional to the current, is not observed, with the noise expressed via the increased temperature and thus the thermal noise.

We are concerned here with quantum Hall effects, observed in a two dimensional electron gas (2DEG) under a strong magnetic field B. The energy spectrum of a 2DEG is then consisting of highly degenerate Landau levels with a degeneracy per unit area $d=B/\phi_0$, with $\phi_0=h/e$ the flux quantum (h being Plank's constant). Whenever the magnetic field is such that an integer number ν (the *filling factor)* of Landau levels are occupied, that is $\nu=n_s/d$ equals an integer (n_s being the areal density of the 2DEG), the longitudinal conductance of the 2DEG vanishes while the Hall conductance equals $\nu e^2/h$ with very high accuracy. This is the celebrated *integer quantum Hall* effect (IQHE) [3,4]. A similar behavior of the conductance is found at *fractional* filling factors, namely, when the filling factor equals a rational fraction with (mostly) an odd denominator $2p+1$, being known as the *fractional quantum Hall* effect (FQHE) [4,5]. In contrast to the IQHE, which is well understood in terms of non-interacting electrons, the FQHE cannot be explained in such terms and is believed to result from interactions among the electrons, brought about by the strong magnetic field. Laughlin's theoretical prediction of the existence of fractionally charged *quasiparticles*, each having a fractional charge $e^*=e/(2p+1)$, e.g., $e/3$, $e/5$ and $e/7$, put forward in order to explain the FQHE effect [6], is very counterintuitive.

Halperin was the first to suggest edge channels transport to explain the conduction mechanism in the IQHE [7]. According to this successful model current flows along the edges of the sample, with electrons performing classical chiral 'skipping orbits'. In a quantum mechanical language, current flows at the crossing of the Landau levels with the chemical potential, near the edges of the sample. In the IQHE regime, electron-correlations are weak and currents are presumed to flow in non-interacting 1d like channels. Alternatively, in the FQHE regime, electron correlations are strong, leading the edge channels to behave as 1d chiral Luttinger liquid [8]. Due to the chirality, backscattering in wide samples is minimized and edge channels propagate void of backscattering for long distances.

Edge channels produce shot noise (called sometimes 'excess noise') on top of the ubiquitous thermal noise ('Johnson-Nyquist' noise), with the latter being a property of any conductor and is independent of its microscopic properties and the electron charge. Its spectral density $S_T=4k_BTg$, with k_B the Boltzman's constant, T the temperature, and g the conductance, originates from microscopic current fluctuations (with no current flowing) due to the finite temperature of the electrons. The so called quantum shot noise differs from the classical one reflecting the noise free property of the emitting reservoir (due to its fermionic nature) [2,9,10]. This was first demonstrated in the simplest mesoscopic system: a ballistic constriction in the 2DEG. The constriction was formed by Quantum Point Contact (QPC) [11], being two closely separated metallic gates evaporated on the surface of the heterostructure embedding the 2DEG (see inset in Fig. 7.1). At zero temperature the contribution to the shot noise of the p'th propagating channel in the constriction is:

$$S_i = 2e^*Vg_p t_p(1-t_p) \ , \tag{7.1}$$

where S_i stands for the low frequency ($f<<eV/h$) spectral density of current fluctuations, V the applied source-drain voltage, g_p the conductance of the fully transmitted p'th channel, and t_p the effective transmission coefficient of the p'th channel. This reduces to the well known *classical* poissonian expression for shot noise (above) when $t_p<<1$ ('Schottky equation').

7.2. Noise of Fractional Edge Channels

Recent theoretical studies of shot noise in the FQHE regime, based on the chiral Luttinger liquid model, are applicable only to the *Laughlin fractional states*, e.g., $\nu=1/3$, 1/5, etc. [8], namely, when the edge current is carried by only one channel. They predict a spectral density of shot noise very much like in Eq. (7.1), with two limits [12]:

$$\begin{aligned} S_{T=0} &= 2e^*Vg_p(1-t_p) = 2e^*I_r \quad ; \quad t_p \approx 1 \\ S_{T=0} &= 2eVg_pt_p = 2eI_t \quad ; \quad t_p \approx 0 \end{aligned} , \tag{7.2}$$

where I_r and I_t are the reflected and transmitted average currents, respectively. The most important outcome of Eq. (7.2) is the prediction that the tunnelling of quasiparticles e^* contributes to the shot noise at weak reflection ($t_p \approx 1$). No easily accessible formulation for an arbitrary transmission exists [12].

One can gain insight into the characteristics of the expected shot noise in the FQHE regime by considering the composite fermions (CF) model [13]. In its simplest form the model identifies the fractionally filled first electron Landau level, $\nu=p/(2p+1)$, as an integer p filled Landau levels of CFs ($\nu_{CF}=p$). Each CF consists of an electron and *two* magnetic flux quanta opposing the original magnetic field attached to it. Hence, the net magnetic field sensed by the CFs is $B\text{-}2n_sh/e$ (equals to zero at $\nu=1/2$). Under this weaker effective magnetic field the CFs can be approximated by weakly interacting quasiparticles, flowing in separate and nearly non-interacting edge channels, hence, justifying the application of the above mentioned formulae for stochastic back scattering of particle and a poissonian shot noise. When the QPC partly pinches off and the conductance is in a transition between two different FQHE plateaus of the series $p/(2p+1)$, only the highest edge channel is being partitioned; the other, lower lying channels, are fully transmitted. Consequently, in Eqs. (7.1) and (7.2), p designates the CF edge channel that is being partitioned. If in the transition region between $\nu=1/3$ and an insulator the values are $p=1$, $g_1=g_Q/3$ and $t_1=3g/g_Q$, then, in the transition region between $\nu=2/5$ and $\nu=1/3$ they are $p=2$, $g_2=(2/5\text{-}1/3)g_Q$, and $t_2 = \dfrac{g/g_Q - 1/3}{2/5 - 1/3}$. Here g is the total two-terminal conductance and $g_Q=e^2/h$ the quantum conductance.

A more general expression for shot noise of non-interacting particles [14], applicable at finite temperatures, is:

$$S = 2e^{*}V\, g_p t_p (1-t_p)\left[coth(\frac{e^{*}V}{2k_B T}) - \frac{2k_B T}{e^{*}V} \right] + 4k_B Tg \ , \qquad (7.3)$$

with a finite noise at zero applied voltage, $S_T=4k_BTg$. When $V>V_T\sim 2k_BT/e^{*}$ the noise approaches the linear behavior predicted by Eqs. (7.1) and (7.2). This expression was utilized for charge determination over a wide range of conditions (transmission and temperature); as described later. It should be noted that a different approach to determine charge, e.g., based on visualization of localized quasiparticles [15] and resonant tunnelling of quasiparticles into an isolated island (or an impurity) [16].

7.3. Experimental Considerations

In order to realize shot noise measurements we formed constrictions in the plane of a 2DEG, embedded in the heterostructure, some 100nm below the surface. Carrier densities are around $n_s=(1\text{-}2)\cdot 10^{11}\text{cm}^{-2}$ and nobilities $\mu=(2\text{-}30)\cdot 10^{6}\text{cm}^{2}/\text{V-s}$ at 4.2K. Samples were cooled in a dilution refrigerator to an electron temperature 10-50mK. Current fluctuations, generated in the constriction, were fed into a *LC* resonant circuit with center frequency ~1MHz and bandwidth of 30-100kHz. The fluctuations were amplified by a nearby extremely low noise, home made, preamplifier, cooled to 4.2K. The preamplifier has a voltage noise as low as 2.5 $10^{-19}\text{V}^2/\text{Hz}$ and a current noise of some $(4\text{-}10)\cdot 10^{-29}\text{A}^2/\text{Hz}$. Outside the cryostat the amplified signal was fed into an additional amplifier followed by a spectrum analyzer. Since the accurate magnitude of the noise signal is important, a careful calibration of the total gain is routinely done by utilizing a calibrated current noise source. Measuring the thermal noise generated by the conduction of the constriction (or by the quantum conductance in a multi-terminal configuration) as function of the inverse conductance $1/g$, both the electron temperature, via $T=(\delta S_T/\delta g)/4k_B$, and the current noise of the amplifier (extracted from the extrapolated total noise to zero inverse conductance) were deduced (as shown in Fig. 7.1).

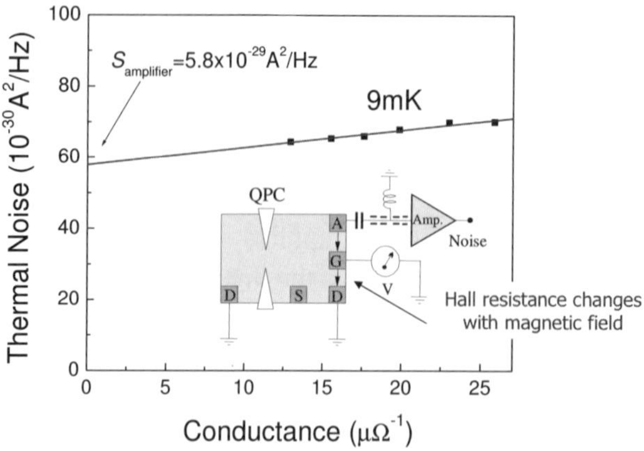

Fig. 7.1. A general configuration of the noise measurement system with the measured shot noise at equilibrium at the input of the preamplifier as a function of the output conductance. The output conductance is independent of the transmission of the QPC and is controlled via the magnetic field. The measured noise is a sum of thermal noise $4k_BTg$ and the constant noise of the amplifier.

7.4. Shot Noise at ν=1/3

Shot noise measurements as a function of the current through the constriction were performed first in the absence of magnetic field. The results, after calibration and subtraction of the amplifier's noise, are shown in Fig. 7.2. The transmission of the lowest laying quasi 1d channel was simply deduced from the measured conductance normalized by $2e^2/h$ (the factor 2 accounts for spin degeneracy). Our data almost perfectly fits the expected noise from electrons predicted by Eq. (7.2) using the independently measured electron temperature [17].

At high magnetic field (10-14Tesla), the two-terminal conductance exhibits Hall plateaus at fractional fillings. We start by exploring filling factor ν=1/3 with $g_{1/3}$=$e^2/3h$. Figures 7.3a and 3b present typical differential conductance of the constriction as function of the applied voltage at electron temperatures T=23mK and T=120mK, for different backscattering potential strengths (determined by the split-gate voltage of the QPC). At the lower temperature, even a relatively weak backscattering potential (transmission t=$g/g_{1/3}$~0.7 at high bias) almost fully reflects the current at zero applied voltage.

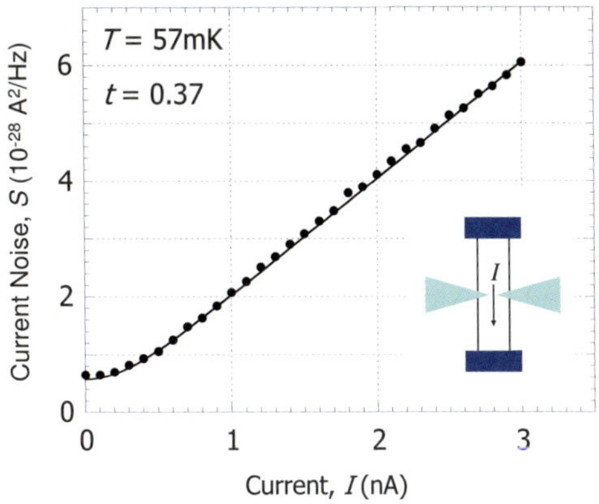

Fig. 7.2. Shot noise as a function of the current I through the constriction without an applied magnetic field. The solid line is the noise given by Eq. (7.3) with the temperature deduced independently (as in Fig. 7.1).

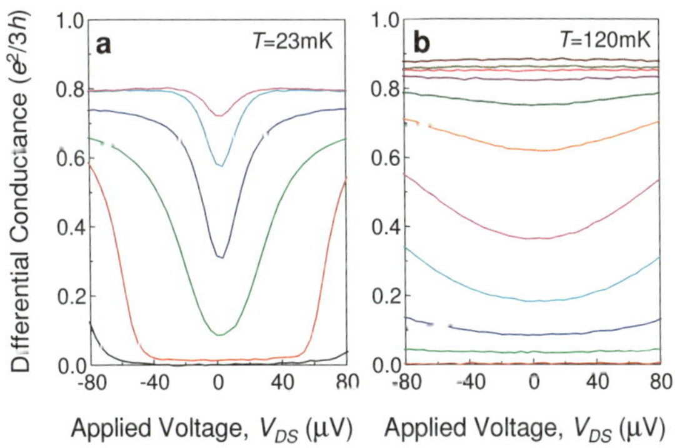

Fig. 7.3. The differential conductance of the constriction at ν=1/3 as function of V_{DS}. **a.** Measured at 23mK for a few backscattering potential strengths; from bottom to top: QPC split-gate voltage changes from -83mV to -0.5mV. **b.** Similar data measured at 120mK; from bottom to top: QPC split-gate voltage changes from -131mV to -11mV.

We turn now to the shot noise in this regime. The conductance and shot noise at different temperatures are plotted as function of the applied voltage across the constriction in Fig. 7.4. At 23mK the conductance is highly non-linear and the noise is non-poissonian. The noise has only the qualitative features of the CLL, predicted by Fendley *et al.* [18]. The quantitative disagreement may be related to a deviation of the scattering potential and its voltage dependence from the assumed idealized impurity [19], or due to added Coulomb interactions [20]. As the temperature increased the differential conductance turned to be only weakly dependent on current and the shot noise reduced. At sufficiently higher temperature, here 120mK, shot noise turned poissonian, with a definite charge $e^*=e/3$ (see also Ref. 21), verifying Laughlin's prediction of fractionally charged quasiparticles (however, flowing in edge channels).

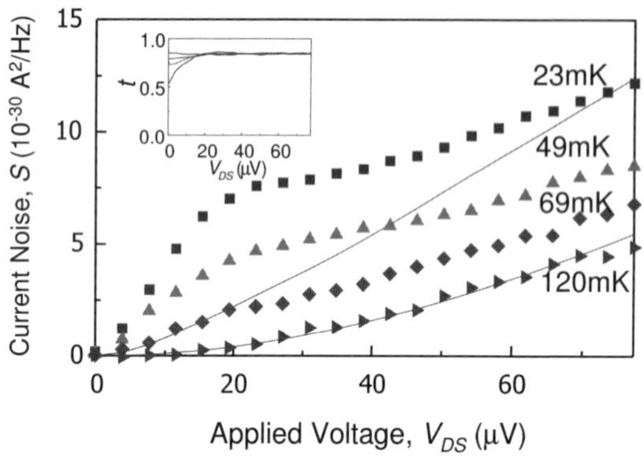

Fig. 7.4. Shot noise and differential conductance at ν=1/3 for different temperatures. Shot noise due to a weak backscattering potential measured at various temperatures: 23, 49, 69 and 120mK. The thin solid lines are the expected shot noise for non-interacting quasiparticles with charge $e/3$ at 23mK (upper line) and 120mK (lower line). Inset: The transmission of the constriction at the same temperatures, with the top most line at 120mK and the lowest one at 23mK.

What happens at an even lower electron temperature, at T~9mK? This temperature is very close to the lattice temperature and was achieved after 'cold grounding' of most of the ohmic contacts in the sample (cooling the electron via Wiedemann-Franz- conduction) [22]. As was shown before, a relatively weak backscattering potential at ν=1/3, with a saturated transmission, $t=g/g_{1/3}$~0.7, led to rather strong backscattering in the linear regime (near zero excitation). Moreover, both the voltage and temperature dependences of the differential conductance were positive – as expected from the CLL model ('valley-like' behavior). However, when the QPC split-gate potential was even smaller (in absolute terms), making backscattering weaker, $t=g/g_{1/3}$>0.85, the dependence of the

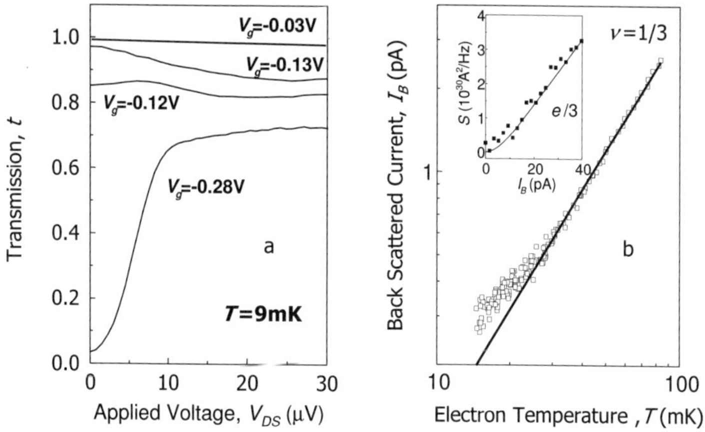

Fig. 7.5. Typical dependence of the transmission coefficient and backscattered current on bias and temperature at ν=1/3. **a.** At different QPC split-gate voltages. When the QPC is very weakly pinched off (split-gate voltage -0.03V, t~0.97) the transmission has a very weak (negative) dependence on the applied bias; opposite to CLL behavior. **b.** The backscattered current as a function of electron temperature measured with AC 10μV RMS. The curve can be fitted with a single slope. Inset: The shot noise generated by the weakly pinched constriction as function of the backscattered current, at an electron temperature of 9mK. Noise is classical and quasiparticles charge is e/3.

conductance on excitation voltage and temperature reversed sign (see Fig. 7.5, a 'mound-like' behavior), contradicting the CLL prediction. In these two regimes the charge determined at low excitation voltage is very large (approaching e), turning to e/3 at higher excitation voltage. Only in the extremely weak backscattering regime (see Fig. 7.5a, V_g=-0.03V), with conductance again independent of excitation voltage the noise was poissonian again, leading to a charge $e^*=e/3$ – very much like at elevated temperatures.

7.5. Shot Noise at ν=2/5 and ν=3/7

What about the charge of quasiparticles in higher CF channels? This measurement is important since it distinguishes between the charge and the conductance. Figure 7.6a shows, as an example, the linear two terminal conductance as the constriction is being pinched for a bulk-filling factor ν_{bulk}=2/5 (fluctuations are due to instabilities).

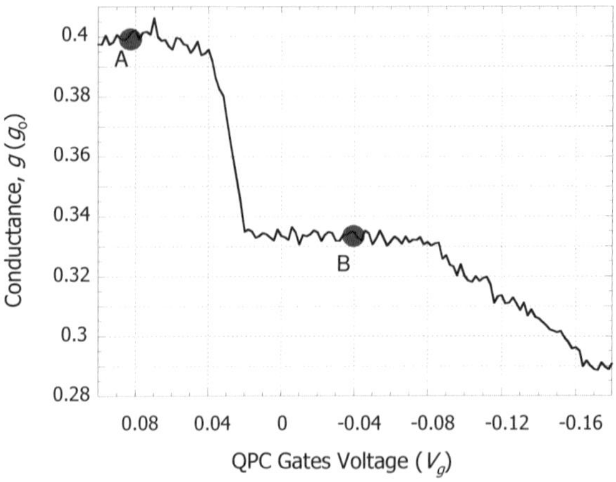

Fig. 7.6a. The conductance as function of the QPC split-gate voltage (affecting the constriction width). Two conductance plateaus, $2e^2/5h$ and $e^2/3h$, are observed.

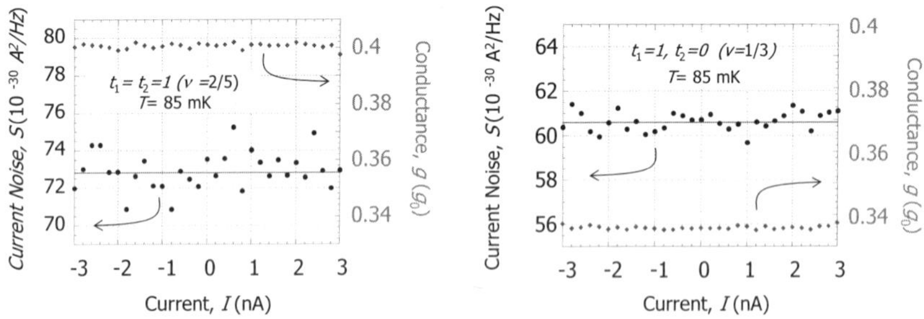

Fig. 7.6b. Noise measured at the two plateaus, at points A and B (Fig. 7.6a). t_1 and t_2 refer to the transmission coefficients of the first and second LL of CFs (ν=1/3 and ν=2/5, respectively). The noise is constant and does not depend on the current, accounting only for the thermal and stray noise; hence, the shot noise is zero. The conductance is shown for reference (by + signs).

Figure 7.6b show the measured noise, as function of the transmitted current in the 2/5 plateau (in the CF language, t_2=t_1=1) and in the 1/3 plateau (t_2=0, t_1=1). As expected, no excess noise (above the thermal and amplifier noises) was measured in each of the plateaus. Since $\left\langle \delta(i_1)^2 \right\rangle = 0$ and $\left\langle \delta(i_1+i_2)^2 \right\rangle = \left\langle \delta(i_1)^2 \right\rangle + \left\langle \delta(i_2)^2 \right\rangle + 2\left\langle \delta(i_1 \cdot i_2) \right\rangle = 0$, the last two terms in the last expression must each be zero independently (unlikely that they cancel each other), leading to the conclusion that the two currents flowing in the two CF channels are independent (last term).

Setting the QPC split-gate voltage to t~0.98 the temperature dependence of the backscattered current, shown in Fig. 7.7a, can be fitted with two distinct slopes in $\log(I_B)$ *vs* $\log(T)$ – with a crossover at T~45mK. The noise was measured across the full temperature range with two examples plotted in Fig. 7.7b. Note that with $t_{eff} = \dfrac{t \cdot g_{2/5} - g_{1/3}}{g_{2/5} - g_{1/3}} = 6t - 5$, t_{eff} is substantially smaller than t, except for t very close to unity. The two solid lines in Fig. 7.7b, which agree very well with the data, are the calculated shot noise with charges q=2e/5 at T~9mK and q=e/5 at T~82mK. The

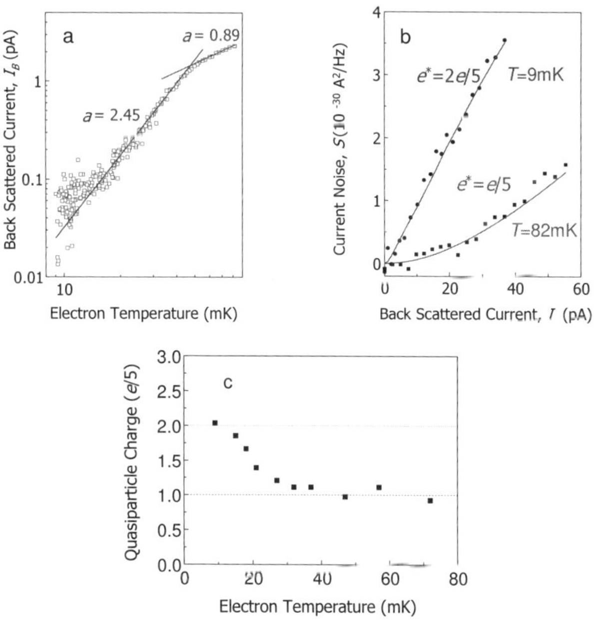

Fig. 7.7. Backscattered current and charge at ν=2/5. **a.** Backscattered current as function of the electron temperature. Two distinct slopes are observed with a transition temperature of about 45mK. **b.** Shot noise at two different temperatures. The backscattered quasiparticle charge is 2e/5 at 9mK and e/5 at 82mK. The constriction was tuned to reflect some 2% of the 2/5 channel at the two temperatures. **c.** The evolution of the quasiparticle charge with temperature.

charge changes smoothly with temperature over a region of some 20mK (Fig. 7.7c). While the scattered charge $q=e/5$, which was measured at the *high* temperature region, had been measured already before [23], a charge $q=2e/5$, measured at the lowest temperature, was unexpected. Apparently, at the lowest temperature, in the very weak backscattering regime, one can view the noise as resulting from partitioning of two quasiparticles, $e/5$ each, at each backscatter event (*bunching*).

Similarly, at $\nu=3/7$ the charge approaches $3e/7$ (maximum measured charge was ~$2.4e/7$) at the lowest available temperature (Fig. 7.8). Presumably an even lower temperature is needed in order to *bunch* three $e/7$ quasiparticles to $3e/7$ charges. This unexpected bunching might be related to the fractional exchange statistics of the quasiparticles.

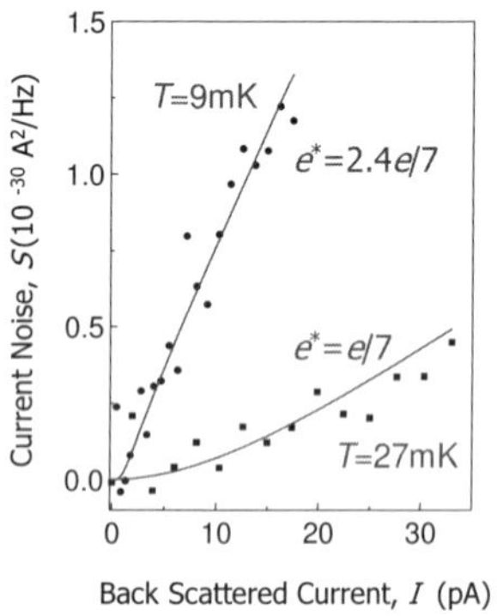

Fig. 7.8. Shot noise at two different temperatures at $\nu=3/7$. The backscattered quasiparticle charge is ~$2.4e/7$ at 9mK and $e/7$ at 27mK. The constriction was tuned to reflect the 3/7 channel by some 2%.

7.6. Shot Noise in the Strong Backscattering Limit

In the strong backscattering limit the quasiparticles charge is expected to be different (Eq. (7.2)). As the constriction is being pinched the *I-V* characteristic becomes nonlinear (g and t depend on current), even at an elevated temperature. Measurements were done at filling factor $\nu=1/3$. To account for the non-linearity, the energy independent Eq. (7.3) was modified, by approximating the integral over energy in Ref. 14.

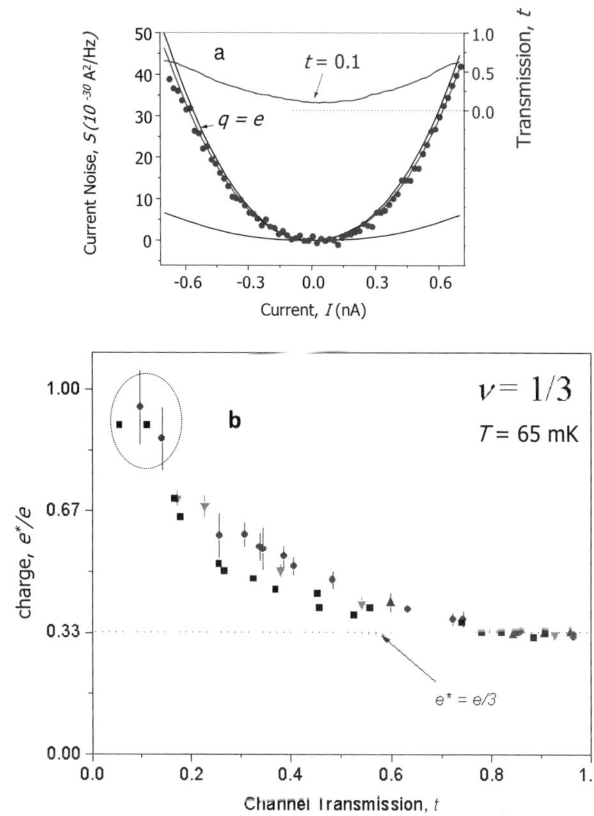

Fig. 7.9. Bunching of $e/3$ quasiparticles at ν=1/3 by a relatively pinched constriction. **a.** Noise and conductance for t~0.1, demonstrating bunching of quasiparticles to charge e. **b.** The evolution of the charge in three different devices as function of the transmission of the constriction (squares, circles, and triangles stand for different samples).

Moreover, the dependence of the conductance on the current was attributed only to t, while the charge e^* was assumed to be constant for a fixed QPC gate voltage. Replacing the integration over energy by a sum over discrete points and substituting t in terms of g and e^* in Eq. (7.3), we get [24]:

$$S\,(I) = 2e^* I \frac{1}{N}\sum_{i=1}^{N}\left(1 - \frac{(g_i/g_0)}{(e_i^*/e)}\right)\left[coth\left(\frac{e^* V}{2k_B T}\right) - \frac{2k_B T}{e^* V}\right] + 4k_B T g\,. \quad (7.4)$$

Here i runs over the measured points (N) up to current I, g_i is the differential conductance at each point, and the term 1-t is being replaced by 1-$\tilde{t}$, with $\tilde{t}$ the effective transmission coefficient of the *flux* of *quasiparticles* e^* and not the current. The *coth* term, however, rising from integration over the Fermi function, was inserted in 'by hand'. The noise expression now contains a single fitting parameter e^*. In other words, for each width of the constriction we find a fitting quasiparticle charge e^* and consequently the channel transmission t (via g).

In Fig. 7.9a the noise and the fitted charge at ν=1/3 are shown for t=0.1 at the lowest applied voltage. The fitted charge is nearly e. The dependence of e^* on t is summarized in Fig. 7.9b [24]. Measurements taken on a few samples collapse into one curve. The charge changes smoothly from e/3 for weak reflection (large t) to around e for strong reflection (at t~0.1) - as expected.

7.7. Evolution of Charge with Partitioning

We discuss now low temperature (~10mK) shot noise measurements in the non-linear regime, where the transmission is 'mound-like' or 'valley-like'. Shot noise and transmission were measured as function of the excitation voltage V_{sd} and split-gate voltage V_g for bulk filling factors ν=5/2, ν=7/3 and ν=1/3. In the weak and strong backscattering regimes, in all three filling factors, both conductance and spectral density exhibited two distinct regions in the excitation voltage. At low excitation voltage, the conductance, with either 'mound' or 'valley' like behaviors, was accompanied with a large slope in the spectral density, which corresponds to a large charge at all filling factors. At higher excitation voltage a rather 'linear characteristic' and a significantly lower charge were obtained.

Focusing on the low excitation voltage range an interesting dependence of the deduced quasiparticle charge on the average transmission was found. For ν=5/2, at high, 'mound-like', transmissions (t~0.9), the charge was substantially higher than the theoretically predicted quasiparticle charge of e^*=e/4 (Fig. 7.10a, and see Ref. 25 for previous measurements and a description of this even-denominator fraction). At intermediate values of the transmission (t~0.4-0.9), with the transmission almost independent of the applied voltage, the charge was very close to $e^*=e/4$ (Fig. 7.10b). At lower, 'valley-like', transmissions, the charge increased towards $e^*=e$ (Fig. 7.10c). The latter result is expected since the filling factor within the potential barrier approached $\nu=2$; enabling thus only backscattering of electrons. Figure 7.10d summaries the low-excitation-charge evolution as function of the effective transmission $\langle t_{5/2-2} \rangle$, as measured on four different samples. A similar behavior was obtained also for the $\nu=7/3$ case. The quasiparticle charge evolution as a function of

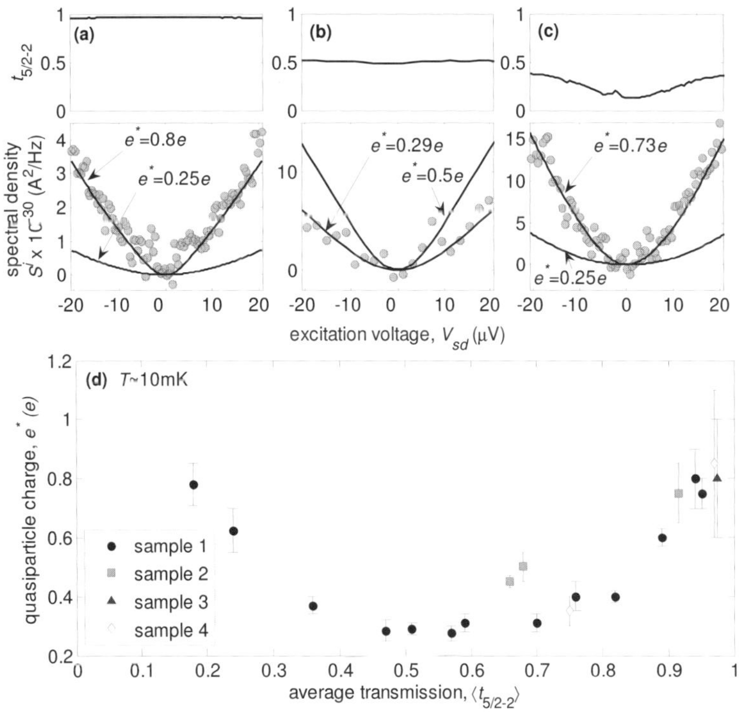

Fig. 7.10. Measurements at $\nu=5/2$ state of the transmission probability and the shot noise in the small V_{sd} range, measured at T~10mK. **(a) - (c)** Spectral density for a few transmissions and the predicted spectral densities (using Eq. 1). **(d)** The evolution of quasiparticle charge as a function of the effective (average) transmission probability (of the 5/2 channel with the lower lying channel 2), $\langle t_{5/2-2} \rangle$, measured on four different samples.

the average effective transmission, $\langle t_{7/3-2} \rangle$, resembles that in the ν=5/2 state, with e^*~e/3 obtained only at the intermediate transmission range - where the transmission is nearly voltage independent (not shown).

For the ν=1/3 state, with measurements performed in the high range of the $t_{1/3}$, with 'mound-like' transmission, and in a somewhat lower transmission which was rather flat. Again, a bimodal quasiparticle charge was observed in the 'mound-like' transmission regime, with e^*=e in the small excitation voltage range and e^*~e/3 at a higher excitation voltages. For the flat transmission, the charge fitted well e^*=e/3 over a wide range of excitation voltage (not shown). The lower transmission range ($t_{1/3}$<0.3), the backscattered charge approached e - as already discussed above [24].

Temperature dependence of the noise measured at high transmissions was studied at bulk filling factor ν=5/2, as described in Fig. 7.11. While the longitudinal resistance of the ν=5/2 state increased weakly with temperature, the resultant backscattering was less than 0.1% at 85mK. As evident from data, the differential transmission coefficient became less sensitive to the applied voltage and the quasiparticle charge reduced. The charge approached e^*=e/4 and saturated thereafter as the temperature reached ~75mK and beyond (Fig. 7.11).

Our measurements present a rather complex evolution of the determined backscattered quasiparticle charge. It roughly portrays two extreme values of the quasiparticle charge. One small - being close to the 'expected value' - at intermediate backscattering probabilities *or* at higher energies (excitation voltage or temperature), in which the transmission is usually linear. The other one is large, approaching e^*=e, for extremely weak backscattering and in the limit of low excitation voltage and temperature, in which the transmission is found to be non linear. This enhancement of the charge can be attributed to either the scattering of an integer multiple of a smaller fundamental charge (*bunching*), or to the fundamental charge being larger than expected. The first assumption is strengthened by the observation of the lower charge at higher excitation voltage or temperature; suggesting that bunching may take place at low energies, and its dissemination into individual fundamental quasiparticle charges at higher energy.

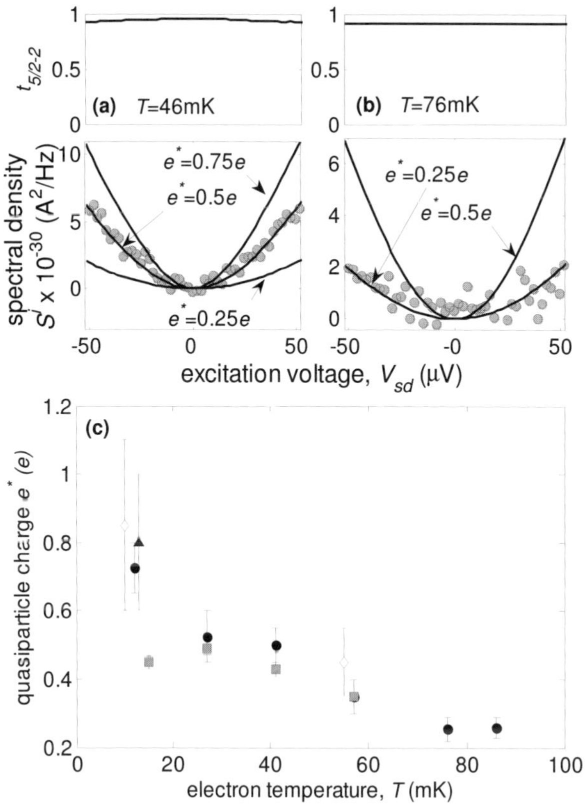

Fig. 7.11. The dependence of the backscattered charge on temperature measured at weak backscattering in the ν=5/2 state. Two examples of the measured data: **a** at T=46mK with e^*~0.5e, and **b** at T=76mK with e^*~e/4. **c** Evolution of the backscattered charge as a function of temperature. At temperatures higher than ~40mK a significant reduction of the charge is observed.

7.8. Shot Noise of Composite Edge Channels

Transport is more complicated if there are one or more counter-propagating edge channels, as is the case of the so called *hole conjugate quantum Hall states*, ν=2/3, 3/5, etc. In the case of ν=2/3, which we describe briefly here, a clean sample devoid of any impurities is expected to support two charged modes: one with conductance of e^2/h - carrying electrons, and a counter-propagating one with conductance $(1/3)e^2/h$, carrying $e/3$ fractional charges [26]. For a smooth edge potential, the two counter-propagating modes will have different momenta (the difference being proportional to the enclosed flux), and hence unlikely to equilibrate. However, in the presence of random inter-channel scattering the momentum of each channel need not be conserved, allowing thus equilibration (and the emergence of a single charge mode) and a universal quantization of Hall conductance $(2/3)e^2/h$. In addition, a *neutral* counter propagating channel (namely, carrying only energy but no charge) is expected to exist [26,27]. Recently, there has been a resurgence of theoretical investigations of this and similar neutral edge channels [28]; ignited by the hypothesis of the much anticipated non-abelian ν=5/2 fractional state, which is expected to carry a neutral Majorana mode [29].

Figure 7.12a shows a plot of the transmission as function of the applied split-gate voltage to the QPC, measured on a typical sample. Scanning the gate voltage revealed a prominent plateau at t=1/2, suggesting a local filling factor in the constriction ν_C=1/3. Adopting the assumption that the electron density drops gradually near edges of the constriction, the ν_C=1/3 plateau seems to confirm that 1/3 edge channel traverses the constriction without backscattering, while the 2/3 edge channel is fully reflected. Under such circumstances the shot noise for t=1/2 should be zero. However, surprisingly, this was not the case here. As shown in Fig. 7.12b, the measured shot noise is *finite*, suggesting a different picture of edge reconstruction in the ν=2/3 case. Since this behavior repeated itself in all the measured samples, we must adopt the notion of a single chiral-composite-edge channel, with transmission $0<t<1$, determined by the split-gate voltage.

The charge over the entire range of the transmission (from very close to unity to t~0.3) fits excellently to a value $e^*=(2/3)e$ (not shown). Upon depleting the constriction further the differential transmission became highly current dependent and the spectral density developed two distinct slopes: at the low range of current the quasiparticle charge was *e* while at higher currents the charge dropped to approximately (2/3)*e*. Since the transmission increases with current (expected in a chiral Luttinger liquid), the charge reverts back to that of the quasiparticle. As the electron temperature increased the charge, measured at $1<t<0.4$, evolved smoothly from (2/3)*e* to *e*/3. This observation, reminiscent of the behavior of the quasiparticle charge in the 2/5 and 3/7 states [22], may suggest that quasiparticles of charge *e*/3 carry the current in the 2/3 state, however, *bunching* takes place at the low temperature regime.

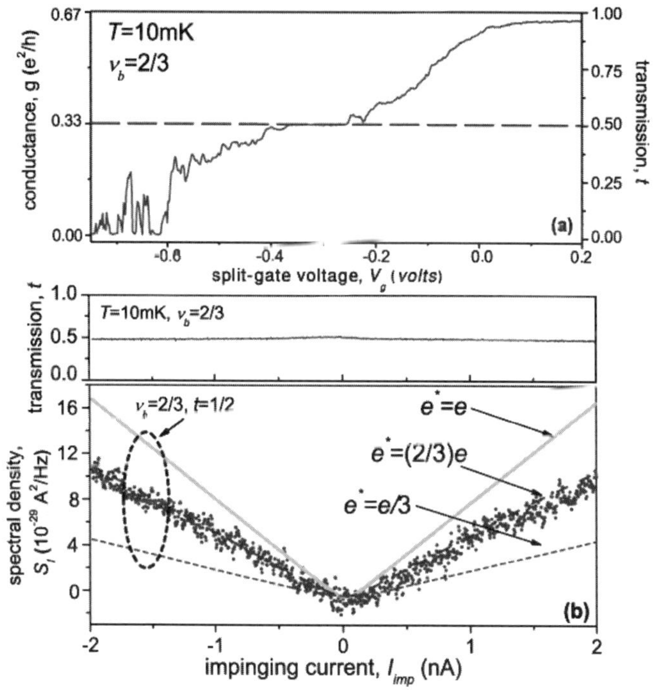

Fig. 7.12. Conductance and spectral density at v=2/3 at electron temperature 10mK. **a.** Conductance g and transmission t of the constriction as a function of split-gate voltage. Note the appearance of a prominent plateau at $g=e^2/3h$ (t=1/2). **b.** Upper panel - dependence of the transmission (zero bias t=1/2) on injected electron energy. Lower panel - spectral density S_I at this value of transmission. The dots are the measured data points. Shown is the expected spectral density for transmission t=1/2, temperature T=10mK and quasiparticle charge $e^*=e$, (2/3)e (dotted line behind the dots) and e/3 (lowest dotted line).

7.9. Shot Noise of Interfering Electrons

Every interferometer must generate phase dependent shot noise since it splits the incoming electrons into at least two paths, or alternatively, its transmission is in general smaller than one. It is intriguing to measure simultaneously the discrete nature of particles and their interference due to their wave behavior. In a symmetric interferometer, where the incoming beam is split equally between two paths, one can express the transmission t_{int}=0.5(1+μ cosφ_{AB}), with μ the visibility and φ_{AB} the AB phase (the Aharonov Bohm phase), one gets for shot noise:

$$S = 0.5eI_{imp}(1-\mu^2 \cos^2 \phi_{AB}) \ , \tag{7.5}$$

with I_{imp} the impinging current. Hence, the shot noise contains only the second AB harmonic.

We developed a two-path electron interferometer based on edge channels transport: a Mach-Zehnder electron interferometer [30]. A single edge channel is split and later recombines after enclosing an AB flux. Changing the area of the interferometer, under the quantizing magnetic field, changes the AB phase leading to interference oscillations in the conductance (Fig. 7.13a). The noise measured under such conditions revealed indeed a second AB harmonic (Fig. 7.13b).

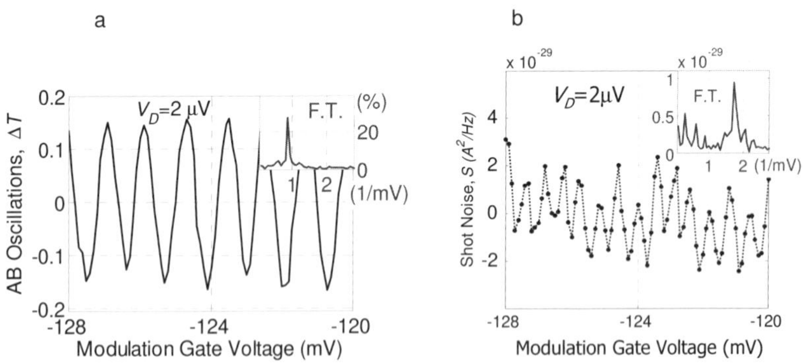

Fig. 7.13. Aharonov-Bohm oscillations of conductance and noise in an electronic Mach-Zehnder interferometer. **a.** The AB oscillations in the conductance as function of the voltage on a gate that changes the area of the interferometer. **b.** Shot noise - second harmonic of the AB oscillations in the conductance.

Acknowledgements

The work described in this review had been carried out by (in a chronological order), M. Reznikov, R. de-Picciotto, E. Comforti, T. Griffith, and Y-C. Chung, N. Ofek, A. Bid, M. Dolev, and Y. Gross. The material was grown by H. Shtrikman and V. Umansky, and the electron beam lithography done by D. Mahalu. The work enjoyed partial support by the Israeli Science Foundation (ISF), the Minerva Foundation, the German Israeli Foundation (GIF), the German Israeli Project Cooperation (DIP), the US-Israel Bi-National Science Foundation, and the European Research Council under the European Community's Seventh Framework Program (FP7/2007-2013)/ERC Grant agreement #227716.

References

1. W. Schottky, Ann. Phys. (Lp) **57**, 541 (1918).
2. M. Reznikov *et al.*, Superlattices and Microstructures **23**, 901 (1998).
3. K. von Klitzing, G. Dorda, & M. Pepper, Phys Rev. Lett., **45**, 494 (1980).
4. T. Chakraborty & P. Pietilainen, *The Quantum Hall Effects*, Springer-Verlag, 1995.
5. D. C. Tsui, H. L. Stormer, & A. C. Gossard, Phys. Rev. Lett., **48**, 1559 (1982).
6. R. B. Laughlin, Phys. Rev. Lett. **50**, 1395 (1982); R. B. Laughlin, Int. J. Mod. Phys. **B 5**, 1507 (1991).
7. B. Halperin, Phys. Rev. B **25**, 2185 (1982).
8. X. G. Wen, Phys. Rev. B **41**, 12838 (1990).
9. V. K Khlus, Sov. Phys. JETP **66**, 1243 (1987); G. B. Lesovik, JETP Lett. **49**, 592 (1989).
10. B. Yurke & G. P. Kochanski, Ohys. Rev. **B 41**, 8184 (1990).
11. M. Reznikov *et al.*, Phys. Rev. Lett. **75**, 3340 (1995); A. Kumar, L. Saminadayar, & D. C. Glattli, Phys. Rev. Lett. **76**, 3806 (1996).
12. C. L. Kane & M. P. A. Fisher, Phys. Rev. Lett. **72**, 724 (1994); C. de Chamon, D. E. Freed, & X. G. Wen, Phys. Rev. B **51**, 2363 (1995); A. Fendley, W. W. Ludwig, & H. Saleur, Phys. Rev. Lett. **75**, 2196 (1995); A. Fendley & H. Saleur, Phys. Rev. B **54**, 10845 (1996).
13. J. K. Jain, Phys. Rev. Lett. **63**, 199 (1989); For a Review see B. Halperin, in *New perspectives in quantum Hall effect*, S. Das Sarma & A. Pinzuk, A. (Eds.), Wiley & Sons, 1997.
14. Th. Martin & R. Landauer, Phys. Rev. B **45**, 1742 (1992).
15. J. Martin *et al.*, Science **305**, 980 (2004).
16. V. J. Goldman & B. Su, Science **267**, 1010 (1995); J. D. F. Franklin *et al.*, Surf. Sci. **361**, 17 (1996).
17. M. Reznikov *et al.*, Phys. Rev. Lett. **75**, 3340 (1995).

18. A. Fendley, W. W. Ludwig, & H. Saleur, Phys. Rev. Lett. **75**, 2196 (1995); A. Fendley & H. Saleur, Phys. Rev. B **54**, 10845 (1996).
19. X. G. Wen, Phys. Rev. B **41**, 12838 (1990).
20. Rosenow & B. I. Halperin, Phys. Rev. Lett. **88**, 096404 (2002).
21. R. de-Picciotto *et al.*, Nature **389**, 162 (1997); L. Saminadayar *et al.,* Phys. Rev. Lett. **79**, 2526 (1997).
22. Yunchul Chung, M. Heiblum & V. Umansky, Phy. Rev. Lett. **91**, 216804-1 (2003).
23. M. Reznikov *et al.*, Nature **399**, 238 (1999).
24. T. G. Griffiths *et al.*, Phys. Rev. Lett. **85**, 3918 (2000).
25. M. Dolev *et al.* Nature **452**, 829 (2008).
26. C. L. Kane & M. P. A. Fisher, Phys. Rev. B **51**, 13449 (1995).
27. C. L. Kane, M. P. A. Fisher, & J. Polchinski, Phys. Rev. Lett. **72**, 4129 (1994).
28. B. Rosenow & B. I. Halperin, Phys. Rev. B **80**, 155305 (2009).
29. C. Nayak *et al.*, Rev. Mod. Phys. **80**, 1083 (2008).
30. Yang Ji *et al.*, Nature **422**, 415 (2003); I. Neder, *et al.*, Phys. Rev. Lett. **96**, 016804 (2006).

Chapter 8

Fabry-Perot Interferometer and Non-Abelian Quantum Hall States

R. Ilan

Department of Condensed Matter Physics, Weizmann Institute of Science, Rehovot 76100, Israel

E. Grosfeld

Department of Physics, University of Illinois, 1110 W. Green St., Urbana IL 61801-3080, U.S.A.

K. Schoutens

Institute for Theoretical Physics, University of Amsterdam, Valckenierstraat 65, 1018 XE Amsterdam, the Netherlands

A. Stern

Department of Condensed Matter Physics, Weizmann Institute of Science, Rehovot 76100, Israel

This chapter reviews proposed transport experiments in the quantum Hall regime in a system designed to form the solid state analog of the Fabry-Perot interferometer. We review studies of that system for a series of quantum Hall states in the second Landau Level called the Read-Rezayi states. We focus here on two types of experiments. The first is a lowest order interference experiment of the fundamental quasi-holes of the system. The second is a Coulomb blockade based measurement which probes features in the flow of electrons through a closed part of the system. Both experiments were suggested as ways to answer a single question: do these quantum Hall states support excitation obeying non-Abelian statistics? After introducing the concept of non-Abelian statistics, reviewing these proposed experiments and the predictions made for their results, we also comment on some of the assumptions made during the analysis and the outcome of their violation.

8.1. Non-Abelian Quasi-particles and the Quantum Hall Effect

The term non-Abelian statistics is used to describe the quantum statistics of particles whose exchange does not result in a simple multiplicative phase to the wavefunction describing their collective state. The first condition for particles to obey Non-Aeblian statistics is that the ground state of a system containing a set of them is degenerate, and is separated from the rest of the spectrum by an energy gap. In such a case, an interchange of two such particles will result in a unitary transformation acting on the set of degenerate states. When two such operators, corresponding to two different interchanges, are non-commuting, the statistics of the particles is termed non-Abelian. Non-Abelian statistics is a key ingredient in the design of quantum gates for topologically protected qubits [1–4], therefore, an experimental proof of the existence of non-Abelian quasiparticles would be a major step towards an implementation of topological quantum computation.

Almost two decades ago, it was suggested by Moore and Read [5] that quasi-holes in a quantum Hall system at filling fraction $\nu = 5/2$ may actually obey non-Abelian statistics. Later on, this prediction was extended to include a whole series of quantum Hall states at $\nu = 2 + \frac{k}{k+2}$ with $k = 1, 2, 3...$ by Read and Rezayi [6]. The idea that quasiparticles obeying non-Abelian statistics may be present in condensed matter systems that are accessible in a laboratory is very exciting. To this day, however, there is no experiment that confirms that prediction.

At present, there are several suggestions for experiments that may allow an observation of signatures of such statistics. Some most prominent ones involve an experimental setup known as the two-path or Fabry-Perot interferometer (FPI). In this chapter, we wish to introduce two examples for such experiments, and discuss the possible difficulties that may be involved in executing such experiments. The introductory parts in this paper are based mainly on Refs. [4] [7], while Sec. 8.4 presents a new perspective on lowest order interference experiments that has not been published elsewhere. We start by introducing the experimental setup, namely the quantum Hall analog of the FPI.

8.2. The Fabry Perot Interferometer

In the context of the quantum Hall effect, the Fabry-Perot interferometer, discussed in details first by Chamon *et al.* [8], is a Hall bar (for concreteness, lying along the x-axis, between $0 < y < w$) perturbed by two constrictions, known as quantum point contacts (say, at $x = \pm d/2$). Current flows chirally along the edges, towards the left along the $y = w$ edge and towards the right along the $y = 0$ edge. The right moving edge is put at a chemical potential difference μ relative to the left moving one. In the absence of the point contacts there is no back scattering of current, and thus when the bulk is in a quantum Hall state the two-terminal conductance (the ratio of the current difference to the voltage difference between the two edges) is quantized. The point contacts introduce an amplitude for inter-edge tunnelling, which backscatters some of the current flowing along the right moving edge. The probability for back-scattering is directly reflected in the voltage between the edges, i.e., in the two terminal conductance.

The presence of two point contacts makes the probability for back-scattering the quantum Hall analog of "the interference screen". If the tunnelling amplitudes introduced by the two point contacts are t_1, t_2 respectively, then the back-scattered current is, to lowest order in these amplitudes, proportional to $|t_1 + t_2|^2$. To this lowest order, the Fabry-Perot interferometer is the QHE analog of the two slit experiment. The relative phase between the two amplitudes may be varied in three principal

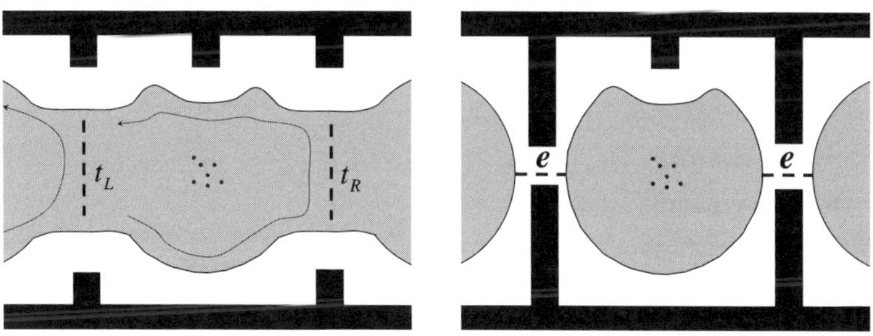

Fig. 8.1. The quantum Hall based Fabry Perot interferometer. The gray region is a gapped quantum Hall fluid, and there are n quasiholes localized in the area between the two quantum point contacts. The figure on the left presents the setup for a lowest order interference experiment, and two possible tunneling paths for an incoming edge quasihole are marked. The right figure shows the Fabry Perot interferometer in the limit of strong quasihole backscattering. Transport through the dot is done via electron (e) tunneling. The area S of the dot may be varied using a side modulation gate.

ways: by varying the magnetic field, by varying the electron density and by varying the area of the "cell" defined by the two edges and two point contacts. The latter is implemented by a side gate that "moves the walls" of the cell in the region it operates on. At least within a crude approximation, a variation of the area of the cell does not vary the filling factor in the bulk. Thus, it varies the number of electrons in the cell (which is the product of density and area), but does not introduce quasi-particles into the bulk. The variation of the electron density and/or magnetic field lead to a more complicated effect. Electrostatic interaction tends to fix the electronic charge density to be such that it exactly neutralizes the positive background. When this interaction dominates, a variation of the bulk Landau filling from that of the quantum Hall state leads to the introduction of localized quasi-holes or quasi-particles. Such a variation may be created either by varying the magnetic field or by varying the positive background density (usually done by applying a voltage to a gate).

Given this picture, the Fabry-Perot interferometer gives us a device in which quasi-holes that are flowing on the edge split into partial waves that interfere, the interference pattern is measured by measuring the backscattered current as a function of three possible parameters, and the area enclosed by the interfering trajectories enclose localized quasi-holes. It then makes sense to expect the dependence of the back-scattered current on the area of the cell, the magnetic field and the density to reflect the statistics of the quasi-particles.

We now turn to introduce the series of non-Abelian states called Read-Rezayi states, and the basic theoretical concepts that are needed in order to understand their construction.

8.3. Non-Abelian Statistics, the Read-Rezayi States and $\mathbb{Z}_k$ Parafermions

The many-body wave function describing a quantum Hall system with a set of quasiholes localized at the coordinates $\mathbf{R} = (R_1, ..., R_n)$ is $|\psi(\mathbf{R})\rangle$. Then, the kinetic energy that dictates the dynamics of $\mathbf{R}$ will include a Berry vector potential

$$\mathbf{A_B} = \mathrm{Im}\langle\psi(\mathbf{R})|\nabla_\mathbf{R}|\psi(\mathbf{R})\rangle \tag{8.1}$$

When a vortex (i.e, a quasi-hole) traverses a closed loop the wave function accumulates a Berry phase of $\int \mathbf{A_B} \cdot d\mathbf{R}$.

The above argument crucially relies on the assumption that once the positions of the quasi-holes are fixed, there is just one ground state to the system. Under this assumption, which is violated in non-Abelian states, when the parameters in that ground state, the positions $\mathbf{R}$, adiabatically traverse a closed loop, the state of the system must evolve from its initial ground state $|\psi(\mathbf{R})\rangle$ to a final state that differs from the original one by a phase factor only. This is the adiabatic theorem. A calculation then shows that the phase factor is just the Berry phase mentioned above.

What if the assumption does not hold, and rather than having one ground state per each configuration of quasi-holes we have several ground states that differ by some internal quantum numbers? Such a degeneracy may in principle appear accidentally, but here we are interested in the case where the degeneracy is robust as it originates from deep properties of the quantum states.

Independently of its origin, with such a degeneracy, if the initial state is a ground state the adiabatic theorem guarantees that the final state will be a ground state as well, but it does not guarantee that it would be the initial state multiplied by a phase factor. Rather, under these conditions, the geometric vector potential $\mathbf{A}(\mathbf{R})$ becomes a matrix,

$$\mathbf{A_B}_{ij} = \mathrm{Im}\langle\psi_i(\mathbf{R})|\nabla_{\mathbf{R}}|\psi_j(\mathbf{R})\rangle \tag{8.2}$$

with the $|\psi_i\rangle$ being the various degenerate ground states. Then, the effect of a motion of the parameters $\mathbf{R}$'s along a closed loop is not only a phase factor, but rather a unitary transformation that acts within the subspace of degenerate ground states. This unitary transformation is

$$\hat{P}e^{i\int \mathbf{A_B}\cdot d\mathbf{R}}, \tag{8.3}$$

with $\hat{P}$ the path ordering operator.

A degenerate subspace of ground states is just one necessary condition for non-Abelian anyons to occur. The second condition is that, up to an Abelian phase, the unitary transformation (8.3) will depend only on the topology of the closed trajectory taken by the parameters $\mathbf{R}$. When that happens an interchange of two particles is associated with a particular unitary transformation. Different unitary transformations associated with different interchanges do not necessarily commute and the state of the system after a series of interchanges of positions of particles may depend on the order at which these interchanges were carried out. This is the origin of the name non-Abelian statistics.

Read and Rezayi (RR) came up with a series of states, for which one can write analytic ground state wavefunctions that are the zero energy solutions of a $k+1$ body interaction [6]. These wavefunctions, written as correlators in a conformal field theory (CFT), were generalizations of the Pffafian wavefunction written earlier by Moore and Read [5]. Since the Pffafian is the BCS state for a superconductor with $p_x + ip_y$ pairing symmetry, the generalized wavefunctions can intuitively be understood as describing condensations of larger clusters of particles, of size k.

Following the rationale of flux attachment as a means to map problems of electrons in a magnetic field onto composite particles at a different magnetic field, let us consider a flux attachment transformation that attaches $\alpha = 1/\nu$ flux quanta to each electron, with ν the filling of the partially filled topmost Landau level. That makes the composite particles experience zero magnetic field on average. Thus, the composite particles carry one electron charge and α flux quanta, and are subjected to zero average magnetic field. For an odd α the electrons are turned into bosons at zero average magnetic field. This viewpoint of the Laughlin fractions (as well as the $\nu = 1$ completely filled lowest Landau level), commonly known as composite boson theory [9], views the Laughlin fractions as Bose condensates, characterized - as all condensates - by the dissipationless flow of currents. For an even α the electrons are transformed into composite fermions at zero average field. These then may form a p-wave super-conductor. The p-wave super-conductor is again a condensate. This time the condensate is formed by Cooper-pairs, clusters of two fermions that, effectively, group together to form a boson.

For a non-integer α the flux attachment turns electrons into anyons at zero average magnetic field. For these anyons to form condensates, where current flows with no dissipation, they have to form effective bosons, made of clusters of an integer number k of anyons. If the phase accumulated when one anyon encircles another is $4\pi/k$, the phase accumulated by an anyon encircling a cluster of k anyons will be 4π, and that accumulated by two clusters encircling one another will be a multiple of 4π. The cluster will then be an effective boson. Thus, for forming a condensate of this sort, the flux attachment transformation should attach $(1 + \frac{2}{k})$ flux quanta to each electron, which would make the electronic filling factor $k/(k+2)$.

When the Laughlin fractions $\nu = 1/m$ (with m an odd integer) are described as condensates of bosons, the Laughlin quasi-particles are the quantized vortices in this condensate. To form a quantized vortex at the point $\mathbf{R}$ in a superfluid of bosons we need to award each boson an angular momentum of $\hbar$ relative to the point $\mathbf{R}$. When the boson is made of two fermions, the relative angular momentum per fermion is $\hbar/2$. When the fermion goes around the vortex, then, it accumulates a phase of π. In analogy, when the boson is made of k anyons, each anyon gets, on average, an angular momentum of $\hbar/k$ and an anyon that goes around a vortex accumulates a phase of $2\pi/k$.

The arguments of the last two paragraphs tell us that there should be a description of the $\nu = k/(k+2)$ fractional quantum Hall state as a condensate of effective bosons made of clusters of k anyons. Within this description an anyon that goes around another would accumulate a phase of $4\pi/k$ while an anyon that goes around a vortex in the condensate would accumulate a phase of $2\pi/k$. Our considerations do not tell us, however, the information we need about a collection of vortices: what is the phase accumulated when vortices go around one another, what are the degenerate ground states, if any, that many vortices form when they are distant from one another, and how these degenerate ground states lead to non-Abelian statistics, if at all. Indeed, for the case of $\nu = 5/2$, one can use the machinery of super-conductivity, particularly the Bogoliubov-deGennes equations for a p-wave super-conductor, to give us that information. The tools that will helps us to analyze the clustering of anyons into effective bosons and the states formed by vortices in the condensate of these bosons will be parafermionic conformal field theories.

In order to account for the charge and statistics of the electrons as well as the quasiholes, one has to compose operators describing those physical elements of the system using two CFT's. The first is the chiral boson theory, used to describe Abelian states such as the Laughlin states. This theory is usually associated with the charge mode of the edge, and the Laughlin - Jastrow factor of the bulk wavefunctions. The second field theory is associated with neutral modes. The fields in this theory are labeled by two quantum numbers Φ^l_m, with $l \in \{0, 1, \ldots, k\}$. The integer m is known as the $\mathbb{Z}_k$ charge of the field Φ^l_m and is defined modulo $2k$. The fields are subjected to the following identifications $\Phi^l_m = \Phi^l_{m+2k} = \Phi^{k-l}_{m-k}$ and $l + m = 0 \bmod 2$.

The creation operators of both electrons and quasiholes in the bulk are then products of two factors. The first is a vertex operator, $e^{i\alpha\varphi(z)}$, with $\alpha = \sqrt{(k+2)/k}$ for the electron, and with $\alpha = 1/\sqrt{k(k+2)}$ for the quasihole . The second factor is one of the fields in the parafermionic theory; the electron creation operator is given by $\psi_1 e^{i\sqrt{\frac{k+2}{k}}\varphi}$, where $\psi_1 = \Phi^0_2$

is one of the parafermionic currents, while the quasihole creation operator is $\sigma_1 e^{i\sqrt{\frac{1}{k(k+2)}}\varphi}$, where σ_1 is one of the parafermionic primary fields (see below).

The fusion rules for the parafermionic CFT are [10, 11]

$$\Phi^{l_\alpha}_{m_\alpha} \times \Phi^{l_\beta}_{m_\beta} = \sum_{l_\gamma=|l_\alpha-l_\beta|}^{\min\{l_\alpha+l_\beta,2k-l_\alpha-l_\beta\}} \Phi^{l_\gamma}_{m_\alpha+m_\beta}. \tag{8.4}$$

The operator product expansion (OPE) is given by

$$\Phi^{l_\alpha}_{m_\alpha}(z)\Phi^{l_\beta}_{m_\beta}(w) = \sum_{l_\gamma} C_{\alpha\beta\gamma}(z-w)^{\Delta h}\, \Phi^{l_\gamma}_{m_\alpha+m_\beta}(w) \tag{8.5}$$

where the fields appearing on the right hand side are determined by Eq. (8.4), $C_{\alpha\beta\gamma}$'s are constants, and $\Delta h = h^{l_\gamma}_{m_\alpha+m_\beta} - h^{l_\beta}_{m_\beta} - h^{l_\alpha}_{m_\alpha}$. As a consequence of that relation, when a field $\Phi^{l_\alpha}_{m_\alpha}$ goes around a field $\Phi^{l_\beta}_{m_\beta}$ and their fusion is to a field $\Phi^{l_\gamma}_{m_\alpha+m_\beta}$, the phase generated is $2\pi\Delta h$. The conformal dimension of the field Φ^l_m is given by

$$\begin{aligned} h^l_m &= h_l + \frac{(l-m)(l+m)}{4k}, \text{ for } -l < m \le l \\ h^l_m &= h_l + \frac{(m-l)(2k-l-m)}{4k}, \text{ for } l < m \le 2k-l\,. \end{aligned} \tag{8.6}$$

Note here, that the parafermionic operator that is used in the representation of the quasi-hole operator, σ_1, is fundamentally different than the parafermion used to construct the electron operator. The fusion product of two σ_1's yields two terms, while the fusion product of any operator with ψ_1 yields only one. This essential difference is the origin of the non trivial non-Abelian statistics of the quasi-holes.

8.4. Lowest Order Interference

Lowest order interference is observed in a FPI in the limit where a single quasihole does not tunnel between opposite edges more than once. The tunneling process introduces a finite value for the longitudinal conductance. The measured backscattered current, injected from the left along the lower edge and collected on the left at the upper edge, interferes only through two trajectories: quasiholes entering from the left along the lower edge are either being backscattered at the left QPC, or transmitted at the left QPC and reflected from the right one. The path difference between the two trajectories forms a closed loop around the island confined between the

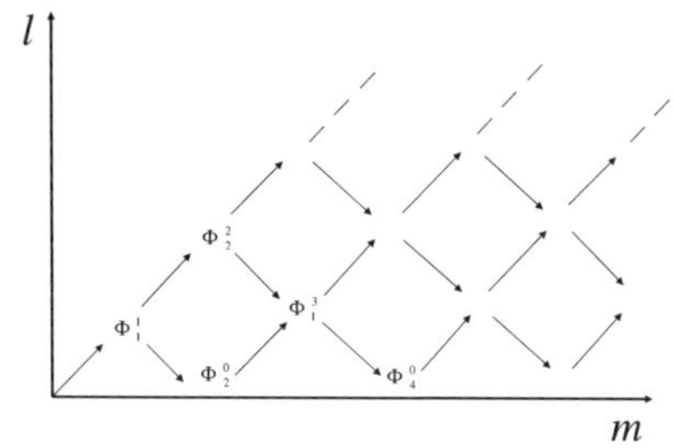

Fig. 8.2. Bratteli diagram for $\mathbb{Z}_k$ parafermionic theory. Propagation in the m direction is done by multiplying the fields in the vertices (m, l) by another spin field $\sigma_1 = \Phi_1^1$. The top diagram (a) shows some examples of this procedure. The bottom diagram (b) is the same diagram, with different labeling of the vertices. The numbers represent the number of paths possible to take from the point $(0, 0)$ to some point (l, m). The top floor of the Bratteli diagram always has $l = k$.

contacts, which may contain localized quasiholes in the bulk of the sample. For the sake of discussing lowest order interference, we assume that there is no hopping of quasiholes between the bulk and the edge. We will review the effect of bulk-edge coupling later on.

The two trajectories sketched in Fig. 8.1 are associated with two partial waves, $|\Psi_L\rangle$ and $|\Psi_R\rangle$. While these partial waves differ by global phases originating from the total magnetic field through the island (due to the Aharonov-Bohm effect), their overlap also encodes information on the mutual statistics of the quasiholes in the system. As mentioned in Sec. 8.2, this is attributed to the fact that the path difference between the two trajectories of the edge quasiholes forms a closed loop encircling localized quasiholes. Altogether, the back-scattered current will be of the form

$$I_{bs} \propto |t_L|^2 + |t_R|^2 + 2\,\mathrm{Re}\,\{t_L^* t_R \langle\psi_L|\psi_R\rangle\}\,. \tag{8.7}$$

The overlap $\langle\psi_L|\psi_R\rangle$ is the main object of interest.

8.4.1. *Lowest order interference - path counting approach*

In this section we re-derive from physical considerations the lowest order interference term for the RR states. This result was originally derived in Ref. [12] using the Modular S matrix of the two CFT's discussed above. Naturally, the results of the two derivations agree, and we find the following derivation to be more intuitive. We stress here that there is an underlying assumption for this calculation, which is the requirement for a flow of many edge quasiholes, which is not essential in previous calculations. The S-matrix approach predicts that a single excitation encircling the bulk excitation should result in the correct interference term. In that sense, perhaps, the S-matrix approach is more rigorous.

To obtain the interference term of a system with n localized bulk quasiholes in a RR state, consider the OPE of an incoming quasihole on the edge with the result of the fusion product of n localized quasiholes in the bulk whose parafermionic part is the spin field σ_1. The result of fusing all the bulk quasiholes together can be obtained by fusing all the bosonic contributions into the single bosonic operator $e^{i\frac{n}{\sqrt{k(k+2)}}\varphi}(0)$, and fusing all the σ_1 fields into a single field using Eq. (8.4).

The fusion of n spin fields σ_1 is most easily considered by using a "Bratteli diagram". The Bratteli diagram (see Fig. 8.2) is a graph with two coordinates corresponding to the two quantum numbers of the fields Φ^l_m. For a particular value of m, all the vertices represent the different results, called fusion channels, of fusing m fields of the type Φ^1_1. These will in turn represent the internal states of the island in our experimental system.

The n quasiholes in the bulk fuse into a superposition of the form $\sum_a \Phi^a_n e^{i\frac{n}{\sqrt{k(k+2)}}\varphi}(0)$, where for all a, the condition $n+a = 0 \bmod 2$ must be satisfied. Let us consider each of these possible fusion channels separately.

Let us start with the case where the number of quasiholes in the bulk is even, and $a = 0$. According to Eq. (8.4), the OPE of an incoming edge quasihole operator with the bulk fusion channel $\Phi^0_n e^{i\frac{n}{\sqrt{k(k+2)}}\varphi}$ has only one term

$$\Phi^1_1 e^{i\frac{1}{\sqrt{k(k+2)}}\varphi}(z)\Phi^0_n e^{i\frac{n}{\sqrt{k(k+2)}}\varphi}(0) \sim \\ z^{\alpha+\gamma}\,\Phi^1_{n+1} e^{i\frac{n+1}{\sqrt{k(k+2)}}\varphi}(0) \tag{8.8}$$

where $\alpha = h^1_{n+1} - h^0_n - h^1_1 = -\frac{n}{2k}$, and the exponent γ is contributed due to the vertex operator, $\gamma = \frac{n}{k(k+2)}$. The state of the system is therefore not a superposition state, $|\Psi_L\rangle = |1\rangle$. The state $|1\rangle$ represents a field at the floor of the Bratteli diagram where $l = 1$.

The state $|\Psi_R\rangle$ differs from $|\Psi_L\rangle$ by multiplicative phases, $|\Psi_R\rangle = e^{i\beta}e^{i\phi_B}e^{i\phi_0}|\Psi_L\rangle$, with $\phi_B = 2\pi\gamma$ and $\phi_0 = 2\pi\alpha$. (Note that β is the phase contributed by the Aharonov Bohm flux and dynamics, and is not unique to a non-Abelian quantum Hall system).

The overlap between these two partial waves is given by $\langle\psi_L|\psi_R\rangle_\psi =$

$e^{i\beta}e^{i\phi_B}e^{i\phi_0}$, and therefore the interference term is given by

$$2|t_L^* t_R| \cos\left(\frac{2\pi n}{k(k+2)} - \frac{\pi n}{k} + \beta\right) \tag{8.9}$$

When a is non-zero, the OPE with an incoming quasihole has two terms. The boson vertex operators has the same OPE as before, $e^{i\frac{1}{\sqrt{k(k+2)}}\varphi}(z)e^{i\frac{n}{\sqrt{k(k+2)}}\varphi}(0) \sim z^\gamma e^{i\frac{n+1}{\sqrt{k(k+2)}}\varphi}(0)$, multiplying the OPE of the spin fields

$$\Phi_1^1(z)\Phi_n^a(0) \sim z^{\alpha_1}\Phi_{n+1}^{a-1}(0) + z^{\alpha_2}\Phi_{n+1}^{a+1}(0)$$

where $\alpha_1 = h_{n+1}^{a-1} - h_n^a - h_1^1 = -\frac{a+2}{2(k+2)} - \frac{n}{2k}$, and $\alpha_2 = h_{n+1}^{a+1} - h_n^a - h_1^1 = \frac{a}{2(k+2)} - \frac{n}{2k}$. Note that both α_1 and α_2 given here are correct up to an addition of an integer. However, since we are calculating phases, these integers may add integer multiples of 2π to these phases and are therefore insignificant. This time, the state of the system is a superposition state, i.e., $|\Psi_L\rangle = c_{a-1}|a-1\rangle + c_{a+1}|a+1\rangle$, where $|a-1\rangle$ represents a field at the $a-1$ floor of the Bratteli diagram, and so on. The state $|\Psi_R\rangle$ will differ from $|\Psi_L\rangle$ by a global phase, and a relative phase between the two terms $|a-1\rangle$ and $|a+1\rangle$, $|\Psi_R\rangle = e^{i\beta}e^{i\phi_B}\left(e^{i\phi_{a-1}}|a-1\rangle + e^{i\phi_{a+1}}|a+1\rangle\right)$, with $\phi_B = 2\pi\gamma$ as before, and $\phi_{a-1} = 2\pi\alpha_1$, $\phi_{a+1} = 2\pi\alpha_2$. The overlap between the two partial waves is given by

$$\langle\psi_L|\psi_R\rangle = e^{i\beta}e^{i\phi_B}\left(|c_{a-1}|^2 e^{i\phi_{a-1}} + |c_{a+1}|^2 e^{i\phi_{a+1}}\right). \tag{8.10}$$

The two constant amplitudes $|c_{a-1}|^2$ and $|c_{a+1}|^2$, represent the probabilities that the fusion of $n+1$ spin fields results in the field Φ_{n+1}^{a-1} or Φ_{n+1}^{a+1}, correspondingly. In order to find these probability we again use the Bratteli diagram. We note that the probabilities are given by the asymptotic values of the number of paths on the Bratteli diagram that lead to the field Φ_n^{a-1} or Φ_n^{a+1} (normalized by the sum of paths that lead to either of these fields). Some intuition on why the use of these asymptotic values is correct in this case may be gained from the following arguments. For simplicity, we may assume that the total number of quasiholes in the system, edge and bulk combined, is $2mk$, where m is some large integer. Since we require that all quasiholes fuse to the identity field, the probability that $n+1$ quasiholes fuse to $\Phi_{n+1}^{a\pm1}$, given that the n quasiholes fused to Φ_n^a, is equal to the probability that the remaining $2mk-n-1$ quasiholes on the edge fused to $\Phi_{2mk-n-1}^{a\pm1}$. In the limit of large m, these probabilities are found by counting the number of paths leading to the fields $\Phi_{2mk-n-1}^{a\pm1}$ on the Bratteli diagram of the edge.

Finding the number of paths leading to a certain vertex of the Bratteli diagram with coordinates (l,m), denoted $D(l,m)$, is done by writing a simple recursion relation [13] for $D(l,m)$. Looking at the Bratteli diagram, it is easy to conclude that

$$D(l,m) = D(l-1,m-1) + D(l+1,m-1). \tag{8.11}$$

Note that the above relation holds only when $D(l,m) \equiv 0$ if the point (l,m) does not appear on the diagram. This recursion relation can also be written in a matrix form

$$\vec{D}(m) = M\vec{D}(m-1) = M^m\vec{D}(0) \tag{8.12}$$

where the components of the vector $\vec{D}(m)$ are $D(l,m)$, with $0 \leq l \leq k$.

The matrix M is a $(k+1)\times(k+1)$ matrix of the form

$$M_{ij} = \delta_{i,j+1} + \delta_{i+1,j} = \begin{pmatrix} 0\,1\,0 & . & . & . \\ 1\,0\,1 & & & \\ 0\,1\,. & & & \\ . & & . & \\ . & & & . \end{pmatrix}. \tag{8.13}$$

Since this matrix is a symmetric matrix it can be diagonalized by an orthogonal transformation, $M = ODO^T$. In order to find the diagonalizing matrix, note that M resembles a tight binding Hamiltonian of a system containing $k+3$ sites, with boundary conditions forcing the wavefunction to be zero at the site 0 and the site $k+2$. It is therefore easy to diagonalize M: the normalized eigenvectors are

$$(v_q)_j = \frac{1}{N_q}\sin\left(\frac{\pi q j}{k+2}\right), \tag{8.14}$$

$$N_q = \left(\sum_{j=1}^{k+1}\sin^2\left(\frac{\pi q j}{k+2}\right)\right)^{1/2},$$

and the corresponding eigenvalues are

$$\lambda_q = 2\cos\left(\frac{\pi q}{k+2}\right) \tag{8.15}$$

with $1 \leq q \leq k+1$. The matrix M^m can therefore be written as

$$(M^m)_{ij} = \sum_q \frac{(\lambda_q)^m}{N_q^2}\sin\left(\frac{\pi q i}{k+2}\right)\sin\left(\frac{\pi q j}{k+2}\right). \tag{8.16}$$

The initial vector we use in order to find $D(l,m)$, is $\vec{D}(1) = (0,1,0,0,0,....)$. Operating with the matrix M^{m-1}, we get

$$D(l,m) = \sum_q \frac{(\lambda_q)^m}{N_q^2} \sin\left(\frac{\pi q}{k+2}\right) \sin\left(\frac{\pi q(l+1)}{k+2}\right) \tag{8.17}$$

In the limit of large m, the main contribution comes from the largest eigenvalues of the matrix M, given by $\lambda_1 = -\lambda_{k+1} = 2\cos\left(\frac{\pi}{k+2}\right)$. Therefore, the asymptotic values for the the coefficients of the superposition (8.10) as a function of k and a are given by

$$|c_{a-1}|^2 = \frac{\sin\left(\frac{\pi a}{k+2}\right)}{\sin\left(\frac{\pi a}{k+2}\right) + \sin\left(\frac{\pi(a+2)}{k+2}\right)} \tag{8.18}$$

$$|c_{a+1}|^2 = \frac{\sin\left(\frac{\pi(a+2)}{k+2}\right)}{\sin\left(\frac{\pi a}{k+2}\right) + \sin\left(\frac{\pi(a+2)}{k+2}\right)} \tag{8.19}$$

We may now use the coefficients $|c_{a-1}|^2$ and $|c_{a+1}|^2$ to find the correct overlap (8.10). We find that the interference term has the same phase and general form of Eq (8.9), but this time it is multiplied by an amplitude which is a function of k and a

$$I_{bs} \propto |t_L|^2 + |t_R|^2 + 2|t_L^* t_R| A_k^a \cos\left(\frac{2\pi n}{k(k+2)} - \frac{\pi n}{k} + \beta\right)$$
$$A_k^a = \frac{\cos\left(\frac{\pi(a+1)}{k+2}\right)}{\cos\left(\frac{\pi}{k+2}\right)} \tag{8.20}$$

Considering Eq. (8.20), we find that when k is even, there will always be a possible state of the dot, $\Phi_n^{k/2}$, for which there is no interference at all, as the pre-factor of the interference term is set to zero. For example, if $k = 2$, the field Φ_n^1 exists for odd n as the only possible internal state of the dot. In that case, for all odd n, there will be no interference, as predicted before in Refs. [14, 15]. For all other cases there will always be a finite amplitude for the interference term, with a phase that depends on the number of quasiholes in the bulk and an amplitude which depends on the particular internal state of the dot.

The above prediction, first derived particularly for the $k = 2$ case corresponding to $\nu = 5/2$ in Refs. [14, 15], and later generalized to include the entire RR series in Ref. [12], shows that the fact that the quasi-holes in the system obey non-Abelian statistics is intimately related to a suppression of their lowest order interference term in the conductance of the interferometer.

Surprisingly, although the derivation of the above result heavily relies on the unique structure of the Hilbert space of a system of quasiparticles through the use of the Bratteli diagram, it turns out that similar features are obtained for a set of Abelian states at the same filling fractions [16]. These Abelian states, known as the generalized Halperin states [17], can show a suppression of the interference term, and at $\nu = 5/2$ an interference measurement will imitate perfectly the Moore Read state. For further discussion see Ref. [16] as well as Sec. 8.6.3.

8.5. Coulomb Blockade Regime

In the limit of strong back-scattering, when the two point contacts on either side of the island are almost closed, the number of electrons in the dot confined between them is quantized to an integer. Quasihole or quasiparticle tunneling into and out of this region is forbidden, and the only way to transport charge through it is by tunneling of electrons. Low-voltage low-temperature conductance through the dot is suppressed due to charging energy, except at "Coulomb blockade peaks", when the ground state energy of the dot with N_e electrons is degenerate with its ground state energy with $N_e + 1$ electrons. Coulomb blockade peaks may be probed by measuring the conductance through the dot as a function of a magnetic field B and the dot's area S, since a variation of S with fixed N_e violates charge neutrality with the positive background [18]. Peaks in the conductance appear for those values of the area S and magnetic field B, for which the following equation

$$E(N_e, S, B) = E(N_e + 1, S, B) \tag{8.21}$$

is satisfied for some integer N_e representing the total number of electrons inside the dot, where $E(N_e, S, B)$ is the energy of the edge modes for a dot of area S in a magnetic field B and containing N_e electrons.

The edge energies of the RR states are stored in a bosonic charged edge mode (a chiral Luttinger liquid) and a neutral edge mode. With $N_e = 0$ defined to be the number of electrons for a dot with area S_0, the energy associated with the charge mode for N_e electrons on the edge is [18]

$$E_c(N_e) = \frac{\pi v_c}{\nu L}\left(N_e - \nu\frac{B_0(S - S_0)}{\phi_0}\right)^2. \tag{8.22}$$

where ϕ_0 is the magnetic flux quantum. The magnetic field $B = B_0$ is that in which there are no quasiholes in the bulk, ν is the filling fraction of the partially filled topmost Landau level, v_c is the velocity of the charged edge mode, and L is the perimeter of the quantum dot (we take $\hbar = 1$). For Abelian Laughlin states, where there is only a single bosonic mode, Eq. (8.21) reduces to $E_c(N_e) = E_c(N_e + 1)$, and the area separation ΔS between its solutions for consecutive values of N_e is $\Delta S = e/n_0$, where n_0 is the charge density inside the dot. The value of ΔS in this case is independent of the magnetic field. For non-Abelian states, this is not the case.

Let us start with a dot where the number of electrons is divisible by k. By setting the magnetic field to a certain value for which the filling factor is a bit less than $k/(k+2)$, quasiholes may be introduced into the bulk, while their counterparts of opposite charge and topological charge are introduced onto the edge. The contribution of the charged part of the edge quasiparticles influences the boundary conditions of the chiral boson theory [18]. The parafermionic part of the edge will be taken from the fusion product of n copies of the field σ_{k-1}, the spin field that makes the edge quasiparticles.

The bulk quasiholes may have several fusion channels as dictated by Eq. (8.4). According to the fusion channel Φ_{bulk} of the bulk quasiholes, i.e, of n copies of the field σ_1, the fusion channel Φ_{edge} of the n copies of σ_{k-1} on the edge will be fixed by the requirement that $\Phi_{bulk} \times \Phi_{edge} \sim \mathbf{1}$. Had the system with localized quasiholes been infinite with no edge, all fusion channels of the bulk quasiholes would be have been degenerate in energy. This is the essential ingredient which causes these quasiholes to have non-Abelian statistics [5, 6]. The presence of the edge and the excitations living on it will lift the degeneracy of these fusion channels.

Every fusion channel of the edge quasi-particles can easily be re-written as a fusion product of some spin field σ_q with j copies of the fundamental parafermion ψ_1. Consequently, the ground state energy of an edge state carrying that particular excitation will be given by its conformal dimension [7]

$$E_\psi(j,q) = h_q + \begin{cases} \frac{j(k-j-q)}{k} & \text{for } j \leq k-q \\ \\ \frac{(k-j)(j+q-k)}{k} & \text{for } j > k-q \end{cases} \tag{8.23}$$

in units of $2\pi v_n/L$. The spacings between Coulomb blockade peaks is easily calculated using Eq. (8.21) which assumes the form

$$\Delta S = \frac{e}{n_0} + \frac{eL\nu}{2n_0\pi v_c}\left[E_\psi(N_e+2) - 2E_\psi(N_e+1) + E_\psi(N_e)\right]. \quad (8.24)$$

Any tunneling event corresponds to an addition of one unit of electron charge to the system, along with an application of $\psi_1^\dagger = \psi_{k-1} \sim \psi_1^{k-1}$ to the edge modes. Using (8.23) one finds that the pattern of Coulomb blockade peaks may be described as follows: If $q \neq 0$, the peaks bunch into alternating groups of q and $k-q$ peaks. The spacing that separates peaks within a group is given by

$$\Delta S_1 = \frac{e}{n_0}\left(1 - \nu\frac{v_n}{v_c}\frac{2}{k}\right), \quad (8.25)$$

while the spacing that separates two consecutive groups is

$$\Delta S_2 = \frac{e}{n_0}\left(1 + \nu\frac{v_n}{v_c}\left(1 - \frac{2}{k}\right)\right). \quad (8.26)$$

If $q = 0$, the peaks bunch into groups of k. The spacing that separates peaks within the group is again given by (8.25), while the spacing between two consecutive groups is

$$\Delta S_2 = \frac{e}{n_0}\left(1 + \nu\frac{v_n}{v_c}\left(2 - \frac{2}{k}\right)\right). \quad (8.27)$$

The periodicity of the peaks is always k. However, if k is even, the peak structure may also be periodic with a period $k/2$, provided that $q = k/2$.

The conclusion is therefore that transport of electrons through the quantum dot, as surprising as it may sound, may actually show signatures for the presence of non-Abelian quasiparticles. This is due to the fact that the edge spectrum is sensitive to the topological charge of the bulk.

8.6. Imperfections of the Fabry Perot Interferometer and their Possible Effects

The predictions of the outcome of the two experiments described in this chapter strongly rely on several assumptions that may in fact not hold in real experiments. We now turn to review some of them and the alternative predictions that are available assuming they do not hold.

8.6.1. *Bulk-edge coupling and lowest order interference*

One of the fundamental assumptions made in analyzing the FPI both in the limit of lowest order interference, and the CB limit, is that there is no coupling between the bulk and the edge. For the purpose of lowest order interference this means that there in no hopping of any kind between the bulk and the edge. However, since the bulk quasi-holes contain localized zero energy states that could in principle be occupied, in a real experiment there is always the possibility that tunneling of neutral particles between the bulk and the edge will take place.

An analysis of such perturbation for a general state of the RR series is hard. However, for the most studied state of that series, which luckily is also the most experimentally relevant state, such analysis is possible and was carried out in Refs. [19–21].

Using the analogy of the 5/2 state to a p-wave superconductor, the bulk quasiholes can be viewed as vortices in the superconductor, and are accompanied by an electric charge $e/4$ due to the Hall conductance. These vortices have a fermionic zero energy bound state at their core, which is a Majorana state [22]. Two such Majorana operators, associated with two different vortices, can be combined into a complex fermion to form a two-level system. The state of this two-level system can be changed by moving a third quasihole around one of the vortices. Depending on the occupancy of this two level system, there are two possible interference terms that are mutually shifted by a π phase. Assuming that the state of the two level system in the bulk is robust and fixed, a particular interference term will be chosen. Alternatively, when the number of vortices in the bulk is odd, no interference term is expected to appear [14, 15], since one of the bulk quasiholes forms a two level system with the interfering edge quasihole, for which the state is completely random, thereby dephasing the overall signal [14].

Note that the same predictions could be inferred from Eq. (8.20). For $k = 2$, the amplitude A_k^a may obtain three possible values $A_2^0 = 1, A_2^2 = -1, A_2^1 = 0$, implying that depending on the state of the bulk interference may be completely suppressed. Recall that the label a may obtain odd values only when the number of quasiholes is odd (see Sec. 8.4). However, for the purpose of this discussion it will be more convenient to think in terms of the p-wave analogy.

Refs. [20, 21] analyzed the influence of a tunnel coupling between bulk and edge-Majorana states on the visibility and phase of interference oscillations. The zero temperature behavior of the interferometer was studied for an interference loop encircling one or two localized quasiholes, as a function of the source-drain voltage and the strength of coupling between the Majorana modes of the impurities and the neutral modes of the edge.

If the bulk houses two Majorana states, and each is coupled to a different edge state, the two-level system formed by the two bulk Majorana states is coupled to a continuum (the edge), and its states acquire a life-time τ. Refs. [20, 21] show that the effect of the coupling depends on the voltage V used to probe the interference. When $eV\tau/\hbar \ll 1$ the bulk-edge coupling is strong, and the bulk quasi-particles are effectively annexed by the edge. When the converse is true, the bulk-edge coupling is weak. In that case, if τ is longer than the duration of the experiment, its effect may be neglected. However, if that is not the case, this coupling leads to a smearing of the even-odd effects that we discussed above.

8.6.2. *Bulk-edge relaxation*

When bulk quasi-particles are coupled to the edge, one may also consider the effect of inelastic relaxation, where hopping of neutral modes is accompanied by an energy transfer to an outside thermal bath, e.g., by an emission of a phonon. Such processes, it turns out, have far reaching consequences for Coulomb Blockade experiments [7].

If the area of the dot is varied slowly with respect to the relaxation rate, then each electron has time to relax onto the lowest energy state by changing the fusion channel of the edge quasiparticles and the bulk quasiholes. Bunching of peaks will still be observed for all the RR states with $k \geq 3$. However, the dependence of the pattern on the number of quasiholes in the bulk is suppressed. The periodicity of this structure depends on k: if k is even, the periodicity is $k/2$ and the pattern is of groups of $k/2$ peaks, while for odd k the periodicity is k, and the pattern is of alternating groups of $(k-1)/2$ and $(k+1)/2$ peaks.

For the case $k = 2$, it was found in Ref. [7] that when inelastic bulk-edge relaxation is allowed, the Coulomb blockade pattern will be the same as for Abelian fractional quantum Hall states, with a uniform spacing between peaks. This means that for slow variation of the area of the dot, the even odd effect predicted in Ref. [14] is smeared out. This smearing is a consequence of the parafermionic edge mode (which for $k = 2$ is a Majorana fermion mode) staying unpopulated throughout the experiment. Whenever the addition of an electron into the dot attempts to populate this mode by a fermion, the bulk-edge relaxation mechanism makes the fermion tunnel inelastically into the bulk, where its presence does not involve any energy cost.

8.6.3. *The uniqueness of interferometry signatures*

Recently, Bonderson et al. [23] showed that the zero temperature Coulomb blockade peak patterns that are predicted for the non-Abelian Read-Rezayi series of states in Ref. [7, 18] are identical to the ones predicted for the Abelian series of multi-component Halperin states. This identity of the predictions is, however, based on a very limiting assumption regarding the Abelian states.

The Abelian multi-component states of $\nu = 2 + \frac{k}{k+2}$ are composed of k flavors of electrons, and their similarity to the non-Abelian Read-Rezayi states holds only under a full symmetry between all the flavors. A breaking of that symmetry affects the Coulomb blockade and interference patterns and distinguishes the Abelian and non-Abelian states. In the $k = 2$ case, corresponding to the $\nu = 5/2$ state, the two flavors of electrons are likely to be the two spin states. The symmetry between the two states would then be broken by the Zeeman coupling of the spin to the magnetic field, or by spin-orbit coupling. In contrast, the patterns predicted for the non-Abelian states enjoy the insensitivity characteristic of these states to local perturbations.

As it turns out, for the most prominent candidate for a non-Abelian state, $\nu = 5/2$, the same identity of patterns holds also for Fabry-Perot interference experiments, as well as for finite temperature Coulomb blockade experiments [16]. Largely, this trend holds also for the more complicated states at $\nu = 2 + \frac{k}{k+2}$ with $k > 2$, although subtle differences do exist in the details of the suppression factors of the interference signals.

8.7. Conclusion

In this chapter we reviewed two proposed experiments to detect non-Abelian statistics in the quantum Hall effect using a Fabry Perot interferometer. Lowest order interference and Coulomb blockade are both affected by the presence of quasi-particles and their statistics in this setup. We have presented the conditions under which such measurements show these features clearly, and discussed some of the possible difficulties that may emerge when these conditions are not met. Although predications may change due to bulk-edge coupling or relaxation, some non trivial signatures should survive in an experiment.

To conclude, it is a pleasure for us to dedicate this article to our teacher and colleague, Prof. Yoseph (Joe) Imry, for his seventieth birthday. When condensed matter physicists deal with highly mathematical subjects, as we try to do in this work, it is imperative that they keep close connection to observables and experiments. There is no better advocator of this principle than Joe, and we feel blessed having had the opportunity to learn it from him.

Acknowledgment

We acknowledge financial support from the US-Israel Binational Science Foundation, the Minerva foundation, Microsoft's Station Q, the foundation FOM of the Netherlands and the INSTANS programme of the ESF.

References

1. A. Y. Kitaev, Fault-tolerant quantum computation by anyons, *Ann. Phys.* **303**, 2–30, (2003).
2. S. D. Sarma, M. Freedman, and C. Nayak, Topologically protected qubits from a possible non-abelian fractional quantum hall state, *Phys. Rev. Lett.* **94**, 166802, (2005).
3. C. Nayak, S. H. Simon, A. Stern, M. Freedman, and S. D. Sarma, Non-abelian anyons and topological quantum computation, *Rev. Mod. Phys, in press.* (2007).
4. A. Stern, Anyons and the quantum hall effect - a pedagogical review, *Annals of Physics.* **323**(1), 204–249 (January, 2008). doi: 10.1016/j.aop.2007.10.008.
5. G. Moore and N. Read, Nonabelions in the fractional quantum hall effect, *Nucl. Phys. B.* **360**, 362–396, (1991).
6. N. Read and E. Rezayi, Beyond paired quantum hall states: Parafermions and incompressible states in the first excited landau level, *Phys. Rev. B.* **59** (12), 8084–8092 (Mar, 1999). doi: 10.1103/PhysRevB.59.8084.
7. R. Ilan, E. Grosfeld, K. Schoutens, and A. Stern, Experimental signatures of non-abelian statistics in clustered quantum hall states, *Phys. Rev. B.* 79(24): 245305, (2009).
8. C. de C. Chamon, D. E. Freed, S. A. Kivelson, S. L. Sondhi, and X. G. Wen, Two point-contact interferometer for quantum hall systems, *Phys. Rev. B.* **55**(4), 2331–2343, (1997). doi: 10.1103/PhysRevB.55.2331.
9. S. C. Zhang, T. H. Hansson, and S. Kivelson, Effective-field-theory model for the fractional quantum Hall effect, *Phys. Rev. Lett.* **62**(1), 82–5, (1989).
10. A. B. Zamolodchikov and V. A. Fateev, Nonlocal (parafermion) currents in two dimensional conformal quantum field theoryand self dual critical points

in z_n symmetric statystical systems, *Zh. Eksp. Teor. Fiz.* **89**, 380–399 (Aug, 1985).
11. D. Gepner and Z. Qiu, Modular invariant partition functions for parafermionic field theories, *Nucl. Phys.* **B285**, 423, (1987).
12. P. Bonderson, K. Shtengel, and J. K. Slingerland, Probing non-abelian statistics with quasiparticle interferometry, *Phys. Rev. Lett.* **97**(1), 016401 (Jul, 2006). doi: 10.1103/PhysRevLett.97.016401.
13. J. K. Slingerland and F. A. Bais, Quantum groups and nonabelian braiding in quantum hall systems, *Nucl. Phys. B.* **612**, 229, (2001).
14. A. Stern and B. I. Halperin, Proposed experiments to probe the non-abelian $\nu = 5/2$ quantum hall state, *Phys. Rev. Lett.* **96**(1), 016802 (Jan, 2006). doi: 10.1103/PhysRevLett.96.016802.
15. P. Bonderson, A. Kitaev, and K. Shtengel, Detecting non-abelian statistics in the $\nu = 5/2$ fractional quantum hall state, *Phys. Rev. Lett.* **96**(1), 016803 (Jan, 2006). doi: 10.1103/PhysRevLett.96.016803.
16. A. Stern, B. Rosenow, R. Ilan, and B. I. Halperin, Interference, coulomb blockade, and the identification of non-abelian quantum hall states, *arXiv:0912.4394v1.* (2009).
17. X. G. Wen and A. Zee, Classification of abelian quantum hall states and matrix formulation of topological fluids, *Phys. Rev. B.* **46**(4), 2290–2301 (Jul, 1992). doi: 10.1103/PhysRevB.46.2290.
18. R. Ilan, E. Grosfeld, and A. Stern, Coulomb blockade as a probe for non-abelian statistics in read-rezayi states, *Phys. Rev. Lett.* **100**(8), 086803, (2008).
19. B. J. Overbosch and X.-G. Wen, Dynamical and scaling properties of $\nu = 5/2$ interferometer, *arXiv:0706.4339.* (2007).
20. B. Rosenow, B. I. Halperin, S. H. Simon, and A. Stern, Bulk-edge coupling in the non-abelian nu = 5/2 quantum hall interferometer, *Physical Review Letters.* **100**(22), 226803, (2008).
21. B. Rosenow, B. I. Halperin, S. H. Simon, and A. Stern, Exact solution for bulk-edge coupling in the non-abelian nu = 5/2 quantum hall interferometer, *Physical Review B.* 80(15):155305, (2009).
22. N. Read and D. Green, Paired states of fermions in two dimensions with breaking of parity and time-reversal symmetries and the fractional quantum hall effect, *Phys. Rev. B.* **61**(15), 10267–10297 (Apr, 2000). doi: 10.1103/PhysRevB.61.10267.
23. P. Bonderson, C. Nayak, and K. Shtengel, Coulomb blockade doppelgangers in quantum hall states, *arXiv:0909.1056.* (2009).

Chapter 9

Molecular Conduction Junctions: Intermolecular Effects

Arie Landau[1], Leeor Kronik[2] and Abraham Nitzan[1]

[1]*School of Chemistry, the Sackler Faculty of Science, Tel Aviv University, Tel Aviv, 69978, Israel*

[2]*Department of Materials and Interfaces, Weizmann Institute of Science, Rehovoth, 76100, Israel*

Intermolecular interactions can affect the conduction properties of molecular junctions in several ways: Direct and through-substrate electronic interactions affect the spectral properties (density of states) of the conducting junction, intermolecular electrostatic interactions affect the positioning of molecular electronic energies and thereby the nature of interface polarization. Such interactions also influence the screening properties of the junction and consequently the electrostatic potential profile across the biased junction. Other consequences include effects on junction mechanical properties that can be manifested by a different temperature dependence of conduction for a single molecule and for a molecular layer, as well as effects on optical response that may be important, for example, for the junction response associated with light induced switching. This article discusses some of these effects and their implications for the performance of molecular junctions.

9.1. Introduction

The electronic conduction properties of metal-molecule-metal junctions are determined by the electronic properties of the metal and molecular constituents, the bonding between them, the junction structure and configuration, external electrostatic (gate) fields and environmental parameters such as temperature. The relationship of function and structure offers routes for controlling the junction operation but also results in uncertainties about performance and stability. Characterization of such relationships is therefore central to the study of molecular conduction junctions.[1] One such issue that was already addressed in several experimental and theoretical papers is the dependence of transport on the number of molecules involved in the conduction process. In particular, a comparison between electronic transport of a single molecule junction and that of a junction comprising the corresponding molecular monolayer is of interest. While a common practice is to assume linear scaling with molecular coverage, i.e. to associate the conduction-per-molecule observed in monolayer junctions with that of the single molecule junction, direct experimental and numerical examinations yield mixed observations on this point. On the one hand, clear evidence of linear scaling is indicated by the results of Cui et al[2] and Xu and Tao[3] in junctions containing 1-5 n-alkane molecules (see Fig. 9.1).[a] Similarly, Kushmerik and coworkers[4] have found that the current-voltage curves of monolayers of isonitrile oligo(phenylene ethynylene) molecules measured in the cross wire technique with varying contact areas can be scaled to a single curve by dividing with different integers in the range 1-1000, suggesting that the scaled curve corresponds to a single molecule junction. Supporting evidence is obtained by comparing STM and cross wire conduction measurements on a series of saturated and conjugated molecules,[5] where similar current-voltage traces where obtained by using a multiplication factor (of the order of 10^3) that may account for the different number of molecules in these junctions. On the other hand, the conduction measured in some single molecule junctions was found to be several orders of magnitudes larger than the conduction-per-molecule mentioned in the corresponding monolayers.[6] In a more recent study, Selzer et al[7] have noticed that the small bias conduction per molecule of a molecular layer is similar to that of the corresponding single molecule; however, the differential conductance of the latter increases considerably more rapidly with bias and can become a thousand times larger than the

[a] In Ref. 3 also 4,4' bipyridine was used with similar results. Note that the conduction values reported by Xu and Tao in these papers are about an order of magnitude larger than those of Cui et al; the difference probably resulting from unrelated technical factors.

per-molecule layer conductance (see Fig. 9.2, left panel). Furthermore, the single molecule (1-nitro-2,5-di(phenylethynyl-4'-mercapto)benzene) junction shows temperature dependence with transition to activated conduction at $T > 100\text{K}$, while the conduction of the corresponding molecular layer remains temperature independent up to room temperature (Fig. 9.2, right panel). Obviously, the effect of intermolecular interactions depends on the nature of the adsorbed molecules. A recent scanning tunneling spectroscopy study[8] of monolayers of 3,4,9,10-perylenetetracarboxylicacid-dianhydride (PTCDA) on a silver substrate showed a remarkable electronic band dispersion (see Fig. 9.3) associated with the layer electronic structure, indicating a free electron behavior with an effective mass of $m_{eff} = 0.47m_e$ (m_e is the free electron mass). This surface electron delocalization indicates strong intermolecular coupling that has been attributed by the authors to mostly through-metal interaction. The existence of such strong intermolecular coupling suggests that conduction in this system will not scale linearly with that of single-molecule junctions based on the same molecule.

Theoretical studies of this issue also lead to varied results. Early studies by Magoga and Joachim[9] concluded that linear scaling of the conduction $g(N)$ with the number N of identical molecules connecting the leads exists in the form

$$g(N) = Ng_{eff}(1), \tag{9.1}$$

where the effective single molecule conduction $g_{eff}(1)$ differs from the corresponding single molecule property $g(1)$ because of intermolecular interactions. The latter can be direct interactions between neighboring molecules or interactions resulting from their mutual coupling to the leads. Obviously, however, N has to be large enough for Eq. (9.1) to hold. In the non-resonant small bias regime $g_{eff}(1) > g(1)$, because intermolecular interaction brings molecular states density closer to the Fermi energy of the lead. Yaliraki and Ratner[10] have made a similar observation by comparing the conduction of two parallel molecular wires to that of a single wire. On the other hand, Lang and Avouris[11] have studied the low bias conduction of a junction containing two carbon wires connecting jellium leads as a function of the interwire separation.

A non-monotonic distance dependence, with an overall conduction increase at larger separation was found and rationalized in terms of the effect of the interwire interaction on the molecular density of states at the Fermi energy. The magnitude of the effect should depend on the adsorbate molecules. Indeed, Kim[12] et al have studied the conduction of junctions based on single molecules and molecular layers of Hexane-dithiolate between Au(111) electrodes and found that the effect of intermolecular coupling on conduction is negligible in these junctions.

Effects of intermolecular interactions on properties of molecular junctions can arise from several sources. Clearly, direct interactions between the molecules may affect the transport properties of a molecular layer, as can indirect interactions mediated by the underlying substrate.

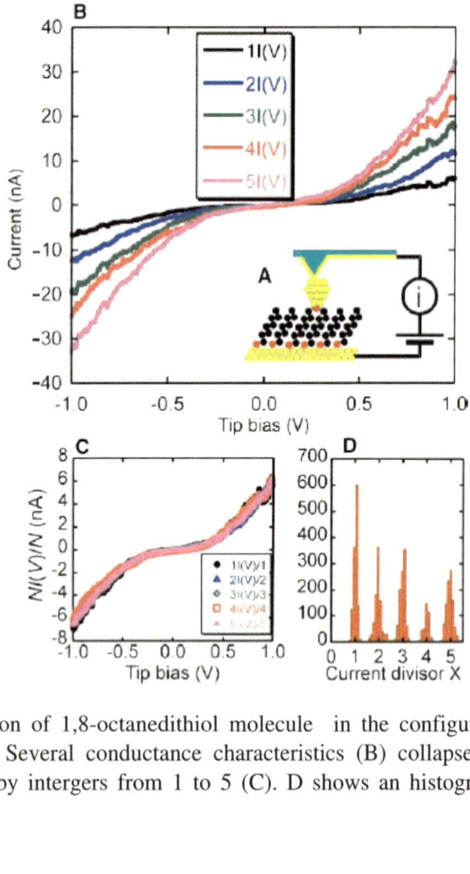

Fig. 9.1. Conduction of 1,8-octanedithiol molecule in the configuration shown in A. (after Cui et al[2]). Several conductance characteristics (B) collapse into a single line following scaling by intergers from 1 to 5 (C). D shows an histogram of conductance signals.

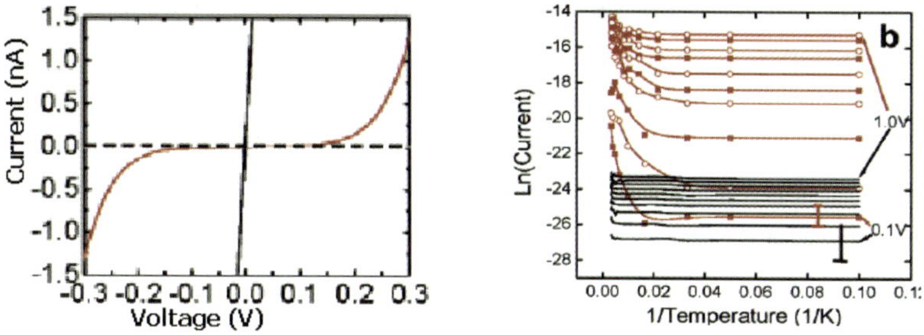

Fig. 9.2. Conductance of 1-nitro-2,5-di(phenylethynyl- 4'-mercapto) benzene between gold leads. Left: Red – single molecule; black – molecular layer. Dashed black is the molecular layer signal normalized per molecule. Right: Temperature and voltage dependence of conduction. Red – single molecule; black – molecular layer per molecule. From Selzer et al.[7]

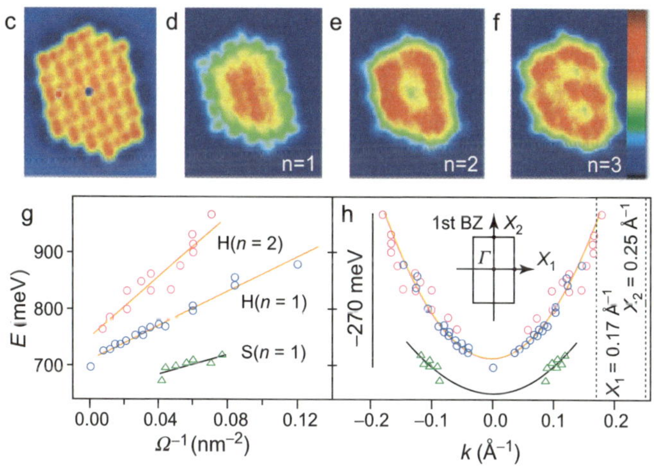

Fig. 9.3. Top panels: Scanning tunneling images of monolayers of 3,4,9,10 perylenetetracarboxylicacid-dianhydride on a silver substrate, showing different electronic states in the confined island geometry. Bottom Panels: The size dependent energy (left) can be translated into an electronic band dispersion (right) associated with the layer electronic structure, indicating a free electron behavior with an effective mass of $m_{eff} = 0.47 m_e$. From Temirov et al.[8]

Such interactions may directly affect transport, but may also influence transport properties by affecting the molecular configuration and its response to the imposed bias potential and to temperature change. Electrostatic effects are particularly noteworthy, having at least two important effects. First, the way an imposed bias is distributed across the junction depends on the lateral dimension of the bridging layer in a way that reflects the layer screening properties.[13,14] In turn, the way by which the bias potential falls along the junction has important implications on the junction transport properties, see, e.g. Refs. 15 and 16. Secondly, polar adsorbate molecules, or molecules that form polarized adsorption sites because of charge transfer to/from the substrate are expected to show strong intermolecular interaction effects,[17-20] as discussed in Sec. 9.3.

Finally, intermolecular interactions affect the manifestations of other transport properties that bear on the performance of molecular conduction junctions. For example, junction stability is related to its structural response to the imposed field and ensuing current, as well as to its temperature rise during conduction. The latter depends on the junction ability to dissipate excess energy, which in turn depends on its immediate environment. In particular, large differences may exist between a junction that can dissipate heat only through the contacting electrodes (as with a single molecule connecting metal leads in vacuum) or also through lateral interactions with neighboring molecules (a situation associated with molecular layers and junctions embedded in solvent environments). As another example, optical control, a desired property in molecular devices, relies on the system optical response which is expected to be a sensitive function of the nature and structure of its molecular interface.

This paper focuses on several of the above mentioned aspects of single molecule versus molecular island and layer behavior of molecular junctions. In Sec. 9.2, we briefly outline a simple generic model for the effect of molecular island size on the potential distribution along a biased junction. In Sec. 9.3, we review theoretical calculations of charge transfer at the interface of semiconductors and polar molecules. In Sec. 9.4, we present a generic study of effects associated with short range intermolecular and molecule-substrate interactions on the transport properties of single molecules and the corresponding molecular islands and layers. We use a minimal model of single-level molecules connecting electrodes characterized by simple cubic lattices in order to examine the effect of intermolecular interactions on the junction spectral function and conduction properties. Section 9.5 concludes.

9.2. Electrostatics: The Bias Potential Distribution along the Junction

Understanding the potential bias distribution along a molecular junction is necessary for understanding and predicting its conduction behavior. A demonstration of the importance of this issue to the understanding of the current-voltage behavior of molecular junctions was first presented by Datta and coworkers.[15,21] They have shown, within a simple Extended-Hückel (EH) model for α,α'-xylyl dithiol bridging between two gold leads, that the potential profile (imposed on the molecule as input to the EH calculation) had a profound effect on quantitative as well as qualitative aspects of the calculated current-voltage characteristic. Since then, the evaluation of the electrostatic potential along the junction has become an integral part of a full scale self consistent calculation.[22] On the experimental side, Bachtold et al[23] have used scanned probe microscopy of both single-walled and multi-walled carbon nanotubes (SWNTs and MWNTs) to measure the potential drop along such nanotubes connecting between two gold electrodes. They found an approximately linear drop of the potential in a MWNT of diameter 9nm while for a metalic SWNT bundle of diameter 2.5nm the potential was flat beyond the screening regions at the tube edges. A theoretical analysis of this behavior was provided in Ref. 24. While these experiments cannot be related directly to the calculations discussed above (the nanotube length is a few microns and impurity and defect scattering may be effective as is most certainly the case in the MWNT measurement), the flat potential seen in the metallic SWNT measurement is in fact a remarkable observation implying a very long mean free path (>1μm) for electrons in these room temperature structures.

As indicated by the above discussion, the computational methodology for evaluating the potential distribution on a specified biased molecular wire is available. Here we focus on a generic issue pertaining to our theme: the dependence of this distribution on the wire lateral cross-section. For a single molecule junction the cross-section radius is of order of 1-2Å and it increases linearly with the number of molecules connecting between the leads. To see the effect of this lateral wire width on the screening that affects the potential distribution, Nitzan et al[13] have employed a maximum screening model: The molecular wire is modeled as a cylinder characterized by the molecular length L and cross-section diameter (molecular wire thickness) σ, perpendicular to and connecting between two planar metal electrode surfaces. The cylinder is oriented parallel to the z axis, with its axis going through the origin in the xy plane. For simplicity, the electrodes are taken to be blocking, so there is no current through the junction, and the potentials at the wire-electrode interface are set to be Φ_1 and Φ_2 at $z=-L/2$ and $z=L/2$, respectively. It is further assumed that screening is described by a Thomas-Fermi model

$$\nabla^2\Phi = -4\pi\rho(\mathbf{r}) = -\lambda^{-2}\Phi \tag{9.2}$$

with the molecular screening length λ taken to be a model parameter. Finally, taking the direction perpendicular to the metal electrode planes to be z, and $\mathbf{r}_{||}$ to denote the lateral direction, a solution is sought in the form

$$\Phi(\mathbf{r}_{||}, z) = F(\mathbf{r}_{||})\Phi(0,0,z)$$

where the function $F(\mathbf{r}_{||})$ is chosen to describe the lateral confinement of the charge distribution in the molecular cylinder. This leads to

$$\Phi(0,0,z) = \Phi_0(z) - \frac{\Delta\Phi}{\pi} \times \sum_{\substack{n=0 \\ (even)}}^{\infty} \frac{(-1)^{n/2}}{n} \frac{F_n}{1+F_n} \sin\left(\frac{2\pi n}{L} z\right) \tag{9.3}$$

where (with Φ_1 and Φ_2 the potentials at the two boundaries)

$$\Phi_0(z) = \frac{1}{2}(\Phi_2 + \Phi_1) + \frac{z}{L}\Delta\Phi \quad ; \qquad \Delta\Phi = (\Phi_2 - \Phi_1) \tag{9.4}$$

and

$$F_n = \frac{1}{\lambda^2} \frac{d^2\mathbf{k}}{(2\pi)^2} \frac{\hat{F}(\mathbf{k}_{||})}{k_{||}^2 + (2\pi n/L)^2} \; ; \quad \hat{F}(\mathbf{k}_{||}) = \int d^2\mathbf{r}_{||} e^{-i\mathbf{k}_{||}\cdot\mathbf{r}_{||}} F(\mathbf{r}_{||}) \tag{9.5}$$

Using for the confinement function the model form

$$F(\mathbf{r}_{||}) = e^{-(\mathbf{r}_{||}/2\sigma)^2} \tag{9.6}$$

(note that this function is not normalized and describes a charge density at the cylinder axis independent of the width σ, as desired when the cylinder models a molecular assembly of a given diameter) leads to the results shown in Fig. 9.4,[13] where $\Phi_1 = -\Delta\Phi / 2$, $\Phi_2 = \Delta\Phi / 2$ are used.

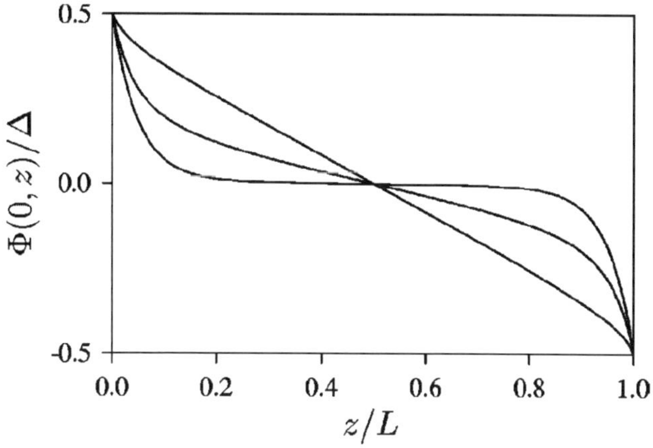

Fig. 9.4. The potential profile along a molecular wire computed from Eqs. (9.3) - (9.6) is shown for a screening length $\lambda/L = 0.05$. The thickness parameter σ/L takes the values 0.0125, 0.05, and 0.5 from the almost linear behavior to the voltage profile containing almost a plateau. From Nitzan et al.[13]

These results show that the wire thickness, determined by the number of molecules connecting in parallel between the conducting leads, does constitute an important generic attribute that determines the general behavior of the potential bias distribution along a molecular wire. Obviously, a Thomas-Fermi screening model is not a good representation of electrostatic screening in molecular systems. However, it suffices for demonstrating this point. Indeed, detailed molecular level calculations[14] of the potential profile in molecular junctions show similar behavior.

9.3. Molecule - Substrate Charge Transfer

At thermodynamic equilibrium, uniformity in the electronic chemical potential in a system comprising a conducting substrate and an adsorbed molecule, molecular island or a molecular layer is achieved by (partial) electronic charge transfer between the molecular adsorbate and the underlying substrate. This charge transfer results in a contribution to the surface dipole density, which is manifested as a change in the substrate work function but should also affect the potential profile in a biased molecular junction. Obviously, the energetic balance that determines the magnitude and direction of the transferred charge depends strongly on the electrostatics within the created dipole layer. For this reason we may expect a strong dependence of this process on the number and density of the molecules involved.

A demonstration of this expectation was recently made in calculations of molecules and molecular monolayers adsorbed on silicon.[18,19] Two types of systems were considered. In one a molecular monolayer (calculations with several benzene derivatives were done) is adsorbed on Si (111) surface.[18] The substrate is represented by a slab with thickness of at least six Si atomic layers embedded in a 3-dimensional periodic supercell. Density functional theory is used in the local density approximation where the Kohn-Sham equations are solved using a plane wave basis set.

The other system type considered is a Si cluster (with several tens of Si atoms) with one adsorbed molecule. When such a cluster is large enough this is a model for a single molecule adsorbed on Si surface. The dipole per adsorbed molecule was computed and compared.

Results of these calculations are shown in Figs. 9.5 (slab/monolayer system) and 9.6 (cluster with single adsorbate molecule), which depict, for several benzene derivatives, the dipole moment per adsorbed molecule against the dipole moment of the isolated (gas phase) molecule. The most important difference between the results displayed in these figures is the fact that the slope of the lines in Fig. 9.5 are smaller than one, indicating that the dipole per molecule in an adsorbate layer is smaller than the corresponding isolated molecule case, while the slopes in Fig. 9.6 are larger than 1. This indicates different charge reorganization mechanisms that result in a larger surface dipole in the case of a single adsorbed molecule and a smaller one in the molecular layer. The slab/molecular-layer observation can be rationalized in terms of the tendency to increase system stability by reducing dipole-dipole repulsion in the molecular layer case,[20] and is consistent with the dependence of the slopes in Fig. 9.5 on the surface coverage.

Calculations show that charge migration between the molecules and the substrate is very small in this case (which is consistent with the electric field suppression outside the layer[20]) and dipole reduction is achieved by depolarization of the molecules themselves. In the cluster/single molecule case the surface dipole enhancement is caused mainly by (cluster size dependent) electron exchange with the Si substrate in the direction that depends on the electron withdrawing/donating property of the molecule. This implies that the net polarization should be a strong function of coverage – a prediction confirmed in recent experimental work.[25] It is interesting to note[20] that electron transfer in the opposite direction is expected in the complementary case, where molecules are *absent* from a molecular monolayer, a phenomenon that often occurs naturally.[26] Indeed, the superposition principle tells us that the electric

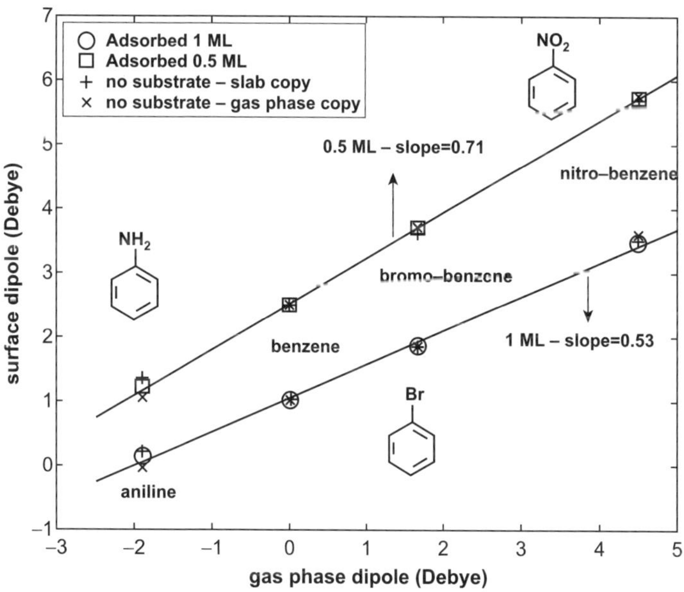

Fig. 9.5. Surface dipole per molecule as a function of gas-phase molecular dipole for a self assembled monolayer (SAM) of benzene derivatives with various functional groups,[18] for both 1 ML and 0.5 ML coverage. For each coverage, dipoles were computed for a SAM/substrate configuration, for a SAM with the substrate removed, but with molecules in their relaxed surface-adsorbed geometry, and for a SAM with the substrate removed and the molecules in their relaxed gas-phase geometry. For ease of comparison all curves were shifted such that the dipole moment of the benzene SAM is the same.

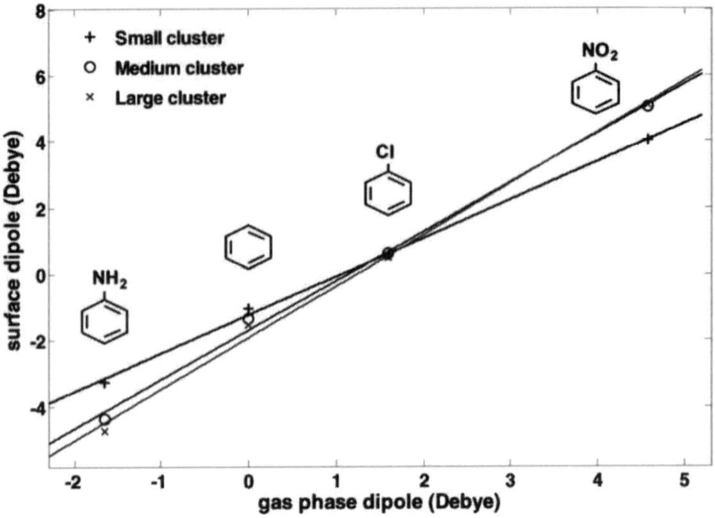

Fig. 9.6. Same as Fig. 9.5 for a single molecule adsorbed on silicon clusters of different sizes.[19] Using linear regression, slopes of 1.15, 1.48, and 1.55 are found for the "small" (1 atom), "medium" (14 Si atoms), and "large" (38 Si atoms) clusters, respectively.

field near a pinhole of an otherwise perfect dipolar monolayer is simply the difference between the electric field of the perfect monolayer and the electric field of the "missing domain". Therefore the electrostatic field of a single-molecule hole in such a monolayer, beyond the field suppression distance (of the order of the intermolecular distance in the layer), is very similar to that of a single dipole in the *opposite* direction, implying similar reversal in the direction of charge transfer.

9.4. Spectral and Conduction Properties of Molecules and Molecular Layers

When a junction comprises several molecules connecting in parallel between the two leads, intermolecular effects arise from direct as well as from substrate mediated interactions. Here we examine the consequences of such interactions on the spectral and conduction properties of the molecular junction, using a simple tight binding model for the leads as well as for the molecular conductor. The following discussion focuses on

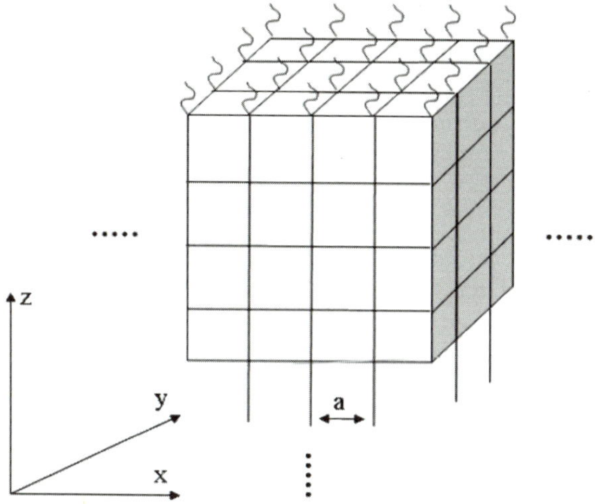

Fig. 9.7. Schematic cartoon of a simple 1:1 ML (curly lines represent molecules) adsorbed on a metallic electrode (grid points represent metallic-atoms). The Z direction is normal to the surface, X and Y are parallel to it, a is the unit distance. The cubic lattice is $\left(n_x = -\infty...\infty, n_y = -\infty,..\infty, n_z = 0...\infty\right)$ and the molecular layer occupies the sites $\left(n_x, n_y, 0\right)$.

metal-molecule-metal junctions. An equivalent analysis of junctions involving semiconductor electrodes was recently published.[27] Our aim is to compare the spectral properties (molecule projected density of states (DOS)) of, and the conduction associated with, a molecular monolayer (ML, Fig. 9.7)), a finite molecular island (IL) and a single molecule (SM) chemisorbed on a metal surface. The metal is described by a tight binding model defined on a simple-cubic cell structure characterized by a lattice constant, a, which is semi-infinite in the Z direction perpendicular to the metal surface. Periodic boundary conditions are used in the X and Y directions (see Fig. 9.7). Both metal and molecular system are described by simple nearest-neighbor (nn) tight-biding models. Each metallic-atom is represented by a single orbital $\left|n_x, n_y, n_z\right\rangle$, coupled to its nearest neighbor orbitals. A similar description is also used for the molecular system and for the molecule–metal interaction. The model is characterized by the molecular and metal site energies, ε_m and ε_t, respectively, and by the interaction between neighboring metal sites,

$V^{(t)}$, neighboring molecules, $V^{(m)}$, and nearest-neighbor metal-molecule interaction $V^{(mt)}$. The model Hamiltonian $\hat{H} = \hat{H}^{(m)} + \hat{H}^{(t)} + \hat{H}^{(mt)}$ has the usual nn tight binding form where, e.g., the molecule-metal interaction is given by

$$\hat{H}^{(mt)} = V^{(mt)} \sum_{n_x} \sum_{n_y} \Big[\big| n_x n_y (n_z = 0) \big\rangle \big\langle n_x n_y (n_z = 1) \big| + h.c. \Big]. \quad (9.7)$$

The sum over the lateral indices n_x, n_y is over the occupied sites of the molecular adsorbate, Z=0, plane. More than one coupling parameter $V^{(mt)}$ may be needed to describe more complex layer structures.

It is convenient to replace the local basis $\left| n_x, n_y, n_z \right\rangle$ by a basis in which Bloch wavefunctions are used in the X, Y directions:

$$\left| \theta_x \theta_y n_z \right\rangle = \frac{1}{\sqrt{N_x N_y}} \sum_{n_x = -(N_x/2)}^{(N_x/2)-1} \sum_{n_y = -(N_y/2)}^{(N_y/2)-1} e^{i(\theta_x n_x + \theta_y n_y)} \left| n_x n_y n_z \right\rangle, \quad (9.8)$$

where $\theta_u = k_u a$, and $k_u = 2\pi j / N_u$ $(j = 1, 2, ..., N_u;\ u = x, y)$. In what follows we also use the notation $\boldsymbol{\theta}_{xy} = (\theta_x, \theta_y)$. For a pure metal this transformation diagonalizes the Hamiltonian, yielding the energy eigenvalues

$$E_t(\boldsymbol{\theta}_{xy}) = \varepsilon_t + 2V^{(t)} (\cos\theta_x + \cos\theta_y) \quad (9.9)$$

The spectral and transport properties we seek can be obtained from the surface Green function (GF), $G^{(ts)}$, and the associated surface self energy (SE), $\Sigma^{(ts)}$. Details of the calculation are given in Ref. 28 and only an outline is given below. For the bare metal surface these are given by

$$G_{\boldsymbol{\theta}_{xy}}^{(ts)}(E) = \left(E - E_t(\boldsymbol{\theta}_{xy}) - \Sigma_{\boldsymbol{\theta}_{xy}}^{(ts)}(E) \right)^{-1} \quad (9.10)$$

and

$$\Sigma_{\boldsymbol{\theta}_{xy}}^{(ts)} = V^{(t)} G_{\boldsymbol{\theta}_{xy}}^{(ts)} V^{(t)} \quad (9.11)$$

Eqs. (9.10) and (9.11) can be solved for the metal surface GF iteratively. In addition, the surface spectral density function of the metal is defined as

$$\Gamma^{(ts)}_{\boldsymbol{\theta}_{xy}}(E) = -2\,\mathrm{Im}\left[\Sigma^{(ts)}_{\boldsymbol{\theta}_{xy}}(E)\right] \tag{9.12}$$

Similarly, for a simple[b] adsorbed monolayer (top layer in Fig. 9.7) the GF is,

$$G^{(ml)}_{\boldsymbol{\theta}_{xy}}(E) = \left(E - E_m\left(\boldsymbol{\theta}_{xy}\right) - \Sigma^{(ml)}_{\boldsymbol{\theta}_{xy}}(E)\right)^{-1} \tag{9.13}$$

where

$$E_m\left(\boldsymbol{\theta}_{xy}\right) = \varepsilon_m + 2V^{(m)}\left(\cos\theta_x + \cos\theta_y\right) \tag{9.14}$$

and

$$\Sigma^{(ml)}_{\boldsymbol{\theta}_{xy}} = V^{(mt)} G^{(ts)}_{\boldsymbol{\theta}_{xy}} V^{(mt)} \; ; \quad \Gamma^{(ml)}_{\boldsymbol{\theta}_{xy}}(E) = -2\,\mathrm{Im}\left[\Sigma^{(ml)}_{\boldsymbol{\theta}_{xy}}(E)\right] \tag{9.15}$$

When the molecular layer is in contact with two metals, L and R, with the same simple 1:1 adsorption geometry, the ML SEs combine additively, $\Sigma^{(ml)}_{\boldsymbol{\theta}_{xy}} = \Sigma^{(ml,L)}_{\boldsymbol{\theta}_{xy}} + \Sigma^{(ml,R)}_{\boldsymbol{\theta}_{xy}}$, where

$$\Sigma^{(ml,K)}_{\boldsymbol{\theta}_{xy}} = V^{(mt,K)} G^{(ts,K)}_{\boldsymbol{\theta}_{xy}} V^{(mt,K)} ; \quad K = L, R\,. \tag{9.16}$$

In the following we also require the position space (XY) representation of these functions. In particular the local site GF

$$G^{(ml)}_{\mathbf{n}_{xy},\mathbf{n}_{xy}}(E) = \frac{1}{4\pi^2}\int_0^{2\pi}\int_0^{2\pi} d\theta_x d\theta_y G^{(ml)}_{\boldsymbol{\theta}_{xy}}(E)\,, \tag{9.17}$$

is related to the local density of states (DOS) per molecule

$$\rho^{(ml)}(E) = -\frac{1}{\pi}\mathrm{Im}\left[G^{(ml)}_{\mathbf{n}_{xy},\mathbf{n}_{xy}}(E)\right]. \tag{9.18}$$

Equivalent expressions can be obtained when instead of a homogeneous molecular layer we have a molecular island of J molecules (J=1 corresponds to a single adsorbed molecule) adsorbed on the metal

---

[b] By "simple" we mean that each adsorbed molecule is coupled to only one metal surface atom, and vice versa.

surface and interacting with N surface atoms. Denote the corresponding interaction elements by $V_{nj}^{(mt)}$; $n=1,..,N;\ j=1,..,J$. Note that there is a one-to-one correspondence between n and a position $\mathbf{n}_{xy}$ on the metal surface. However, we have dropped the requirement of "simplicity" as defined above[b] with respect to the adsorbate island or ML. The molecular island Hamiltonian, $\hat{H}^{(il)}$, is a combination of a diagonal part, $\varepsilon_m \mathbf{1}_J$, where $\mathbf{1}_J$ is a unit matrix of order J, and a non-diagonal part associated with the interactions $V^{(m)}$ between island molecules. The island GF (a $J \times J$ matrix in the position representation) is now given by

$$G^{(il)}(E) = \left(E - \hat{H}^{(il)} - \hat{\Sigma}^{(il)}\right)^{-1} \tag{9.19}$$

where the island SE is given by

$$\Sigma_{j,j'}^{(il)} = \sum_{n}^{N} \sum_{n'}^{N} V_{jn}^{(mt)} G_{n,n'}^{(ts)} V_{n'j'}^{(mt)} \ ; \quad \Gamma_{j,j'}^{(il)}(E) = i\left(\Sigma^{(il)} - \Sigma^{(il)\dagger}\right)_{j,j'} \tag{9.20}$$

and may be easily evaluated for any finite island once the metal surface GF (Eq. (9.10) has been calculated. Again, when the molecular island connects between two metals the self-energy in (9.19) is the sum $\Sigma^{(il,L)} + \Sigma^{(il,R)}$, where

$$\Sigma_{j,j'}^{(il,K)} = \sum_{n}^{N} \sum_{n'}^{N} V_{jn}^{(mt,K)} G_{n,n'}^{(ts,K)} V_{n'j'}^{(mt,K)}; \quad K = L, R. \tag{9.21}$$

With these results we can compare spectral properties of infinite and finite molecular layers, down to a single molecule. For example, a relationship between the per molecule DOS associated with a single adsorbed molecule, $\rho^{(sm)}(E)$, and the corresponding function of a molecular layer $\rho^{(ml)}(E)$ is obtained in the form[28]

$$\frac{\rho^{(sm)}(E)}{\rho^{(ml)}(E)} = \frac{\operatorname{Im}\left[\left(E - \varepsilon_m - \frac{1}{4\pi^2} \int_0^{2\pi} \int_0^{2\pi} d\theta_x d\theta_y \Sigma_{\boldsymbol{\theta}_{xy}}^{(ml)}(E)\right)^{-1}\right]}{\operatorname{Im}\left[\frac{1}{4\pi^2} \int_0^{2\pi} \int_0^{2\pi} d\theta_x d\theta_y \left(E - E_m(\boldsymbol{\theta}_{xy}) - \Sigma_{\boldsymbol{\theta}_{xy}}^{(ml)}(E)\right)^{-1}\right]} \tag{9.22}$$

Next, consider conduction properties of these molecular systems when placed between two metal electrodes. According to the Landauer formula,[29] conduction in the linear response limit is given by $\left(e^2 / \pi\hbar\right)\mathcal{T}\left(E_F\right)$ where the transmission coefficient $\mathcal{T}(E)$ is given in terms of the GF and the SE of the subsystem comprising the molecular bridge

$$\mathcal{T}(E) = \mathrm{Tr}_{\mathrm{bridge}}\left[\Gamma^{(L)}(E)G(E)\Gamma^{(R)}(E)\left(G(E)\right)^{\dagger}\right] \tag{9.23}$$

$\mathrm{Tr}_{\mathrm{bridge}}$ stands for a trace over the states of the bridging system. For a finite molecular island this trace is easily evaluated in the representation of local island states *j*, using Eqs. (9.19) and (9.20). In particular, for a single molecule junction this yields

$$\mathcal{T}^{(sm)}(E) = \Gamma^{(sm,L)}(E)G^{(sm)}(E)\Gamma^{(sm,R)}(E)\left(G^{(sm)}(E)\right)^{\dagger} \tag{9.24}$$

For the simple molecular layer the transmission *per molecule* (denoted $\mathcal{T}^{(ml)}(E)$ below) is given by $\left\langle \mathbf{n}_{xy}\middle|\Gamma^{(L)}(E)G(E)\Gamma^{(R)}(E)\left(G(E)\right)^{\dagger}\middle|\mathbf{n}_{xy}\right\rangle$, where $\mathbf{n}_{xy}$ corresponds the lateral position of any one molecule on the ML. This leads to

$$\mathcal{T}^{(ml)}(E) = \frac{1}{4\pi^2}\int_0^{2\pi}\int_0^{2\pi} d\theta_x d\theta_y \Gamma_{\boldsymbol{\theta}_{xy}}^{(ml,L)}(E)G_{\boldsymbol{\theta}_{xy}}^{(ml)}(E)\Gamma_{\boldsymbol{\theta}_{xy}}^{(ml,R)}(E)\left(G_{\boldsymbol{\theta}_{xy}}^{(ml)}(E)\right)^{\dagger} \tag{9.25}$$

Once the transmission function (9.24) or (9.25) has been obtained, we can also compute the current through the junction using the Landauer formula,

$$I(\Phi) = \int dE\left(f_L(E) - f_R(E)\right)\mathcal{T}^{(M)}(E,\Phi) \tag{9.26}$$

where *M* stands for s single molecule (SM), a molecular island (IL) or a molecular monolayer (ML), and where $f_L(E)$ and $f_R(E)$ are the Fermi functions of the left and right electrode, respectively.

$$f_K(E) = \left(1 + \exp\left(\frac{E - \mu + e\Phi^K}{k_B T}\right)\right)^{-1} \; ; \; K = L, R, \qquad (9.27)$$

where μ is the unbiased Fermi energy, Φ^K is the potential on the electrodes $K = L, R$, and k_B and T are the Boltzmann constant and the temperature, respectively. In the calculation reported below we set $\Phi^L = 0$ and denote $\Phi^R = \Phi$. The latter assignment is expressed by shifting all metal sites energies on the right by $-e\Phi$. The molecular site energy is then taken as $\varepsilon_m(\Phi) = \varepsilon_m(0) - Se\Phi$ where the shift parameter $0 \le S \le 1$ reflects a particular assumption about the way by which the bias voltage falls along the molecular bridge.

Figures 9.8-9.10 show some results based on this model, using model parameters that reflect orders of magnitude of physical observables such as metallic band widths, computed interaction energies between adsorbate molecules, and lifetimes of excess electrons on molecules adsorbed on metal surfaces (see Ref. 28).

Figure 9.8 depicts results obtained for the transmission coefficient (Eq. (9.23)) per molecule, through a molecular monolayer with (white circles) and without (black circles) direct intermolecular interactions. Also shown is the transmission coefficient through a single adsorbed molecule. The difference between the two latter results stems from the fact that molecules interact with each other, even in the absence of direct interactions, through their mutual interactions with the underlying metal.

The most significant effect of intermolecular interactions on the transmission coefficient is the considerable broadening of the transmission resonance associated with the molecular level. With our choice of parameter the direct intermolecular interaction plays a dominant role in this broadening, even though through metal interaction has a non-negligible effect, as suggested in Ref. 8.

Figure 9.9 shows the consequence of this broadening on the current-voltage characteristics of the junction, computed using (9.26) with the molecular-shift parameter S=0.5. When the molecular site-energy is well within the Fermi window, the SM current is higher than the current per molecule through a ML, because part of the broadened layer spectral

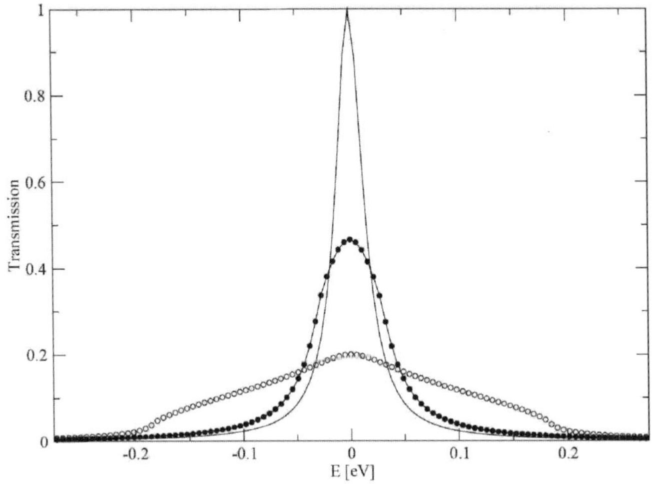

Fig. 9.8. White circles: Transmission (per molecule) as a function of energy through a ML, calculated with model parameters $V^{(t)}$ = 0.03a.u.~0.82 eV, $V^{(m)}$ = 0.0035 a.u. ~ 0.095 eV, $V^{(mt)}$ = 0.004 a.u. ~ 0.11 eV and $\varepsilon_m = \varepsilon_t = 0$. Black circles: Same, except with $V^{(m)}$ = 0. Full line: The transmission coefficient through a single molecule using the same parameters ($V^{(m)}$ is irrelevant in this case).

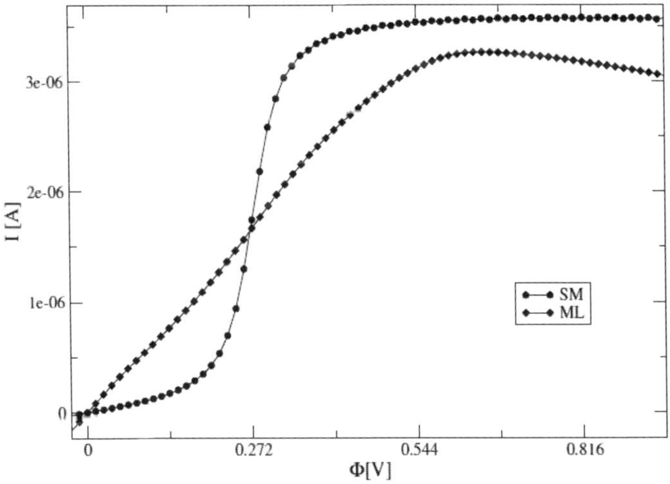

Fig. 9.9. Current I (per molecule) as function of the bias voltage Φ via a SM (circles) and ML (diamonds), with the molecular level set at ε_m = –0.136eV. ε_t, taken as the energy origin is also the Fermi energy in this calculation (corresponding to a monovalent metal). In this system $I(\Phi) = I(-\Phi)$ and only the $\Phi > 0$ side is shown.

density is outside that window. In the opposite case the molecular resonance may lie outside the Fermi window while the layer band may be broad enough to 'spill' into the window. The layer conductance will be higher in this case. Thus, in the case where $\varepsilon_m = -0.136$ eV and $S = 0.5$ the molecular level enters the Fermi window at Φ = 0.272V, and for lower bias the ML current is higher than its SM counterpart. This is reversed above $\Phi = 0.272\,\text{V}$, where the SM spectrum is better contained within the Fermi window. The decrease in the ML current at higher voltages is a manifestation of the fact that the broadened layer spectrum starts to 'feel' the metallic band edge so that the tail of the molecular band that extends beyond this edge cannot contribute to conduction. Again, level broadening arguments make it clear that such an effect will occur in a molecular layer at lower voltages than in the SM case.

Next, consider the scaling of conduction with the number of conducting molecules. Experimentally, such considerations pertain to two possible situations: (A) an essentially infinite molecular layer is engaged by a probe (or probes) that connect to a varying number of molecules, and (B) the junction involves molecular islands of varying sizes. In both cases we find that linear scaling appears only beyond some characteristic island size that depends on the strength of intermolecular interactions. Figure 9.10 demonstrates this behavior in case B. To ease the computational effort a two-dimensional junction model is considered in which the surface and the adsorbed molecular layers are represented by 1-dimensional rows of sites. The molecules are taken to couple more strongly to one of the electrodes ("substrate") then to the other (probe). Also, the tight binding parameter for the 2-dimensional "metal" is taken to be large enough to yield a sufficiently large metal bandwidth. Our results show strong dependence on the molecular island size N, which saturates to linear scaling behavior (and converge to the ML results) only for island exceeding $N \sim 30$ molecules in this 2-dimensional model. Interestingly, the discrete spectrum of an island comprising a small number of molecule leads to a distinct structure in the current-voltage characteristics. While this is an obvious possibility, it is usually disregarded in theoretical analysis of experimental "single molecule" junctions.

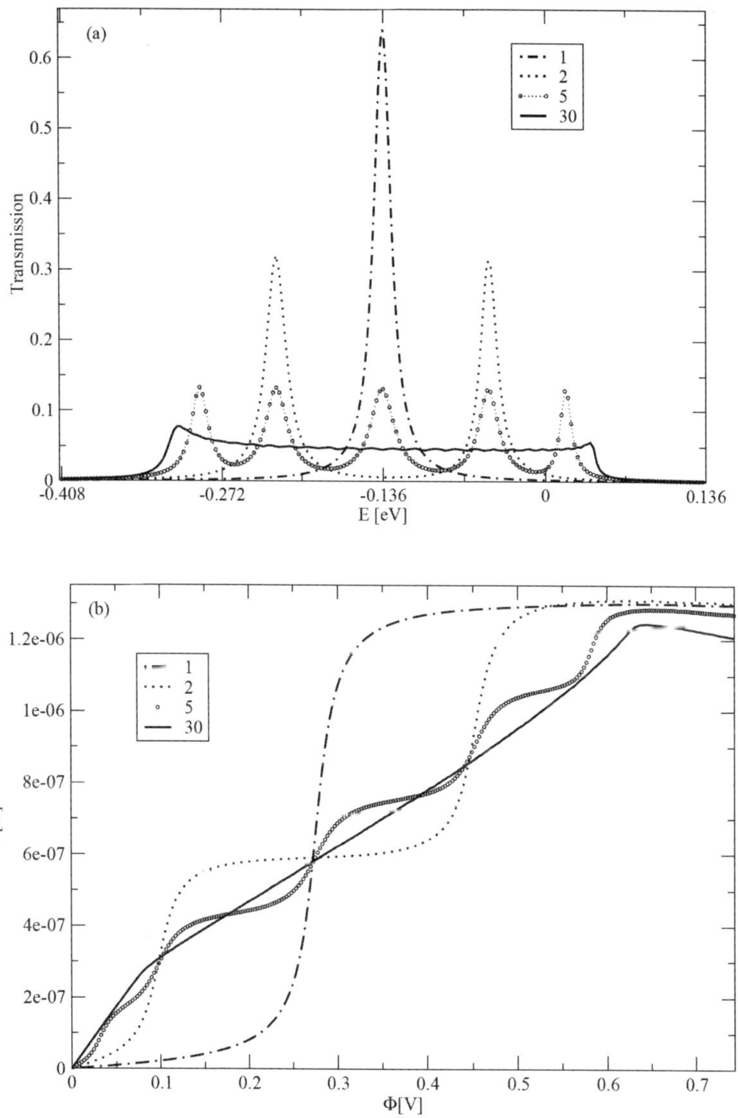

Fig. 9.10. The transmission coefficient (a) and the current-voltage characteristics (b) of 2-dimensional junctions comprising molecular islands of different sizes (expressed by the number of molecules N). The parameters used in these calculations are $V^{(t)} = 1.1$ eV, $V^{(m)} = 0.095$ eV, $\varepsilon_m - \varepsilon_t = -0.136$ eV, $V^{(mt1)} = 0.11$ eV (substrate) , $V^{(mt2)} = 0.055$ eV (probe) and $S = 0.5$.

9.7. Summary and Conclusions

The conduction properties of molecular junctions depend on structural and interaction parameters. In this review we have focused on effects associated with intermolecular interactions, as may be revealed in observations of the dependence of conduction on the number of conducting molecules; in the extreme case, comparing single molecules junctions to junctions based on molecular layers. In this review we have focused on several factors that affect this dependence: electrostatic screening, electrostatic interactions affecting molecule-substrate charge transfer, and intermolecular interactions affecting broadening of transmission resonances.

The individual or collective nature of molecular conduction often comes under discussion with respect to observation of apparent scaling of conduction with the number of molecules involved. We have pointed out, and demonstrated in model calculations, that such observations may depend on the nature of the molecular system. Simple linear scaling of the single molecule behavior is expected only beyond some molecular cluster size that depends on the intermolecular interaction. In particular, in our simplified 2-dimensional calculation, and based on our (physically motivated) choice of short range interaction parameters, we have estimated this linear size to be of the order of ~30 molecular sites. Again we emphasize that the conduction per molecule in these linear scaling regime may be quite different from that of the corresponding single molecule junction.

Finally, we point out that other collective molecular effects can influence molecular conduction and related transport properties. Of considerable interest are mechanical properties associated with the temperature dependence of molecular conduction (see Fig. 9.2) and heat conduction effects[30] that besides their intrinsic significance are importantly manifested in junction stability properties.

Acknowledgements

We thank David Cahen and Ron Naaman for many helpful discussions. This research was supported by grants from the Israel Science Foundation, the USA-Israel Binational Science Foundation, the Germany-Israel

Foundation and the Schmidt Minerva Center for Supramolecular Chemistry. This paper is dedicated to Yosef Imry, an inspiring leader of our field.

References

1. H. B. Akkerman and B. de Boer, J. Phys.: Cond. Mat. 20 (1), 013001 (2008).
2. X. D. Cui, A. Primak, X. Zarate, J. Tomfohr, O. F. Sankey, A. L. Moore, T. A. Moore, D. Gust, G. Harris, and S. M. Lindsay, Science 294, 571 (2001).
3. B. Xu and N. J. Tao, Science 301 (5637), 1221 (2003).
4. J. G. Kushmerick, J. Naciri, J. C. Yang, and R. Shashidhar, Nano Lett. 3 (7), 897 (2003).
5. A. S. Blum, J. G. Kushmerick, S. K. Pollack, J. C. Yang, M. Moore, J. Naciri, R. Shashidhar, and B. R. Ratna, The Journal of Physical Chemistry B 108, 18124 (2004).
6. A. Salomon, D. Cahen, S. M. Lindsay, J. Tomfohr, V. B. Engelkes, and C. D. Frisbie, Advanced Materials 15, 1881 (2003).
7. Y. Selzer, L. Cai, M. A. Cabassi, Y. Yao, J. M. Tour, T. S. Mayer, and D. L. Allara, Nano Letters 5, 61 (2005).
8. R. Temirov, S. Soubatch, A. Luican, and F. S. Tautz, Nature 444 (7117), 350 (2006).
9. M. Magoga and C. Joachim, Phys. Rev. B 59 (24), 16011 (1999).
10. S. N. Yaliraki and M. A. Ratner, J. Chem. Phys. 109 (12), 5036 (1998).
11. N. D. Lang and P. Avouris, Phys. Rev. B 62, 7325 (2000).
12. Y.-H. Kim, J. Tahir-Kheli, P. A. Schultz, W. A. Goddard, and Iii, Physical Review B (Condensed Matter and Materials Physics) 73 (23), 235419 (2006).
13. A. Nitzan, M. Galperin, G.-L. Ingold, and H. Grabert, J. Chem Phys. 117 (23), 10837 (2002).
14. G. C. Liang, A. W. Ghosh, M. Paulsson, and S. Datta, Phys. Rev. B 69, 115302 (2004).
15. W. D. Tian, S. Datta, S. H. Hong, R. Reifenberger, J. I. Henderson, and C. P. Kubiak, J. Chem. Phys. 109 (7), 2874 (1998).
16. M. Galperin, A. Nitzan, S. Sek, and M. Majda, J. Electroanalytical Chem. 550-551, 337 (2003).
17. D. Cahen, R. Naaman, and Z. Vager, Advanced Functional Materials 15 (10), 1571 (2005); G. Heimel, L. Romaner, E. Zojer, and J. L. Bredas, Nano Lett. 7, 932 (2007); D. Cornil, Y. Olivier, V. Geskin, and J. Cornil, Adv. Function. Mat. 17 (7), 1143 (2007); L. Romaner, G. Heimel, C. Ambrosch-Draxl, and E. Zojer, Adv. Funct. Mater 18, 3999 (2008).
18. A. Natan, Y. Zidon, Y. Shapira, and L. Kronik, Phys. Rev. B 73 (19), 193310 (2006).
19. D. Deutsch, A. Natan, Y. Shapira, and L. Kronik, J. Am. Chem. Soc. 129 (10), 2989 (2007).
20. A. Natan, L. Kronik, H. Haick, and R T. Tung, Advanced Materials 19 (23), 4103 (2007).

21. S. Datta, W. D. Tian, S. H. Hong, R. Reifenberger, J. I. Henderson, and C. P. Kubiak, Phys. Rev. Lett. 79 (13), 2530 (1997).
22. N. D. Lang and P. Avouris, Phys. Rev. Lett. 84 (2), 358 (2000); P. S. Damle, A. W. Ghosh, and S. Datta, Phys. Rev. B 64, 201403 (2001); M. Brandbyge, J. L. Mozos, P. Ordejon, J. Taylor, and K. Stokbro, Phys. Rev. B 65 (16), 165401 (2002).
23. A. Bachtold, M. S. Fuhrer, S. Plyasunov, M. Forero, E. H. Anderson, A. Zettl, and P. L. McEuen, Phys. Rev. Lett. 84, 6082 (2000).
24. N. Mingo, J. Han, M. P. Anantram, and L. Yang, Surface Science 482 (485), 1130 (2001).
25. N. Gozlan and H. Haick, J. Phys. Chem. C 112, 12599 (2008); N. Gozlan, U. Tisch, and H. Haick, J. Phys. Chem. C 112, 12988 (2008); N. Peor, R. Sfez, and S. Yitzchaik, J. Am. Chem. Soc. 130, 4158 (2008).
26. A. Ulman., *An Introduction to Ultrathin Organic Films*. (Academic Press, San Diego, 1991).
27. A. Landau, A. Nitzan, and L. Kronik, Journal of Physical Chemistry A 113 (26), 7451 (2009).
28. A. Landau, L. Kronik, and A. Nitzan, Journal of Computational and Theoretical Nanoscience 5, 535 (2008).
29. R. Landauer, IBM J. Res. Dev. 1, 223 (1957); S. Datta, *Electric transport in Mesoscopic Systems*. (Cambridge University Press, Cambridge, 1995).
30. Z. Liu and B. Li, Phys. Rev. E 76, 051118 (2007).

Chapter 10

Mesoscopic Effects in the Electron-Glass

Z. Ovadyahu

Racah Institute of Physics, The Hebrew University,
Jerusalem, Israel

Electronic transport studies of mesoscopic systems may yield unique information on a variety of physical processes. Here we discuss results of transport measurements on electron-glasses with sizes approaching the mesoscopic regime of hopping conductance. The study reveals fluctuations in the conductance as function of the gate voltage in samples that are smaller than $\approx 100\mu m$, when measured at liquid helium temperatures. The relative amplitude of the conductance fluctuations decreases with the sample volume while, down to sample size of few microns, the non-equilibrium features that are attributed to the electron-glass are indistinguishable from those observed in macroscopic samples. These include the memory-dip, slow relaxation, and simple-aging behavior. The mesoscopic conductance fluctuations, as well as the 1/f noise, appear to be different phenomena, independent of the electron-glass dynamics.

10.1. Introduction

Mesoscopic physics is ordinarily associated with quantum interference effects in samples with lateral size comparable to the phase-coherent length.[1] These effects were extensively studied in diffusive systems, and yielded a wealth of information on basic quantum transport issues as well as on the dynamics of scattering centers.[1,2] Similar effects, albeit not related to quantum interference, can be seen in transport studies of small samples that are in the insulating regime (where the conductance mechanism is by hopping between localized sites).[3] Conductance fluctuations (CF) can be readily observed in Anderson insulating samples, even when their lateral dimensions are considerably larger than the percolation radius that, in terms of ensemble-averaging plays the role of the length scale for mesoscopic behavior in the hopping regime.[4] Indeed, at low temperatures (typically, liquid helium temperatures), samples with L$\leq$100μm exhibit reproducible conductance fluctuation as function of the chemical potential.[5,6] These are observable in the conductance versus gate-voltage G(V_g) scans as a sample-specific, 'noise-like' component that is reproducible upon repeated gate-voltage scans. This pattern however, changes form when the sample is thermally re-cycled.

In this work we review the behavior the electron-glass when the system size approaches the mesoscopic regime of hopping conductance. One of the goals in these studies is to shed light on the interplay between the electronic slow-dynamics and the ionic one, which is presumably associated with the temporal evolution of the CF pattern. The respective feature that characterizes the electron-glass dynamics, is the memory-dip (MD). This is a cusp-like modulation observable as a local minimum at G(V_g^0) when a G(V_g) scan is taken after the sample was allowed to equilibrate[7–9] under V_g^0. This feature does not appear immediately after cool-down from high temperature; its amplitude evolves with time and becomes progressively more conspicuous as the equilibration process proceeds.[8] In contrast, the CF appears as a superimposed structure *immediately* after a cool-down and, as long as the sample remains cold, its rms amplitude does not vary with time. The two types of conductance modulation observable in the G(V_g) scans, namely the CF and the MD are associated with different time scales and different spatial scales; they therefore appear to be distinct physical phenomena.

10.2. Experimental

10.2.1. *Sample preparation*

The In_2O_{3-x} films used in this work were e-gun evaporated on either 100μm glass-slide with Au layer deposited on its back side to act as gate, or on SiO_2 insulating layer (0.5 μm thick) thermally grown on a $\langle 100\rangle$ Si wafer. The Si wafer was boron doped and had resistivity $\leq 2\cdot 10^{-3}\Omega$cm, deep in the degenerate regime. It thus could be used as an equipotential gate electrode for a low-temperature measurement where the In_2O_{3-x} served as the conducting layer in a traditional MOSFET configuration.

The choice of In_2O_{3-x}, the crystalline version of indium-oxide for this study, was motivated by its essentially constant stoichiometry imposed by crystal-chemistry constraints. This, in turn, guarantees a fairly constant carrier concentration n$\approx$(3-6)$\cdot 10^{19}$ cm^{-3}. (The amorphous version of indium-oxide exhibits a large variance in its glassy features depending on its carrier concentration,[9] which might overshadow the size dependence). Lateral dimensions of the samples were controlled using a stainless steel mask (for samples larger than 0.1mm), or optical lithography (for sizes in the range 30-200μm), and e-beam lithography (employing a XL-30 SEM) for samples smaller than 30μm. Samples used in this study had length L and width W that ranged from 2μm to 2mm, and thickness d in the range (35-100)$\pm$5Å. The 'source' and 'drain' contacts to the In_2O_{3-x} film were made from thermally evaporated $\approx$500Å thick gold films. Fuller details of sample preparation and characterization are given elsewhere.[8]

10.2.2. *Measurement techniques*

Conductance measurements were performed using two terminal ac technique, employing ITHACO-1211 current preamplifier and PAR-124A lock-in amplifier. In the MOSFET-like samples, gate voltage sweeps were affected by charging a 10μF capacitor with a constant current source. All the measurements reported here were performed with the samples immersed in liquid helium at T=4.11K held by a 100 liters storage-dewar. This allowed long term measurements without disturbing the samples as well as a convenient way to maintain a stable bath temperature. These requirements are of particular importance for studies of glassy systems where sample history may influence time dependent measurements, as was demonstrated in previous studies.[10] Care was taken to use small bias in the conductance measurements to ensure linear-response conditions. The only exception was in the 2μm samples where some deviations from Ohms law were observable even at the lowest bias we managed to measure with a reasonable signal to noise ratio. Artifacts associated with measurements outside of the linear response regime are discussed elsewhere.[4]

10.3. Results and Discussion

10.3.1. *Early studies of mesoscopic effects in the Electron-Glass*

Conductance fluctuations in G(V_g) of electron-glasses were already observed in the first report of the anomalous memory-dip in insulating indium-oxide films.[11] These were observed even in millimeter-size samples provided their resistance was sufficiently large ($\geq$100GΩ). In retrospect, the reason for the observability of these mesoscopic effects in such large samples is the inhomogeneous nature of the films. In such highly resistive samples the current-carrying network is a rarefied structure, embedded in a 'dead-wood' matrix thus the actual size of the 'active' part in these samples is much smaller than their geometric extent.

The next experiment, performed few years later, employed lithographically-patterned samples, which made the connection between the system size and the CF visibility much clearer. The original motivation for this experiment was to elucidate the mechanism by which the system gets out of equilibrium when the gate voltage is suddenly changed. For this purpose, samples were equipped with two additional voltage probes, prepared by a two-level, e-beam lithography. A typical sample configuration is shown in Fig. 10.1.

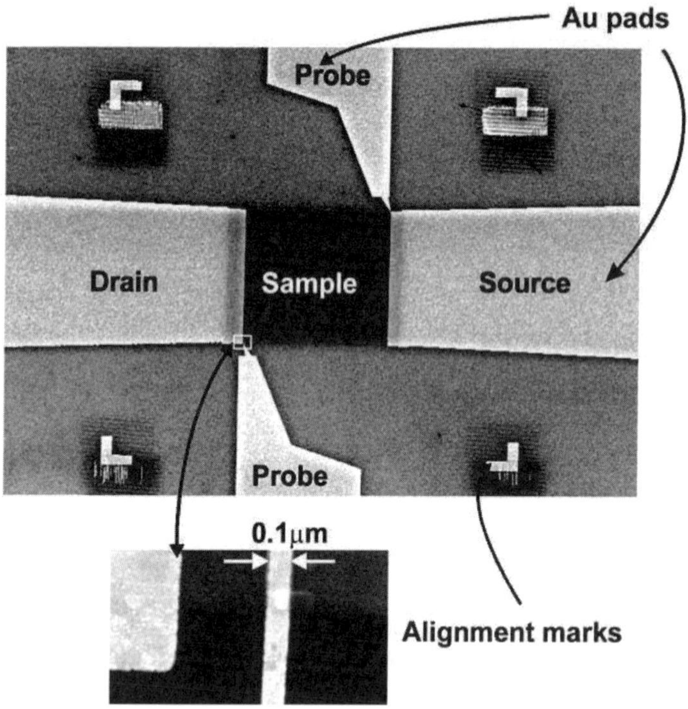

Fig. 10.1. Scanning electron micrograph of a typical sample made for monitoring the gate excitation process (Vaknin, Ph.D. thesis, Hebrew University, 2003). The source, drain, probes and alignment marks are Au films deposited and processed in the first level lithography. The alignment marks are used for the second level lithography (where the indium-oxide is deposited to make contact to the Au source and drain electrodes).

The voltage between a probe and the source (drain) $V^{s,d}_{probe}(V_g)$ was recorded in these experiments while V_g was swept at a constant rate across a voltage interval in either direction. As expected, the magnitude of the *average* $V^{s,d}_{probe}(V_g)$, and its polarity turned out to be proportional to dV_g/dt (since $V^{s,d}_{probe}$ is just a product of the charging current and the resistance of the part of the sample between the probe and the source or drain). The dc component of $V^{s,d}_{probe}$ is essentially identical with the voltage on a constant resistor connected in series with the source as depicted by the dashed lines in Fig. 10.2. It is noteworthy that no significant threshold-bias for insertion (or removal) of charge from the system is detected in this measurements; It might have been expected that a finite bias will be needed to overcome the Coulomb-gap to get charge into or out of the sample. The width of the Coulomb gap for In_2O_{3-x} is ≈6meV, which is ≈3 orders of magnitude larger than the voltages detected in Fig. 10.2. A possible reason for the absence of a Coulomb-gap signature in these experiments may be the related to the scenario discussed by Larkin and Shklovskii.[12]

In addition to the dc component, $V^{s,d}_{probe}$ exhibits a fluctuating part which is fairly reproducible upon subsequent scans, and remarkably, has amplitude that often exceeds the dc voltage. These voltage fluctuations are so conspicuous because the sample probed by $V^{s,d}_{probe}$ is small and thus its transport properties are dominated by mesoscopic effects. The nature of these effects is elucidated in the next subsection.

10.3.2. *Conductance fluctuations in the Electron-Glass*

Figure 10.3 shows $G(V_g)$ scans for two of the studied samples with different lateral dimensions. Both samples exhibit the two types of $G(V_g)$ modulation that we wish to focus on here: A memory-dip, MD centered at V_g=0 Volt (at which the sample was cooled and allowed to equilibrate), and a set of conductance fluctuations, CF. Two $G(V_g)$ traces are shown for each sample; one obtained a short time after cooling the system, and another after longer time of equilibration. Note that during this time the MD amplitude grows while the rms condcutance modulation associated with the CF pattern does not change. Another difference between these two types of modulations is their dependence on the system spatial extent; All other things being equal (e.g., temperature, resistance, sweep rate), the CF amplitude *increases* when the sample size *decreases* while the MD amplitude is found to be size independent (over the studied range of 2μm to 2mm). An example that illustrates this trend may be seen in Fig. 10.3 where the

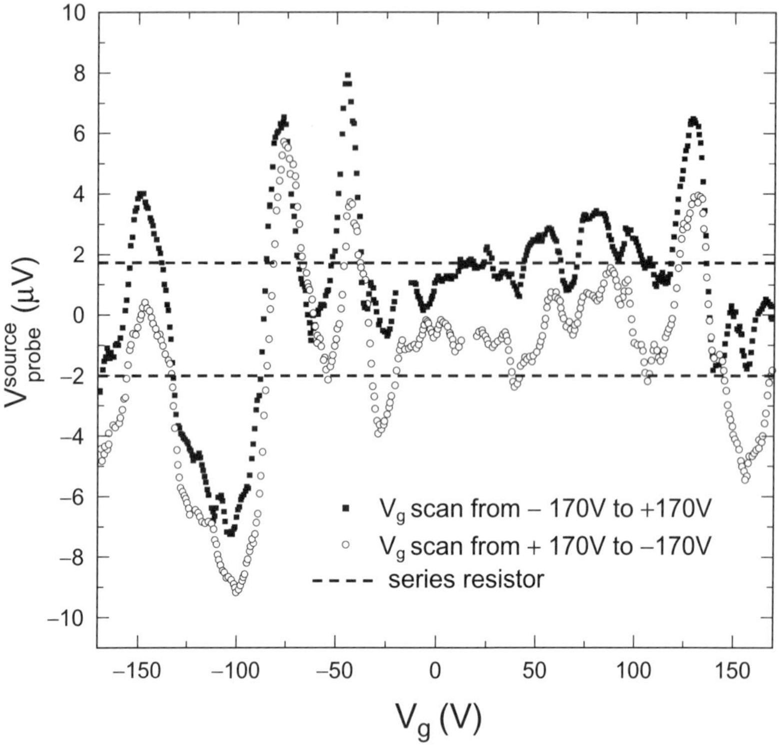

Fig. 10.2. The voltage between the source electrode and the adjacent probe for a typical sample (with 100μm SiO_2 spacer) as function of V_g for the two scan directions. Sweep rate is 0.5V/s. The voltage drop on a series resistor, having the same value as the source-probe resistance ($\approx$150kΩ), is shown for comparison. T=4.11K.

amplitude of the CF relative to the MD is larger in the smaller sample.

In our smallest samples (2x2μm^2), the CF amplitude may be so large as to mask the MD modulation (one needs to resort to averaging to expose the glassy nature of the system in this size regime[5]). Figure 10.4 illustrates a typical small-sample behavior; ΔG/G of the CF is 40-50%, which is rarely the amplitude of the MD in this range of temperatures and resistances. Another important aspect revealed in these experiments concerns the temporal behavior of the MD vis-a-vis that of the CF modulation. The data in Fig. 10.4 depicts G(V_g) of the sample while V_g is continuously swept back and forth across the (-5V,5V) interval starting from equilibrium. Note that the CF pattern shows only relatively small changes during the whole experiment. While there are some deviations from perfect reproducibility of the CF structure, they are always within the level expected from the observed magnitude of the 1/f noise that was studied in these samples.[4] Namely, the deviation, judged by the cross-correlation between a traces taken at time $t_1$and t_2, was always consistent with the magnitude of the noise estimated at $[t_2\text{-}t_1]^{-1}$. By contrast, a continuous sweep of V_g does not allow the system to equilibrate and form the MD at any place along the voltage range. Furthermore, by continuously sweeping the gate voltage, a pre-formed MD is weakened, and eventually completely disappears from the $G(V_g)$ plot.[5]

The disparity in the dynamics between the CF and the MD was illustrated in the experiment described before;[5] Starting from a relaxed state at the cool-down $V_g=V_g^o$, multiple sweeps of the gate voltage over an interval straddling V_g^o yield $G(V_g)$ traces that clearly show that the cusp amplitude decays with time *much* faster than the time it takes the CF pattern to change appreciably.

Before discussing the interplay between the CF and the MD, the origin of the CF in these samples needs to be identified. There is some similarity in behavior between the CF observed in our $G(V_g)$ and the so called 'universal-conductance-fluctuation' (UCF) phenomonon familiar from the weakly-localized (diffusive) regime.[2] Most notably in that the fluctuation pattern produced is like a set of 'fingerprints', characteristic of a particular structural, frozen-in configuration. This can be demonstrated by performing thermal-cycling experiments. A typical result is given in Fig. 10.5 showing two $G(V_g)$ traces for a given sample. These were obtained by the following protocol: After an initial cool-down and equilibration for 12 hours at T=4K, the sample was first warmed up to T≈70K, held there for 10sec, then quench-cooled to 4K, allowed to relax again for 12 hours, and then a $G(V_g)$ sweep was taken. In these instances the $G(V_g)$ curves show a well-developed MD centered at the gate voltage at which the sample equilibrated, in addition to a distinctive set of CF. Each such experiment run yields a *quench-specific* set of CF (Fig. 10.5) while the average value of G is nearly identical; Note that the relative change in the conductance between the two quenches is less than 2% which must be regarded as a minute change - recall that G is exponential with temperature, and in this sample it varies by a factor of ≈70 in the range 4-70K. More importantly, both the shape and magnitude of the associated MD are *unchanged by thermally cycling the sample* (e.g., inset to Fig. 10.5).

The observation that a thermal cycle generates a new CF pattern yet preserves (on average) the value of G *and* the rms CF amplitude, is indeed

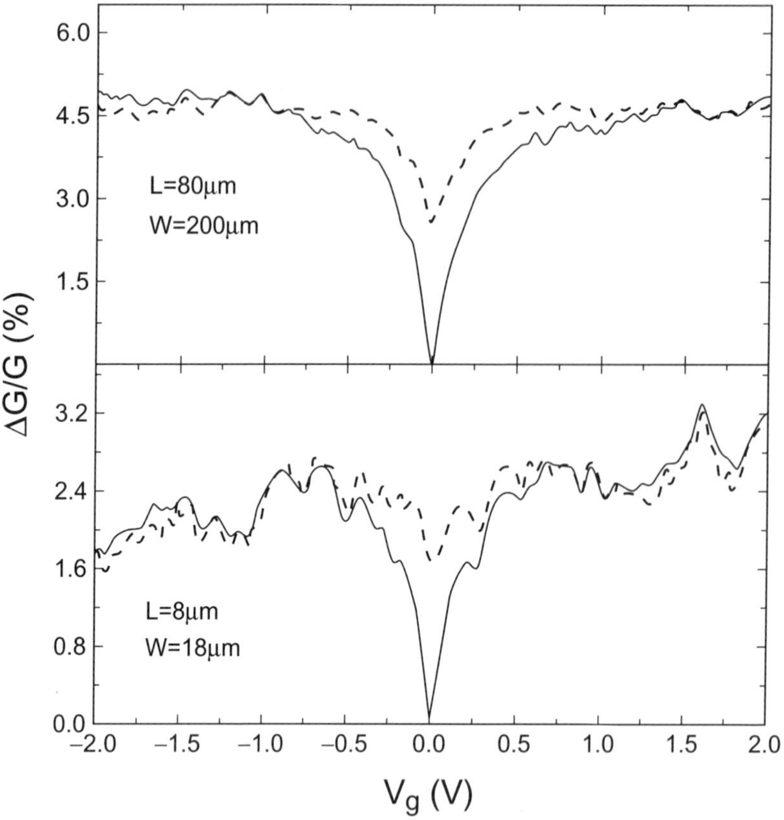

Fig. 10.3. Relative change of conductance versus the gate voltage for two In_2O_{3-x} samples with similar resistances but different lateral dimensions. Upper graph: $R_\square$=34MΩ Lower graph: $R_\square$=25MΩ. The dashed (full) lines are scans taken ≈ 1 min (≈ 5 h) after cool-down from room temperatures. Note that the MD for the smaller sample can be distinguished from the CF only after the longer period of equilibration.

reminiscent of the respective situation in diffusive systems.[2] The underlying mechanism for the phenomenon studied here is different however; The UCF pattern results from a quantum interference effect usually produced by sweeping a magnetic field H that modulates G via the Aharonov-Bohm effect (similar behavior has been observed in gate experiments on diffusive samples, also related to an interference effect[14]). The CF observed here is associated with a mechanism described by Lee[13] to account for experiments on quasi-one dimensional hopping samples.[3] The underlying physics of the mechanism proposed by Lee does not depend on the particular dimension-

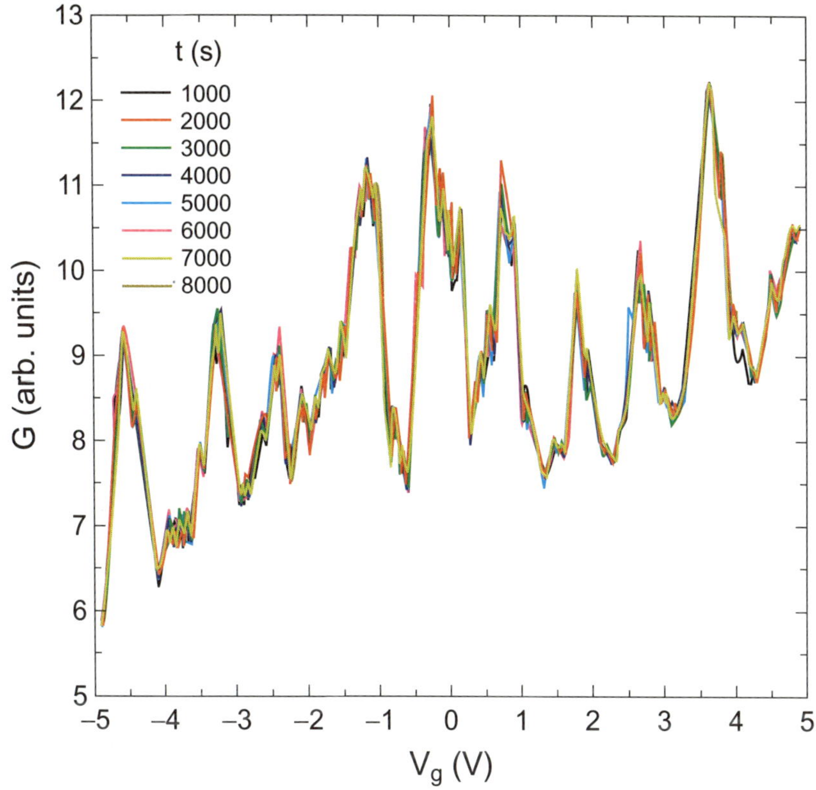

Fig. 10.4. Conductance versus gate voltage for a $2\text{x}2\mu\text{m}^2$ sample and $R_\square$=4MΩ at T=4.11K. Consecutive, 'side-to-side' gate-voltage sweeps were taken with a sweep rate $dV_g/dt=10^{-2}$V/s. Traces are labeled by the time at which the sweep ended.

ality: The current in hopping systems is carried in a percolation-network and controlled by a relatively small number of 'critical' or 'bottleneck' resistors.[15–17] Like any other resistor in the system, the values of these resistors are given by the Miller-Abrahams expressions[18]

$$R_{ij} \propto \exp\left[\frac{2r_{ij}}{\xi} + \frac{|E_i - E_j| + |E_i - \mu| + |E_j - \mu|}{k_B T}\right] \tag{10.1}$$

where r_{ij} is the inter-site distance in space, ξ is the localization length, E_i, E_j are the site energies, and μ is the chemical potential. Therefore the actual resistance of these circuit elements depend (among other things) on

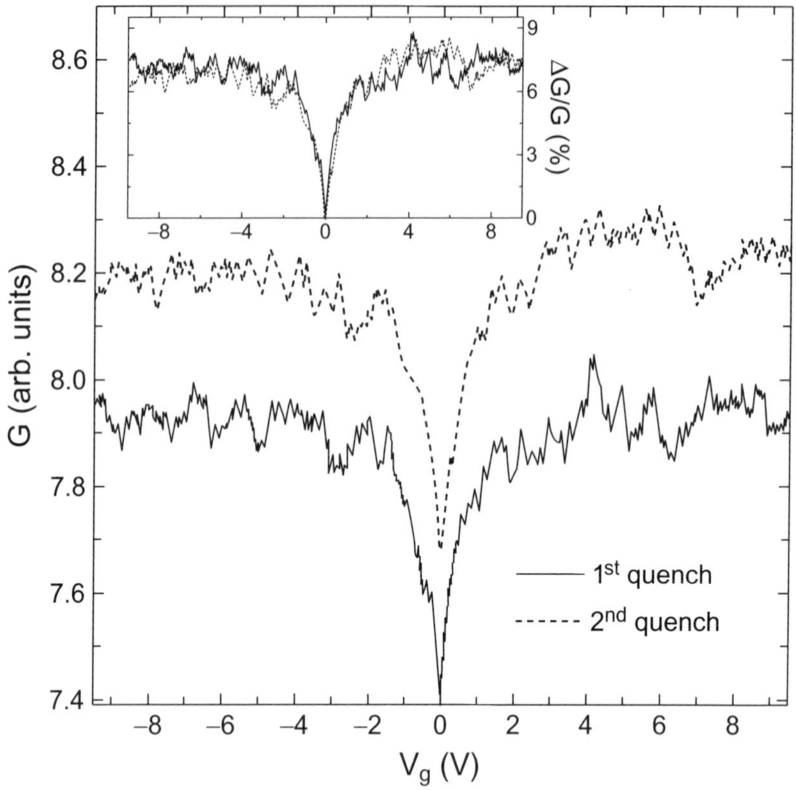

Fig. 10.5. Conductance-fluctuations and memory-dip for a single In_2O_{3-x} sample with width, W=100μm, length L=25μm and sheet resistance $R_\square$=57.5MΩ. Traces were taken after quench-cooling from 70K to 4.1K, after equilibration (see text for detailed protocol). Inset illustrates that, the normalized MD is reproducible while the CF are quench dependent.

the value of the chemical potential. Sweeping the gate voltage causes the chemical potential to change thereby re-shuffling the values of the resistors in the current-carrying network. The CF pattern in this picture reflects the process by which some critical resistors in the percolation network are replaced by other critical resistors as the chemical potential is varied.[3] The fact that by going back and forth with V_g the pattern reproduces itself suggests that there is an underlying backbone structure of sites that are fairly stable, most likely the positions of the ions that control the background potential in the system. This is the analog of the distribution of scattering centers that determine the 'magnetic fingerprints' in the UCF phenomenon.

In both phenomena the underlying structure changes upon thermal recycling, and in both the deviations from perfect reproducibility are associated with non-stationary potential disorder which in turn is believed to be the main cause of the 1/f noise.[19]

The relative magnitude of the CF for a given two-dimensional sample is determined by two factors. The first is related to the basic conductance swing ΔG associated with a fluctuation in the hopping regime, which is of the order of the system conductance G. This is a characteristic feature of the strongly localized regime, (which in the present case of 2D samples mandates a sheet resistance $R_\square$ that fulfills $R_\square \gg \frac{\hbar}{e^2}$). In other words, for a system with size L smaller or equal to the scale relevant for the phenomenon observed, $\Delta G/G$ is of the order unity.[20] This should be contrasted with the 'universal' conductance fluctuation $\Delta G \approx e^2/h$ characterizing the respective situation in weak localization (which holds for $G \gg e^2/h$ and thus $\Delta G/G \ll 1$ is the typical result in this transport regime). This, incidentally is the main reason why the conductance noise and the CF (whether due to an interference effect or an energy-shift mechanism), are much more prominent in the hopping regime than they are in the diffusive one. The second factor that controls the fluctuation amplitude $\Delta G/G$ is how many independent fluctuators contribute to the conductance noise. $\Delta G/G$ is then essentially determined by the square root of the number of critical resistors in the sample $N \approx \frac{L \cdot W}{\mathcal{L}^2} = \frac{A}{\mathcal{L}^2}$ where $\mathcal{L}$ is the percolation radius[15–17] and thus, $\frac{\Delta G}{G} \approx \mathcal{L}A^{-1/2}$. This dependence on the sample area A was confirmed in.[5] A similar dependence on $\mathcal{L}$ was observed for the 1/f noise magnitude,[4] which is consistent with the assumption of a common mechanism for the two phenomena.

It should be noted that $\mathcal{L}$ is disorder-dependent quantity[15–17] and for a fixed temperature, as is the case for the data presented here, the relative amplitude of the CF (as well as the 1/f-noise) increases with disorder. The relative amplitude of the MD also increases with disorder[21] but is independent of the system lateral dimensions. The other feature that clearly distinguishes between these two phenomena is their vastly different temporal behavior as noted above.

To re-iterate, while the phenomenology associated with the CF appears to have several common features with the 1/f noise, the electron-glass effects seem to be controlled by a different spatial scale, and quite a different time scale. Neither the width nor the relative amplitude of the memory-dip change when L is decreased down to the smallest scale used in our experiments. This is in contrast with the behavior of the CF and the 1/f noise that grows as L decreases, making the glassy effects difficult to resolve below a certain sample size.

10.3.3. *Aging experiments with mesoscopic samples*

In addition to the memory-dip discussed above, other characteristic features of the electron glass are observable in the samples that approach the mesoscopic regime. Figure 10.6 illustrates the aging behavior of a 8 by 18μm sample using the 'gate-protocol'.[8,10,22] This protocol involves the following procedure. The sample is allowed to relax for 12 hours at T=4K under a gate voltage V_g^o reaching a near-equilibrium conductance G_0 (which is used as baseline for the aging protocol). Then the gate voltage is quickly changed (typically over 5 seconds) from V_g^o to V_g^n and is held there for t_w after which V_g is re-instated at V_g^o. Figure 10.5 is a plot of the excess conductance (normalized to the equilibrium conductance G_0) measured after V_g is set back to V_g^o. The figure shows such $\Delta G/G$ curves as function of time for different 'waiting-times' t_w (Fig. 10.6a). The same data are plotted as function of t/t_w to illustrate simple-aging behavior (Fig. 10.6b), which is almost identical for the 'mesoscopic' and macroscopic sample (compare with e.g., figure 9 in[10]). The only apparent difference is a more pronounced noise superimposed on the data of the small sample.

It was recently shown by Amir, Oreg, and Imry[23] that this simple form of aging can be accounted for by assuming a broad distribution of relaxation rates (which leads to a logarithmic relaxation[24]) and an additional assumption on the symmetry in dynamics involved in the creation of new MD at V_g^n versus the decay of the old one at V_g^o. The plausibility of these assumptions were discussed in.[25] It should be noted however, that simple aging in this system has been observed to hold even when the above symmetry is *not* obeyed.[22,26]

To summarize, we have shown in this work that the dynamics of the electron-glass has a different time and spatial scale than those associated with the mesoscopic conductance fluctuations. These conclusions are based on the following experimental results:[5]

A) Thermally cycling the sample creates a different pattern of CF while the shape and relative magnitude of the memory-dip remains the same (as long as the average G is not changed by the cycling). B) The CF pattern is fully developed immediately after cooling the sample, while the memory-dip amplitude takes many hours to build. C) Starting from a well-equilibrated sample, continuously sweeping V_g causes the memory-dip amplitude to di-

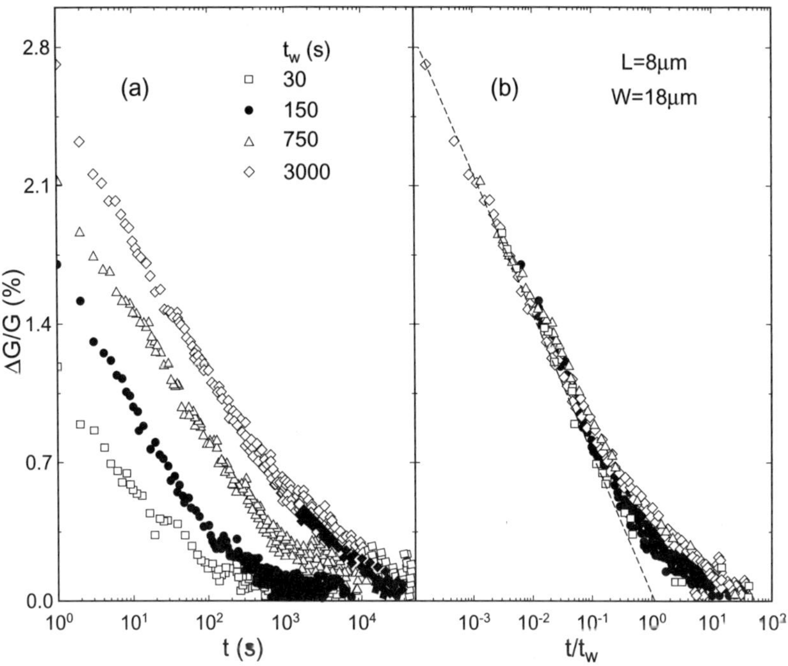

Fig. 10.6. Aging experiment using the gate protocol for a sample with lateral dimensions $L = 8$ μm and $W = 18$ μm and sheet resistance $R_\square = 3.5$ MΩ. T=4.11 K. See text for the protocol details.

minish with a typical time scale of hours while the CF visibility and distinctive fingerprints remain essentially intact for many days.

Inasmuch as the CF pattern is taken as a snapshot of the underlying potential landscape, these findings seem to indicate that the electron-glass dynamics may be treated as essentially independent of the lattice dynamics although in terms of energy exchange (and dissipation) the two systems are coupled. The glassy effects seem to be simply superimposed on the CF and noise without a noticeable interference between these phenomena. For example, Fig. 10.3 demonstrates that the reproducibility of the CF is unaffected by the presence of the memory-dip. This is not a trivial result as there is no reason to assume that the lattice potential is stationary over the time scale associated with the electron glass dynamics. In fact, ions, atoms, and perhaps group of atoms, are presumably mobile even at cryogenic temperatures, thus producing potential fluctuations.[27] These are believed to be the main source of the 1/f noise which includes time scales that overlap those typical of the electron glass dynamics. It is therefore natural to expect that the two phenomena affect one another to some degree. For example, one intuitively expects that the conductance noise will be modified when the system is relaxing from an excited state (or, more generally, when it is taken out of equilibrium). In the present case however, where ΔG in the excited state is typically only few percents of the equilibrium conductance, this effect, is yet to be detected.[28]

An attempt to find a change in the 1/f noise characteristics when the system is deliberately taken out of equilibrium yielded a null result; Such an experiment was carried out by keeping the sample in an agitated state for a sufficiently long time to allow adequate averaging of the noise spectrum.[29] This steady-state, out-of-equilibrium situation was achieved by a continuous sweep of the gate voltage over a certain voltage interval. Data for G(t) were taken after the memory-dip was erased by allowing this pro-

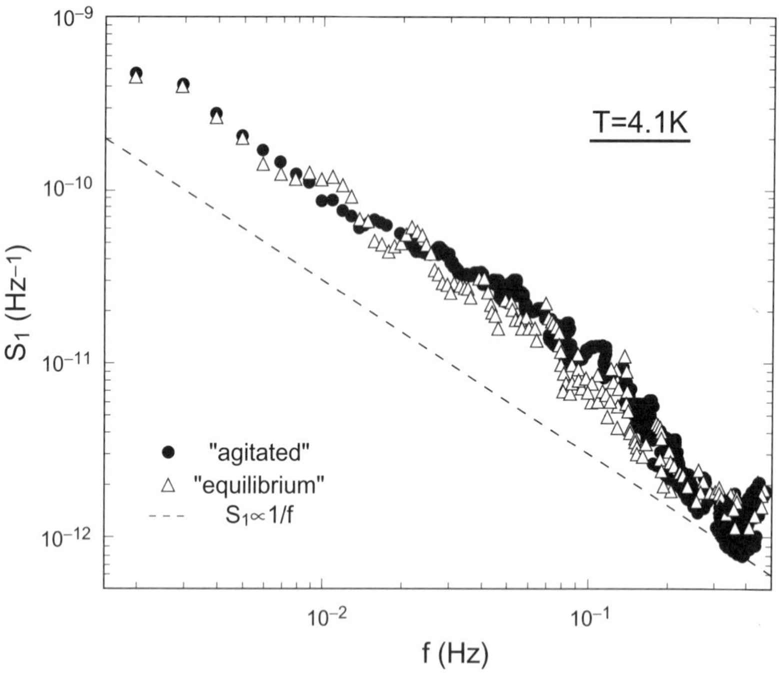

Fig. 10.7. Comparison of the noise power spectrum for the 'agitated' versus 'equilibrium' state of a typical electron glass sample (see text for details). The In_2O_{3-x} sample used here has $R_\square$=17.6MΩ (at 4.1K), thickness d=47Å, and lateral dimensions of LxW=1x1mm^2.

cess to proceed for 10 days. The noise characteristics were compared with those taken with a fixed-gate situation for which data were taken after the sample was allowed to relax (and build a memory-dip). In either case the noise spectrum was obtained by Fourier transforming 25 conductance time-traces each composed of 1024 points taken with 1 second resolution. Figure 10.7 shows the results of a specific experiment that deliberately employed a large-size sample (to minimize mesoscopic effects in the gate-voltage scans). Evidently, within the error of the procedure,[30] the 1/f noise magnitude and spectrum are essentially the same for the 'equilibrium' and the 'agitated' sample. This suggests that the main contribution to the 1/f noise magnitude is *not* related to the glassy nature of the electronic relaxation. Although, some contribution to the 1/f noise from the slow electronic dynamics seems inevitable; it is apparently too small relative to the noise due to ions/atoms dynamic. Indeed, the magnitude of the 1/f noise in these systems increases with disorder and becomes quite large already in the *diffusive* regime, and at room temperatures[30] where no electron-glass effects can be observed.

Finally, throughout the size-range where the memory-dip could be resolved over the background of the CF and the 1/f noise, its amplitude and characteristic width do not depend on sample size. Likewise, the dynamics associated with conductance relaxation is independent of sample size down to 2μm.[5] These findings suggest that the relevant length scale for the electron glass (at T$\simeq$4K) is considerably smaller than the sample sizes used in these studies.

Acknowledgment

Useful discussion with Y. Imry, A. Efros, and M. Pollak are gratefully acknowledged. This research was supported by a grant administered by the US Israel Binational Science Foundation and by the Israeli Foundation for Sciences and Humanities.

References

1. Yoseph Imry, *"Introduction to Mesoscopic Physics"*, Oxford University Press (1997).
2. S. Washburn and R. A. Webb, Reports on Prog. in Phys., **55**, 1311 (1992), and references therein.
3. A. B. Fowler, A. Hartstein, and R. A. Webb, Phys. Rev. Lett., **48**, 196 (1982); W. J. Skocpol, L. D. Jackel, E. L. Hu, R. E. Howard, and L. A. Fetter, Phys.

Rev. Lett. **49**, 951(1982); R. F. Kwasnick, M. A. Kastner, J. Melngailis, and P. A. Lee, Phys. Rev. Lett. **52**, 224 (1984).
4. V. Orlyanchik, V. I. Kozub, and Z. Ovadyahu, Phys. Rev. B **74**, 235206 (2006).
5. V. Orlyanchik, and Z. Ovadyahu, Phys. Rev. B **75**, 174205 (2007).
6. J. Delahaye, T. Grenet and F. Gay, Eur. Phys. J. B **65** ,5 (2008).
7. M. Ben Chorin, Z. Ovadyahu and M. Pollak, Phys. Rev. B **48**, 15025 (1993); G. Martinez-Arizala, D. E. Grupp, C. Christiansen, A. Mack, N. Markovic, Y. Seguchi, and A. M. Goldman, Phys. Rev. Lett., **78**, 1130 (1997). G. Martinez-Arizala, C. Christiansen, D. E. Grupp, N. Markovic, A. Mack, and A. M. Goldman, Phys. Rev. B **57**, R670 (1998); T. Grenet, Eur. Phys. J, **32**, 275 (2003).
8. A. Vaknin, Z. Ovadyahu, and M. Pollak, Phys. Rev. B **65**, 134208 (2002); A. Vaknin, Z. Ovadyhau, and M. Pollak, Phys. Rev. Lett., **84**, 3402 (2000).
9. A. Vaknin, Z. Ovadyhau, and M. Pollak, Phys. Rev. Lett., **81**, 669 (1998).
10. Z. Ovadyahu, and M. Pollak, Phys. Rev. B **68**, 184204 (2003).
11. M. Ben-Chorin, D. Kowal and Z. Ovadyahu, Phys. Rev. B **44**, 3420 (1991).
12. A. I. Larkin and B. I. Shklovskii, phys. stat. sol. (b) **230**, 189 (2002).
13. P. A. Lee, Phys. Rev. Lett., **53**, 2042 (1984).
14. W. J. Skocpol, P. M. Mankiewich, R. E. Howard, L. D. Jackel, and D. M. Tennant, Phys. Rev. Lett. **56**, 2865 (1986).
15. V. Ambegaokar, B. I. Halperin, J. S. Langer, Phys. Rev. **B4**, 2612 (1971)
16. B. I. Shklovskii, A.L. Efros, Sov. Phys. JETP **33**, 468 (1971).
17. M. Pollak, J. Non-Cryst. Solids **11**, 1 (1972).
18. A. Miller and E. Abrahams, Phys. Rev. **120**, 745 (1960).
19. P. A. Lee and A. D. Stone, Phys. Rev. Lett. **55**, 1622 (1985); S. Feng, P. A. Lee, and A. D. Stone, Phys. Rev. Lett. **56**, 1960 (1986).
20. Y. Imry, Europhys. Lett. **1**, 249 (1986); F. P. Milliken and Z. Ovadyahu, Phys. Rev. Lett., **65**, 911 (1990); Z. Ovadyahu, Waves in Random Media, **9**, 241 (1999) and references therein.
21. M. Pollak and Z. Ovadyahu, J. de Physique I France, **7**, 1595 (1997).
22. Z. Ovadyahu and M. Pollak, Phys. Rev. Lett., **79**, 459 (1997).
23. Ariel Amir, Yuval Oreg, and Yoseph Imry, Phys. Rev. Lett. **103**, 126403 (2009). See also: T. Grenet, J. Delahaye, M. Sabra, and F. Gay, Eur. Phys. J. B **56**, 183 (2007).
24. A. Vaknin, Z. Ovadyahu, and M. Pollak, Phys. Rev. B 61, 6692 (2000); Ariel Amir, Yuval Oreg, and Yoseph Imry, Phys. Rev. **B** 77, 165207 (2008).
25. M. Pollak, and Z. Ovadyahu, phys. stat. sol. (c) **3**, 283 (2006).
26. V. Orlyanchik and Z. Ovadyahu, Phys. Rev. Lett., **92**, 066801 (2004).
27. Movements of these objects, commonly referred to as two-state-systems, may be triggered by electronic transitions such as changes in bonding angle. Therefore these are as much 'electronic' as they are 'atomic'. The only distinction we make is based on whether the objects actually contribute to macroscopic dc current or just modulate it. The possible role of ions movement and its relation to the electron glass dynamics was discussed before vis-a-vis the dependence of the dynamics on disorder and magnetic field.[22]

28. Note that re-normalization of hopping rates due to the coupling of electrons to two-state-systems may still be at work and that may affect e.g., the relaxation rate of ΔG see; Brage Golding, M. Neil Zimmerman, and S. N. Coppersmith, Phys. Rev. Lett. **68**, 998 (1992); F. Pérez-Willarda, C. Surgersa, H. Von Löhneysen, P. Pfundstein, Physica E, **22**, 872 (2004); Z. Ovadyahu, Phys. Rev. B **73**, 214208 (2006); Phys. Rev. Lett., **99**, 226603 (2007).
29. Note that the act of sweeping the gate voltage while collecting G(t) data for the noise analysis may also actually contribute to the 'agitated' noise curve (due to 'instrumnetal' reasons). This contribution is probably also overshadowed by the high level of intrinsic 1/f noise.
30. The relaxation of ΔG in a 'one-shot' relaxation experiment does not leave ample time for a meaningful measurement of 1/f-noise while the system is in the 'excited' state; due to the logarithmic law most of the excess conductance, which is small anyhow relative to the equilibrium value, is gone by $\Delta t \approx 100$-500sec. This makes it hard to perform an accurate measurement of the low frequency noise.
31. O. Cohen, Z. Ovadyahu, and M. Rokni, Phys. Rev. Lett., **69**, 3555 (1992); O. Cohen and Z. Ovadyahu, Phys. Rev. B **50**, 10442 (1994).

Chapter 11

Electron Coherence and Dephasing in Metal Films, Wires and Rings

Daniel E. Prober

Departments of Applied Physics and Physics, Yale University
New Haven, CT 06520-8284 USA
E-mail: daniel.prober@yale.edu

The fields of weak localization, quantum interference and dephasing in disordered metal microstructures underwent dramatic advances in the past three decades. We describe the earlier advances, focusing on those at Yale in which Prof. Imry had significant input. We also present a novel physical picture to understand the essential mechanics of dephasing by Nyquist noise.

11.1. Introduction

In the late 1970s and early 1980s, the understanding of electron transport in disordered metal microstructures advanced dramatically. Electron interference effects were predicted and observed. These initially provided theoretical challenges, and also challenged our intuition regarding dissipation and conduction processes. The theoretical understanding developed at that time[1] is quite complete. The early experiments benefitted significantly from discussions between theorists and experimentalists. Much of the initial theory was done in the USSR, and therefore the discussions were not as rapid and personal as in modern times. Professor Imry, then at Tel Aviv University and at IBM, had a significant hand in guiding some of our ideas at Yale University. We therefore present here a brief review of some of that early work. This may also serve as a brief introduction for beginning students.

11.2. Weak Localization and the Aharonov-Bohm Effect

Weak localization is an effect that was first associated with one-dimensional (1d) wires. The prediction by D. Thouless[2] in 1977 that 1d metal wires would all be insulators at a sufficiently low temperature shocked some of us, and stimulated experiments. The first such experiments were published in 1979[3,4] and showed that 1d effects definitely occurred in wires approximately ≤100 nm in diameter,[3] but that 2d weak localization (WL) occurred in wider wires, of order a micron in width.[4] Weak localization is the high temperature precursor of the insulating state predicted for T = 0. WL arises from electron partial waves which can backscatter and travel along time-reversed paths to return to the point of origin, where they interfere constructively (in the simplest case) - see Fig. 11.1b. Since the pathlengths are identical for the time-reversed pairs, the interference is constructive.

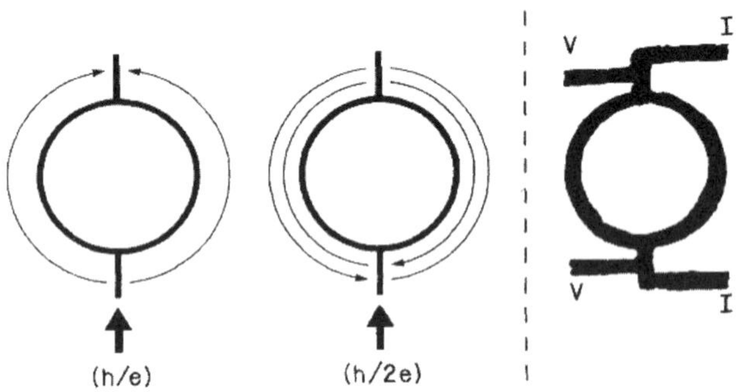

Fig. 11.1. Paths for h/e and h/2e AB interference effects in metal rings; micrograph of a ring; from Ref. 10.

The likelihood of returning to the origin is higher in this coherent case than for the case of classical diffusion, where there is no phase present. Thus, in this quantum coherent case, the electron in Fig. 11.1b gets somewhat stuck at the point of origin. It is weakly 'localized'. A magnetic field couples to and increments the phases of the electron waves, through the magnetic vector potential. For a ring, this gives the Aharonov-Bohm (AB) effect. Subsequent studies of the magnetoresistance of wires[5,6] related more explicitly the effects due to weak localization to the processes by which the waves lose their phase, called dephasing.

The physical understanding of the weak localization mechanism was advanced by the experimental work of Sharvin and Sharvin.[7] They observed the Aharonov-Bohm interference effect in a metal cylinder with thin walls, with a magnetic field oriented along the length of the cylinder. This striking result made clear that electron interference was directly observable, and this led to the search for interference effects in single rings, of diameter order of 1 μm. The AB interference they observed had a magnetoresistance which oscillated with field, with a flux period of $\Delta\phi = h/2e$, with h Planck's constant; this oscillation with field is due to a backscattering process, Fig. 11.1b, where the magnetic flux is encircled twice by the two partial waves.

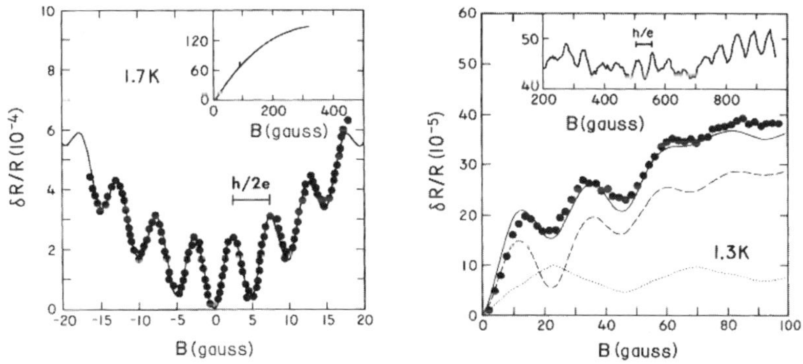

Fig. 11.2. Left: Magnetoresistance (MR) of Al ring showing interference effect with flux period h/2e; inset: high field MR with different scales; Right: MR of Ag ring, at low and high field. At low field, the dashed line is the contribution due to h/2e interference, and the light dotted line that due to h/e AB effect; inset: at high field, only the h/e effect is observed; from Ref. 10.

In 1985 Webb et al. observed interference in a single ring that corresponded to the forward-scattering interference effect.[8] The conductance varied in a periodic fashion with magnetic flux, with a flux period $\Delta\phi = h/e$; see Fig. 11.1a. This effect is analogous to the 'universal conductance fluctuations' [9] also seen at low temperatures, in short metal wires. In the wires, the forward scattering effect is present, but there is no single area, as in the ring, which defines a unique flux period for the magnetoresistance. This forward scattering effect is different from the interference effect that gives backscattering and which gives rise to weak localization. This is because for the forward scattering effect, the two waves never travel on paths that are identical.

Research at Yale in 1985 on micron-size rings[10] demonstrated both the h/e effect and the backscattering effect that is involved in weak localization, shown in Fig. 11.1b, with magnetic flux period $\Delta\phi = h/2e$. In Fig. 11.1a, many pairs of paths add up incoherently (they have random phases) to give the forward-scattering effect. In Fig. 11.1b, only the few special, time-reversed pairs of paths add, but these add coherently, having the same phase vs. magnetic field, because the path lengths of the two time-reversed paths are the same. The data for these rings is shown in Fig. 11.2. The h/e effect is dominant only at high field; at low field, the h/2e interference due to backscattering is dominant. The relative magnitudes depend on the temperature as well as the field.

11.3. Dephasing Mechanisms in 1d and 2d Metallic Systems

The size scale of the rings studied[8,10] was set by the electron phase-coherence length. At *very* low temperatures and energies, the electron wave propagates diffusively due to elastic (impurity) scattering, staying in a state of constant energy. However, at finite temperature, interactions with other electrons and with the lattice lead to loss of phase memory, with a temperature-dependent time scale τ_φ and a length scale $(D\tau_\varphi)^{1/2}$; D is the diffusion constant. Phonon scattering and electron-electron scattering both contribute to the dephasing rate τ_ϕ^{-1}. Dephasing occurs when the pairs of paths that give interference (see Fig. 11.1) either lose their individual phase memories, or the different pairs of paths change their relative phase relations as a function of time (pairs relative to one another), so that dc interference no longer is observed. Research at Yale established that the dephasing rate in the 2d metal films[11] and 1d wires[12] is given by the sum of the electron-phonon (el-ph) process and the electron-electron (el-el) process,

$$\tau_\varphi^{-1} = \tau_{\varphi,\text{el-ph}}^{-1} + \tau_{\varphi,\text{el-el}}^{-1}. \tag{11.1}$$

The electron-electron inelastic scattering that gives dephasing is due to a new mechanism, Nyquist noise. This is different from the inelastic electron-electron scattering with *large* energy exchanges, which in clean metals has a temperature dependent rate $\approx T^2$ with exchanges of energy of order k_BT. Nyquist noise dephasing is much stronger in dirty (diffusive) metals than the dephasing due to electron-electron scattering with large energy exchange.

The ideas on dephasing mechanisms grew out of theory developed to understand weak localization. The mechanisms of dephasing described here are indeed all due to inelastic processes. However, the inelastic time is the characteristic time for an electron to lose energy (equal to k_BT). The inelastic time is, in general, different from the dephasing time.[1]

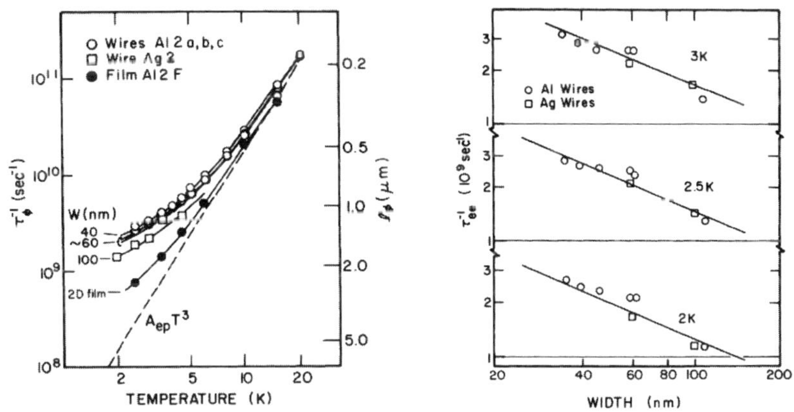

Fig. 11.3. Left: Phase-breaking rate vs. temperature; the electron-phonon contribution, with rate $\approx T^3$, is shown as a dashed line. The electron-electron contribution to the total dephasing rate, $\tau_{\varphi,\text{el-el}}^{-1}$, is shown on the right. Solid lines are theory;[1] from Ref. 12.

We also investigated at Yale the new mechanisms causing dephasing. We therefore measured[12] the weak-localization magnetoresistance of Ag and Al wires vs. temperature, for wires of various widths. The magnetoresistance was analyzed to deduce τ_φ. We extracted $\tau_{\varphi,\text{el-ph}}$ from the data for wide wires at high temperatures. Using Eq. 11.1, we obtained

$\tau_{\varphi,el\text{-}el}$. The results are plotted in Fig. 11.3. The fit to the theory predictions is excellent, for both the magnitude and for the dependence on temperature and wire width.

Recent research in this field has characterized the persistent currents that flow in individual rings at very low temperatures. These persistent currents were first seen in individual rings by Chandrasekhar et al. at IBM,[13] but the magnitude has remained puzzling. Recent work at Stanford[14] and Yale[15] has advanced the understanding, and recent theory[16] points to possible resolution of the puzzling magnitude of the observed effect. Small quantities of magnetic scatterers may play an important role in the experiments. Challenges remain open for future experiments, and this field continues to pose interesting questions.

11.4. Dephasing - a Simple Pedagogical Model

Dephasing of electron waves in metals was not a significant research topic prior to the concepts of weak localization and associated interference effects. Measurements at Yale have confirmed in detail the novel theoretical predictions for electron-electron dephasing in 2d films and 1d wires, discussed above. These experiments demonstrate that at low temperatures, electron-electron dephasing (by Nyquist-Johnson noise) dominates. At higher temperatures, typically above a few K, electron-phonon inelastic scattering dominates. We present below a physical picture that argues for the *plausibility* of the dephasing due to Nyquist noise, which causes a fluctuating electric field **E** in the metal. Our argument is not intended to make a quantitative prediction. Rather, it is intended to be a pedagogical introduction to the rationale for dephasing, and to explain how finite-frequency noise can cause dephasing.

Dephasing is caused by energy exchange with the environment that is different for the two partial waves. We consider the simplest case of a metal film. We draw in Fig. 11.4 the paths of the two electron partial waves, paths 1 and 2, as direct, with one scattering event along each path. Even though the real motion is diffusive, this simplification still captures the essential ideas. The two partial waves originate (split apart) at point A. The energy on path 1 is given by $E_1 = hf_1$. If there is no change in the energies of the two partial waves when they travel to point B where they recombine, they will interfere at this exit point, B. For a given pair of paths the interference can have any phase, but that phase will be constant in time. However, if E_1 changes during propagation from point A to point B, and E_2 is constant, then the two partial waves 1 and 2 arrive at point B having different frequencies, and cannot have dc interference at point B. Dephasing has occurred due to the change of energy E_1 (and frequency f_1).

If the two waves in Fig. 11.4 accumulate a significantly phase difference along the two paths, dephasing will occur. The typical time to traverse from A to B is τ_φ. The phase difference, for a frequency difference of Δf, is $\Delta\varphi \approx (\Delta f\ \tau_\varphi)$. A phase difference of $\Delta\varphi = \pi$ is significant for dephasing. This amount $\Delta\varphi = \pi$ would be accumulated if the two energies differ only by the small amount $\Delta E = h\ \Delta f \approx \pi\ h\ \tau_\varphi^{-1} << k_B T$. The actual phase difference changes vs. time, since we will find below that ΔE varies in time. Thus, dc interference is destroyed. The inequality $\Delta E << k_B T$ in this result means that *small* energy exchanges, $<< k_B T$, can fully dephase the two partial waves; large energy exchanges ($\approx k_b T$) are not necessary to cause dephasing. Multiple small energy exchanges of even smaller energies also dephase.

Energy exchange can occur due to phonon emission or absorption, typically of large energy $\approx k_B T$. But fluctuating electric fields can also dephase the electron waves. Consider a time dependent **E** field in the direction shown in Fig. 11.4, with a single oscillation during the time to propagate from A to B. (Typically, the relevant time scale is the dephasing time, τ_φ, determined self-consistently.) On path 2, the energy E_2 is decreased (compared to zero electric field) since **E** and the velocity $\mathbf{v}_2$ are in the same direction during the first half of the cycle, when $\mathbf{v}_2$ is parallel to **E.** (This causes a **de**crease of energy because the charge is negative.). During the second part of the cycle, in contrast, $\mathbf{v}_2$ is orthogonal to **E**; no energy is added to the wave. On path 1, the energy E_1 is increased, since **E** and $\mathbf{v}_1$ are in opposite directions during the second part of the cycle, and the charge is negative. During the first half of the cycle, no energy is added to that wave. Thus, a significant ΔE and frequency difference, Δf, can occur between the two waves. Because **E** fluctuates in time, the phase difference will also fluctuate for those two paths. The time-dependent **E** field waveform shown will thus dephase the two electron partial waves. A dc electric field will not dephase, by the same argument. Nyquist noise contains components at all frequencies. The most effective frequencies are $f \approx \tau_\varphi^{-1}$.

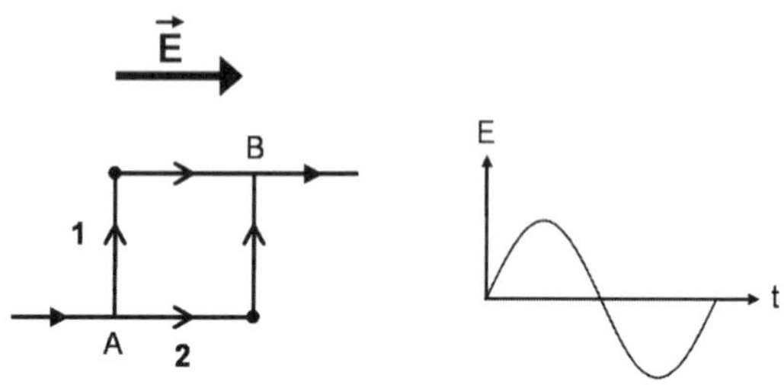

Fig. 11.4. Model of dephasing by a time-varying electric field **E.** An electron enters from left, and two partial waves split apart at point A. These two waves traverse the paths 1 and 2, and then recombine at point B. Dc interference can occur if the two frequencies are the same. The time-varying **E** field shown will cause a time varying phase shift between the two paths. The overall conductance is due to many such pairs of interfering paths, and the **E** field shown will cause different phase shifts for the *different* pairs of paths, leading to loss of dc interference.

11.5. Conclusions

Experiments and theory for electron transport in disordered metal films, wires, and rings have led to a deep understanding of issues of coherence and dissipation in disordered conductors, and revealed new and unexpected interference effects.[17] Dephasing by Nyquist noise was one of the new effects, and plays a key role in all the experiments. We have presented a 'toy' model of this effect which captures the essence of Nyquist dephasing. It does not predict the temperature dependence of the dephasing rate, nor how this rate depends on sample dimension and material parameters. Happily, the full quantitative theory[1] is available to predict all these. The predictions have been carefully tested in our experiments, with very satisfying confirmation of this novel and important theory. Significant further development, and the establishment of mesoscopic physics as an important area of research and understanding, has occurred in the subsequent decades.[17]

Acknowledgments

We thank the many colleagues who have participated in the research with the author at Yale, including N. Giordano, P. Santhanam, V. Chandrasekhar, S. Wind, and M.J. Rooks in the early research described here. In that work we benefitted greatly from theoretical advice from many colleagues, especially Y. Imry and A.D. Stone. We thank B. Reulet, D. Santavicca and J. Chudow for advice on this chapter. The research at Yale was supported by NSF-DMR.

References

1. B.L. Altshuler and A.G Aronov, Electron-Electron Interaction in Disordered Conductors, in *Electron-Electron Interactions in Disordered Systems*, edited by A.L. Efros and M. Pollak, (Elsevier, Amsterdam, 1985).
2. D.J. Thouless, Maximum Metallic Resistance in Thin Wires, *Phys. Rev. Lett.* **39**, 1167 (1977).
3. N. Giordano, W. Gilson, and D.E. Prober, Experimental Study of Anderson Localization in Thin Wires, *Phys. Rev. Lett.* **34**, 725 (1979).
4. G. J. Dolan and D. D. Osheroff, Nonmetallic Conduction in Thin Metal Films at Low Temperatures, *Phys. Rev. Lett.* **43**, 721 (1979).
5. A.E. White, M. Tinkham, W.J. Skocpol, and D.C. Flanders, Evidence for Interaction Effects in the Low-Temperature Resistance Rise in Ultrathin Metallic Wires, *Phys. Rev. Lett.* **48**, 1752 (1982); J.M. Gordon, C.J. Lobb and M. Tinkham, Electron inelastic lifetime and electron-electron attraction strength in Al films, *Phys. Rev.* **B28**, 4046 (1983).
6. P. Santhanam, S.J. Wind, and D.E. Prober, One-Dimensional Electron Localization and Superconducting Fluctuations in Narrow Aluminum Wires, *Phys. Rev. Lett.* **53**, 1179 (1984).
7. D. Yu. Sharvin and Yu. V. Sharvin, Pis'ma Zh. Teor. Eksp. Fiz. 34, 285 (1981); [JETP Lett. 34, 272 (1981).
8. R.A. Webb, S. Washburn, C.P. Umbach, and R.B. Laibowitz, Observation of h/e Aharonov-Bohm Oscillations in Normal-Metal Rings, *Phys. Rev. Lett.* **54**, 2696 (1985).
9. P.A. Lee and A.D. Stone, Universal Conductance Fluctuations in Metals, *Phys. Rev. Lett.* **55**, 1622 (1985).

10. V. Chandrasekhar, M.J. Rooks, S. Wind, and D.E. Prober, Observation of Aharonov-Bohm Electron Interference Effects with Periods h/e and h/2e in Individual Micron-Size, Normal-Metal Rings, *Phys. Rev. Lett.* **55**, 1610 (1985).
11. P. Santhanam and D.E. Prober, Inelastic electron scattering mechanisms in clean aluminum films, *Phys. Rev.* **B29**, 3733 (1984).
12. S. Wind, M.J. Rooks, V. Chandrasekhar, and D.E. Prober, One-Dimensional Electron-Electron Scattering with Small Energy Transfers, *Phys. Rev. Lett.* **57**, 633 (1986).
13. V. Chandrasekhar, R. A. Webb, M.J. Brady, M.B Ketchen, W.J. Gallagher, and A. Kleinsasser, Magnetic Response of a Single, Isolated Gold Loop, *Phys. Rev. Lett.* **67**, 3578 (1991).
14. H. Bluhm, N. C. Koschnick, J. A. Bert, M. E. Huber, K. A. Moler, Persistent Currents in Normal Metal Rings, *Phys. Rev. Lett.* **102**, 136802 (2009).
15. A. C. Bleszynski-Jayich, W. E. Shanks, B. Peaudecerf, E. Ginossar, F. von Oppen, L. Glazman, J. G. E. Harris, Persistent Currents in Normal Metal Rings, *Science* **326**, 272 (2009).
16. H. Bary-Soroker, O. Entin-Wohlman, and Y. Imry, Effect of Pair-Breaking on Mesoscopic Persistent Currents Well Above the Superconducting Transition Temperature, *Phys. Rev. Lett.* **101**, 057001 (2008).
17. Y. Imry, Introduction to Mesoscopic Physics, Second Edition (Oxford, 2002); Eric Akkermans and G. Montambaux, Mesoscopic Physics of Electrons and Photons, (Cambridge, 2006).

Chapter 12

The Third Moment of Current Fluctuations in a Tunnel Junction: Experiments in the Classical and Quantum Regimes

B. Reulet, J. Gabelli

Laboratoire de Physique des Solides, CNRS UMR8502, Université de Paris-Sud 11, Orsay, France

L. Spietz* and D. E. Prober

Department of Applied Physics, Yale University, New Haven CT, USA

We report the first experimental data of the third moment of current fluctuations in a tunnel junction. We show that both in the classical and quantum regimes (low or high frequency as compared to voltage), it is given by $S_{I^3} = e^2 I$. We discuss environmental effects in both regimes.

12.1. Introduction

In a seminal article,[1] Levitov and Lesovik asked themselves the following, simple question: what is the probability that n electrons cross a coherent conductor in a time τ ? The final answer came three years later, by a quantum mechanical treatment of the electronic wavefunctions, of the scattering by the sample and even of the charge detector.[2] Despite a fully quantum approach, the result of the calculation was quite disappointing: electrons behave as classical particles. At zero temperature, electrons try to enter the device at a pace eV/h, and for each attempt they succeed with a probability $p = |t|^2$, where t is the transmission amplitude of the electronic wave through the scattering region (e is the electron charge and h Planck's constant). The distribution is binomial, $P(n, \tau) = C_N^n p^n (1-p)^n$ where $N = \tau eV/h$ is the total number of attempts and $C_N^n = N!/[n!(N-n)!]$. Only the long time limit ($N \gg 1$) was considered. The result of ref.[2] implies non-Gaussian current fluctuations in the sample, the simplest measure of this non-Gaussian aspect being given by the existence of a third moment

*Present address: NIST, Boulder CO, USA.

of the transferred number of particles: $\langle \delta n^3 \rangle = p(1-p)(1-2p)N$, with $\delta n = n - \langle n \rangle$. This reflects the asymmetry of the fluctuations around the mean value: the distribution is skewed, one direction is preferred. In the following we will only consider tunnel junctions, which are the simplest system that exhibit shot noise, and for which the second moment has been thoroughly studied.[3]

12.2. The Third Moment in the Classical Regime

For a tunnel junction one has $p \ll 1$ and the probability distribution $P(n, \tau)$ is Poissonian, with $\langle \delta n^3 \rangle = \langle \delta n^2 \rangle = \langle n \rangle$. This translates into current fluctuations having a third moment spectral density $S_{I^3} = e^2 I$ where $I = \langle I \rangle = \langle n \rangle e/\tau$ is the dc current (the spectral density of the variance is $S_{I^2} = eI$). The relation between the fluctuations of the current integrated in the bandwidth from f_1 to f_2 and the spectral density is non-trivial: $\langle \delta I^2 \rangle = 2(f_2 - f_1)S_{I^2}$ (the factor 2 come from positive and negative frequencies) and $\langle \delta I^3 \rangle = 3B^2 S_{I^3}$ with $B = f_2 - 2f_1$ if $f_2 > 2f_1$ and $B = 0$ otherwise (we have confirmed this unusual dependence of the measurement on the frequency cutoffs[4]). As a consequence, the relation between the measured current distribution and the real one is very subtle.[6] We have experimentally confirmed the prediction $S_{I^3} = e^2 I$, see.[4] We show on Fig.

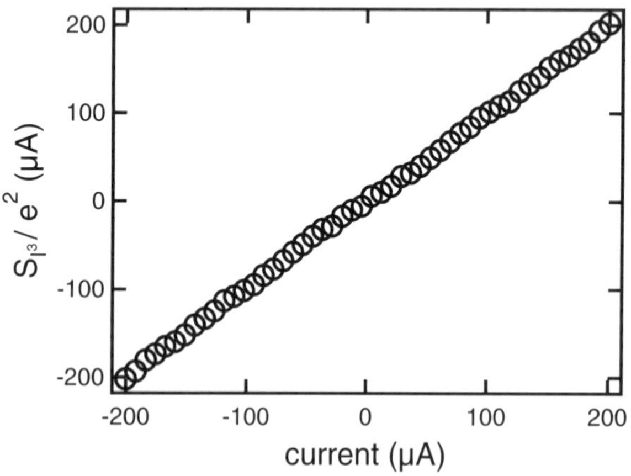

Fig. 12.1. Normalized third moment of current fluctuations S_{I3} vs. dc current measured at temperature $T = 4.2$ K and low frequency ($0.01 - 1$ GHz). It is obtained from the top panel of Fig. 12.2 after subtraction of the environmental contributions, see Eq. (12.1).

12.1 the result for a tunnel junction measured in the bandwidth 0.01-1 GHz at temperature $T = 4.2$ K. At equilibrium ($I = 0$), positive and negative current fluctuations are equiprobable, so $S_{I^3}^{eq} = 0$. Subsequent work has confirmed our result.[7]

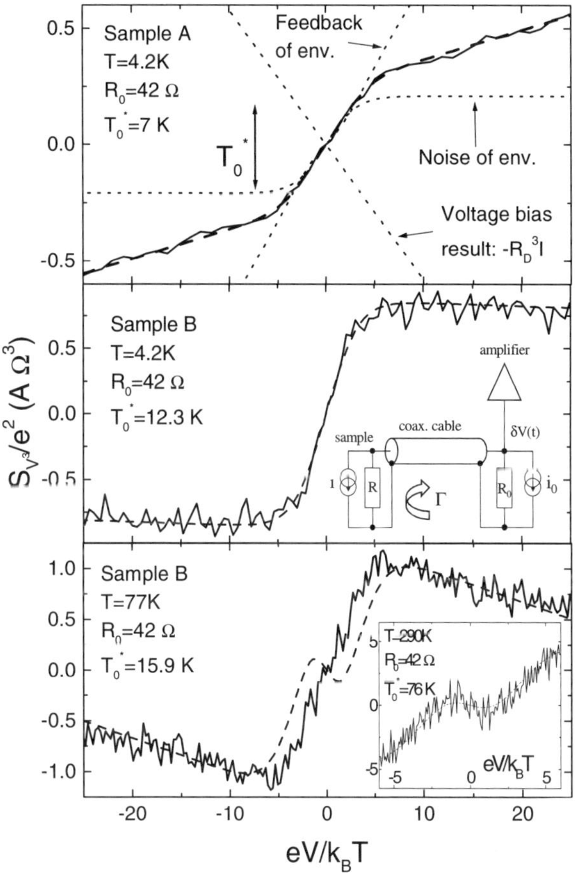

Fig. 12.2. Measurement of $S_{V^3}(eV/k_BT)$ (solid lines). The dashed lines corresponds to the best fit with Eq. (12.1). The dotted lines in the top plot correspond to the different contributions to S_{V^3} (see text). Inset of the middle plot: schematics of the measurement circuit. The effective noise temperature of the environment T_0^* is equal to the real noise temperature times the reflection coefficient Γ of the sample when the cable between the sample and the amplifier is long (which was not the case at room temperature).

Finite temperature affects the transport, so the probability for charge transfer is temperature dependent. In a tunnel junction, the spectral density of the variance (second moment) of current fluctuations is given by: $S_{I^2} = GeV \coth eV/(2k_BT)$, with T the temperature and G the sample's conductance. Noise vs. voltage evolves continuously from the equilibrium, Johnson-Nyquist noise $S_{I^2}^{eq} = 2k_BTG$ at $V = 0$, to the full shot noise $S_{I^2}^{shot} = eI$ at large voltage, $V \gg k_BT/e$ (for a review on S_{I^2}, see[8]). Strikingly, S_{I^3} has a very different behavior: it is totally temperature independent, i.e. thermal fluctuations of current are symmetric. Thus S_{I^3} can be measured at high temperature and/or low voltage to reveal the charge of the carriers, contrary to S_{I^2} which requires $eV > k_BT$. We have measured S_{I^3} from 300 K down to 50 mK and indeed found no temperature dependence.[4,5] We show on Fig. 12.2 measurements at 300 K, 77 K and 4.2 K on different samples. They all indicate $S_{I^3} = e^2I$ independent of temperature, despite the environmental contributions which do depend on temperature (see below).

12.3. Environmental Effects in the Classical Regime

Usual theories consider the fluctuations of charge transfer at a fixed bias voltage. This rarely describes an experimental situation: voltage or current sources have a finite output impedance and voltage/current amplifiers a finite input impedance, in particular at high frequency. This usually does not matter for average quantities or variances: the average current under voltage bias is $\langle I \rangle_V = GV$, the average voltage under current bias is $\langle V \rangle_I = RI$, and the conductance G and the resistance R are inverse of each other: $R = 1/G$. Similarly, voltage fluctuations under current bias (i.e., sample is open at ac) and current fluctuations under voltage bias (i.e., sample is short-circuited at ac) are proportional, $\langle \delta I^2 \rangle_V = G^2 \langle \delta V^2 \rangle_I$, or in terms of spectral densities, $S_{I^2} = G^2 S_{V^2}$. Finite source or amplifier impedances lead to a simple rescaling by a resistance ratio, and the noise of the source or amplifier adds incoherently to that of the sample: $S_{V^2} = R_D^2(S_{I^2} + S_{i_0^2})$ where $R_D = RR_0/(R + R_0)$ is the parallel combination of the sample's and the environmental resistances and $S_{i_0^2}$ is the noise spectral density of the amplifier, for the circuit depicted in the middle inset of Fig. 12.2.

For the third and higher order moments, the story becomes more complicated: in the presence of a finite external impedance, the voltage across the sample fluctuates, which modifies the probability of charge transfer.[9,10] There are two contributions to the voltage fluctuations: noise of the environment (Johnson noise of the external impedance or current noise of the amplifier, i_0), and current fluctuations of the sample itself that are con-

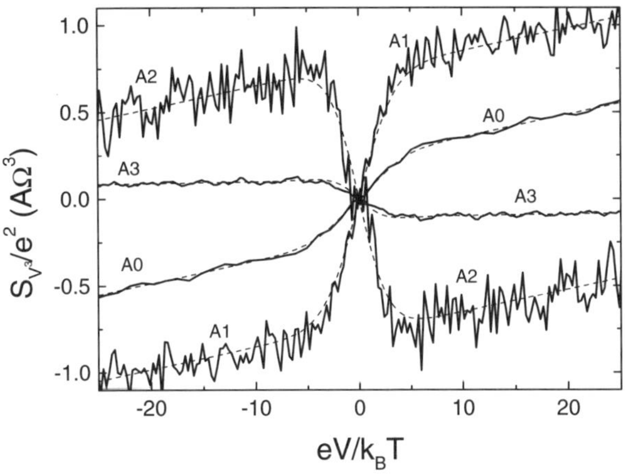

Fig. 12.3. Measurement of $S_{V^3}(eV/k_BT)$ for sample A at T=4.2 K (solid lines). A0: no ac excitation (same as Fig. 1, top panel). A1: with an ac excitation at frequency $\Omega/2\pi$ such that $\cos 2\Omega\Delta t = +1$; A2: $\cos 2\Omega\Delta t = -1$. The ac excitation at frequency Ω simulates the noise of the amplifier sent to the sample, and demonstrates the influence of the environmental noise. The frequency dependence with the delay time Δt explains why T_0 has to be replaced by T_0^* in the environmental contribution. A3: no ac excitation but a 63 Ω resistor in parallel with the sample. This demonstrates the effect of the environmental impedance. The dashed lines corresponds to fits with Eq. (12.1). A1 and A2 differ from A0 only by a term $\propto dS_{I^2}/dV$. A3 is deduced from A0 with no fitting parameter.

verted into voltage fluctuations by the external impedance R_0. The first environmental contribution to S_{I^3} expresses how the amount of noise generated by the sample (the variance S_{I^2}) is modulated by a small fluctuating voltage $R_D i_0$, and is $\propto R_D S_{i_0^2} dS_{I^2}/dV$. It disappears if the environment is perfectly quiet ($i_0 = 0$), but exists even if the external impedance is zero. In contrast in the second contribution, the feedback of the environment, the noise source is the sample itself, so it does not depend on the environmental noise temperature, but the environmental impedance converts the sample current fluctuations into voltage fluctuations, resulting in a contribution to S_{I^3} that is $\propto R_D S_{I^2} dS_{I^2}/dV$. As a result, the spectral density of the third moment of voltage fluctuations measured across the resistance R_0 in parallel with the sample is given by:

$$S_{V^3} = R_D^3 \left[-S_{I^3} + 3R_D S_{i_0^2} \frac{dS_{I^2}}{dV} + 3R_D S_{I^2} \frac{dS_{I^2}}{dV} \right] \qquad (12.1)$$

We have experimentally demonstrated the existence of these two mechanisms and their contribution to S_{V^3}, see Fig. 12.3. Note that the absence of environmental corrections to the average current and to S_{V^2} are only approximate. The feedback mechanism indeed leads to an environmental correction to the dc current, better known as dynamical Coulomb blockade.[5,11,12] Dynamical Coulomb blockade is hardly visible for low impedance samples but is important for the third moment. Environmental corrections to the second moment have not yet been observed.[3]

12.4. The Third Moment in the Quantum Regime

The result of ref.[2] that electrons behave as marbles is valid for $\tau \to +\infty$, i.e. in the low frequency limit $f \ll eV/h$. At high frequency, one may hope that quantum mechanics will dictate the statistics of charge transfers. As a matter of fact, the spectral density of the variance of current fluctuations at finite frequency f is given by $S_{I^2}(V,f) = [S_{I^2}(V+hf/e,0) + S_{I^2}(V-hf/e,0)]/2$ for a tunnel junction. At zero temperature, it is voltage-independent and equal to Ghf as long as $|eV| < hf$, since electrons of energy eV cannot emit photons of greater energy $hf > eV$. The low voltage plateau on S_{I^2} is clearly of quantum nature. We show on Fig. 12.4 the observation of the plateau, in a way clearer than previously reported data[13–15] thanks to the high $hf/(k_BT) = 15$ ratio that we have been able to achieve ($f = 6$ GHz, $T = 20$ mK).

S_{I^2} at finite frequency results from the product of the current at frequency f by the current at frequency $-f$. In classical mechanics, those two are c-numbers; in quantum mechanics they are operators. These operators do not commute, since the current operators taken at two different times do not commute. Thus, the correlators $S_+(f) = \langle \hat{I}(f)\hat{I}(-f)\rangle$, $S_-(f) = \langle \hat{I}(-f)\hat{I}(f)\rangle$, or more generally $S_\alpha(f) = \alpha S_+ + (1-\alpha)S_-$ are all different (but are identical in the classical limit), and one has to face the problem of knowing which experimental setup measures which combination. However, for simple systems S_+ and S_- are related by $S_+ - S_- = Ghf$, so all the correlators differ only by a constant $\propto Ghf$, independent of both temperature and voltage, thus indistinguishable in many experiments (although some setups correspond to a well established value of α[16–18]). This simplification does not occur for higher order cumulants, and the various ways of ordering the operators leads to different voltage dependance.[19–21]

We have performed a measurement of $\langle \delta V(\epsilon)\delta V(f-\epsilon)\delta V(-f)\rangle$ ($\epsilon < k_BT/h \ll f$) using linear amplifiers and square law detectors, from which we deduce $S_{I^3}(\epsilon \to 0, f)$. We observe that $S_{I^3}(0,f) = e^2 I$ regardless of the frequency f, see Fig. 12.5. It is very surprising that S_{I^3} is finite even

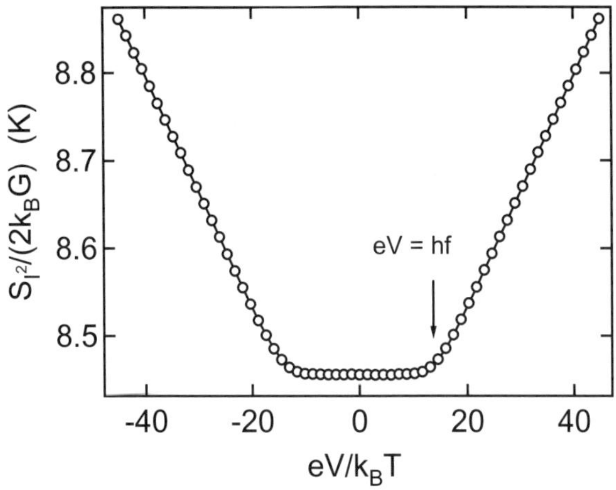

Fig. 12.4. Measurement of $S_{I^2}(V)/(2k_BG)$ at $f = 5.5 - 6.5$ GHz and $T = 20$ mK. The rounding of the curve is not limited by the temperature but by the bandwidth of the measurement (1 GHz is 50 mK). The value at $V = 0$ is $T_N + hf/(2k_B) = T_N + 0.15$K with T_N the amplifier's noise temperature.

though $eV < hf$. This means that S_{I^3} can be non-zero even if no photon of frequency f is emitted by the sample. A non-zero $S_{I^3}(0, f)$ implies the existence of correlations between the low frequency current fluctuations $I(\epsilon \to 0)$ and the high frequency noise $I(f)I(-f)$ i.e., for $eV < hf$, between the low frequency current fluctuations and the high frequency zero point motion of electrons. This phenomenon has never been discussed. Our measurement directly raises the problem of how to relate an experimental setup to a quantum mechanical prediction. Had we used a photo-detector (which clicks each time it absorbs a photon) instead of a linear amplifier followed by a square law detector, we would have obtained $S_{I^3}(0, f < eV/h) = 0$ since no photon would have been detected for $eV < hf$.[22]

12.5. Environmental Effects in the Quantum Regime: The Noise Susceptibility

Since we have seen that environmental effects are crucial to S_{I^3} measured at low frequency, it is necessary to investigate what they become in the quantum regime. The environmental contributions to S_{I^3} involve the modulation of S_{I^2} by a time-dependent voltage v_0. This modulation is given by $\delta S_{I^2} = v_0 \chi$ where χ, the noise susceptibility, describes by how much the

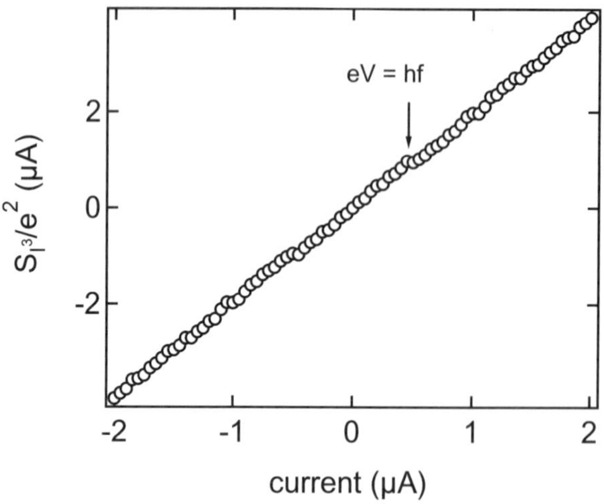

Fig. 12.5. Normalized third moment of current fluctuations $S_{I^3}(0,f)$ measured at very low temperature $T = 27$ mK and high frequency $f = 5.5 - 6.5$ GHz.

noise is modulated by the ac excitation. In the adiabatic limit, i.e. provided the noise "follows" the excitation, $\chi = dS_{I^2}/dV$. This happens as long as the frequency f_0 of the ac excitation is smaller than some characteristic frequency f_c.

In order to have an intuition of f_c, let us first consider the simple case of a macroscopic sample, for which the noise is the equilibrium value $S_{I^2}^{eq} = 2k_BTG$, biased by a voltage $V + v_0 \cos 2\pi f_0 t$. The oscillating voltage creates oscillating Joule heating of the sample, thus oscillating temperature and oscillating noise (measured at frequency $f \gg f_0$). In this case f_c is clearly the inverse of the thermalization time. The linear response of the sample noise temperature to the ac Joule heating defines the noise thermal impedance between the system and its thermal environment, and allows for a direct measurement of thermalization times, e.g. by electron-phonon interaction or by diffusion of hot electrons.[23] This also explains the existence of an intrinsic S_{I^3} in a diffusive wire or a chaotic cavity,[24,25] where the variation of the Joule heating is given by the current fluctuations in the sample times the bias voltage, and the cutoff f_c, proportional to the inverse diffusion time, determines the frequency dependence of S_{I^3}.

This clear picture however does not apply in the quantum regime, for which $f \gg f_0$ is not fulfilled. In that case, we have shown that δS_{I^2} is given by a correlator involving current taken at two different frequencies separated by f_0: $\delta S_{I^2} = \langle \hat{I}(f)\hat{I}(f_0 - f)\rangle + \langle \hat{I}(f - f_0)\hat{I}(-f)\rangle$.[12] We have measured this quantity for $f \sim f_0$, in which case δS_{I^2} measures how low frequency and high frequency currents become correlated by a high frequency excitation.[15] We show on Fig. 12.6 the noise susceptibility χ that characterizes environmental contributions for $f = f_0 = 6$ GHz. It clearly differs from the low frequency result $\propto dS_{I^2}/dV$. From the knowledge of environmental effects in the quantum regime, we can extract the intrinsic quantum third moment of the current fluctuations that we have discussed before.

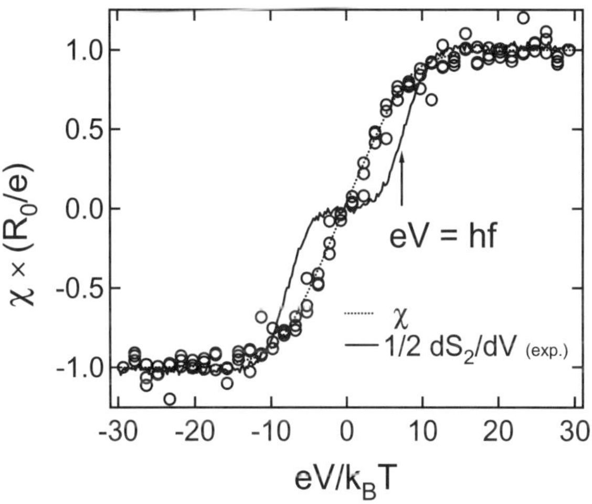

Fig. 12.6. Normalized noise susceptibility χ vs. normalized dc bias. Voltage fluctuations are detected in the range $f = 5.5 - 6.5$ GHz and $|f_0 - f| = 0.01 - 0.5$ GHz, whereas rf excitation is of amplitude V_{ac} at frequency $f_0 = 6$ GHz. Symbols: data for various levels of ac excitation ($eV_{ac}/(hf_0) = 0.85$, 0.6 and 0.42). Dotted lines: theoretical prediction for χ. Solid line: measured $(1/2)dS_2(f)/dV$, as a comparison.

12.6. Conclusion

We have performed the first measurements of non-Gaussian noise in a conductor, both in the classical and quantum regimes. This work opens routes towards the study of higher cumulants of noise in complex devices to extract interesting physics (see, e.g.[6]) as well as it stresses the need for theoretical investigations to model measurements beyond average quantities. That a time-averaged measurement is predicted by the quantum and statistical average of the well suited operator is well established. What correlator (with the appropriate times and time-orderings) describes the measurement of a quantity at different times by a given setup is not yet solved.[26–29]

Acknowledgments

We are very grateful to J. Senzier for participating in the early stage of this work and to C. Wilson for providing us with a tunnel junction used for the first measurements. We acknowledge fruitful discussions with M. Aprili, C. Beenakker, A. Bednorz, W. Belzig, M. Büttiker, A. Clerk, M. Devoret, M. Kindermann, L. Levitov and Y. Nazarov.

References

1. L.S. Levitov and G.B. Lesovik, JETP Lett. **58**, 230 (1993).
2. L.S. Levitov, H.W. Lee and G.B. Lesovik , Electron counting statistics and coherent states of electric current, *J. Math. Phys.* **37**, 4845 (1996).
3. L. Spietz, K. W. Lehnert, I. Siddiqi, R. J. Schoelkopf, Primary Electronic Thermometry Using the Shot Noise of a Tunnel Junction, *Science* **300** 1929 (2003)
4. B. Reulet, J. Senzier and D. E. Prober, Environmental Effects in the Third Moment of Voltage Fluctuations in a Tunnel Junction, *Phys. Rev. Lett.* **91** 196601 (2003)
5. B. Reulet, Higher Moments of Noise, Les Houches Summer School of Theoretical Physics, Session LXXXI, Nanophysics: Coherence and Transport. NATO ASI. H. Bouchiat, Y. Gefen, S. Guéron, G. Montambaux and J. Dalibard editors. Elsevier (2005). (*cond-mat*/0502077).
6. J. Gabelli and B. Reulet, Full Counting Statistics of Avalanche Transport: an Experiment, Phys. Rev. B 80, R161203 (2009).
7. Yu. Bomze, G. Gershon, D. Shovkun, L. S. Levitov and M. Reznikov, Measurement of Counting Statistics of Electron Transport in a Tunnel Junction, *Phys. Rev. Lett.* **95** 176601 (2005)
8. Y.M. Blanter and M. Büttiker, Shot Noise in Mesoscopic Conductors, *Phys. Rep.* **336** 1 (2000)
9. M. Kindermann, Yu. V. Nazarov and C. W. J. Beenakker, Distribution of Voltage Fluctuations in a Current-Biased Conductor, *Phys. Rev. Lett.* **90** 246805 (2003)
10. C.W.J. Beenakker, M. Kindermann and Yu V. Nazarov, Temperature-dependent Third Cumulant of Tunneling Noise, Phys. Rev. Lett. **90**, 176802 (2003).
11. G.-L. Ingold and Yu. V. Nazarov, Charge Tunneling Rates in Ultrasmall Junctions, in *Single Charge Tunneling*, edited by H. Grabert and M.H. Devoret, NATO ASI Series B, Vol. 294, 21 (Plenum, New York, 1992).

12. J. Gabelli and B. Reulet, The Noise Susceptibility of a Photo-excited Coherent Conductor, arXiv:0801.1432 (2008).
13. R. J. Schoelkopf, P. J. Burke, A. A. Kozhevnikov, and D. E. Prober and M. J. Rooks, Frequency Dependence of Shot Noise in a Diffusive Mesoscopic Conductor, *Phys. Rev. Lett.* **78** 3370 (1997).
14. E. Zakka-Bajjani, J. Segala, F. Portier, P. Roche, C. Glattli, A. Cavanna and Y. Jin, Experimental Test of the High-Frequency Quantum Shot Noise Theory in a Quantum Point Contact, *Phys. Rev. Lett.* **99**, 236803 (2007).
15. J. Gabelli and B. Reulet, Dynamics of Quantum Noise in a Tunnel Junction under ac Excitation, *Phys. Rev. Lett.* **100**, 026601 (2008).
16. R. Deblock, E. Onac, L. Gurevich, and L.P. Kouwenhoven, Detection of Quantum Noise from an Electrically-Driven Two-Level System , *Science* **301**, 203 (2003).
17. U. Gavish, Y. Levinson and Y. Imry, Detection of quantum noise, *Phys. Rev. B* **62** R10, 637 (2000).
18. Lesovik and Loosen, On the Detection of Finite-Frequency Current Fluctuations, *Pisma ZhETF* **65**, 269 (1997).
19. A. Galaktionov, D. Golubev, and A. Zaikin, Statistics of current fluctuations in mesoscopic coherent conductors at nonzero frequencies, *Phys. Rev. B* **68**, 235333 (2003).
20. D. S. Golubev, A. V. Galaktionov, and A. D. Zaikin, Electron transport and current fluctuations in short coherent conductors, *Phys. Rev. B* **72**, 205417 (2005).
21. J. Salo, F. W. J. Hekking, and J. P. Pekola , Frequency-dependent current correlation functions from scattering theory, *Phys. Rev. B* **74**, 125427 (2006).
22. J. Gabelli and B. Reulet, High frequency Dynamics and Third Cumulant of Quantum Noise, J. Stat. Mech. P01049 (2009).
23. B. Reulet and D.E. Prober, Noise Thermal Impedance of a Diffusive Wire, Phys. Rev. Lett. **95**, 066602 (2005).
24. S. Pilgram, K. E. Nagaev and M. Büttiker, Frequency-dependent third cumulant of current in diffusive conductors, Phys. Rev. B**70**, 045304 (2004).
25. K. E. Nagaev, S. Pilgram, and M. Bttiker, Frequency Scales for Current Statistics of Mesoscopic Conductors, Phys. Rev. Lett. **92**, 176804 (2004).
26. A. Bednorz and W. Belzig, Formulation of Time-Resolved Counting Statistics Based on a Positive -operator-Valued Measure, Phys. Rev. Lett. 101, 206803 (2008).
27. A. Bednorz and W. Belzig, Quantum tape model of mesoscopic time-resolved current detection, unpublished (arXiv:0906.0038).
28. K. V. Bayandin, A.V. Lebedev and G. B. Lesovik, Measurement of high order current correlators, JETP 106, 117 (2008).
29. S. Bachmann, G.M. Graf, and G.B. Lesovik, Time ordering and counting statistics, unpublished (arXiv:0907.3848).

Chapter 13

Magnetic Avalanches in Molecular Nanomagnets

M. P. Sarachik

Department of Physics, City College of New York, CUNY,
New York, New York, 10031, USA
E-mail: sarachik@sci.ccny.cuny.edu

S. McHugh

Department of Physics, University of California,
Santa Barbara, California, 93106-9530, USA

The magnetization of the prototypical molecular magnet Mn_{12}-acetate exhibits a series of sharp steps at low temperatures due to quantum tunneling at specific resonant values of magnetic field applied along the easy c-axis. An abrupt reversal of the magnetic moment of such a crystal can also occur as an avalanche, where the spin reversal proceeds along a "deflagration" front that travels through the sample at subsonic speed. In this article we review experimental results that have been obtained for the ignition temperature and the speed of propagation of magnetic avalanches in molecular nanomagnets. Fits of the data with the theory of magnetic deflagration yield overall qualitative agreement. However, numerical discrepancies indicate that our understanding of these avalanches is incomplete.

13.1. Background

First synthesized by Lis in 1980,[1] $Mn_{12}O_{12}(CH_3COO)_{16}(H_2O)_4$ (referred to hereafter as Mn_{12}-ac) is a particularly simple, prototypical molecular magnet. Shown in Fig. 13.1(a), the magnetic core of Mn_{12}-ac has four Mn^{4+} (S = 3/2) ions in a central tetrahedron surrounded by eight Mn^{3+} (S = 2) ions. The ions are coupled by superexchange through oxygen bridges with the net result that the four inner and eight outer ions point in opposite directions, yielding a total spin $S = 10$.[2] The magnetic core is surrounded by acetate ligands, which serve to isolate each core from its neighbors in a body-centered tetragonal lattice. A crystalline sample contains $\sim 10^{17}$ or more (nearly) identical, weakly interacting single molecule nanomagnets in (nearly) identical crystalline environments.

While there are very weak exchange interactions between molecules, the exchange between ions within the magnetic core is very strong, resulting in a rigid spin 10 unit that has no internal spin degrees of freedom at low temperatures. To lowest order, the spin Hamiltonian is given by:

$$\mathcal{H} = -DS_z^2 - g_z\mu_B H_z S_z + \ldots + \mathcal{H}'. \tag{13.1}$$

The first term denotes the anisotropy barrier, the second is the Zeeman energy that splits the spin-up and spin-down states in a magnetic field, and the last term, $\mathcal{H}'$, contains all symmetry-breaking operators that do not commute with S_z. For Mn_{12}-ac, $D = 0.548$K, $g_z = 1.94$, and μ_B is the Bohr magneton.

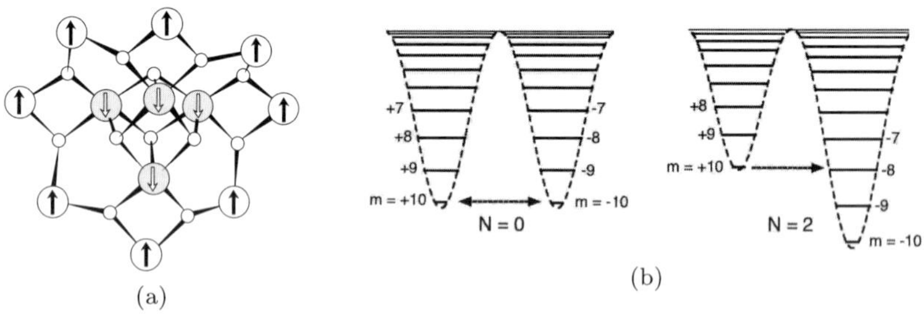

Fig. 13.1. (a) Chemical structure of the core of the Mn_{12} molecule. The four inner spin-down Mn^{3+} ions each have spin $S = 3/2$; the eight outer spin-up Mn^{4+} ions each have spin $S = 2$, yielding a net spin $S = 10$ for the magnetic cluster; the small open circles are O bridges and arrows denote spin. Acetate ligands and water molecules have been removed for clarity. (b) Left: Double-well potential in the absence of magnetic field showing spin-up and spin-down levels separated by the anisotropy barrier. Different spin projection states $|m>$ are indicated. The arrows denote quantum tunneling. Right: Double-well potential for the N=2 step in a magnetic field applied along the easy axis.

As illustrated by Fig. 13.1(b), the spin's energy can be modeled as a double-well potential, where one well corresponds to the spin pointing "up" and the other to the spin pointing "down." A strong uniaxial anisotropy barrier of the order of 66 K yields doubly degenerate ground states in zero field. The spin has a set of energy levels corresponding to different projections, $m = 10, 9, \ldots, -9, -10$, of the total spin along the easy (c-axis) of the crystal.[3–6]

13.2. Spin Reversal by Quantum Tunneling

Slow relaxation below the blocking temperature, $T_B \sim 3$ K, gives rise to hysteresis loops that display steps[7] as a function of magnetic field, H_z, swept along the easy *c*-axis of the crystal.[8,9] Figure 13.2(a) shows the magnetization M as a function of magnetic field $\mu_0 H_z$; the derivative, dM/dH, which reflects the magnetic relaxation rate, is plotted as a function of $\mu_0 H_z$ in Fig. 13.2(b). These steps, characteristic of molecular magnets, can be understood with reference to the double well potential of Fig. 13.1(b): a magnetic field H_z introduces a Zeeman splitting that tilts the potential wells and causes energy levels in the right (left) well to move down (up). Levels in opposite wells align at particular values of magnetic field (dashed lines in Fig. 13.1(b)), allowing the spin to reverse by tunneling. Full (saturation) magnetization is thereby reached in a stepwise fashion, with the detailed form of the steps depending on sweep-rate and temperature.

13.3. Spin Reversal by Avalanches

As first reported by Paulsen and Park,[10] Mn_{12}-ac crystals sometimes exhibit a sudden, complete reversal of magnetic moment during a field-swept measurement. This phenomenon, also observed in other molecular magnets, was attributed to a thermal runaway (avalanche) in which the relaxation of magnetization toward the direction of the field results in the release of heat that further accelerates the magnetic relaxation. Direct measurements of the heat emitted have confirmed the thermal nature of the avalanches. In addition to releasing thermal energy, molecular crystals emit bursts of radiation during magnetic avalanches.[11–13] Once considered events to be avoided, as they interfere with a detailed study of the stepwise process of magnetization, magnetic avalanches became the focus of attention and renewed interest stimulated by the theoretical suggestion that the radiation emitted during an avalanche is in the form of coherent (Dicke) superradiance.[14] Although the issue of coherence of the radiation has yet to be resolved, recent studies have clarified the nature of the avalanche process itself.

13.3.1. *Magnetic Deflagration*

From time-resolved measurements of the local magnetization using an array of micron-sized Hall sensors placed on the surface of Mn_{12}-ac crystals,

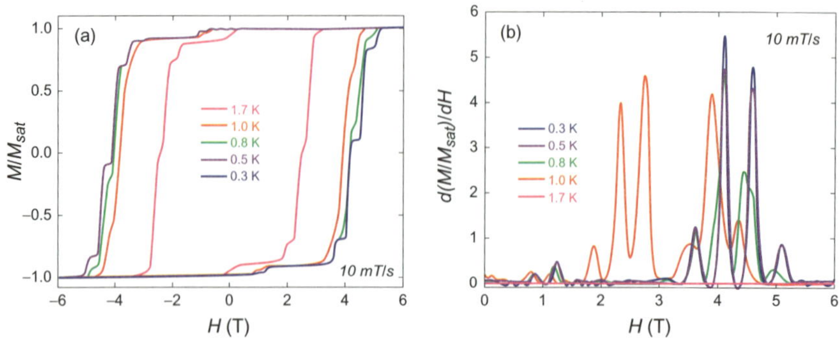

Fig. 13.2. (a) Magnetization of a Mn_{12}-ac crystal normalized by its saturation value as a function of magnetic field applied along the uniaxial c-axis direction at different temperatures below the blocking temperature; the magnetic field was swept at 10 mT/s. (b) The derivative, dM/dH of the data in part (a) as a function of magnetic field.

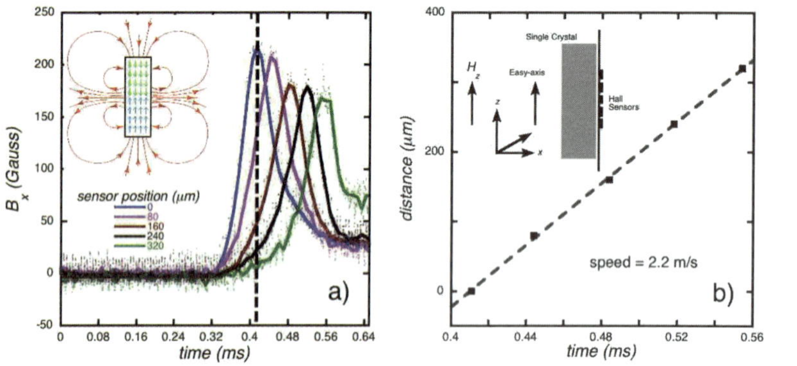

Fig. 13.3. (a) The local magnetization measured as a function of time by an array of micron-sized Hall sensors placed along the surface of the sample. The inset illustrates the "bunching" of magnetic field lines as the deflagration front travels past a given Hall sensor. (b) The sensor position as a function of the time at which the sensor registered the peak; the inset shows the placement of the Hall sensors on the crystal. The propagation speed for this avalanche is 2.2 m/s, approximately three orders of magnitude below the speed of sound.

Suzuki *et al.*[15] discovered that a magnetic avalanche propagates through the crystal at subsonic speed in the form of a thin interface between regions of opposite magnetization. Figure 13.3(a) shows traces recorded during an avalanche by sensors placed in sequential positions near the center of a Mn_{12}-ac sample. The inset is a schematic that illustrates the bunching of field lines at the propagating front that gives rise to the observed peaks.

Figure 13.3(b) is a plot of the sensor position versus the time of arrival of the peak. From these measurements one deduces that the front separating up- and down-spins travels with a constant (field-dependent) speed on the order of 5 m/s, two to three orders of magnitude slower than the speed of sound.

From a thermodynamic point of view, a crystal of Mn_{12} molecules placed in a magnetic field opposite to the magnetic moment is equivalent to a metastable (flammable) chemical substance. A well-known mechanism for the release of energy by a metastable chemical substance is combustion or slow burning, technically referred to as deflagration.[16] It occurs as a flame front of finite width propagates at a constant speed small compared to the speed of sound. For "magnetic deflagration" in Mn_{12}-ac, the role of the chemical energy stored in a molecule is played by the difference in the Zeeman energy, $\Delta E = 2g\mu_B HS$, for states of the Mn_{12}-ac molecule that correspond to $\mathbf{S}$ parallel and antiparallel to $\mathbf{H}$.

13.3.2. *Avalanche Ignition*

Although the probability of a spontaneous avalanche has been shown to be higher at resonant magnetic fields than off-resonance,[17] avalanche ignition is unpredictable and uncontrolled when an external magnetic field is swept back and forth, the experimental protocol generally used to study the steps in the hysteresis loops. Avalanche ignition under these conditions is a stochastic process that depends on factors such as the sweep rate, the temperature, and the quality of the crystal. Controlled ignition of avalanches has now been achieved using surface acoustic waves (which serve to heat the sample),[18] and by using a heater[19] , as described below and in the next section.

McHugh *et al.*[19] employed a resistive wire element as a simple electric heater to trigger avalanches in a manner similar to the work of Paulsen and Park.[10] In these experiments, an external magnetic field is ramped to and held at a fixed value. The heater is then turned on to slowly heat the sample until an avalanche is triggered at a temperature measured by a small thermometer. Avalanches launched by this method occur at well-defined, reproducible ignition temperatures. Figure 13.4(a) shows a typical temperature profile: starting at the base temperature of 300 mK, the temperature gradually rises until an abrupt sharp increase in the temperature signals the ignition of an avalanche. For this avalanche triggered at $\mu_0 H_z = 1.85$ T, the ignition temperature is about 0.6 K.

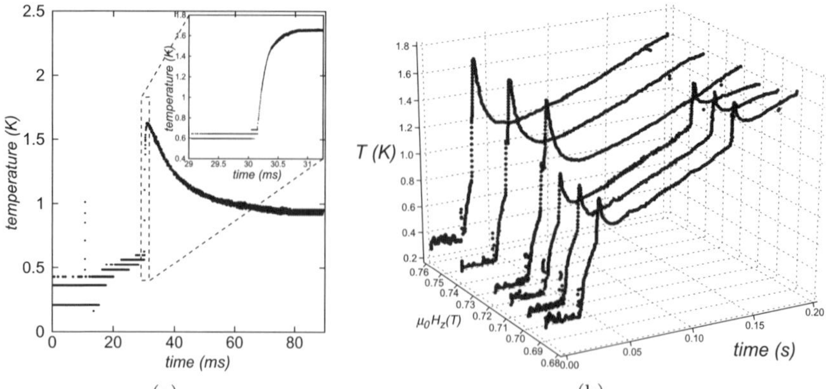

Fig. 13.4. (a) Temperature recorded by a thermometer in contact with a Mn_{12} crystal during the triggering of an avalanche. The heater is turned on at ~ 0.012 s, the temperature then increases slowly until an abrupt rise in temperature at 0.03 s signals the ignition of an avalanche. The inset shows data taken near ignition with higher resolution. The noise at low temperatures derives from digitizing the analog output of the thermometer, which depends weakly on temperature below 0.4 K. (b) Temperature profiles for avalanches of major and minor species triggered at low fields in a Mn_{12} crystal. The two types of avalanches are triggered separately below a sample-dependent magnetic field, while at higher fields ignition of the minor species triggers the ignition of the major species.

Single crystals of Mn_{12}–ac are known to contain two types of molecules. In addition to the primary or "major" species described earlier, as-grown crystals contain a second "minor" species at a level of ≈ 5 percent with lower (rhombohedral) symmetry.[20,21] These faster-relaxing molecules can be modeled by the same effective spin Hamiltonian, Eq. 13.1, with a lower anisotropy barrier of 0.49 K. Avalanches of the each species can be studied in the absence of the other through an appropriate magnetic protocol described in Ref. 22. Interestingly, avalanches are separately triggered by the two species in low magnetic field. As shown in Fig. 13.4(b), at low fields the minor species relaxes prior to and independently of the major species, while above ~ 0.7 T, the major and minor species ignite together and propagate as a single front. It is analogous to grass and trees that can sustain separate burn fronts that abruptly merge into a single front when the grass becomes sufficiently hot to ignite the trees.

Despite the turbulent conditions that one might expect for deflagration (analogous to chemical combustion), quantum mechanical tunneling clearly plays a role, as demonstrated in Fig. 13.5, where the temperature above

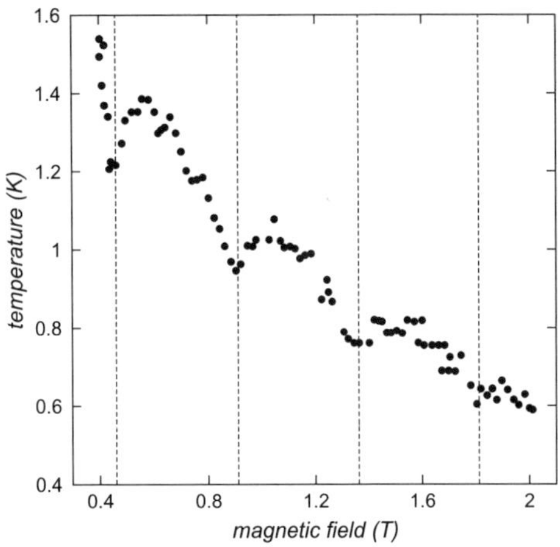

Fig. 13.5. Temperature required to ignite avalanches plotted as a function of magnetic field. The vertical lines denote the magnetic fields where sharp minima occur in the ignition temperature corresponding to tunneling near the top of the anisotropy barrier. The overall decrease in ignition temperature is due to the reduction of the anisotropy barrier as the field is increased.

which ignition occurs is plotted as a function of a preset, constant magnetic field.[19] The temperature required to ignite avalanches exhibits an overall decrease with applied magnetic field, reflecting the fact that larger fields reduce the barrier (see the double-well potential in Fig. 13.1(b)). The role of quantum mechanics is clearly evidenced by the minima observed for the ignition temperature at the resonant magnetic fields due to tunneling when levels cross on opposite sides of the anisotropy barrier.

In the ignition studies described above, the barrier against spin reversal was lowered by applying an external magnetic field, H_z, along the uniaxial c-direction, which serves to unbalance the potential wells and lower the barrier to spin reversal. A transverse field, H_x, also reduces the anisotropy barrier by introducing a symmetry-breaking term, $(g\mu_B H_x S_x)$, to the Hamiltonian, Eq. 13.1, thereby promoting tunneling. Macià *et al.*[17] investigated the threshold for avalanche ignition in Mn_{12}-ac as a function of the magnitude and direction of a magnetic field applied at various angles with respect to the anisotropy axis and as a function of temperature. As the external field is increased at a constant rate from negative saturation to positive values, both H_z and H_x increase, tracing a trajectory in the (H_z, H_x) parameter space; (examples of sweeps starting from zero are shown by the arrows in Fig. 13.6. As shown in Fig. 13.6(a), an avalanche was recorded

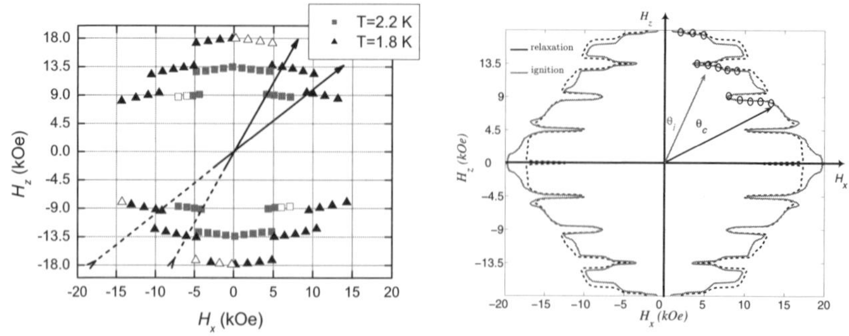

(a) (b)

Fig. 13.6. (a) Angle dependence of metastability measured through the occurrence of avalanches. Squares (triangles) denote parameter values where deflagration occurs for initial temperature 2.2 K (1.8 K). (b) Theoretical calculation for the area of stability against ignition of avalanches (solid curve) and against slow relaxation (dashed curve). Circles denote points where avalanches are predicted to occur at a given angle θ_i within the first quadrant. The angle θ_c denotes the crossing point between areas of slow relaxation and avalanche stability. These results were obtained with T_f as a parameter varying from 6.8 K for $H = 4600$ Oe to 10.9 K for $H = 9200$ Oe. From Macià *et al.*[17]

for each pair (H_x, H_z) denoted by a square (for $T = 2.2$ K) or a triangle (for $T = 1.8$ K). We postpone a detailed explanation of these results and a comparison with theory to a later section.

13.3.3. *Avalanche Speed*

Following the initial studies of Suzuki *et al.*[15] in which avalanches were triggered stochastically in swept magnetic field, Hernández-Mínguez *et al.*[18,23] carried out a systematic investigation of avalanche speeds as a function of a preset, constant magnetic field, $\mu_0 H_z$, for avalanches triggered controllably using surface acoustic waves. From SQUID-based measurements of the total magnetization of a crystal of known dimensions, and the realization that the avalanche propagates as an interface between regions of opposite magnetization,[15] Hernández-Mínguez *et al.*[18,23] deduced the speed of propagation as a function of magnetic field shown in Fig. 13.7. The speed of the avalanches is enhanced at the resonant fields where tunneling occurs, confirming the important role of quantum mechanics, and prompting the authors to name the phenomenon "quantum magnetic deflagration." Similar results have been obtained from local, time-resolved magnetization measurements using micron-sized Hall sensors.

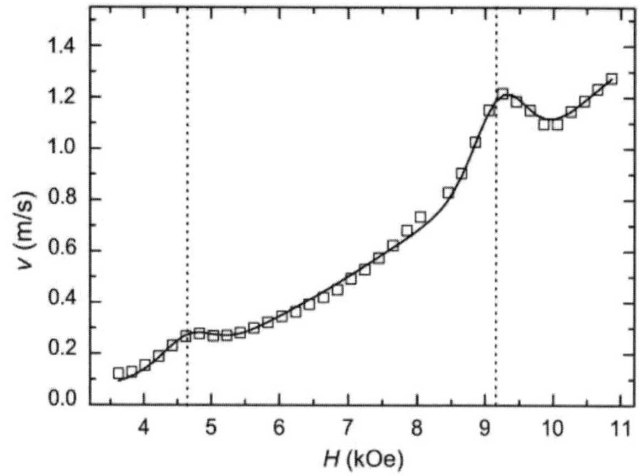

Fig. 13.7. The speed of propagation of the magnetic avalanche deflagration front is plotted as a function of the (fixed) field at which the avalanche is triggered; note the enhancement of propagation velocity at magnetic fields corresponding to quantum tunneling (denoted by vertical dotted lines). From Hernández-Mínguez *et al.*[18]

McHugh *et al.*[24] have reported a detailed, systematic investigation of the speed of magnetic avalanches for various experimental conditions. The speed of propagation of an avalanche is described approximately[15] by the expression, $v \sim (\kappa/\tau_0)^{1/2}\exp[-U(H)/2k_BT_f]$, where U is the barrier against spin reversal, T_f is the flame temperature at or near the propagating front where energy is released by the reversing spins, κ is the thermal diffusivity, and τ_0 is an attempt time. In these studies, avalanches were controllably triggered: (A) in various external fields with fixed (maximum) initial magnetization, so that both U and T_f vary; (B) in fixed external field with different initial magnetization so that the avalanches differ primarily through the amount of energy released, and thus the flame temperature T_f; and (C) where external magnetic fields and initial magnetization are varied and adjusted to hold the energy released, and thus T_f, constant. These parameters will be discussed in more detail in the next section. We note at this point that the energy barrier U and the flame temperature T_f appear only as the ratio U/T_f in the above expression for the velocity. It is therefore convenient to plot the speed of the avalanche as a function of U/T_f as is done in Fig. 13.8(a).[24]

To end this section, we note that Villuendas *et al.*[25] have recently reported a novel approach to studying avalanche dynamics using the magneto-optical Kerr effect to image the magnetization during an

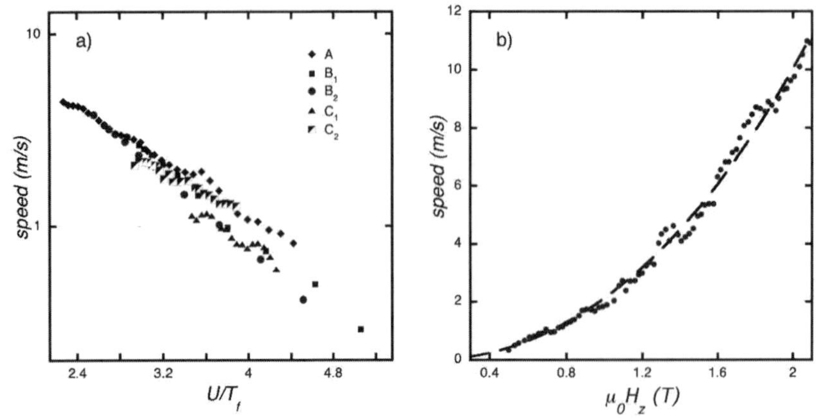

Fig. 13.8. (a) Avalanche speeds for a single crystal with various initial magnetic preparations. A denotes avalanches with $\Delta M/2M_s = 1$; B_1 and B_2 denote data taken at $\mu_0 H_z = 2.2$ T and 2.5 T, respectively; C_1 and C_2 denote avalanches with estimated flame temperatures $T_f \approx 10$ K and 12 K, respectively. (b) Avalanche speeds for different crystal with $\Delta M/2M_s = 1$. The fit requires an unphysical temperature dependence for the thermal diffusivity, $\kappa \propto T^{3.5}$.

avalanche. Although the resolution was quite limited, with further improvements this technique could provide a valuable avenue for studying magnetic deflagration.

13.4. Comparison with the Theory of Magnetic Deflagration

There are two essential ingredients for magnetic deflagration: the metastable spins release energy as they relax to the ground state, and this energy diffuses as heat through the crystal and thermally stimulates the reversal of neighboring metastable spins. Garanin and Chudnovsky[26] developed a comprehensive theory of magnetic deflagration encapsulated in the following equations:[27]

$$C\frac{\partial T}{\partial t} - \nabla \cdot k\nabla T = -\langle E\rangle \frac{\partial n}{\partial t} \tag{13.2}$$

$$\frac{\partial n}{\partial t} = -\Gamma n. \tag{13.3}$$

Equation 13.2 describes the flow of heat through the solid with the relaxing spins, $\frac{\partial n}{\partial t}$, as a source of heat; the thermal conductivity, k, is related to the heat capacity, C, through the thermal diffusivity, κ, as $k = \kappa C$. Equation 13.3 describes the relaxation of the metastable spin density, n, with a thermal relaxation rate Γ given by an Arrhenius law, $\Gamma = \Gamma_0 \exp[-U/k_B T]$; $\langle E \rangle$ is the average amount of heat released per molecule when its spin relaxes to the stable state. For a single relaxing molecule, the energy change from the metastable to stable state is simply the Zeeman energy $\Delta E = 2g\mu_B S B_z$. However, since not all spins necessarily relax during an avalanche, the average energy released per molecule must be introduced:

$$\langle E \rangle = 2g\mu_B S \left(\frac{\Delta M}{2M_s} \right) B_z, \tag{13.4}$$

where M_s is the saturation magnetization and $\Delta M = |M_z - M_s|$ is the change from initial to final magnetization.

Since both k and C are functions of temperature, Eqs. 13.2 and 13.3 are coupled, nonlinear differential equations, which make them very difficult to solve in general. The samples used in the experiments generally have large aspect ratios, so that we can use simple approximations for quasi-one-dimensional avalanche propagation[26] to compare theory with experimental results.

We begin by comparing the theory with experimental results for the ignition temperatures and the stability criteria. Heuristically, a deflagration front can develop when the rate at which energy is released by the relaxing metastable spins exceeds the rate of energy loss through the boundaries of the crystal. This condition can be expressed in terms of a critical relaxation rate,[26]

$$\Gamma_c = \frac{8k(T_0)k_B T_0^2}{U \langle E \rangle l^2}, \tag{13.5}$$

where T_0 is the initial temperature, $\Gamma_c = \Gamma_0 \exp[-U/k_B T_0]$, and l^2 is the characteristic cross section of the crystal. The curves shown in Fig. 13.6(b) are the result of a calculation using Eq. 13.5. Two areas are defined in the (H_z, H_x) parameter space where the spins are expected to be metastable against relaxation: the solid line denotes the region of metastability against relaxation by triggering avalanches while the dashed curve delineates the region of metastability against slow, stepwise relaxation. If the experimental trajectory, denoted by the arrows, crosses the grey solid line first, an avalanche will ignite. If the dashed line is crossed first, the metastable spins will relax slowly without triggering an avalanche. This defines a critical angle θ_c, above which an avalanche cannot occur.

Macià *et al.* measured the ignition threshold by applying an increasing external field at an angle with respect to the crystal. The relaxation rate increases as the field grows until Γ_c is reached and deflagration ignites, as shown in Fig. 13.6(a). For sufficiently large values of H_x, they found that the slow relaxation of the metastable spins occurs before deflagration can ignite. This defines a line in parameter space separating regions where one or the other mode of relaxation occurs, as shown in Fig. 13.6(b). The theory predicts that the transverse field should result in a significant decrease in the magnetization metastability at the resonant fields of H_z. The data recorded in Fig. 13.6(a) confirm this and are consistent with the ignition temperatures of Fig. 13.5. In addition, ignition thresholds were measured at two different temperatures. The area of stability is clearly reduced by the increased initial temperature, as expected.

To summarize, the theory of magnetic deflagration is in excellent agreement with the experiments of McHugh *et al.*, where the critical relaxation rate was reached by varying T_0 with a heater, and the experiments of Macià *et al.*, where the ignition threshold was reached by controlling the barrier U using H_x and H_z.

We now proceed to compare the measured avalanche speeds with the theory of magnetic deflagration. An approximate expression for the speed of the deflagration front is given by[26]

$$v = \sqrt{\frac{3k_B T_f \kappa \Gamma(B, T_f)}{U(B)}}. \tag{13.6}$$

Eq. 13.6 requires that the relaxation rate of the spins at the highest temperature, $\Gamma(U, T_f)$, is significantly slower than the rate at which heat traverses the interface width, κ/δ^2, where δ is the magnetic interface width; this condition is expected to hold for all speeds considered here.

The barrier $U(B_z = \mu_0 H_z)$ is calculated from the effective spin Hamiltonian (Eq. 13.1). The temperature of the front, T_f, can be estimated from the measured heat capacity at zero magnetic field,[28] the calculated contribution of the spins in magnetic field, and the average energy released per molecule:

$$\langle E \rangle = \int_{T_0}^{T_f} C(B_z, T) dT. \tag{13.7}$$

Typical values calculated for T_f range from 7 K to 16 K.

With reference to the theoretical expression for the avalanche speed, Eq. 13.6, if one assumes the thermal diffusivity κ is approximately independent of temperature, or that its temperature dependence is unimportant compared to that of other parameters in the problem, then all measured avalanche velocities should collapse onto a single curve when plotted as a function of $[U(H)/T_f]$. Although an approximate collapse is indeed obtained, as shown in Fig. 13.8(a), there are clear and systematic deviations depending on whether (A): there is full (maximum) magnetization reversal, $\Delta M/2M_s = 1$ (both U and T_f vary); (B_1,B_2): the amount of "fuel" $\Delta M/2M$ is varied for a fixed magnetic field (thus U is held constant); (C_1,C_2): avalanches are triggered such that the product $\Delta M \times H$ remains the same so that the energy released and the flame temperature T_f are held constant. That these different experimental protocols introduce systematic variations, albeit small, suggests that the theory is incomplete.

An attempt to fit to the theory by allowing the thermal diffusivity to vary with temperature as a power law is shown in Fig. 13.8(b) for avalanches of type (A) that involve full magnetization reversal. Note that the enhancements of the velocity at certain values of magnetic field are associated with the tunneling resonances, in agreement with Fig.13.7. The fit with Eq. 13.6 is obtained for a thermal diffusivity that varies with temperature as $\kappa \sim T^{3.5}$. This is distinctly unphysical, as the thermal diffusivity is generally a strongly decreasing function of temperature[29] for these materials. A similar analysis was performed[22] on avalanches of the minor species that yields an even steeper increase of κ with temperature. Regrettably, experimental measurement of the thermal diffusivity of Mn_{12} are not available.

We conclude that, although the theory of deflagration agrees well with the measured conditions for the ignition of avalanches and provides a description of the avalanche velocity that is qualitatively correct, there are detailed discrepancies that suggest that additional factors need to be included in the theory to obtain good quantitative agreement with experiment.

13.5. Outlook

More experimental work is clearly needed. Measurements of the thermal diffusivity would provide an important constraint on the theory, as would a reliable determination of the (local) temperature of the deflagration front. Investigations of the influence of sample shape, size and quality would also be useful. Spatial control of the avalanche ignition points, possibly by optical means, could provide important information. Studies of the shape of the deflagration front, and its character (turbulent or laminar) would be particularly interesting.

The possibility of observing detonation is intriguing. Deflagration is but one type of combustion process. Another, more violent type, is detonation, where heat spreads from the reaction front as a shock wave rather than by diffusion. It is natural to ask whether crystals of molecular magnets can support the magnetic analog of chemical detonation. Decelle et al.[30] have reported intriguing results hinting at this possibility using high external field sweep rates (4 kT/s). The interpretation of these experiments is not entirely clear, and much work remains to be done.

We close by noting that, to the degree that magnetic deflagration resembles chemical deflagration, the magnetic manifestation of this process offers some major advantages for the systematic study of chemical combustion. The magnetic analog is non-destructive, reversible and continuously tunable using an external field. Unlike the chemical process, it is a particularly interesting realization of deflagration in which quantum mechanical tunneling plays an important role.

Acknowledgments

We thank Ferran Macià and Jonathan R. Friedman for their careful reading of the manuscript and for constructive suggestions. Support was provided by NSF Grant No. DMR-00451605.

References

1. T. Lis, *Acta Cryst.* B **36**, 2042-2046 (1980).
2. R. Sessoli, D. Gatteschi, A. Caneschi, and M. A. Novak, *Nature (London)* **365**, 141 (1993).
3. J. R. Friedman, in *Exploring the Quantum/Classical Frontier: Recent Advances in Macroscopic Quantum Phenomena*, ed. J. R. Friedman and S. Han, pp. 219-49 (Nova Science, Hauppauge, NY, 2002)
4. R. Bagai and G. Christou, *Chem. soc. Rev.* **38**, 1011 (2009).
5. D. Gatteschi and R. Sessoli, *Angew. Chem. Int.* **42**, 268 (2003).
6. D. Gatteschi, R. Sessoli and J. Villain, *Molecular Nanomagnets*, (Oxford University Press, 2006).
7. J. R. Friedman, M. P. Sarachik, J. Tejada, and R. Ziolo, *Phys. Rev. Lett.* **76**, 3830 (1996).
8. J. M. Hernandez, X. X. Zhang, F. Luis, J. Bartholome, J. Tejada and R. Ziolo, *Europhys. Lett.*, **35**, 301 (1996).
9. J. M. Hernandez, X. X. Zhang, F. Luis, J. Tejada, J. R. Friedman, M. P. Sarachik and R. Ziolo, *Phys. Rev. B*, **55**, 5858 (1997).
10. C. Paulsen and J.G. Park, in *Quantum Tunneling of Magnetization-QTM'94*, edited by L. Gunther and B. Barbara (Kluwer, Dordrecht, The Netherlands, 1995), pp. 189-207.

11. J. Tejada, E. M. Chudnovsky, J. M. Hernández, and R. Amigo, *Appl. Phys. Lett.* **84**, 2373 (2004).
12. A. Hernández-Mínguez, M. Jordi, R. Amigo, A. García-Santiago, J. M. Hernández, and J. Tejada, *Europhys. Lett.* **69**, 270 (2005).
13. O. Shafir and A. Keren, *Phys. Rev. B* **79**, 180404 (2009).
14. E. M. Chudnovsky and D. M. Garanin, *Phys. Rev. Lett.* **89**, 157201 (2002).
15. Yoko Suzuki, M. P. Sarachik, E. M. Chudnovsky, S. McHugh, R. Gonzalez-Rubio, N. Avraham, Y. Myasoedov, E. Zeldov, H. Shtrikman, N. E. Chakov and G. Christou, *Phys. Rev. Lett.* **95**,147201 (2005).
16. L. D. Landau and E. M. Lifshitz, *Fluid Dynamics* (Pergamon, 1987).
17. F. Macià, J. M. Hernandez, J. Tejada, S. Datta, S. Hill, C. Lampropoulos, and G. Christou, *Phys. Rev. B* **79**, 092403 (2009).
18. A. Hernández-Mínguez, J. M. Hernández, F. Macià, A. García-Santiago, J. Tejada, and P. V. Santos, *Phys. Rev. Lett.* **95**, 217205 (2005).
19. S. McHugh, R. Jaafar, M. P. Sarachik, Y. Myasoedov, A. Finkler, H. Shtrikman, E. Zeldov, R. Bagai and G. Christou, *Phys. Rev. B* **76**, 172410 (2007).
20. A. Caneschi, T. Ohm, C. Paulsen, D. Royal, C. Sangregorio, and R. Sessoli, *J. Magn. and Magn. Matt.* **177**, 1330 (1998); Z. Sun, D. Ruiz, N. R. Dilley, M. Soler, J. Ribas, K. Folting, M. B. Maple, G. Christou and D. N. Hendrickson, *Chem. Commun. (Cambridge)* **19**, 1973 (1999);
21. W. Wernsdorfer, R. Sessoli, D. Gatteschi, *Europhys. Lett.* **47**, 254 (1999).
22. S. McHugh, R. Jaafar, M. P. Sarachik, Y. Myasoedov, A. Finkler, E. Zeldov, R. Bagai, and G. Christou, *Phys. Rev. B* **80**, 024403 (2009).
23. A. Hernández-Mínguez, F. Macià, J. M. Hernández, J. Tejada, and P. V. Santos, *J. Mag. and Mag. Mat.* **320**, 1457 (2008).
24. S. McHugh, B. Wen B, X. Ma, M. P. Sarachik, Y. Myasoedov, E. Zeldov, R. Bagai, and G. Christou, *Phys. Rev. B* **79**, 174413 (2009).
25. D. Villuendas, D. Gheorghe, A. Hernández-Mínguez, F. Macià, J. M. Hernández, J. Tejada, and R. J. Wijngaarden, *Europhys. Lett.* **84**, 67010 (2008).
26. D.A. Garanin and E.M. Chudnovsky, *Phys. Rev. B* **76**, 054410 (2007).
27. See also A. Hernández-Mínguez, F. Macià, J. M. Hernández, J. Tejada, L. H. He, and F. F. Wang, *Europhys. Lett.* **75**, 811 (2006).
28. A. M. Gomes, M. A. Novak, R. Sessoli, A. Caneschi, and D. Gatteschi, *Phys. Rev. B* **57**, 5021(1998).
29. C. Enss and S. Hunklinger, *Low-temperature Physics*, (Springer-Verlag, Berlin, 2005).
30. W. Decelle, J. Vanacken, V. V. Moshchalkov, J. Tejada, J. M. Hernández, and F. Macià, *Phys. Rev. Lett.* **102**, 027203 (2009).

Chapter 14

Mesoscopics and the High T_c Problem

D. J. Scalapino

Department of Physics, University of California, Santa Barbara, CA 93106-9530, USA

E. Berg

Department of Physics, Harvard University, Cambridge, MA 02138, USA

S. A. Kivelson

Department of Physics, Stanford University, Stanford, CA 94305 4045, USA

Mesoscopic physics concerns itself with systems which are intermediate between a single atom and a bulk solid. Besides the many intrinsically interesting properties of mesoscopic systems, they can also provide physical insight into the physics of bulk systems. Here we discuss three examples of this from the field of high temperature superconductivity.

Many interesting electronic materials, especially those that exhibit high temperature superconductivity, lie in an intermediate coupling regime where the strength of the interaction is comparable to the electron bandwidth. The absence of any small parameter makes analytic approaches to such problems difficult. Conversely, in the absence of a small parameter there is every reason to expect the correlation lengths associated with any form of electronic order to be of order 1 in units of the lattice constant (barring an accident which places the system close to a critical point). Thus, with some sensible analysis (and with a little bit of luck), the properties of the system in the thermodynamic limit may be apparent in the properties of mesoscopic systems, even systems that are small enough that they can be studied by essentially exact numerical methods. Here we see how three such studies have provided insight into the high T_c problem.

Following the discovery of the high T_c cuprates there were various suggestions of ways in which the strongly repulsive Hubbard interaction, U, between two electrons on the same site could (paradoxically) produce high temperature pairing in a doped antiferromagnetic insulator. Numerous studies began with the $t-J$ model on a square lattice, which can be thought of as the strong coupling limit of the Hubbard model. Here it is assumed that U is sufficiently large to prevent double occupancy of any site, leaving a one-electron near-neighbor hopping term t and an exchange coupling J. An early argument for pairing was based upon the observation that if one adds two holes to the half-filled (one-electron per site) system, then eight exchange J bonds are broken if the holes were well separated. However, if the two holes are placed on near neighbor sites, only seven exchange J bonds are broken. Thus there is an effective near-neighbor attraction between the holes. This picture however was soon seen as more applicable to the phase separation regime [1]. Alternatively, in the context of the resonance-valence-bond [2] approach, variational calculations using Gutzwiller projected wavefunctions [3] and auxiliary-boson meanfield [4, 5] calculations found a superconducting state with d-wave symmetry in the $t-J$ model. From a more weak-coupling perspective, the idea of spin-fluctuation exchange mediated pairing near an antiferromagnetic instability [6] was also found to lead to d-wave pairing due to the increasingly positive strength of the pairing interaction at large momentum transfer. However, none of these approaches gave a simple, crisp real space picture, especially one that makes clear why d-wave rather than extended s-wave symmetry is preferred.

To address this, Trugman and one of the authors [7] decided to imagine that a 4-site plaquette was extracted from the lattice. The "undoped" groundstate of the $t-J$ model on a plaquette with 4 electrons is a singlet having a wave function

$$|\psi_0(4)\rangle = (\Delta^+_{12}\Delta^+_{34} - \Delta^+_{23}\Delta^+_{41})\ |0\rangle. \tag{14.1}$$

Here, $\Delta^+_{ij} = (c^+_{i\uparrow}c^+_{j\downarrow} - c^+_{i\downarrow}c^+_{j\uparrow})/\sqrt{2}$ creates a singlet pair on sites ij and we have numbered the sites of the plaquette in a clockwise manner. This state is odd under a $\pi/2$ rotation. The two-electron groundstate

$$|\psi_0(2)\rangle = N(c^+_{2\downarrow}c^+_{1\uparrow} + c^+_{4\downarrow}c^+_{1\uparrow} + \cdots)\ |0\rangle \tag{14.2}$$

has spin 0 and is invariant under a $\pi/2$ rotation. Therefore the pairfield annihilation operator that connects the zero-hole (4-electron) and two-hole (2-electron) groundstates of the 2×2 plaquette must transform as $d_{x^2-y^2}$.

The same calculation can be performed for the Hubbard model on a single plaquette; while the wave-functions are somewhat more complex, in this case, the symmetry of the 2 and 4 electron ground-states are invariant for any U in the range $0 < U < \infty$. As Carlson et al. [8] noted, it showed the robustness of the d-wave character of the pairing in $t - J$ and Hubbard models.

Of course studies of a 4-site model could not say anything about the possibility of superconducting order. However, it turned out that studies of 2-leg $t - J$ and Hubbard ladders yielded important insights concerning the character of the superconducting groundstate. The study of 2-leg ladders was motivated by a simple picture based upon the case in which the rung exchange interaction J_r is large compared to the near neighbor leg exchange J_ℓ. In this limit, for the undoped Heisenberg ladder, spin singlets tend to form on the rungs leading to a spin gapped groundstate. Then when holes are added, where $J_r > t$, they would occupy sites on either side of a rung so as to break only one exchange rung coupling. A measure of the spatial correlation of these rung hole pairs would then allow one to probe the superconducting order. Based on this large J_r/J_ℓ picture, it was initially a surprise when numerical calculations [9] showed that at half-filling the spin gap persisted to small values of J_r/J_ℓ. In addition, for the hole doped ladder, despite the fact that $t > J_r$, the equal time pairfield-pairfield correlations appeared to have a power law decay, indicative of quasi-long-range "superconductivity". In later work [10, 11], it was understood that the ladder would have a spin gap at half-filling for any finite $J_r/J_\ell > 0$, and that the groundstate of the doped ladder is a Luther-Emery [12] liquid. Furthermore, the pair structure is d-wave-like in the sense that the rung and leg pairfield amplitudes have opposite signs. We now also know that, in the limit as the length of the ladder tends to infinity, the $t-J$ ladder has perfect Andreev reflection in response to an externally applied pairfield at one end of the ladder [13]. The 2-leg $t - J$ and Hubbard ladders now represent some of the best understood models of strongly correlated electron systems.

Admittedly, since the plaquette and the ladder are, respectively, zero and one dimensional systems, neither can support a superconducting phase with a finite transition temperature. However, in many cases it is possible to analyze the phase diagram of a higher dimensional system constructed as an array of *weakly* coupled mesoscale structures, starting from the exact numerical solution of the isolated structure, and treating the coupling between clusters in the context of perturbation theory [14]. Studies of arrays of weakly coupled two-leg ladders [15, 16] and plaquettes [17, 18] (the "checkerboard Hubbard model") lead to rather complex phase diagrams with many competing phases, even where the above analysis shows strong superconducting correlations on the isolated cluster. Nonetheless, among those phases there are robust regions of d-wave, or d-wave-like superconductivity.

As a final example of insights gained from studies of small systems, we turn to calculations on a 2-leg ladder model of an Fe-pnictide superconductor [19]. Figure 14.1 shows the typical Fermi surfaces of the Fe-pnictide materials in an unfolded (1 Fe/cell) Brillouin zone. There are two-hole Fermi surfaces α_1 and α_2 around the Γ point and two-electron Fermi surfaces β_1 and β_2 around $(\pi, 0)$ and $(0, \pi)$. The symbols indicate the dominant d-orbital contributing to the Bloch state on the indicated portion of the Fermi surfaces. In weak coupling, RPA [20, 21] and functional renormalization group [22] calculations suggest that the pairing arises from the scattering of time-reversed-pairs from the d_{xz}-dominated states on the α_1 Fermi surface to paired states with the same orbital character on the β_2 Fermi surface, and from the analogous processes involving pairs in the d_{yz} dominated states on the α_1 and β_1 Fermi surfaces. This is illustrated in Fig. 14.1 for the $d_{xz} - d_{xz}$ pair scattering.

In order to use numerical methods to study these processes in the intermediate to strong coupling limit, the problem needs to be simplified. If we accept that the type of scattering processes shown in Fig. 14.1 capture the essential physics, we can focus exclusively on pair scattering involving two bands and only one orbital. The resulting two-leg Hubbard ladder retains the d_{xz} states along two cuts through the 2d BZ, $\mathbf{k} = (k_x, 0)$ which passes through the α_1 Fermi surface, and $\mathbf{k} = (k_x, \pi)$ which passes through β_2. This reduces the problem to that of the two-leg Hubbard ladder shown in Fig. 14.2a which can then be studied using the numerical density matrix renormalization group (DMRG) [23]. The one-electron hopping parameters $t_1 = -0.32$, $t_3 = -0.57$ in units of t_2, were chosen to reproduce the density functional bandstructure [24] near the Fermi surface for $k_y = 0$ and π. The repulsion U between two electrons in the same orbital was varied in the range 3–4 in units of t_2.

In the undoped, one electron per site, limit one finds the expected spin gapped groundstate. By applying a magnetic field to one of the end sites of the ladder, the resulting expectation value of the spin appears as shown in Fig. 14.2b. Here one sees "stripe"-like $(0, \pi)$ spin correlations which decay with a slow exponential due to the spin gap. Hole doping the system and applying an external pairfield on the end rung of the ladder, one obtains

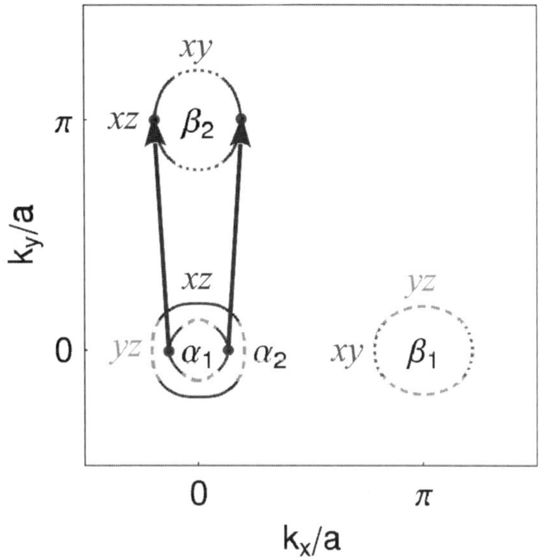

Fig. 14.1. The Fermi surfaces for a five-orbital tight binding model of the Fe-pnictides. The main orbital contributions to the Bloch states are indicated: d_{xz} (solid line), d_{yz} (dashed line) and d_{xy} (dotted line). The arrows illustrate the type of $d_{xz} - d_{xz}$ inter-Fermi surface scattering processes that lead to pairing in the spin-fluctuation-exchange calculations.

the pairfield singlet amplitudes illustrated in Fig. 14.2c. Here a pairfield of unit strength was applied to the left end rung and Fig. 14.2c shows the strength of the induced pairfields $\langle \Delta_{ij} \rangle$ ten sites to the right. The relative positive sign of the pairfield on the rung and diagonal and the negligible value of the pairfield on the legs is expected if the gap changes sign between the α_1 and β_2 Fermi surfaces [20, 21, 25].

The stripe-like SDW pattern of the spin correlations in the undoped system as well as the structure of the pairfield are consistent with what is found in the RPA calculations [20, 21]. However, what we found most interesting was the relationship between the Fe-ladder and the previously studied 2-leg cuprate ladder. This is illustrated in Fig. 14.3. Here in Fig. 14.3a, every other rung has been twisted by 180° and the phase of the d_{xz}-orbit has been changed by π on each of the sites of the twisted rung. In this way, the rung one-electron hopping matrix element remains t_2, but the leg and diagonal hoppings are changed to $-t_3$ and $-t_1$, respectively. The dominant hoppings on the twisted Fe-ladder are along the legs and rungs with only a weak diagonal hopping. The spin correlations shown in Fig. 14.3b, obtained

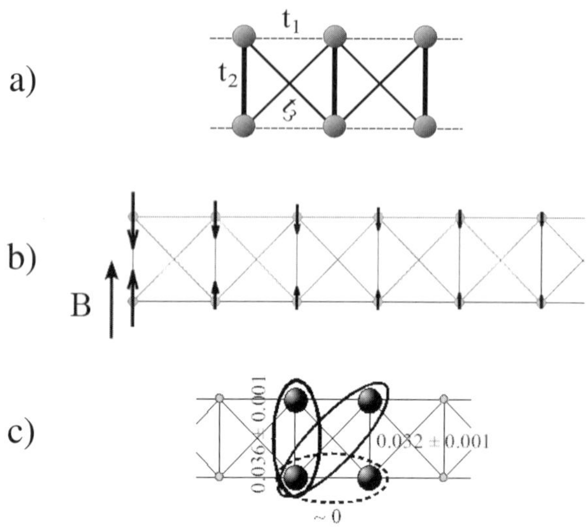

Fig. 14.2. a) An Fe two-leg ladder with $t_1 = -0.32$, $t_3 = -0.57$ and $U = 3$ in units where $t_2 = 1$. These hopping parameters were chosen to fit the DFT calculation of the bandstructure for cuts with $k_y = 0$ and π.; b) The spin structure $\langle S^z(\ell_x, \ell_y)\rangle$ induced on the undoped Fe ladder when an external magnetic field is applied to the lower left hand site.; c) The singlet pairfield $\langle \Delta_{ij}\rangle$ induced across a rung, along a diagonal and along a leg at a distance 10 sites removed from the end of a 32×2 Fe-ladder with a unit external pairfield applied to its end rung.

by twisting every other rung of Fig. 14.2b, look just like the spin gapped AF correlations of the previously studied 2-leg Hubbard cuprate ladder. Because of the twist and the phase change $e^{i\pi} = -1$ of the orbitals on the sites of the twisted rungs, the pairfield correlations take on the d-wave-like form shown in Fig. 14.3c. In short, the twist maps $(\pi, 0)$ magnetic and sign-changing s-wave pairing correlations on the Fe-ladder into (π, π) magnetic and d-wave-like pairing correlations in the cuprate ladder! Finally, it turns out that the ratio of the leg-to-rung hopping 0.57 obtained from the fit to the Fe-pnictide DFT bandstructure is near the value which was previously found [26] to give the slowest pairfield decay for a cuprate ladder. Thus this Fe-ladder turns out to simply be a twisted version of the cuprate 2-leg Hubbard ladder with parameters near those which are optimal for pairing. This provides a direct link between the physics of these two materials.

Now, as noted by Joe in his book *Introduction to Mesoscopic Physics* [27], "the interest in studying systems in the intermediate size range between microscopic and macroscopic is not only in order to understand the macroscopic limit. Many novel phenomena exist that are intrinsic to

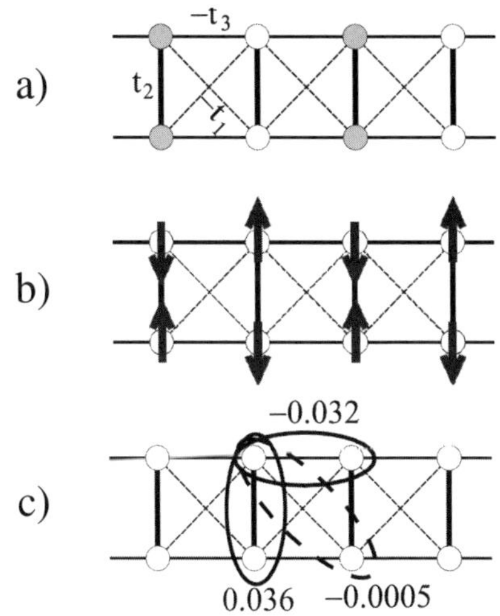

Fig. 14.3. a) Here every other rung of the ladder shown in Fig. 14.2a has been twisted by 180° and the phases of the orbitals denoted by the shaded sites have been changed by $e^{i\pi} = -1$.; b) The spin expectation values of Fig. 14.2b for the twisted ladder show the spin gapped (π, π) antiferromagnetic behavior of the familiar cuprate ladder.; c) The induced pairfield correlations of Fig. 14.2c become the familiar d-wave-like pairing correlations seen for models of the cuprate ladders.

mesoscopic systems." Here we have only touched on some examples where strongly correlated mesoscopic models have been introduced in the hope that they can provide some insight into the macroscopic high T_c problem. It is natural to ask whether there aren't novel mesoscopic phenomena as well. Indeed, there are. For example, the difference between the even- and odd-legged Heisenberg ladders in which the even-leg ladders have a spin gap while the odd-leg ladders are gapless is a mesoscopic width effect [15]. It is also known that while the doped 2-leg ladder goes into a Luther-Emery phase [12], it takes a finite doping to bring the 3-leg ladder into this phase [28, 29]. Ladders also appear in the striped phase of the cuprates and, a better understanding of the mesoscopic properties of multi leg ladders may shed light on the recently proposed π-phase shifted d-wave stripes [30].

Acknowledgments

It is a pleasure to contribute to this volume in honor of Yoseph Imry's 70th birthday. We know Joe as a scientist, teacher, co-worker and friend. DJS met Joe in the early seventies when Joe first came to UCSB. He remembers how they often talked about how one might gain insight into the basic physics of a macroscopic system from calculations on small subsystems and how these subsystems had their own interesting features. All of us are eager to recognize how much we have learned from Joe's work, over the years concerning the intrinsic, subtle and beautiful physics one finds in the mesoscopic world. This work was supported in part by the National Science Foundation under the grant PHY05-51164 at the KITP. DJS acknowledges the Center for Nanophase Materials Science at ORNL, which is sponsored by the Division of Scientific User Facilities, U.S. DOE. EB was supported by the NSF under grants DMR-0705472 and DMR-0757145 at Harvard. SAK was supported, in part, by the NSF grant number DMR-0758356 at Stanford.

References

1. V.J. Emery, S.A. Kivelson and H.Q. Lin, *Phys. Rev. Lett.* **64**, 475 (1990).
2. P.W. Anderson, *Science* **235**, 1196 (1987).
3. C. Gros, R. Joynt and T.M. Rice, *Z. Phys. B* **68**, 425 (1987).
4. A.E. Ruckenstein, P.J. Hirschfeld and J. Appel, *Phys. Rev. B* **36**, 857 (1987).
5. G. Kotliar and J. Liu, *Phys. Rev.* **38**, 5142 (1988).
6. N.E. Bickers, D.J. Scalapino and R.T. Scalettar, *Int. J. Mod. Phys.* **B1**, 687 (1987).
7. D.J. Scalapino and S.A. Trugman, *Philos. Mag. B* **74**, 607 (1996).
8. E.W. Carlson, V.J. Emery, S.A. Kivelson and D. Orgad, Chap. 21, Superconductivity **Vol. 1**, eds. K.H. Bennemann and J.B. Ketterson (Springer, 2006).
9. E. Dagatto, J. Riera and D. Scalapino, *Phys. Rev. B* **45**, 5744 (1992).
10. For a review see E. Dagatto and T.M. Rice, *Science* **271**, 618 (1996).
11. S. Chakravarty, *Phys. Rev. Lett.* **77**, 4446 (1996).
12. A. Luther and V.J. Emery, *Phys. Rev. Lett.* **33**, 589 (1974).
13. A.E. Feiguin, S.R. White, D.J. Scalapino and I. Affleck, *Phys. Rev. Lett.* **101**, 217001 (2008).
14. D.J. Scalapino, Y. Imry and P. Pincus, *Phys. Rev. B* **11**, 2042 (1975).
15. H. Tsunetsugu, M. Troyer and T.M. Rice, *Phys. Rev. B* **49**, 16078 (1994).
16. E. Arrigoni, E. Fradkin and S.A. Kivelson, *Phys. Rev. B* **69**, 214519 (2004).
17. E. Altman and A. Auerbach, *Phys. Rev. B* **65**,104508 (2002).
18. H. Yao, W-F. Tsai and S. A. Kivelson, *Phys. Rev. B* **76**, 161104 (2007).
19. E. Berg, S.A. Kivelson and D.J. Scalapino, *New J. Phys.* **11**, 085007 (2009).
20. K. Kuroki, S. Onari, R. Arita, H. Usui, Y. Tanaka, H. Kontani and H. Aoki, *Phys. Rev. Lett.* **101** 087004 (2008).
21. S. Graser, T.A. Maier, P.J. Hirschfeld and D.J. Scalapino, *New J. Phys.* **11** 025016 (2009).

22. F. Wang, H. Zhai, Y. Ran, A. Vishwanath, and D.-H. Lee, *Phys. Rev. Lett.* **102**, 47005 (2009).
23. S.R. White, *Phys. Rev. Lett.* **69**, 2863 (1992).
24. C. Cao, P.J. Hirschfeld and H.-P. Cheng, *Phys. Rev. B* **77**, 220506(R) (2008).
25. I.I. Mazin, D.J. Singh, M.D. Johannes and M.H. Du, *Phys. Rev. Lett.* **101**, 057003 (2008).
26. R.M. Noack, N. Bulut, D.J. Scalapino and M.G. Zacher, *Phys. Rev. B* **56**, 7162-7166 (1997).
27. Y. Imry, *Introduction to Mesoscopic Physics*, Oxford University Press (1997).
28. T.M. Rice, S. Haas, M. Sigrist and F. Zhang, *Phys. Rev. B* **56**, 6569 (1997).
29. S.R. White and D.J. Scalapino, *Phys. Rev. B* **57**, 3031 (1998).
30. E. Berg, E. Fradkin, S.A. Kivelson and J.M. Tranquada, cond-mat/arXiv:0901.4826.

Chapter 15

An Effective Theory of Pulse Propagation in a Nonlinear and Disordered Medium in Two Dimensions

Georg Schwiete

Department of Physics, Texas A&M University, College Station, TX 77843-4242, USA

Alexander M. Finkel'stein

Department of Physics, Texas A&M University, College Station, TX 77843-4242, USA and Department of Condensed Matter Physics, The Weizmann Institute of Science, 76100 Rehovot, Israel

We develop an effective theory of pulse propagation in a nonlinear *and* disordered medium. The theory is formulated in terms of a nonlinear diffusion equation. Despite its apparent simplicity this equation describes novel phenomena which we refer to as "locked explosion" and "diffusive" collapse. The equation can be applied to such distinct physical systems as laser beams propagating in disordered photonic crystals or Bose-Einstein condensates expanding in a disordered environment.

15.1. Introduction

In recent years, novel experimental techniques made possible first observations of wave-packets evolving in the presence of random scatterers and nonlinearities. In a number of optical experiments, a laser beam was sent into a nonlinear optical medium with a random refractive index, and the beam profile in the transverse direction(s) was monitored on the opposite side of the sample,[1,2] for an illustration see Fig. 15.1(a). In a second class of experiments, atoms forming a Bose-Einstein condensate were released from a trap and subjected to a disorder potential during the expansion,[3–8] see Fig. 15.1(b). The experiments were inspired by the idea that in these setups, unlike for transport experiments in electronic systems, one can *visualize* the phenomenon of Anderson localization, whereby a wave-packet

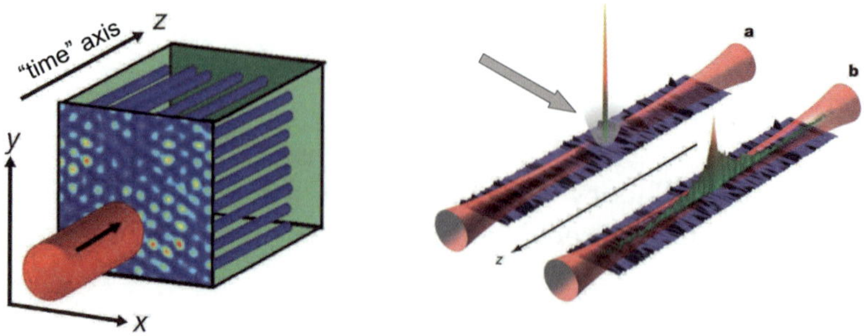

(a) (b)

Fig. 15.1. (a) In the experiment of Ref. 1 a laser beam (in red) is sent onto a disordered nonlinear photonic crystal. Inside the crystal, the evolution of the smooth envelope of the electric field in the transverse (xy) plane is approximately described by the nonlinear Schrödinger equation (NLSE). In this equation, the z-coordinate plays the role of "time". The intensity profile of the outgoing beam is measured at the opposite side of the crystal. Adapted from Schwartz et al., Ref. 1. (b) In the experiment of Ref. 7 an atomic Bose-Einstein condensate is released from a small trap (displayed in white) and subsequently expands along a one-dimensional channel, formed by an external confining potential (in red). During the expansion, the atoms are subjected to a disorder potential (in blue). After a while, the motion comes to a halt and the condensate becomes localized. The experiment was performed at small densities, so that the nonlinear term in the Gross-Pitaevskii equation can be considered as small. Adapted from Billy et al., Ref. 7.

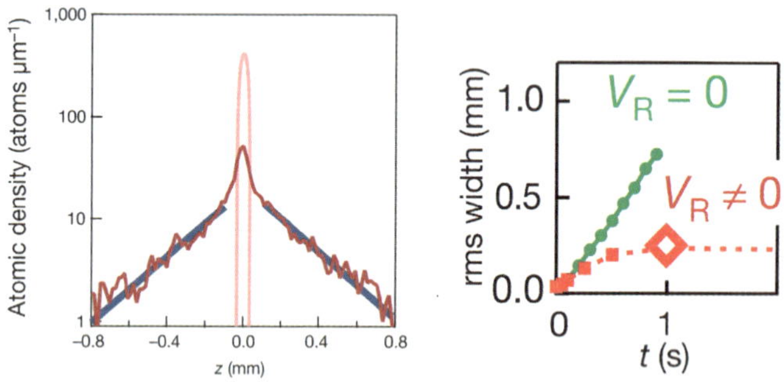

Fig. 15.2. On the left hand side, the profile of the condensate at the final stage of the experiment of Ref. 7 is displayed on a logarithmic scale. The tails of the wave function decay exponentially. This is interpreted as evidence for Anderson localization. On the right hand side the rms width of the condensate is plotted as a function of time. In the presence of a disorder potential (red line), the width approaches a constant value, while in the absence of disorder (green line), the condensate expands ballistically. Adapted from Billy et al., Ref. 7.

or quantum particle is confined within a finite volume as a result of multiple scattering on a random potential (Figs. 15.2,15.3,15.4). The evolution of the injected wave-packet in both experiments can be described by the non-linear Schrödinger equation (NLSE), in the context of atomic Bose-Einstein condensates referred to as the Gross-Pitaevskii Equation (GPE). This equation differs from the linear Schrödinger equation by an additional cubic term and is used as a paradigmatic description for nonlinear waves. The nonlinearity is a consequence of interactions between particles in the case of atomic condensates and of a change in the refractive index in response to the electric field (Kerr effect) in the case of laser beams.

Motivated by these experiments we derive, starting from the GPE/NLSE, a kinetic equation that describes the evolution of an injected wave-packet in a weakly disordered nonlinear medium in two dimensions. Analysis of this equation reveals a rather nontrivial picture: Irrespective of the sign of the nonlinearity the mean square radius of the wave-packet changes linearly in time, $\partial_t \left\langle r^2 \right\rangle \propto E_{tot}$, where E_{tot} is the total energy of the wave-packet. For a repulsive nonlinearity the initial change of the profile displays features of an explosion, although the overall size of the wave-packet is growing slowly as in ordinary diffusion. For an attractive nonlinearity, the radius can either grow or decrease, depending on the sign of E_{tot}. In particular, for $E_{tot} < 0$ we predict a slow "diffusive" collapse as the radius of the wave-packet shrinks towards zero.

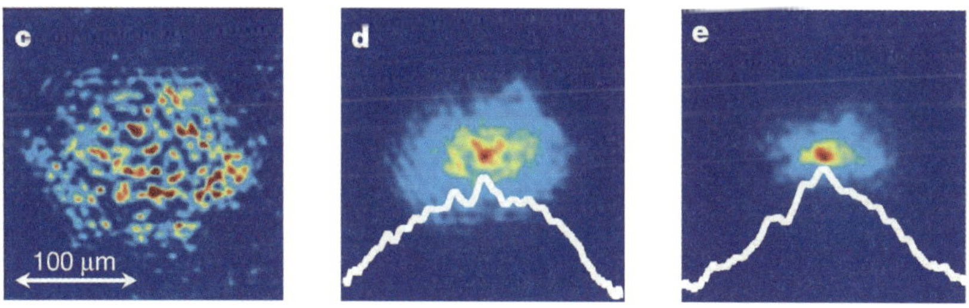

Fig. 15.3. The intensity distribution of the outgoing laser beam in the experiment of Ref. 1 is shown in a regime for which the nonlinearity is negligible. The disorder level increases from left to right, starting from the clean lattice. All distributions are averaged over different disorder configurations. The white lines display the logarithm of the intensity as a function of the transverse coordinate. The authors of Ref. 1 interpret the distribution in the middle as the result of diffusion, and the distribution on the right as a signature of Anderson localization. Courtesy of Schwartz et al., Ref. 1.

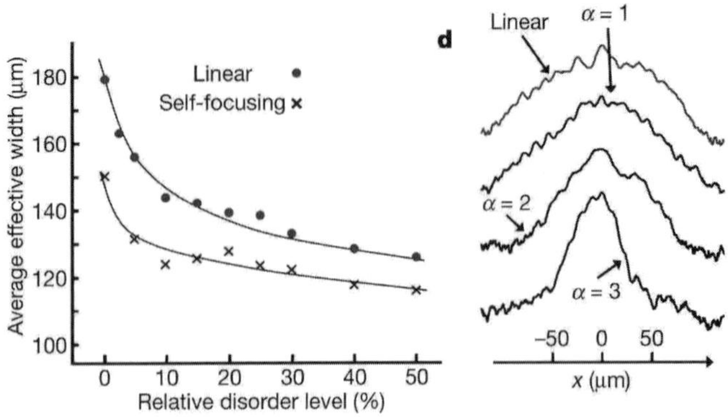

Fig. 15.4. The average width of the outgoing laser beam as a function of the disorder level in the experiment of Ref. 1 is compared for the linear case and in the presence of a self-focusing nonlinearity (left panel). The right panel compares (the logarithm of) the averaged intensity profile for a fixed disorder strength as a function of the transverse coordinate. The parameter α is a dimensionless measure for the strength of the nonlinearity. The main result is that the self-focusing nonlinearity promotes localization. Courtesy of Schwartz et al., Ref. 1.

In this paper we will mostly use the language related to the GPE, but also indicate below how to translate to a language more suitable for optical experiments. The GPE describes the evolution of a wave-function Ψ:[9]

$$i\partial_t\Psi(\mathbf{r},t) = -\frac{1}{2m}\nabla^2\Psi(\mathbf{r},t) + u(\mathbf{r})\Psi(\mathbf{r},t) + \lambda|\Psi(\mathbf{r},t)|^2\Psi(\mathbf{r},t), \quad (15.1)$$

where we set $\hbar = 1$. For positive (negative) λ this equation contains a repulsive (attractive) self-consistent potential $\lambda|\Psi(\mathbf{r},t)|^2$. This corresponds to a nonlinearity of the de-focusing (self-focusing) type. The static disorder potential $u(\mathbf{r})$ is the source of randomness in the equation. For simplicity we choose for our calculation a Gaussian white noise potential with correlation function $\langle u(\mathbf{r})u(\mathbf{r}')\rangle = \delta(\mathbf{r}-\mathbf{r}')/(m\tau)$ [for a discussion of averaging for speckle potentials see, e.g., Ref. 10]. The angular brackets denote averaging over disorder configurations and τ is the scattering time.

The NLSE used in optics is derived in the so-called paraxial approximation,[11] and describes the evolution of the smooth envelope of the electric field. The main propagation direction of the laser beam, say the z-direction, plays the role of time in the NLSE, see Fig. 15.1(a). In this sense, the disorder potential which results from random variations of the refractive index is static when it is z-independent. For example, the two-dimensional (2d) transverse evolution of a pulse is studied in a 3d sample. The intensity of the beam is proportional to $|\Psi(\mathbf{r}, z)|^2$. In the NLSE, the mass m in the GPE is replaced by the wave vector $k = \omega/c$, where ω is the frequency of the carrier wave and c the velocity of light in the medium.

For a condensate released from a confining harmonic oscillator potential, as is typical for experiments on cold atomic gases, the GPE *without disorder* can be solved exactly.[12,13] During an initial stage the potential energy originating from the nonlinearity is almost entirely converted into kinetic energy. This period of violent acceleration is followed by a second stage, during which the nonlinearity is no longer essential. Expansion in the presence of disorder in the two-dimensional case was recently addressed in reference.[14] In this paper it has been assumed that for repulsive nonlinearity an initial ballistic stage is not affected by disorder, while the subsequent diffusive expansion is not affected by the nonlinearity, thereby separating the two effects. In contrast, we are interested here in the *interplay* of disorder and nonlinearity, both attractive and repulsive. This is especially interesting in 2d, as it is known that for linear wave propagation and weak disorder there is an extended diffusive regime preceding localization on (exponentially) large length scales. It is this regime that we address.[15]

Since the system is far out of equilibrium, we choose to work with a kinetic equation. The derivation of the kinetic equation proceeds as follows. We use methods of classical statistical field theory to derive a functional integral expression for the disorder averaged density.[16,17] The formalism involves a doubling of the degrees of freedom, similar to the Keldysh or closed-time-path approaches for quantum systems,[17] where two fields are introduced on forward and backward time-contours. Instead of averaging over a statistical ensemble in the initial state, we assume that the wave-function at the initial time is known. Averaging is performed over disorder configurations. Scattering on impurities is included on the level of the self-consistent Born approximation. While interference (weak localization) corrections are not covered by this approximation, it allows for a consistent description of diffusion in the presence of nonlinearity. The nonlinearity is treated by introducing a self-consistent potential $\vartheta(\mathbf{r}, t)$. In this way it is possible to include interaction effects in a non-perturbative way, which is crucial for the problem at hand. To obtain the kinetic equation for the density in the diffusive limit, we assume that the initial wave-function sets a momentum scale p_0 characterizing the main part of the momentum distribution, so that the weak disorder condition $p_0 l \gg 1$ is fulfilled, where $l = p_0\tau/m$ is the mean free path. We further assume that the density varies smoothly on scales of l, in particular that the size of the condensate is much larger than the mean free path. Both of these conditions can be met simultaneously. The phase of Ψ, which is related to the momentum, may change rapidly, while the amplitude, which determines the density, may vary smoothly. Even if the density does not satisfy the smoothness condition initially, it is natural to expect that in the case of an expansion it will become sufficiently smooth after some time.[18] The derivation of the kinetic equation will be presented elsewhere.[19]

Starting from Eq. (15.1), the outlined steps lead to the following kinetic equation in the diffusive regime

$$\begin{aligned}&\partial_t\tilde{n}(\mathbf{r},t,\varepsilon)-\nabla(D_{\varepsilon-\vartheta}\nabla\tilde{n}(\mathbf{r},t,\varepsilon))+\partial_t\vartheta(\mathbf{r},t)\partial_\varepsilon\tilde{n}(\mathbf{r},t,\varepsilon)\\&=\delta(t)\,F(\varepsilon-\vartheta(\mathbf{r},0),\mathbf{r})\end{aligned}\tag{15.2}$$

where

$$F(\varepsilon,\mathbf{r})=\int\frac{d^2qd^2p}{(2\pi)^4}\,F(\mathbf{p},\mathbf{q})\,\exp(i\mathbf{qr})\,2\pi\delta(\varepsilon-\varepsilon_{\mathbf{p}}),\tag{15.3}$$

and $F(\mathbf{p},\mathbf{q})=\Psi_0(\mathbf{p}+\mathbf{q}/2)\Psi_0^*(\mathbf{p}-\mathbf{q}/2)$ is determined by the initial wave function Ψ_0; $\varepsilon_{\mathbf{p}}=p^2/(2m)$ is the kinetic energy, and $D_\varepsilon=\varepsilon\tau/m$ the diffusion coefficient. The equation should be supplemented with the self-consistency relation for the potential $\vartheta(\mathbf{r},t)=2\lambda n(\mathbf{r},t)$, where $n(\mathbf{r},t)=\int d\varepsilon/(2\pi)\;\tilde{n}(\mathbf{r},t,\varepsilon)$. Despite its apparent simplicity it is a rather complicated nonlinear integro-differential equation. The equation effectively sums an infinite series of diagrams of the type shown in Fig. 15.5. It is a peculiarity of the perturbation theory for a classical field equation such as the GPE that no closed loops arise,[20] making it quite distinct from the related problem in interacting electron systems. The relation between certain blocks appearing in diagrammatic perturbation theory and the corresponding terms in the kinetic equation is visualized in Fig. 15.6.

The physics described by this equation is essentially classical. Imagine first that the potential ϑ does not depend on time. Consider now a particle diffusing with total energy ε on the background of a smoothly varying potential ϑ, see Fig. 15.7 for illustration. If scattering events are frequent enough, the diffusion coefficient is determined by the kinetic energy $\varepsilon_{\mathbf{p}}=\varepsilon-\vartheta$ that varies locally in space. If the potential additionally varies in time, the particle may change its total energy. If on the other hand the potential depends on time *only*, the kinetic energy does not change. Therefore, it is expected that a purely time-dependent potential has no effect on the density. This observation is related to the fact that in the original GPE a

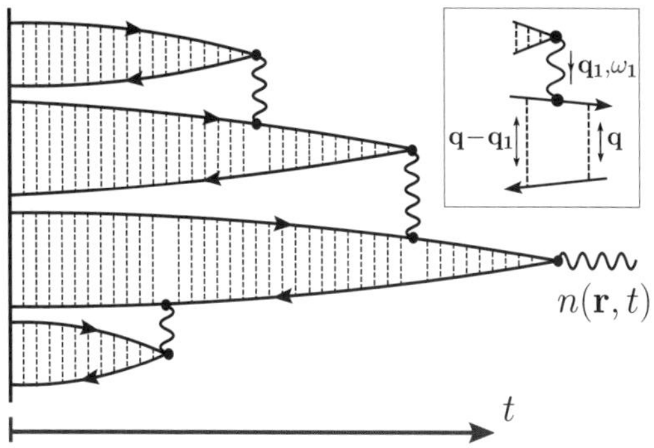

Fig. 15.5. Graphical illustration of the kinetic equation, Eq. (15.2). The injection process takes place on the left. Solid lines are disorder averaged particle propagators. Ladders graphically represent density diffusion. The particular way of disorder averaging is justified for $\varepsilon_{kin}\tau \gg 1$; ε_{kin} is defined below Eq. (15.6). The wavy lines account for the nonlinearity in Eq. (15.1). The block magnified in the inset gives rise to the terms $\nabla(D_\vartheta \nabla \tilde{n}(\mathbf{r}, t, \varepsilon))$ and $\partial_t \vartheta(\mathbf{r}, t)\partial_\varepsilon \tilde{n}(\mathbf{r}, t, \varepsilon)$ in the kinetic equation, Eq. (15.2).

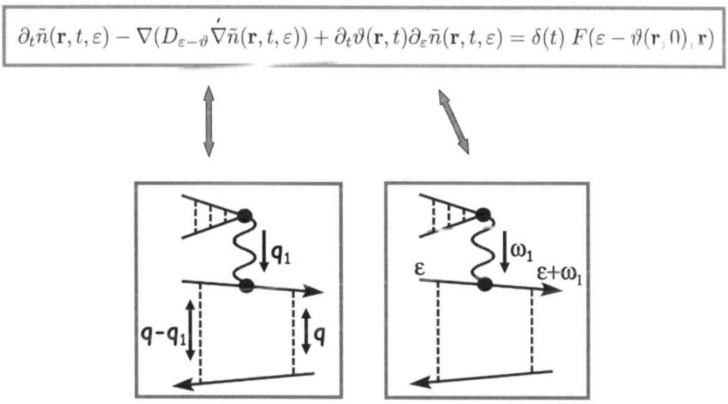

Fig. 15.6. This figure illustrates the correspondence between different terms in the kinetic equation and a certain block in perturbation theory.

purely time dependent potential $V(t)$ may be removed by a suitably chosen gauge-transformation,

$$\Psi(\mathbf{r}, t) \to \Psi(\mathbf{r}, t)\ \exp\left(-i \int_{t_0}^{t} dt' V(t')\right), \tag{15.4}$$

that does not affect the density $|\Psi(\mathbf{r}, t)|^2$. Indeed, we can make this point obvious in Eq. (15.2) by shifting the energy variable so that it will correspond to the kinetic energy instead of the total energy, $n(\mathbf{r}, \varepsilon, t) =$

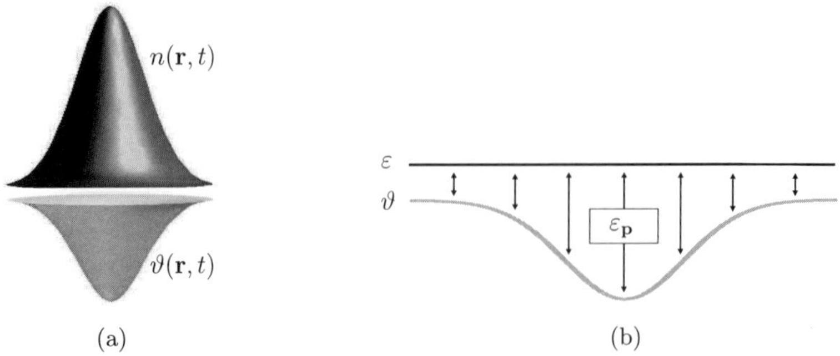

Fig. 15.7. The density distribution $n(\mathbf{r},t)$ creates its own self-consistent potential $\vartheta(\mathbf{r},t)$. We illustrate the attractive (focusing) case. In a classical analogy, the variable ε in Eq. (15.2) can be interpreted as the total energy and ϑ as the potential energy of a diffusing particle. Correspondingly, the diffusion coefficient is determined by the kinetic energy $\varepsilon_{\mathbf{p}} = \varepsilon - \vartheta$.

$\tilde{n}(\mathbf{r}, \varepsilon + \vartheta(\mathbf{r},t), t)$. Expressed in the new coordinates the equation reads

$$\partial_t n(\mathbf{r},\varepsilon,t) - [\nabla - \nabla\vartheta_{\mathbf{r},t}\partial_\varepsilon]\, D_\varepsilon\, [\nabla - \nabla\vartheta_{\mathbf{r},t}\partial_\varepsilon]\, n(\mathbf{r},\varepsilon,t) = \delta(t)\, F(\varepsilon,\mathbf{r}). \quad (15.5)$$

An equation for the density $n(\mathbf{r},t)$ can be obtained by integrating Eq. (15.5) in ε

$$\partial_t n(\mathbf{r},t) - \frac{\tau}{m}\nabla^2\left(\bar{\varepsilon}(\mathbf{r},t) + \lambda n^2(\mathbf{r},t)\right) = \delta(t)\, n(\mathbf{r},0), \quad (15.6)$$

where

$$\bar{\varepsilon}(\mathbf{r},t) = \int \frac{d\varepsilon}{2\pi}\, \varepsilon n(\mathbf{r},t,\varepsilon) \equiv \varepsilon_{kin}(\mathbf{r},t) n(\mathbf{r},t). \quad (15.7)$$

It can be written in the compact form $\partial_t n - \nabla^2(D_{eff} n) = \delta(t) n$ when defining an effective space and time-dependent diffusion coefficient $D_{eff} = (\varepsilon_{kin} + \lambda n)\tau/m$. The apparent simplicity of these equations is however deceiving. They are not closed equations for the density evolution, since the kinetic energy $\bar{\varepsilon}$ depends on the nonlinearity and needs to be determined separately via Eq. (15.5). Nevertheless, we arrive at the conceptually important result that in the diffusive regime the nonlinearity effectively introduces a density dependence of the diffusion coefficient. It seems clear that a closed form solution of the nonlinear equations for arbitrary initial conditions cannot be found. In order to make progress we will rely on two approaches: the use of conservation laws and the study of solvable limiting cases. When combined, they will enable us to arrive at a qualitative picture both for repulsive and attractive nonlinearity.

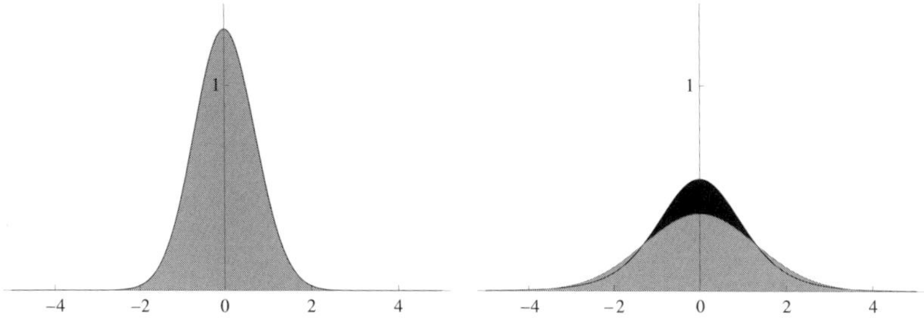

Fig. 15.8. This figure compares two different *linear* diffusion processes, one with a fixed diffusion coefficient (gray) and one with *energy*-dependent diffusion coefficient D_ε and an initial energy distribution of the form $C\exp(-\varepsilon/\varepsilon_0)$, where C and ε_0 are constants. On the left hand side the initial density distribution is shown, which is chosen to be identical for both processes. After a while, the density that corresponds to energy dependent diffusion is larger both for small distances and for large distances from the origin when compared to the diffusion process with a fixed diffusion constant D_{ε_0}. This reflects the fact that both energies $\varepsilon \ll \varepsilon_0$ and $\varepsilon \gg \varepsilon_0$ are present in the initial energy distribution, leading to slow and fast diffusion respectively.

First we briefly discuss the linear case, $\vartheta = 0$. In the absence of nonlinearity, $n(\mathbf{r}, \varepsilon, t)$ evolves independently for each energy ε. In this limit, Eq. (15.5) has the obvious solution

$$n(\mathbf{r}, \varepsilon, t) = \frac{\Theta(t)}{4\pi D_\varepsilon t} \int d\mathbf{r}_1 \, \mathrm{e}^{-(\mathbf{r}-\mathbf{r}_1)^2/(4D_\varepsilon t)} F(\varepsilon, \mathbf{r}_1). \qquad (15.8)$$

Here, for each energy ε diffusion is determined by the corresponding diffusion coefficient D_ε, and should be weighted according to the energy distribution in the injected wave-packet. The linear case was discussed in Ref. 14. Starting from a broad energy distribution centered around some ε_0, both tails and central part of the density in the long time limit are more pronounced compared to diffusion at fixed energy ε_0, because the tails are determined by energies $\varepsilon > \varepsilon_0$, while the central part is dominated by energies $\varepsilon < \varepsilon_0$, see Fig. 15.8.

Next we turn to the nonlinear case. We will make use of the conservation laws for particle number and energy. By integrating Eq. (15.6) over $\mathbf{r}$, we obtain that the particle number (or normalization) $\int d\mathbf{r} \; n(\mathbf{r}, t) = N$ is fixed in time. An equation for $\overline{\varepsilon}(\mathbf{r}, t)$ can be derived by first multiplying Eq. (15.5) by ε before integrating in this variable. Then by combining the equation for $n(\mathbf{r}, t)$ with the equation for $\overline{\varepsilon}(\mathbf{r}, t)$ we find that the energy

$$E_{tot} = \int d\mathbf{r} \; \left(\overline{\varepsilon}(\mathbf{r}, t) + \lambda n^2(\mathbf{r}, t)\right) \qquad (15.9)$$

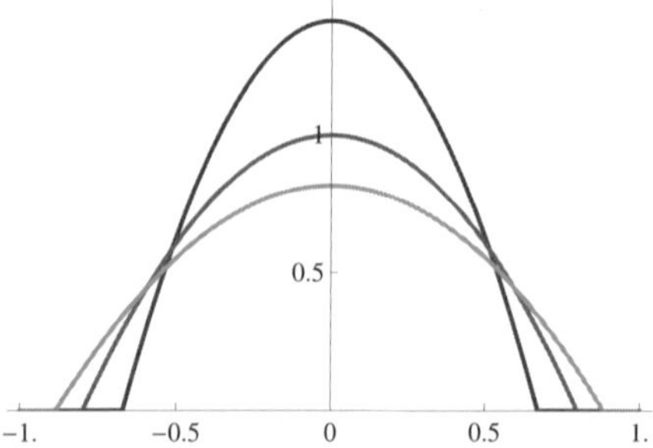

Fig. 15.9. This figure shows the time-evolution of the Barenblatt-solution of the porous medium equation. It has the shape of an inverted parabola. It is worth noting that the wave-front has a finite spatial derivative at the boundary of the distribution. For this solution, the mean radius squared evolves rapidly as $\langle r^2 \rangle \propto t^{1/2}$ and correspondingly the density at $r = 0$ drops as $n(0,t) \propto t^{-1/2}$.

is constant in time. Remarkably, this conservation law completely determines the time evolution of the mean radius squared of the wave-packet, $\langle r^2 \rangle \equiv \int d\mathbf{r}\ r^2\ n(\mathbf{r},t)/N$. Indeed, multiplying Eq. (15.6) by r^2 and subsequently integrating in $\mathbf{r}$ one obtains that $\partial_t \langle r^2 \rangle = 4D_{\varepsilon_{tot}}$, where $\varepsilon_{tot} = E_{tot}/N$. The linear dependence of the mean square radius on time during the whole evolution is guarded by energy conservation. This is one of the central results of this paper. When compared to the linear case, the effective diffusion coefficient $D_{\varepsilon_{tot}}$ is reduced for attractive and enhanced for repulsive nonlinearities.

In the following we discuss more specifically the repulsive and attractive cases. For the repulsive nonlinear case it is instructive to consider a situation in which the second term on the RHS of Eq. (15.6) dominates. The equation $\partial_t n = \nabla^2 n^2$, which one obtains after simple rescaling, is an example of the famous porous medium equation (PME).[21] For the $2d$ case the solution describing the evolution of a delta-function pulse $M\delta(\mathbf{r})$ is given by $n(\mathbf{r},t) = (C - r^2/(16t^{1/2}))/t^{1/2}$, where $C^2 = M/(8\pi)$.[22,23] This solution is often referred to as Barenblatt's solution. It conserves the normalization $\int d\mathbf{r}\ n(\mathbf{r},t) = M$ but, unlike ordinary diffusion, it is nonzero only in a finite region of space, see Fig. 15.9. The special importance of Barenblatt's solution in the theory of the PME is related to the fact that, roughly speaking, any solution starting from a sufficiently benign initial pulse with weight M is eventually well-approximated by Barenblatt's solution with the same weight.[21]

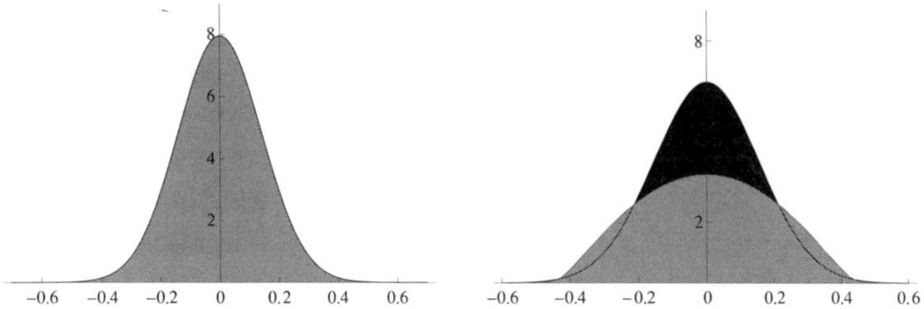

Fig. 15.10. This figure compares a conventional diffusion process (black) to the time evolution described by the porous medium equation (gray) in two spatial dimensions. The same initial distribution is chosen for both processes and displayed on the left hand side. One can see, that the density at the center drops much faster for the solution of the porous medium equation. Simultaneously, the mean radius grows much faster, so that the total density is conserved.

For Barenblatt's solution, the mean radius squared evolves as $\langle r^2\rangle \propto t^{1/2}$ and the density at $r = 0$ drops as $n(0,t) \propto t^{-1/2}$. At short times this solution describes a much faster "explosive" evolution than the source-type solution of the diffusion equation, for which $\langle r^2\rangle \propto t$ and $n(0,t) \propto t^{-1}$, see Fig. 15.10. At first sight there seems to be a contradiction. If one injects a bell-shaped pulse with a large potential energy, it appears that the potential part of the effective diffusion coefficient $D_{eff} = (\varepsilon_{kin} + \lambda n)\tau/m$ dominates. Therefore, naively, one would assume that the initial evolution is "explosive", while our exact result $\langle r^2\rangle = r_0^2 + 4D_{\varepsilon_{tot}}t$ rules out this possibility. This puzzle can be resolved in the following way. The explosion takes place only in the central part of the density distribution, which has only a small weight when calculating $\langle r^2\rangle$. Right in the center, for $\mathbf{r} = 0$, Eq. (15.6) can be written as

$$\partial_t n = \frac{\tau}{m}\left[\nabla^2\overline{\varepsilon} + 2\lambda n \nabla^2 n\right] \tag{15.10}$$

for $t > 0$; we consider here a rotationally symmetric distribution with $\nabla n(0,t) = 0$ and $\nabla^2 n(0,0) < 0$. For sufficiently large λn, the potential part is dominant and leads to a fast initial decrease of the density before either $\lambda n \nabla^2 n$ becomes small or $\nabla^2\overline{\varepsilon}$ becomes positive as a consequence of the outward-flow of the kinetic energy. Away from the center, where the density and correspondingly the term $2\lambda n\nabla^2 n$ are small, Eq. (15.6) takes the form

$$\partial_t n \approx \frac{\tau}{m}\left[\nabla^2\overline{\varepsilon} + 2\lambda(\nabla n)^2\right] \tag{15.11}$$

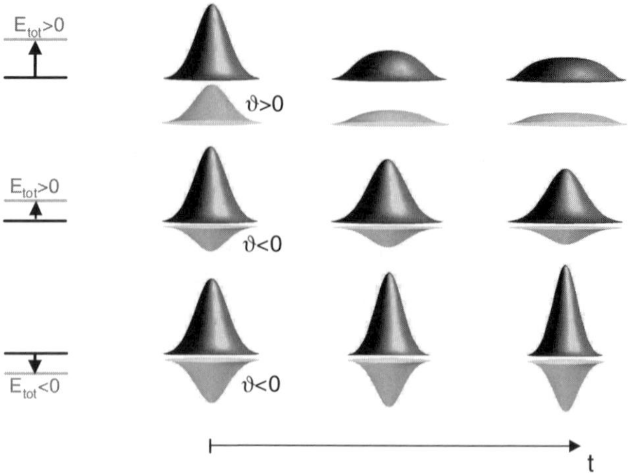

Fig. 15.11. Time-evolution of the disorder averaged density n sketched qualitatively for the three relevant cases. According to the relation $\partial_t \langle r^2 \rangle = E_{tot}\tau/m$, the radius grows for $E_{tot} > 0$ and shrinks for $E_{tot} < 0$. The condition $E_{tot} > 0$ can be realized for positive (first line) or negative potential $\vartheta = \lambda n$ (second line), while $E_{tot} < 0$ can be realized only for negative ϑ (third line). In the repulsive case a rapid drop of the density distribution is expected in the center for large $\vartheta = \lambda n$, while $\langle r^2 \rangle$ grows only linearly in t ("locked explosion"). For $E_{tot} < 0$ a "diffusive" collapse is expected. In the cases with $\vartheta < 0$ a fragmentation of the cloud may occur. The evolution is determined by the initial density *and* energy distributions.

for $t > 0$. For the PME it is the second term that determines the propagation of the boundary. For Eq. (15.6), however, the large kinetic energy outside the center leads to $\nabla^2\overline{\varepsilon} < 0$ for intermediate distances, and this prevents the term $2\lambda(\nabla n)^2$ from dominating. It is therefore an inversion of the distribution of kinetic energy compared to that of the density that does not allow for an explosive expansion and leads to a linear dependence of $\langle r^2 \rangle$ on t. A sketch of a typical density evolution expected for this "locked explosion" is presented in the first line of Fig. 15.11.

We now discuss general features of wave-packet dynamics in the disordered and nonlinear medium.[24] For an expanding wave-packet, i.e. $E_{tot} > 0$, the overall potential energy related to the nonlinearity is converted into kinetic energy. As a result, the total kinetic energy increases in the repulsive case and decreases in the attractive case. Correspondingly, during the course of the expansion localization effects can be expected to be weakened for repulsive nonlinearity and enhanced for attractive nonlinearity. In particular, for an attractive (self-focusing) nonlinearity the slowing down and eventual localization of the injected pulse (not considered here) occurs at smaller distances than in the linear case as observed in the experiment.[1] [Regarding the role of interference effects for localization three main factors need to be accounted for. As long as the evolution does not come to a halt, the time-dependent potential leads to dephasing, which weakens localization effects. When the wave-packet becomes broader, longer paths become available for interference and at the same time the kinetic energy decreases, whereby $\varepsilon\tau$ decreases. The latter two effects support localization.]

The attractive case is richer than the repulsive one (see Fig. 15.2), because the total energy may also be negative, $E_{tot} < 0$. Then the mean radius squared would become equal to zero after a finite time. This corresponds to a celebrated phenomenon in nonlinear physics, the collapse.[25,26] Here it is realized for the diffusive system. To the best to our knowledge, this "diffusive" collapse has not been discussed in the literature. Since our reasoning is based on a diffusive kinetic equation and thus assumes frequent scattering, the linear decrease of $\langle r^2 \rangle$ only holds as long as the radius of the cloud exceeds the mean free path. [In the *clean case* the virial theorem for the NLSE in 2d[25,26] states that the *second* time derivative of $\langle r^2 \rangle$ is proportional to the total energy. In this article we describe diffusive motion and correspondingly obtain a different time dependence for the size of the cloud.]

Even for $E_{tot} > 0$ the collapse can play a role when the nonlinearity is attractive, if part of the cloud has a *negative energy*, while the remaining part expands. As a result one can expect a fragmentation of the cloud. If a part of the cloud with a *positive but small energy* lags behind, this fragment may have a strong tendency to localize. One may expect that this kind of localized or collapsing fragment generically remains from an expanding cloud with $E_{tot} > 0$ but attractive nonlinearity.

To conclude, the nonlinear diffusion equation discussed in this paper contains rich physics that invites further theoretical and experimental investigations.

Acknowledgements

The research was supported by the Minerva Foundation. We thank H. U. Baranger, A. Belyanin, C. Di Castro, G. Falkovich, Y. Lahini, C. A. Müller, V. L. Pokrovsky, Y. Silberberg, J. Sinova and M. D. Spector for their interest in the work.

References

1. T. Schwartz, G. Bartal, S. Fishman, and M. Segev, Transport and Anderson localization in disordered two-dimensional photonic lattices, *Nature.* **446**, 52, (2007).
2. Y. Lahini, A. Avidan, F. Pozzi, M. Sorel, R. Morandotti, D. N. Christodoulides, and Y. Silberberg, Anderson localization and nonlinearity in one-dimensional disordered photonic lattices, *Phys. Rev. Lett.* **100**, 013906, (2008).
3. D. Clément, A. F. Varón, M. Hugbart, J. A. Retter, P. Bouyer, L. Sanchez-Palencia, D. M. Gangardt, G. V. Shlyapnikov, and A. Aspect, Suppression of transport of an interacting elongated Bose-Einstein condensate in a random potential, *Phys. Rev. Lett.* **95**, 170409, (2005).
4. C. Fort, L. Fallani, V. Guarrera, J. E. Lye, M. Modugno, D. S. Wiersma, and M. Inguscio, Effect of optical disorder and single defects on the expansion of a Bose-Einstein condensate in a one-dimensional waveguide, *Phys. Rev. Lett.* **95**, 170410, (2005).
5. J. E. Lye, L. Fallani, M. Modugno, D. S. Wiersma, C. Fort, and M. Inguscio, Bose-Einstein condensate in a random potential, *Phys. Rev. Lett.* **95**, 070401, (2005).
6. T. Schulte, S. Drenkelforth, J. Kruse, W. Ertmer, J. Arlt, K. Sacha, J. Zakrzewski, and M. Lewenstein, Routes towards Anderson-like localization of Bose-Einstein condensates in disordered optical lattices, *Phys. Rev. Lett.* **95**, 170411, (2005).
7. J. Billy, V. Josse, Z. Zuo, A. Bernard, B. Hambrecht, P. Lugan, D. Clement, L. Sanchez-Palencia, P. Bouyer, and A. Aspect, Direct observation of Anderson localization of matter waves in a controlled disorder, *Nature.* **453**, 893, (2008).
8. G. Roati, C. D'Errico, L. Fallani, M. Fattori, M. Zaccanti, G. Modugno, M. Modugno, and M. Iguscio, Anderson localization of a non-interacting Bose-Einstein condensate, *Nature.* **453**, 895, (2008).
9. L. Pitaevskii and S. Stringari, *Bose-Einstein Condensation.* (Clarendon, Oxford, 2003).
10. R. C. Kuhn, O. Sigwarth, C. Miniatura, D. Delande, and C. A. Müller, Coherent matter wave transport in speckle potentials, *New J. Phys.* **9**, 161, (2007).
11. Y. R. Shen, *Principles of Nonlinear Optics.* (Wiley, 1984).
12. Y. Castin and R. Dum, Bose-Einstein condensates in time dependent traps, *Phys. Rev. Lett.* **77**, 5315–5319, (1996).
13. Y. Kagan, E. L. Surkov, and G. V. Shlyapnikov, Evolution of a Bose-condensed gas under variations of the confining potential, *Phys. Rev. A.* **54**, R1753, (1996).
14. B. Shapiro, Expansion of a Bose-Einstein condensate in the presence of disorder, *Phys. Rev. Lett.* **99**, 060602, (2007).
15. I. García-Mata and D. Shepelyansky, Delocalization induced by nonlinearity in systems with disorder, *Phys. Rev. E.* **79**, 026205, (2009).

16. P. C. Martin, E. D. Siggia, and H. A. Rose, Statistical dynamics of classical systems, *Phys. Rev. A.* **8**, 423, (1973).
17. A. Kamenev, *Proceedings of the Les Houches Summer School, Session LXI*, chapter 3, p. 177. Elsevier, Amsterdam, (2005).
18. We demand $\omega\tau \ll 1$ and $lq \ll 1$ for typical frequencies ω and momenta q characterizing the average density distribution $n(\mathbf{q},\omega)$. The nonlinearity can be strong, but gradients should not be too large, $lq\vartheta(\mathbf{q},\omega)/\varepsilon_0 \ll 1$ and $\omega\tau\vartheta(\mathbf{q},\omega)/\varepsilon_0 \ll 1$, for typical kinetic energies ε_0.
19. G. Schwiete and A. M. Finkel'stein, *in preparation.*
20. J. Zinn-Justin, *Quantum Field Theory and Critical Phenomena.* (Oxford University Press, 2002).
21. J. L. Vazquez, *The Porous Medium Equation. Mathematical Theory.* (Oxford University Press, 2006).
22. Y. B. Zeldovich and A. S. Kompaneets, In ed. P. I. Lukirskii, *Collection of Papers Dedicated to the 70th Anniversary of A. F. Ioffe*, pp. 61–72. Izd. Akad. Nauk. SSSR, Moskau, (1950).
23. G. I. Barenblatt, On some unsteady motions of a liquid or a gas in a porous medium, *Prikl. Mat. Mekh.* **16**, 67–78, (1952).
24. M. Hartung, T. W. C. Müller, K. Richter, and P. Schlagheck, Coherent backscattering of Bose-Einstein condensates in two-dimensional disorder potentials, *Phys. Rev. Lett.* **101**, 020603, (2008).
25. S. N. Vlasov, V. A. Petrishchev, and V. I. Talanov, Average description of wave beams in linear and nonlinear media (the method of moments), *Radiophys. Quantum Electron.* **14**, 1062, (1971).
26. C. Sulem and P. L. Sulem, *The Nonlinear Schrödinger Equation: Self-Focusing and Wave Collapse.* (Springer, Berlin, 1999).

Chapter 16

Fluctuations Effects in Small Superconducting Rings

Georg Schwiete

Department of Physics, Texas A&M University, College Station, TX 77843-4242, USA

Yuval Oreg

Department of Condensed Matter Physics, The Weizmann Institute of Science, 76100 Rehovot, Israel

We review and extend theoretical studies of the contribution of fluctuating Cooper pairs to the persistent current in superconducting rings subjected to a magnetic field. For sufficiently small rings, in which the coherence length ξ exceeds the radius R, mean field theory predicts a full reduction of the transition temperature to zero near half-integer flux through the ring. We show that nevertheless a very large current is expected to persist in the ring as a consequence of Cooper pair fluctuations that do not condense near the quantum critical point. For larger rings with $R \gg \xi$, we calculate analytically the susceptibility in the critical region of strong fluctuations and show that it reflects the competition of two interacting complex order parameters. We summarize the main physical results and experimental implications. Details of the calculations are presented in a number of appendices.

16.1. Introduction

The study of superconducting fluctuations has already a long history, for a comprehensive review see Ref. 1. When approaching the superconducting phase from the metallic side, for example by lowering the temperature T, precursors of superconductivity reveal themselves long before the superconducting state is fully established. In this regime, electrons form Cooper pairs only for a limited time. Being charged objects themselves, the Cooper pairs participate in charge transport. At the same time the density of states of the unpaired electrons is reduced. These simple qualitative arguments already indicate that superconducting fluctuations can affect both transport and thermodynamic properties of the metal *outside* the superconducting phase. Detailed studies of these effects have been conducted in different contexts.[1,2] It is well known, for example, that fluctuation effects are more pronounced when the effective dimensionality of the superconductor is reduced or in the presence of disorder.

In bulk superconductors the transition temperature T_c can be partially or even entirely suppressed by various phair-breaking mechanisms, most notably by applying a magnetic field or introducing magnetic impurities. An additional pair-breaking mechanism can become effective in doubly connected superconductors like superconducting rings or cylinders, when they are threaded by a magnetic flux ϕ. In this case one observes so-called Little-Parks oscillations,[3] the transition temperature T_c is *periodically* reduced as a function of ϕ. Due to the periodicity, it is immediately evident that this effect is qualitatively different from the mere suppression of superconductivity by a magnetic field in bulk superconductors. The period of the oscillations is equal to 1 as a function of the reduced flux $\varphi = \phi/\phi_0$, where the superconducting flux quantum is $\phi_0 = \pi/e$,[4] see Fig. 16.1. The maximal T_c reduction occurs when φ takes half-integer values.

The magnitude of the T_c reduction is size-dependent. It is convenient to measure the ring radius R in units of the zero-temperature coherence length ξ and to define $r = R/\xi$. A representative mean field phase diagram is displayed in Fig. 16.1 for two rings of different size. As we see in Fig. 16.1, mean field theory predicts a moderate T_c reduction for moderately small rings with $r \gtrsim 1$. Most strikingly, it also shows that for very small rings for cylinders with $r < 0.6$ the transition temperature is expected to be equal to zero in a finite interval close to half-integer fluxes. Correspondingly, a flux-tuned quantum phase transition is expected to occur in these rings or cylinders at a critical flux φ_{c0}. The mean field transition line can be found from Eq. (16.15) to be discussed below.

As is well known, superconducting rings threaded by a magnetic flux ϕ support a dissipationless persistent current.[2,6] In this article, we study theoretically the *fluctuation* persistent current in different regions of the phase diagram, both for rings with $r \gtrsim 1$ and moderate T_c reduction and for rings with $r \lesssim 0.6$ with strong T_c suppression. In particular, we will study in detail the large fluctuation persistent current I which occurs even at fluxes for which T_c is reduced to zero while the system has a finite resistance. A short account of the most important results of this study has already been presented in Ref. 5. Here we take the opportunity to discuss

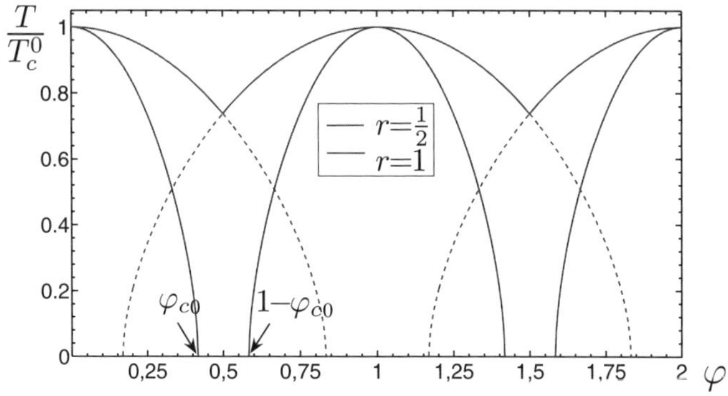

Fig. 16.1. Mean field phase diagram. $T_{c\varphi}$ separates the metallic (high T) and the superconducting (low T) phase as a function of the flux $\varphi = \phi/\phi_0$ through the ring. For sufficiently small rings with effective radius $r = R/\xi < 0.6$ mean field theory predicts a full reduction of T_c for fluxes between $\varphi_{c0} \approx 0.83r$ and $1 - \varphi_{c0}$ near $\varphi = 1/2$. The transition line reflects the condition $\mathcal{L}_{00}^{-1} = 0$, cf. Eq. (16.14). The dotted lines give $T_{n\varphi}$ defined below Eq. (16.5) for $n \in \{0, 1, 2\}$. For $r \gtrsim 1$ it is well approximated by the formula $T_{c\varphi} \approx T_{c0}(1 - \varphi^2/r^2)$. The phase diagram is periodic in φ with period 1 for vanishing ring thickness. Courtesy of Schwiete et al., Ref. 5.

the results in a broader context, extend them to different regions of the phase diagram and to include details of the derivations.

This study is motivated by recent experiments that are significant to our understanding of fluctuation phenomena in superconductors with doubly-connected geometry. Strong Little-Parks oscillations for cylinders with $r < 0.6$, for which T_c is reduced to zero near half-integer flux, have been observed in a transport measurement on superconducting cylinders.[7] Koshnick et al.[8] measured the persistent current in small superconducting rings with $r \gtrsim 1$, for the smallest rings under study T_c was reduced by approximately 6%. Due to experimental difficulties, it has so far unfortunately not been possible to measure the persistent current in superconducting rings close to the flux-tuned quantum critical point. We discuss the possibility that a combination of different pair-breaking mechanisms could be used to make progress in this direction. For the sake of completeness and for the convenience of readers unfamiliar with the microscopic theory of pair-breaking transitions, the classical GL functional, or their relation to each other, we summarize important aspects of this (standard) material within the article and in appendix A.

The literature on superconductivity in systems with doubly connected geometry is extensive. We would like to mention a number of works, where related phenomena have been discussed. The possibility of finding complete suppression of superconductivity near half integer flux in small superconducting rings was pointed out by de Gennes.[9,10] A detailed study of the fluctuation persistent current in rings for which T_c is reduced to zero by magnetic impurities (at any value of the flux) can be found in Refs. 11 and 12. In Refs. 13 and 14 the perturbative fluctuation contribution to the conductivity of long superconducting cylinders near a flux-tuned quantum critical point was discussed as a particular example for transport near a pair-breaking transition. Fluctuation effects in moderately small superconducting rings ($r \gtrsim 1$) were studied in Refs. 15 and 16 with particular emphasis on the regime of strong fluctuations near the transition. The phase diagram of superconducting cylinders was considered in Ref. 17 taking into account the interplay of pair-breaking effects caused by the flux on the one hand and the magnetic field penetrating the walls (of finite width) on the other hand. The role of the back-action effects causes by the self-induction of cylinders and rings was discussed in this context in Refs. 18 and 19. While we will discuss finite thickness effects, we will generally assume that the self-induction and the persistent currents of the *rings* under study in this article are sufficiently small so that back-action effects can be neglected.

The article is organized as follows. Section 16.2 is devoted to the discussion of the persistent current in rings with $r \gtrsim 1$. We first describe the conventional Ginzburg-Landau approach with a particular emphasis on the zero dimensional limit, which plays a crucial role for the analysis of rings examined in the experiment of Ref. 8. We then present results on the fluctuation persistent current close to the thermal transition, including the regime of strong fluctuations in the immediate vicinity of the transition. In Sec. 16.3 we study in detail the fluctuation persistent current in different regions of the phase diagram for rings with $r \lesssim 0.6$, including the vicinity of the quantum critical point. In Sec. 16.4 we discuss different ways to reduce T_c to zero in rings with finite width or by introducing magnetic impurities.

Most of the technical details are relegated to a series of appendices. In addition, appendix A sketches basic aspects of the microscopic theory of pair-breaking transitions, which is the convenient language for discussing rings with a strong reduction of T_c. A much more comprehensive discussion can be found in Ref. 1. Before presenting details, however, we summarize the main results of our analysis.

16.1.1. *Main Results*

I. Regime with $r = R/\xi < 0.6$: For $r < 0.6$ the mean field T_c vanishes in a finite interval of fluxes $(\varphi_{c0}, 1 - \varphi_{c0})$ and one would naively expect a small normal state persistent current. We find, however, that close to the critical mean field line (see Fig. 16.1) there is a parametrically large enhancement of the persistent current due to quantum fluctuations that decays slowly away from that line. The magnitude of the normal persistent current is $I_N \sim \frac{1}{\phi_0}\frac{D}{R^2}\frac{1}{\log g}$, where D is the diffusion coefficient and g is the dimensionless ring conductance.[20,21] Our calculations show that the persistent current due to pair fluctuations near φ_{c0} is parametrically larger and at low $T \ll T_{c\varphi=0}$ given by $[T_{c\varphi=0} \equiv T_c^0]$

$$I_{\rm FL} \approx -\frac{T_c^0}{\phi_0}\frac{1}{\varphi_{c0}}\frac{\xi}{R}\log\left(\frac{1}{\Delta\varphi}\right), \tag{16.1}$$

where $\Delta\varphi \equiv (\varphi - \varphi_{c0})/\varphi_{c0}$ measures the distance to the critical flux φ_{c0}. When increasing T the persistent current initially grows before going through a maximum at finite T, where it can considerably exceed the result of Eq. (16.1) [see Fig. 16.3]. Since $r^{-1} = \xi/R$ is a number of order 1 and $\frac{D}{R^2} = \frac{8}{\pi}\frac{T_c^0}{r^2}$ for a weakly disordered superconductor, we find an enhancement factor of $\log(g)\log(1/\Delta\varphi)$. For rings with $r < 0.6$, the fluctuation persistent current is discussed in detail in Sec. 16.3 at vanishing as well as at finite temperatures and for different values of the flux φ. In particular, in Fig. 16.2 we display the persistent current as a function of the flux at temperatures $T > T_c^0$ [in the Gaussian regime]. In Fig. 16.3 the temperature dependence of the persistent current in the vicinity of the quantum critical point is shown for different fluxes $\varphi \gtrsim \varphi_{c0}$ and in Fig. 16.4 the whole temperature regime up to $T = T_c^0$ is covered.

Our results are obtained for the case when the flux acts as a pair-breaking mechanism. Other pair-breaking mechanisms, e.g. magnetic impurities or a magnetic field penetrating the ring itself will lead to similar results. They cause a reduction of T_c to zero, the pair fluctuations, however, lead to a parametric enhancement of the persistent current in the normal state. Ref. 11 suggests that a similar mechanism due to magnetic impurities is related to the unexpectedly large persistent current in noble metal rings.[22,23]

A metallic state with small but finite resistance was observed experimentally in superconducting cylinders[7] with $\varphi \approx 1/2$. Further studies will be needed to clarify the relation to our findings, where a large persistent current is caused by pair fluctuations that are unable to condense.

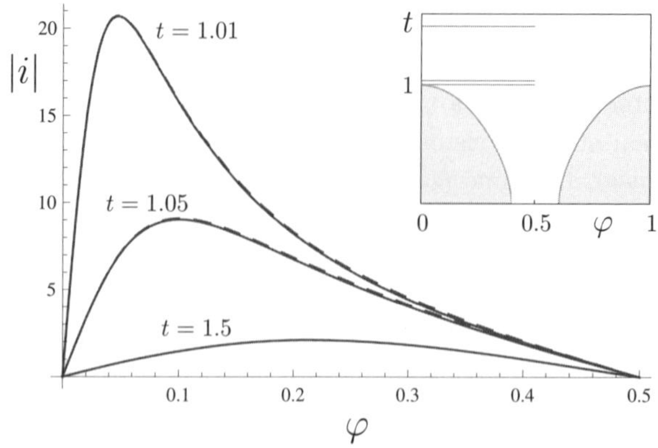

Fig. 16.2. The modulus of the fluctuation persistent current $i = I/[T_c^0/\phi_0]$ (see Eq. 16.16) is shown for a ring with $\varphi_{c0} = 0.4$, corresponding to $r \approx 0.48$, for three temperatures above T_c^0 $[t = T/T_c^0]$ (solid lines). The dashed lines show the thermal ($\omega = 0$) part of i_s, see Eq. (16.22), which gives a very good approximation. The current decreases as the temperature grows. At the same time the shape changes considerably as explained in the text. In the inset the phase diagram is displayed, indicating the temperatures in question.

In Sec. 16.4 we discuss how the use of rings with an annulus of finite width or the addition of magnetic impurities could make possible the observation of the quantum phase transition in experiments on superconducting *rings*.

II. Regime with $r > 1$: The case $r > 1$ is suitable for the description of the experiments on persistent currents by Koshnick et al.[8] Previously this experiment has been interpreted using a one-dimensional Ginzburg-Landau (GL) theory to describe the order parameter fluctuations.[24,25] Following these lines one has to resort to numerical methods[24] in order to describe the critical region close to T_c, where fluctuations proliferate.

Our key observation is that part of the rings in the experiment allow for a description using a suitable generalization[15] of the $0d$ Ginzburg-Landau theory. Indeed, following an expansion of the order parameter field $\psi(\vartheta)$ in terms of angular momentum modes ψ_n, a simple physical picture arises in the limit $\sqrt{g} \gg r$.[26] Two of the modes compete with each other close to half-integer flux, while at the same time both of them strongly fluctuate in the critical regime close to T_c.

Formally, the competition arises due to the quartic term in the GL functional that induces an interaction between the modes[27] and reveals itself

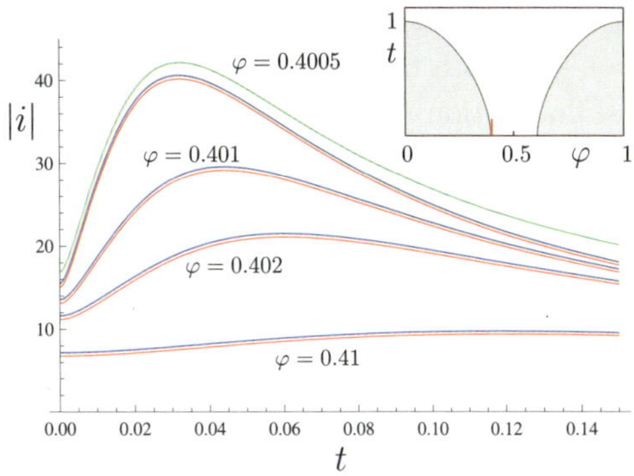

Fig. 16.3. The modulus of the persistent current $i = I/[T_c^0/\phi_0]$ (see Eq. 16.16) in the vicinity of the quantum critical point is shown in red as a function of the reduced temperature $t = T/T_c^0$ for four different fluxes φ close to $\varphi_{c0} = 0.4$. The inset shows the mean field phase diagram indicating the region relevant for this plot. The persistent current decreases when moving away from the quantum critical point. It displays a pronounced maximum at low but finite temperatures. Thermal fluctuations grow with increasing temperature, but for fixed flux the system moves further away from the critical line. At vanishing temperatures the persistent current is still large and entirely caused by quantum fluctuations. Also shown is i_s (of Eq. (16.17)) in blue, which provides a very good approximation. The small difference between i and i_s is i_{ns} (Eq. D.6). In this temperature regime i_{ns} can be obtained from Eq. (16.28). The green line was calculated from the approximate result of Eq. (16.23), which provides a handy estimate for the magnitude of i and the position of the maximum near the quantum critical point.

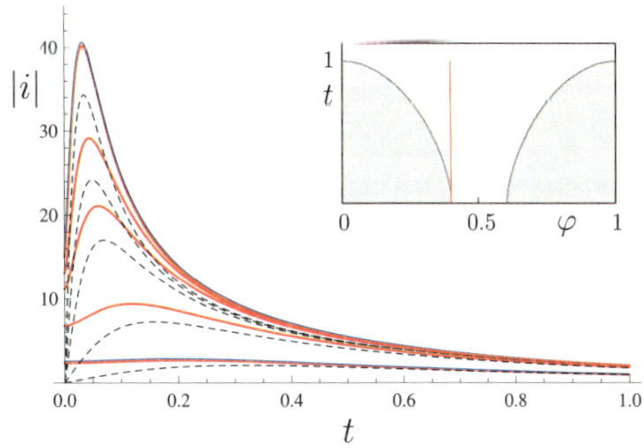

Fig. 16.4. The modulus of the persistent current $i = I/[T_c^0/\phi_0]$ (calculated numerically according to Eq. 16.16) is shown in red as a function of the reduced temperature $t = T/T_c^0$ for three different fluxes φ close to $\varphi_{c0} = 0.4$, $\varphi = 0.4005$, 0.401, 0.402 0.41 and 0.45 from top to bottom. The inset shows the mean field phase diagram indicating the region relevant for this plot. Also displayed is the thermal ($\omega = 0$) contribution to the current, which goes to zero at vanishing temperatures (black dashed line). Thermal fluctuations become increasingly important as the temperature grows. For two examples we indicate the singular contribution i_s (of Eq. (16.17)) in blue, which gives a very good approximation for all parameters. The small difference between i and i_s is i_{ns}.

in the experiment mostly in the slope of the persistent current, i.e. in the susceptibility $\chi = -\frac{\partial I}{\partial \phi}$. With this insight χ can be calculated analytically even in the critical fluctuation regime.

As an example, denoting the susceptibility right at T_c and vanishing flux by χ_0 and at $T_{c,\varphi=1/2}$ by $\chi_{1/2}$, we find

$$\chi_{1/2}/\chi_0 \approx -2.7\sqrt{g}/r. \tag{16.2}$$

Experimentally, a strong enhancement of the magnetic susceptibility near $\varphi = 1/2$ compared to $\varphi \approx 0$ was observed and Eq. (16.2) demonstrates that it is controlled by the parameter $\sqrt{g}/r$. If it is large, the current rapidly changes sign as a function of the flux at half-integer flux, leading to a saw-tooth like shape of i_φ. The full T dependence of $\chi_{\varphi=1/2}$ is given in Eq. (16.10). For the smallest rings in Ref. 8, $\sqrt{g} \approx 33r$.[28]

16.2. Thermal Transition for Rings with $r \gtrsim 1$

In this section we will discuss the description of rings with only a moderate suppression of T_c, i.e. rings for which $r \gtrsim 1$. For these rings the superconducting transition occurs at a finite temperature and fluctuations can be described with the help of the classical GL functional in which the order parameter field is static. In the imaginary time formalism this amounts to neglecting order parameter field components with finite Matsubara frequency in the functional, see appendix A. This simplified description is valid close to the transition line in the $T_c - \varphi$ phase diagram.

16.2.1. *The classical GL functional*

To set the stage for our subsequent discussion of superconducting fluctuations in rings, let us recall some general properties of the classical GL functional

$$\mathcal{Z} = \int D(\psi, \psi^*) \exp\left(-\mathcal{F}/T\right) \tag{16.3}$$

$$\mathcal{F} = \mathcal{F}_N + \int d\mathbf{r} \left(a|\psi(\mathbf{r})|^2 + \frac{b}{2}|\psi(\mathbf{r})|^4 + \frac{1}{4m}\left|(-i\nabla - 2e\mathbf{A}(\mathbf{r}))\psi(\mathbf{r})\right|^2 \right) \tag{16.4}$$

$\mathcal{F}_N$ describes the the normal (non-superconducting) part of the free energy and will not be discussed further. At the mean field level, the sign change of the quadratic form in ψ signals the onset of the superconducting phase.

Therefore one should find the lowest eigenvalue of the corresponding operator $a + \frac{1}{4m}(-i\nabla - 2e\mathbf{A}(\mathbf{r}))^2$. For translationally invariant systems without magnetic field, the system becomes homogeneously superconducting at a transition temperature T_c^0 determined by the sign change of a alone, since gradients in the kinetic part of the operator vanish for the spatially homogeneous order parameter configuration. This observation motivates the parametrization $a = \alpha T_c^0 \varepsilon$, where $\varepsilon = \frac{T - T_c^0}{T_c^0}$. A characteristic length scale, the (zero temperature) coherence length, can be identified $\xi = 1/\sqrt{4m\alpha T_c^0}$. The microscopic theory for disordered superconductors gives rise to the following relations $\alpha^2/b = 8\pi^2\nu/7\zeta(3)$, $\xi^2 = \pi D/8T_c$ where D is the diffusion coefficient. The normalization of ψ allows for a certain arbitrariness, this is why only the ratio of α^2 and b is fixed.

The quartic part of the functional stabilizes the system once it is tuned below the transition temperature. Above this temperature, the quartic term gives only a small contribution to thermal averages, except in the very vicinity of the transition, the so-called Ginzburg region. As long as one stays outside of this region on the metallic side, one can restrict oneself to a quadratic (Gaussian) theory (i.e. neglect the quartic term), which is much easier to handle theoretically, but becomes unreliably close to the transition where the quadratic theory becomes unstable. Importantly, even above the transition temperature, in the normal region of the mean field phase diagram, the average of $|\psi|^2$ with respect the functional $\mathcal{F}$ is finite. It means that at temperatures above the transition temperature T_c^0 to the superconducting state, when the system is still metallic, pairs of electrons can form. An analytic computation of the functional integral necessary to obtain $\mathcal{Z}$ is in general not possible and one has to resort to approximation schemes. A notable exception is the so-called zero-dimensional limit. We want to discuss this case briefly in the following, because it helps to illustrate the role of Gaussian and non-Gaussian fluctuations and moreover will turn out to be instrumental for the theoretical description of fluctuations in small superconducting rings.

16.2.2. *Zero-dimensional limit*

When the spatial extension of a superconductor is of the order of the coherence length, the order parameter does typically not vary appreciably across the sample. In this case the spatially constant order parameter field configuration ψ_0 gives the major contribution to the functional integral that determines $\mathcal{Z}$ and one can exclude components with nonzero momenta (the momenta are quantized for a finite system). Remarkably, in this case the Ginzburg-Landau functional can be evaluated exactly. The free energy takes the form $\mathcal{F}_0 = a|\psi_0|^2 + \frac{b}{2V}|\psi_0|^4$, where V is the volume of the superconducting sample, and the partition function is obtained by evaluating an ordinary integral over the complex field ψ_0: $\mathcal{Z} = \int d\psi_0 d\psi_0^* \exp(-\mathcal{F}_0/T)$. The simple calculation, described in appendix B, yields the result $\mathcal{Z} = \frac{\pi\sqrt{\pi}}{\alpha Gi}\exp(x^2)\,\mathrm{erfc}(x)$. Here, $x = \varepsilon/Gi$ compares the relative deviation of the temperature from the critical one $\varepsilon = (T - T_c^0)/T_c^0$ to the zero-dimensional Ginzburg number $Gi = \sqrt{2b/\alpha^2 T_c V}$.[26] The temperature interval for which $|x| \lesssim 1$ is often referred to as the Ginzburg region or region of strong fluctuations. The function $\mathrm{erfc}(x) = \frac{2}{\sqrt{\pi}}\int_x^\infty dt\ \mathrm{e}^{-t^2}$ is the complementary error function.[29] Figure 16.5 sketches the temperature dependence of of $\langle|\psi_0|^2\rangle = 2f(x)/\alpha Gi$ as calculated from $\mathcal{Z}$, where $f(x) = \frac{\exp(-x^2)}{\sqrt{\pi}\mathrm{erfc}(x)} - x$. We see that for $x \gg 1$, i.e. for temperatures far above T_c^0 the result merges with the result obtained from the Gaussian theory, namely $\langle|\psi_0|^2\rangle = 1/\varepsilon$. The latter one in turn diverges when approaching T_c^0 from above, which is an unphysical consequence of the Gaussian approximation. Far below T_c^0, for $x \ll -1$, $\langle|\psi_0|^2\rangle$ agrees with the result of a *simple* mean field analysis, for which one determines ψ_0 approximately from the condition that $\mathcal{F}_0$ is stationary, $\delta\mathcal{F}_0/\delta\psi_0|_{\psi_{MF}} = 0$. This yields $\langle|\psi_0|^2\rangle \sim \langle|\psi_{MF}|^2\rangle = -2x/Gi\alpha$. Neither the Gaussian nor the simple mean field approximation can account for the fluctuations in the region of strong fluctuations.

16.2.3. *GL functional in the ring geometry*

After this preparation, we turn to the description of small superconducting rings. When the superconducting coherence length $\xi(T) = \xi/\sqrt{\varepsilon}$ and the magnetic penetration depth $\lambda(T)$ are much larger than the ring thickness, the system is well described by a one-dimensional order parameter field ψ, albeit with periodic boundary conditions.[6] In order to account for these boundary conditions, it is convenient to introduce angular momentum modes as $\psi(\vartheta) = \frac{1}{\sqrt{V}}\sum_n \psi_n\ \mathrm{e}^{in\vartheta}$. $V = 2\pi R S_\perp$ is the volume of the ring, $S_\perp$ the cross-section of the wire forming the ring. The vector potential can be chosen as $\mathbf{A} = \mathbf{B}\times\mathbf{r}/2$, the integration as $\int d\mathbf{r} \to S_\perp R\int d\phi$. Then, the free energy functional takes the form

$$\mathcal{F} = \sum_n a_{n\varphi}|\psi_n|^2 + \frac{b}{2V}\sum_{nmkl}\delta_{n+k,l+m}\psi_n\psi_m^*\psi_k\psi_l^* \tag{16.5}$$

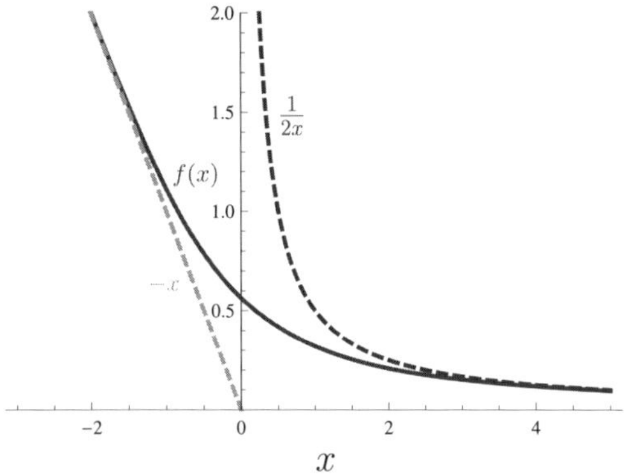

Fig. 16.5. In the zero-dimensional limit the squared modulus of the order parameter $\langle|\psi_0|^2\rangle = 2f(x)/\alpha Gi$ is proportional to the function $f(x)$, which is displayed as a solid line. $x = \varepsilon/Gi$ compares the relative deviation ε of the temperature from the critical one to the zero-dimensional Ginzburg number Gi. The regime of strong fluctuations is characterized by the condition $|x| \lesssim 1$. Also shown are the functions $1/(2x)$ and $-x$, which correspond to the Gaussian and mean field approximations in the 0d limit. Asymptotically, i.e. for large or low temperatures, they provide good estimates, but not close to the transition.

where $a_{n\varphi} = a + (n-\varphi)^2/2mR^2$. Let us make three important observations. First, the free energy functional is flux dependent. As a consequence a persistent current can flow in the ring. Second, the functional is periodic in the reduced flux φ with period 1. Correspondingly, the same is true for all thermodynamic quantities derived from the the GL functional. This property holds strictly speaking only in the idealized limit of a one-dimensional ring. In reality, the external magnetic field also penetrates the superconductor and provides an additional mechanism for the suppression of superconductivity. We will come back later to this point in Sec. 16.4. The third observation is that now the kinetic energy of the Cooper pairs vanishes only when the reduced flux φ takes integer values. Otherwise it gives a finite contribution to the quadratic part of the functional and therefore the transition takes place at a temperature $T_c(\varphi)$ that is in general reduced with respect to T_c^0 of the bulk material. Let us parameterize $a_{n\varphi} = \alpha T_c^0 \varepsilon_{n\varphi}$. Then $\varepsilon_{n\varphi} = (T - T_{n\varphi})/T_c^0$ and a_n change sign at a temperature $T_{n\varphi} = T_c^0[1-(n-\varphi)^2/r^2]$. This temperature can loosely be interpreted as the transition temperature of the nth angular momentum mode ψ_n.[30] The mean field transition for the ring occurs at $T_{c\varphi}$ that is equal to the maximal T_n for given φ, i.e. at the point where the first mode becomes superconducting when lowering the temperature (cf. Fig. 16.1).

The parameter $\sqrt{g}/r \approx 1/5r^2 Gi$ has been used when stating our results in the introduction. Its relevance is now easily understood. $1/r^2$ is a measure for the typical spacing between the transition temperatures T_n for different modes, since e.g. $[T_0(\varphi) - T_1(\varphi)]/T_c^0 = (1 - 2\varphi)/r^2$ (compare Fig. 16.1). This spacing can be compared to the typical width of the non-Gaussian fluctuation region, Gi. Only in this region fluctuations are strong.

Approximation schemes If the spacing is large and one is interested in the region of strong fluctuations close to the transition temperature $T_c(\varphi)$, an effective theory including only one angular momentum mode ψ_n for $n \sim \varphi$ is applicable, $\mathcal{F}_n \sim a_n|\psi_n|^2 + \frac{b}{2V}|\psi_n|^4$. This is so, since in this case the temperature $T \sim T_c(\varphi)$ lies far above the individual transition temperatures $T_m(\varphi)$ $(m \neq n)$ of all other modes ψ_m and they will give only a small contribution when calculating observables. This is very convenient, because in this case one comes to the zero-dimensional limit of the GL functional already mentioned in Sec. 16.2.2. The partition function based on this free energy functional can be calculated exactly and all thermodynamic quantities derived from it.

Clearly, for half-integer values of the flux the spacing between two adjacent modes always goes to zero and due to this degeneracy at least two modes are required for the description. Let us choose for definiteness the example $0 < \varphi < 1$, then one may work with

$$\mathcal{F}_{01} = \sum_{i=0,1} a_i|\psi_i|^2 + \frac{b}{2\text{Vol}}\left(|\psi_0|^4 + |\psi_1|^4 + 4|\psi_0|^2|\psi_1|^2\right). \tag{16.6}$$

In this situation the parameter $\sqrt{g}/r \approx 1/5r^2 Gi$ is still useful, because if it is large, then additional modes need not be taken into account and a two-mode description is valid. To summarize the discussion so far, for large $\sqrt{g}/r \approx 1/5r^2 Gi$, a single mode description is appropriate when discussing fluctuations in the Ginzburg region near integer fluxes φ and a two-mode description near half-integer fluxes.

If $\sqrt{g}/r$ is not exceedingly large, the small contribution of the remaining modes can easily be accounted for by using the Gaussian approximation for them. One should, however, not forget that the presence of the dominant mode(s) can influence the effective transition temperatures of all others via the quartic term. As an example, let us write the resulting effective action for the modes with $n \neq 0$ assuming that $\varphi \sim 0$ and the mode with $n = 0$

is the dominant one, $\mathcal{F}^{eff}_{n\neq 0} = \sum_{n\neq 0}(a_n + \alpha^2 Gi^2 T_c^0 |\psi_0|^2)|\psi_n|^2$. Essentially the same argument was first presented in Ref. 15.

If temperatures are sufficiently high $T \gg T_c^0(1+Gi)$ the quadratic term may be dropped altogether and one may work with a purely Gaussian theory $\mathcal{F}_n \approx \sum_n a_{n\varphi}|\psi_n|^2$. In the next section we will put these approximation schemes into effect when calculating the persistent current.[26]

16.2.4. *Persistent current*

The persistent current I is found from the free energy $F = -T \ln \mathcal{Z}$ by differentiation $I = -\partial F/\partial\phi$. The normalized current is given by

$$i = I/(T_c^0/\phi_0) = \sum_{n=-\infty}^{\infty} \frac{2\alpha}{r^2}(n-\varphi)\left\langle |\psi_n|^2 \right\rangle . \tag{16.7}$$

The averaging is performed with respect to the functional $\mathcal{F}$ in Eq. (16.5). Just as the free energy functional $\mathcal{F}$, the persistent current i is periodic in the flux φ with period one. Since it is also an odd function of φ, the persistent current vanishes when φ takes integer or half-integer values.

Case $\varphi \approx 0$: As pointed out in the previous Sec. 16.2.3, the most important contribution in the regime of non-Gaussian fluctuations close to integer fluxes comes from the angular momentum mode ψ_n with the highest transition temperature $T_{n\varphi}$. One may then approximate Eq. (16.5) by a single-mode and calculate with $\mathcal{F}_n = a_n|\psi_n|^2 + \frac{b}{2V}|\psi_n|^4$.[16] This is the 0d limit of the GL functional[31] already introduced above. In this limit, Eq. (16.7) gives

$$i_n = 4\Lambda(n-\varphi)f(x_n) \quad \text{for } \varphi \approx n. \tag{16.8}$$

Here $x_n = \varepsilon_n/Gi$ and $\Lambda \equiv 1/r^2 Gi \approx 5\sqrt{g}/r$ and the function f was introduced before in Sec. 16.2.2 and displayed in Fig. 16.5. It is worth emphasizing the high degree of universality implied by this result: All persistent current measurements will fall on the same curve, if the persistent current – measured in suitable units $i = I/(T_c^0/\phi_0)$ – and the reduced temperature $\varepsilon_\varphi = (T - T_{c\varphi})/T_c^0$ are scaled as $i \to i\frac{r}{\sqrt{g}}$, $\epsilon_\varphi \to \epsilon_\varphi\, r\sqrt{g}$. This relation can serve as a valuable guide in characterizing different rings in experiments.

It is possible to make contact with the Gaussian and the mean field results using the asymptotic expansion of the conjugated error function $\sqrt{\pi}x\mathrm{erfc}(x) \sim \exp(-x^2)(1 - 1/(2x^2))$ for $x \to \infty$ and the limit $\mathrm{erfc}(x) \to 2$ for $x \to -\infty$, see Fig. 16.5. Far above T_c one obtains as a limiting case the Gaussian result for a single mode $i_n \approx 2(n-\varphi)/r^2\varepsilon_{n\varphi}$, that can also be obtained directly by neglecting the quartic term in the GL functional. In this form, however, it is of limited use, since for the temperatures in question one should sum the contribution of all modes. Indeed, in this case one can use the relation $\langle|\psi_n|^2\rangle \sim T_c^0/a_n$ and perform the sum in Eq. (16.7) to obtain[32]

$$i(\varphi) = \frac{-2\pi\sin(2\pi\varphi)}{\cosh(2\pi\sqrt{\varepsilon}r) - \cos(2\pi\varphi)} \tag{16.9}$$

One of the main features of this result besides the periodicity in φ is the exponential decay of the persistent current as a function of temperature for $\varepsilon > 1/(2\pi)^2r^2$. Turning back to Eq. (16.8), we see that far below T_c one recovers the mean field result $i_{MF} \equiv \frac{-4}{r^2Gi^2}\varepsilon_{n\varphi}(n-\varphi)$, which gives an estimate for the persistent current in the superconducting regime. In this approximation the current grows linearly with $|T - T_c|$ and as soon as $|\varepsilon| \gg Gi + (n-\varphi)^2/r^2$ (i.e. $|x_n| \gg 1$) one expects a sawtooth-like behavior as a function of the flux (i.e. linear dependence from $\varphi = n - 1/2$ to $\varphi = n + 1/2$ passing through zero at integer n) with a discontinuous jump at half integer φ. Both the Gaussian and the mean field result are reliable only outside the region of strong fluctuations, the persistent current i_n in Eq. (16.8) covers this region and interpolates smoothly between them.

Case $\varphi \approx 1/2$: A very interesting situation occurs at half integer values of φ. The transition temperatures for two modes become equal. In the vicinity of this point in the phase diagram the two dominant modes influence each other, their *coupling* becomes crucial. We discuss the case $\varphi \approx 1/2$ for definiteness, and use the form of the free energy functional already displayed in Eq. (16.6).

Calculation of the persistent current in the presence of the coupling requires a generalization of the approach used for the single mode case.[15] Explicit formulas for the persistent current are derived and displayed in appendix C for the sake of completeness. In Fig. 16.6 we display the persistent current i_2 as calculated from Eq. (16.6) for three different temperatures, $T_{c\frac{1}{2}} < T < T_c^0$ (Little-Parks regime), $T = T_{c\frac{1}{2}}$ and $T < T_{c\frac{1}{2}}$. We compare it to the mean field result as well as to i_{20} obtained by neglecting the coupling $|\psi_0|^2|\psi_1|^2$ in Eq. (16.6).

Above $T_{c\varphi}$ ($\varepsilon = -0.05$ in Fig. 16.6), in the region where the mean field result vanishes near half-integer flux, the persistent current is *purely fluctuational.* We deduce from Fig. 16.6 that the coupling of the modes is

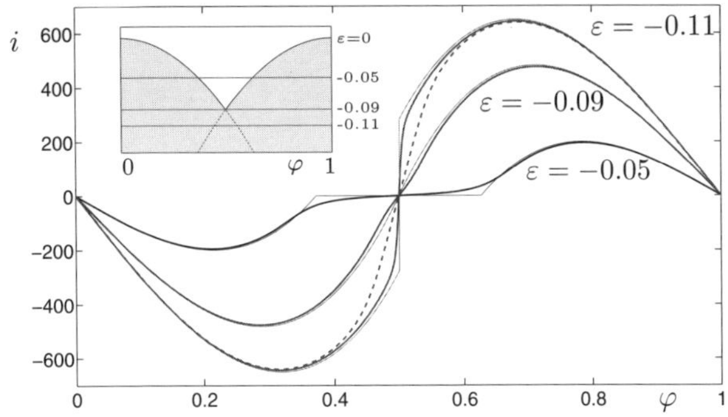

Fig. 16.6. The persistent current $i = I/(T_c^0/\phi_0)$ as a function of the flux φ. Parameters are $r = R/\xi = 1.66$, $\Lambda = 1/(r^2 Gi) \approx 5\sqrt{g}/r = 50$, $\varepsilon = (T - T_c^0)/T_c^0$. The transition for $\varphi = 1/2$ occurs at $\varepsilon = -0.09$. Full lines: i_2 calculated with $\mathcal{F}_{01}$ of Eq. (16.6), it takes into account two modes and the interaction between them. We compare i_2 to two approximations, which neglect this interaction. Dotted lines: The mean field approximation i_{MF} [discussed before Eq. (16.6)] and dashed line: i_{20} calculated with $\mathcal{F}_{01}$ of Eq. (16.6) *without* coupling.[33] Inset: MF phase diagram, superconducting region in grey. Courtesy of Schwiete et al., Ref. 5.

crucial for $\chi(1/2)$, but not for the overall shape when $T > T_{c\frac{1}{2}}$. However, just below $T_{c\frac{1}{2}}$ ($\varepsilon = -0.11$ in Fig. 16.6) the coupling is essential. The mean field result is not applicable as it gives an infinitely sharp jump in the persistent current at half-integer flux. The result without coupling of the modes, i_{20}, gives a finite slope, but it is far from the full current i_2 that includes the mode coupling. The coupling drives the current i_2 towards the mean field approximation i_{MF} which includes only one mode. This occurs because for a repulsive coupling the dominant mode suppresses the subdominant one.[27] If mode $n = 0$ is dominant then the coupling adds a mass term $(2b/V)\left\langle|\psi_0^2|\right\rangle|\psi_1|^2$ to mode $n = 1$ and effectively reduces its T_c. It has become obvious that the susceptibility at $\varphi = 1/2$ is a very sensitive measure for the interplay of superconducting fluctuations on the one hand and the competition of the two angular momentum modes on the other hand. We turn to this quantity now.

16.2.5. *Susceptibility for $\varphi \approx 1/2$*

We will now discuss in more detail the slope at half-integer flux, which is most sensitive to the coupling between the modes below, and to the non-Gaussian fluctuations close to T_c. Differentiating the expression[34] for i_2

(see appendix C) one obtains

$$\overline{\chi}_{\varphi=1/2} = \frac{\chi}{T_c^0/\phi_0^2} = 4\Lambda\, g_1(x) - 4\Lambda^2\, g_2(x)\,, \tag{16.10}$$

where $x \equiv x_0(1/2) = x_1(1/2) = \frac{\varepsilon}{Gi} + \frac{1}{4}\Lambda$.[26] Note that $x = 0$ at the mean field transition for $\varphi = 1/2$, i.e. the condition $x = 0$ defines $T_{c,1/2} = T_c^0(1 - 1/4r^2)$. The dimensionless smooth functions g_n are defined by the relations

$$g_1(x) = \frac{1}{2J(x)} e^{\frac{1}{3}x^2} \operatorname{erfc}(x) - \frac{2x}{3} \tag{16.11}$$

$$g_2(x) = \frac{3}{2\sqrt{\pi}J(x)}\, e^{-\frac{2}{3}x^2} - \frac{3x}{2J(x)} e^{\frac{1}{3}x^2} \operatorname{erfc}(x) - 1, \tag{16.12}$$

where $J(x) = \int_x^\infty dt\, e^{\frac{1}{3}t^2} \operatorname{erfc}(t)$. In Fig. 16.7 we compare these functions to the Gaussian result for just the two modes under consideration, which one can obtain by asymptotic expansion as $g_1(x) \to 1/x$ and $g_2(x) \to 1/2x^2$ for $x \to \infty$. Exact results can be given for the functions g_n at the transition, i.e. for $x = 0$. These give a useful estimate for the magnitude inside the fluctuation region, $g_1(0) = \sqrt{\pi}/[2\sqrt{3}\text{arctanh}\,(1/\sqrt{3})] \approx 0.78$ and $g_2(0) = \sqrt{3}/[2\text{arctanh}\,(1/\sqrt{3})] - 1 \approx 0.315$. For large $\Lambda = 1/r^2 Gi \approx 5\sqrt{g}/r$ one can neglect the first term in Eq. (16.10). Then one obtains $\overline{\chi}_{1/2} = -\,4\Lambda^2\, g_2(x)$. For the susceptibility close to integer flux one easily obtains $\overline{\chi}_0 = 4\Lambda f(x_0)$ from Eq. (16.8). Comparing to the expression for $\overline{\chi}_{1/2}$, we find Eq. (16.2). This is the strong enhancement of $\chi_{1/2}$ compared to χ_0 observed in the experiment.[8]

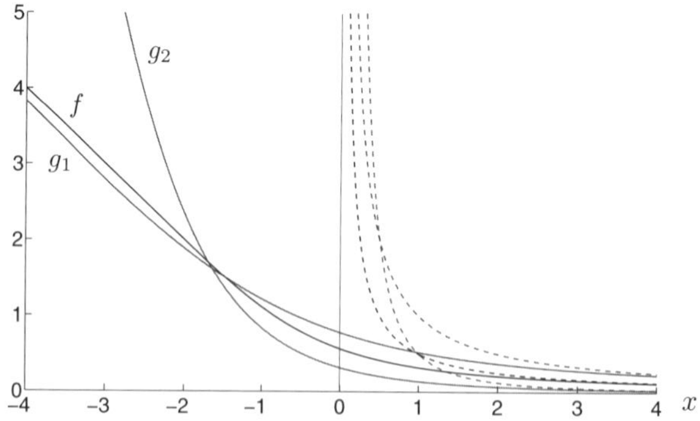

Fig. 16.7. Functions g_2, g_1 and f with respective asymptotic behavior $1/2x^2$, $1/x$ and $1/2x$ for $x \to \infty$.

16.3. Gaussian Fluctuations for Rings with $r < 0.6$

So far we have discussed the limit of moderately small rings with $r = R/\xi > 1$. For these rings the transition temperature is only weakly suppressed at finite flux and the phase transition occurs at a finite temperature. We will now discuss even smaller rings with $r < 1$. For such rings the theoretical description based on the classical GL functional used so far is not valid in large regions of the phase diagram, as it is applicable only in a relatively small temperature interval close to T_c^0. For rings with $r < 1$, however, the transition near half-integer flux can occur at temperatures far below T_c^0, or even at vanishing temperatures. The theoretical description for these rings can be developed in close analogy to the general theory of pair-breaking transitions. In this section we will first determine the mean field transition line and then calculate the contribution of Gaussian fluctuations to the persistent current outside the superconducting regime. Elements of the underlying microscopic approach and its relation to the classical GL description are summarized in appendix A in order to make the paper self-contained.

16.3.1. *Mean field transition line*

The partition function is conveniently formulated in terms of an integral over a complex order parameter field Δ as $\mathcal{Z} = \int D(\Delta, \Delta^*) \exp(-S)$, where

$$\mathcal{S} = T \sum_{\omega} \sum_{n} \Delta^*(n,\omega) \mathcal{L}^{-1}(n,\omega) \Delta(n,\omega). \tag{16.13}$$

The field Δ – unlike the field ψ in the classical GL functional – is dynamical, $\omega_m = 2\pi m T$ is a bosonic Matsubara frequency. Referring for more details to appendix A, let us note here that the field ψ_0 is proportional to the static component $\Delta(\omega_n = 0)$. A neglect of fields with nonzero Matsubara frequencies can be justified for the thermal transition, where it leads to the classical GL functional, but is not justified near the quantum phase transition. In full analogy to previous considerations, Δ has been expanded in terms of angular momentum modes. The fluctuation propagator $\mathcal{L}$ fulfills

$$(\nu \mathcal{L})^{-1}(n,\omega) = \ln\left(\frac{T}{T_c^0}\right) + \psi\left(\frac{1}{2} + \frac{\alpha_n + |\omega|/2}{2\pi T}\right) - \psi\left(\frac{1}{2}\right). \tag{16.14}$$

Here we introduced the pair-breaking parameter $\alpha_n = \varepsilon_T (n - \varphi)^2/2$ for the problem under consideration and ν is the density of states at the Fermi energy. ψ is the Digamma function, (not to be confused with the field ψ_n).

As usual, the mean field transition occurs when $\mathcal{L}^{-1}(n,\omega)$ changes sign first for arbitrary n and ω. Assuming that $|\varphi| < 0.5$, this happens for $n = 0$, $\omega = 0$, so that the condition for the mean field transition reads $\mathcal{L}^{-1}(0,0) = 0$. It defines an implicit equation for $T_c(\varphi)$

$$\ln\left(\frac{T_c(\varphi)}{T_c^0}\right) = \psi\left(\frac{1}{2}\right) - \psi\left(\frac{1}{2} + \frac{\alpha_0(\varphi)}{2\pi T_c(\varphi)}\right) \tag{16.15}$$

The larger α_0, the lower values of $T_c(\varphi)$ are required to fulfill this relation. Eventually, for $\alpha_0 = 2\pi T_c^0 \exp(\psi(1/2)) = \pi T_c^0/2\gamma_E$ the transition temperature $T_c(\varphi)$ vanishes, i.e. one reaches a (flux-tuned) quantum critical point. This happens for the critical flux $\varphi_c = \pi r/(2\sqrt{2\gamma_E})$, $\gamma_E \approx 1.78$.[4] Due to the flux-periodicity of the phase diagram, a quantum transition can only be observed in the ring geometry if $\varphi_c < 1/2$ which implies $r < \sqrt{2\gamma_E}/\pi \approx 0.6$. Notice that this critical value of $r = R/\xi$ is less restrictive than a naive application of the quadratic approximation valid for $r \gg 1$ would suggest. (The latter would give $1 - (1/2r)^2 = 0 \Rightarrow r = 1/2$.)

16.3.2. *Persistent Current*

The formula for the persistent current $I = \frac{T}{\phi_0}\partial_\varphi \ln \mathcal{Z}$ in the Gaussian approximation reads

$$I = -\frac{T}{\phi_0}\sum_{n,\omega}\frac{\partial\mathcal{L}^{-1}(n,\omega)}{\partial\varphi}\mathcal{L}(n,\omega) \tag{16.16}$$

In order to gain a good qualitative and quantitative understanding of the flux and temperature dependence, the magnitude and relevance of the scales involved in the problem, we will in the following derive simpler expressions for the persistent current and discuss various limiting cases. A particular emphasis will be put on the analysis of the persistent current in the vicinity of the quantum critical point. This approach will be corroborated by a direct numerical evaluation of (16.16). When doing so, some care needs to be exercised in order to correctly deal with the slow convergence properties for large values of $|n|$ and $|\omega|$. As before, we find it convenient to consider the dimensionless quantity $i = I/(T_c^0/\varphi_0)$ and for definiteness analyze fluxes φ in the interval $(0, 0.5)$. Since the persistent current is periodic $i(\varphi+1) = i(\varphi)$ and odd $i(\varphi) = -i(-\varphi)$ in the flux, this is sufficient to infer the persistent current for arbitrary fluxes.

As is shown in appendix D, the persistent current can be written as the sum of two contributions, $i = i_s + i_{ns}$. The rationale behind this decomposition is the following: The singular part i_s diverges on the transition line

$\varphi_c(T)$. Outside the Ginzburg region on the normal side of the transition, i_s is still strongly flux and temperature dependent. It describes by far the dominant contribution to the fluctuation persistent current in the entire normal part of the phase diagram. The nonsingular part i_{ns}, on the other hand, displays a smooth flux dependence and is much smaller. i_s and i_{ns} have opposite signs. We will mostly discuss i_s, for explicit formulas for i_{ns} we refer to appendix D. The expression for i_s reads

$$i_s = -2\pi t \sum_{\omega} \frac{\sin(2\pi\varphi)}{\cosh(2\pi z_s) - \cos(2\pi\varphi)} \tag{16.17}$$

Here, z_s is defined as the unique positive solution of the equation

$$\psi\left(\frac{1}{2}\right) - \ln\left(\frac{T}{T_c^0}\right) = \psi\left(\frac{1}{2} + \frac{|\omega| - \varepsilon_T z^2}{4\pi T}\right), \tag{16.18}$$

for which the argument of the digamma function on the right hand side is positive.

In order to find a more convenient expression for z_s it is useful to define the function $\alpha_c(T)$ describing the phase boundary in the α_0–T phase diagram, i.e. $\alpha_c(T) = \alpha_0(\varphi_c(T))$,

$$\ln\left(\frac{T}{T_c^0}\right) = \psi\left(\frac{1}{2}\right) - \psi\left(\frac{1}{2} + \frac{\alpha_c(T)}{2\pi T}\right). \tag{16.19}$$

One immediately reads off

$$z_s = \sqrt{(|\omega| - 2\alpha_c(T))/\varepsilon_T}, \tag{16.20}$$

Here, $\alpha_c(T)$ in Eq. (16.19) is allowed to become negative as soon as $T > T_c^0$. Note that z_s is always real for $T > T_c^0$. This is no longer true for $T < T_c^0$. In this regime it is instructive to use the fact that $\alpha_c(T)$ and the temperature dependent critical flux $\varphi_c(T)$ are closely related, $\varepsilon_T \varphi_c^2(T) = 2\alpha_c(T)$. As a result,

$$z_s = \varphi_c(T)\sqrt{|\omega|/2\alpha_c(T) - 1}, \quad (T < T_c^0). \tag{16.21}$$

Now z_s is real for $\omega > 2\alpha_c(T)$, but becomes purely imaginary for $\omega < 2\alpha_c(T)$. The fact that z_s becomes imaginary for small $|\omega|$ and $T < T_c^0$, but not for larger $T > T_c^0$, is intimately related to the occurrence of the phase transition. Indeed, the denominator in the expression for i_s vanishes for $\varphi = \varphi_c(T)$ at $\omega = 0$, signaling the onset of the superconducting regime. For reference, recall that $\alpha_{c0} \equiv \alpha_c(0) = \pi T_c^0/2\gamma_E \approx 0.88 T_c^0$.

Fluctuation persistent current for $T > T_c^0$ Let us first make contact with the Gaussian result for larger rings $r \gtrsim 1$ stated before in Eq. (16.9). For $T \sim T_c^0$, when the logarithm on the left hand side of Eq. (16.19) is small, one can expand the Digamma function on the right hand side. When keeping only the most dominant term in the sum, the term with vanishing Matsubara frequency, one easily finds $z_s \sim \sqrt{\varepsilon}r$ and in this way reproduces the result of Eq. (16.9) after setting $t \approx 1$. The restriction to $\omega = 0$ is justified in this case, because $z_s(\omega_n) \sim \sqrt{|\omega_n|/\varepsilon_T} = \pi r\sqrt{n}/2$ is real and larger than one for finite Matsubara frequencies, and the denominator in the expression for i_s becomes large, thereby strongly suppressing the contribution of finite $n \neq 0$.

Now we turn to the smaller rings with $r < 0.6$, for which a quantum phase transition takes place at zero temperature. We first analyze the flux dependence of i for a given temperature $T \gtrsim T_c^0$. As long as φ_{c0} and correspondingly r do not become very small, the same argument used for the rings with $r \gtrsim 1$ is applicable here. For an estimate, $z_s(\omega_1) \sim \sqrt{|\omega_1|/\varepsilon_T} = \pi r/2$ becomes equal to $1/2$ only for $r < 1/\pi$, which corresponds to $\varphi_{c0} \approx 0.27$. Interestingly, the same parameter $\pi r/2$ determines the relevance of the nonsingular contribution in this case (see appendix D). As long as this parameter does not become considerably smaller than one, it is therefore safe to concentrate on the $\omega = 0$ term of the singular contribution only. Whenever it is justified to use only this term near T_c^0, then it is also justified for larger temperatures (as can be seen by comparing z_s to z_0 defined in appendix D).

To summarize this somewhat technical discussion, for not too small rings with $\varphi_{c0} \gtrsim 0.3$ we obtain a good description for the entire temperature range $T > T_c^0 + Gi$ by the formula[26]

$$i \approx -2\pi t \frac{\sin(2\pi\varphi)}{\cosh(2\pi\sqrt{-2\alpha_c(T)/\varepsilon_T}) - \cos(2\pi\varphi)}, \quad T > T_c^0 + Gi \quad (16.22)$$

The formula given above remains valid for $T \lesssim T_c^0$ for fluxes, for which the ring is in the normal regime. In Fig. 16.2 we display the flux dependence of the persistent current for different temperatures. At small fluxes, the persistent current is proportional to φ. For temperatures close to T_c^0, this leads to a rapid increase of $|i|$ for small φ. As the flux increases, however, $T_c(\varphi)$ decreases rapidly for the small rings under consideration here. Therefore the distance to the critical line in the phase diagram grows as φ increases while the temperature is kept constant. This is why the current subsequently drops. For larger temperatures, the situation is different. As the temperature increases, $\cosh(2\pi z_s) \approx \exp(2\pi z_s)$ grows and the flux dependence is determined by the numerator in the expression for i_s, Eq. (16.17). As a consequence the shape becomes more sinusoidal. For an estimate we can use that for $T \gtrsim T_c^0$, $z_s \sim \sqrt{\varepsilon} r$ (see Eq. (16.9). The exponential decay of i and transition to a sinusoidal shape therefore starts at $\varepsilon \approx 1/(2\pi r)^2$. The persistent current at these high temperatures results from a combined effect of many angular momentum modes and is therefore much less sensitive to the rapid drop of the transition line than for temperatures $T \sim T_c^0$.

Fluctuation persistent current near the quantum critical point The situation is quite different in the low temperature limit $T \ll T_c^0$ for rings with $r < 0.6$, to which we will turn now. Indeed, for very low temperatures a restriction to the thermal fluctuations, namely those with $\omega = 0$, is not justified as we will see now.

We discuss the vicinity of the critical line for $T \ll T_c^0$. Here one can expand the denominator in the general expression for i_s in small $\Delta\varphi_T = [\varphi - \varphi_c(T)]/\varphi_c(T)$ as well as small $|\omega|/\alpha_c(T)$. In this way one obtains the approximate relation[26]

$$i_s \sim -\frac{2\varphi_c(T)}{\gamma_E \varphi_{c0}^2} h(\Delta\varphi_T, t), \tag{16.23}$$

where the dimensionless function h is defined as

$$h(\Delta\varphi_T, t) = \pi T \sum_{\omega} \frac{1}{|\omega| + 4\alpha_c(T)\Delta\varphi_T}. \tag{16.24}$$

The same expression could be obtained directly from the initial formula for i (upon identifying $\varphi \approx \varphi_c(T)$), when considering the most singular angular momentum mode $n = 0$ only. It is worth noting, however, that for fixed n the sum in ω is ultraviolet divergent (even *before* expanding in $|\omega|/\alpha_c(T)$). When proceeding in this way a cut-off has therefore to be introduced by hand. In contrast, our formula (Eq. 16.17) for i_s immediately reveals that terms in the sum with $\omega > 2\alpha_c(T)$ are suppressed, since z_s becomes real and $\cosh(2\pi z_s)$ grows rapidly for larger Matsubara frequencies. We can therefore perform the sum with logarithmic accuracy and choose $\omega = 2\alpha_c(T)$ as the upper cut-off. Put in different words, the upper limit for the $|\omega|$-summation is effectively provided by mutual cancelations between different angular momentum modes.

The result of the described procedure is

$$h(\Delta\varphi_T, t) = \frac{1}{2s} + \psi\left(1 + \frac{s}{2\Delta\varphi_T}\right) - \psi(1+s), \tag{16.25}$$

where $s = \frac{2\Delta\varphi_T}{t}\frac{\alpha_c(T)}{\pi T_c^0} \approx \frac{\Delta\varphi_T}{\gamma_E t}$. The first term in the expression for h is the classical $\omega = 0$ contribution to the sum. These thermal fluctuations are proportional to the temperature and correspondingly vanish for $T \to 0$. This does, however, not imply the vanishing of the persistent current in this limit. In order to see this more clearly, let us display the asymptotic behavior of the function h:

$$h \sim \begin{cases} \frac{1}{2s} + \ln\frac{1}{2\gamma_E t} & \Delta\varphi_T \ll t \ll 1 \\ \ln\frac{1}{\Delta\varphi_T} & t \ll \Delta\varphi_T \ll 1 \end{cases}. \tag{16.26}$$

Even at vanishing temperatures, a flux dependent contribution to the persistent current $i_s \propto \ln(1/\Delta\varphi)$ remains. As pointed out in the introduction, this contribution, which is entirely due to quantum fluctuations, can become quite large close to the transition.

Turning to finite temperatures next, it is important that $\Delta\varphi_T$ is T-dependent itself and therefore, in order to reveal the full T-dependence of i_s, one should first find the transition line $\alpha_c(T)$. We display the persistent current near the quantum critical point in Fig. 16.3, also comparing the different approximations and showing the contribution of the classical zero frequency part in the sum of Eq. (16.24).

The maximum of $|i|$ at finite T is a result of two competing mechanisms. As T grows from zero, thermal fluctuations become stronger. At the same time the distance to the critical line becomes larger for fixed fixed φ, which eventually leads to a decrease of $|i|$. With the help of the following analytic approximation, $\varphi_c^2(T) \sim \varphi_{c0}^2(1 - 2\gamma_E^2 t^2/3)$ for $T \ll T_c^0$, one can obtain an estimate for the position of the maximum $\varphi_m(T)$ in the $\varphi - T$ phase diagram,

$$\frac{\varphi_m(T) - \varphi_{c0}}{\varphi_{c0}} = \left(1 - \sqrt{1 - 16\gamma_E t/3}\right)\frac{\gamma_E t}{4} - \frac{1}{3}\gamma_E^2 t^2. \tag{16.27}$$

When increasing the flux at fixed temperature the persistent current decays logarithmically away from the transition. It should be kept in mind, however, that the persistent current vanishes at $\varphi = 0.5$ due to symmetry reasons. The validity of the approximations leading to Eq. 16.25 is restricted to small $\Delta\varphi_T$, for larger $\Delta\varphi_T$ the full expression for i_s should be used (see Fig. 16.3).

In appendix D it is shown that at low temperatures $T \ll T_c^0$ the non-singular contribution to the persistent current i_{ns} can be written as

$$i_{ns} = 2\pi t \sum_\omega \int_{-\infty}^{\infty} dx\, \frac{H(\varphi_{cT}\sqrt{\mathrm{e}^x + |\omega|/2\alpha_{cT}}, \varphi)}{x^2 + \pi^2} \tag{16.28}$$

where $H(x,\varphi) = \sin(2\pi\varphi)/(\cosh(2\pi x) - \cos(2\pi\varphi))$. One can perform the integration in ω at zero temperature

$$i_{ns} = \frac{\varepsilon_T}{T_c^0}\int_0^\infty dy \;\; H(\sqrt{y},\varphi)\Big(1 + (2/\pi)\arctan\big[\ln(y/\varphi_{c0}^2)/\pi\big]\Big) \quad (16.29)$$

It is obvious, that i_s and i_{ns} have opposite signs. For a comparison with the singular contribution near the critical point it is instructive to calculate the zero temperature value of i_{ns} right at the critical flux, $i_{ns}(\varphi_{c0} = 0.45) = 0.17$ and $i_{ns}(\varphi_{c0} = 0.4) = 0.43$. Since i_{ns} is also a monotonously decreasing function of φ and vanishes at $\varphi = 0.5$ we conclude that it is numerically small for all fluxes of our interest, $|i_{ns}| \ll |i_s|$. The same remains true at finite temperatures, see Figs. 16.3, 16.4.

The intermediate temperature regime for rings with $r < 0.6$ We display the temperature dependence of the persistent current in the entire temperature interval $0 < T < T_c^0$ in Fig. 16.4. For comparison, the thermal $\omega = 0$ contribution is also shown. One can see, that at low temperatures nonzero Matsubara frequencies give a sizable contribution to the persistent current. When increasing the temperature the thermal $\omega = 0$ contribution becomes increasingly important. In Fig. 16.4 we compare the numerically obtained current i to the approximation i_s. Obviously, it provides a very good approximation in the entire temperature range.

16.4. Discussion

As mentioned previously, the maximal reduction of T_c at finite flux in the experiment of Ref. 8 was about 6 %. Unfortunately, it has so far not been possible to measure the persistent current close to the quantum phase transition. For this type of experiment one would need *both* sufficiently small rings [in order to fulfill the condition $r < 0.6$] *and* a measurement device that allows measuring the persistent current at the comparatively strong magnetic fields necessary to generate a flux of $\phi \approx \phi_0/2$ threading the small area πR^2. (In the experiment at Yale[35] the use of cantilever required a strong magnetic field that most probably reduces almost totally the contributions of pair fluctuation.)

We suggest two possible strategies to relax these conditions. If the size of the ring is the main problem, one can try to use wider rings, because in this case the magnetic field penetrating the annulus of the ring helps to suppress superconductivity. A first consequence is that the critical flux

φ_{c0} is reduced and correspondingly the condition $\varphi_{c0} < 0.5$ for observing the quantum phase transition in the ring geometry is less restrictive, the ring radius R is allowed to be larger. This effect, however, is rather small, as we will see below. A second effect is that the phase diagram is no longer periodic. It can then be advantageous to consider the transition at $\phi \approx 1.5\phi_0$ or even higher fluxes, because then the effect of the magnetic field itself (as opposed to the flux) is stronger (see Fig. 16.8 below). This approach requires measurements at high magnetic fields. If in turn the main problem lies in measuring at high magnetic fields, then the addition of magnetic impurities can help. Magnetic impurities reduce T_c^0 itself and thereby also reduce φ_{c0}. We will now discuss the two mentioned effects in more detail.

Rings of finite width So far we considered the idealized case for which the width w of the ring (in radial direction) is vanishingly small. Next we discuss corrections to this result, resulting from a finite width. While doing so, we will still assume that the width is much smaller than both the coherence length ξ and the penetration depth λ. The first assumption implies that the order parameter field does not vary appreciably as a function of the radius r, the second assumption implies that the magnetic field is almost constant as a function of r.

In appendix E we sketch, following a method outlined in Ref. 14, the derivation of the following result. For a ring of finite width the pair-breaking parameter acquires a correction, $\alpha_{0n} = \alpha_{0n}^{(0)} + \alpha_{0n}^{(1)}$. Here, $\alpha_{0n}^{(0)} = \frac{D}{2R^2}(n-\varphi)^2$ is the width-independent part used so far and the leading corrections in w/R are[17]

$$\alpha_{0n}^{(1)} = \frac{D}{2R^2}\frac{w^2}{4R^2}\left(\varphi^2 + \left[\frac{1}{3} + \frac{w^2}{20R^2}\right]n^2\right) \tag{16.30}$$

The relation $\xi^2 = \pi D/(8T_c^0)$ can be used to express the result through T_c^0 and r. Most importantly, the width-dependent correction $\alpha_{0n}^{(1)}$ is not a function of $n - \phi/\phi_0$. Correspondingly, the phase diagram is no longer periodic in φ. The physical interpretation of this result is simple. Since superconductivity is already weakened by the magnetic field penetrating the annulus of the ring, T_c can be suppressed at a smaller flux compared to a ring of vanishing width. For the sake of brevity, we will write only the leading correction in the following.

Let us examine some of the consequences. For the transition line we should now solve an equation analogous to Eq. (16.15), where we now $\alpha_0(\varphi)$

should be replaced by $\alpha_{0,m} = \min_n(\alpha_{0n})$. Let us first consider the regime of small suppression $r \gg 1$ and small fluxes $\varphi \ll 1$. Then we can use $T_c(\varphi) \sim T_c^0$, $\alpha_{0,m} = \alpha_{0,0}$ and for $\alpha_{0,0}(\varphi) \ll T_c(\varphi)$ one can approximately calculate the reduction of the transition temperature. At vanishing flux, there is no magnetic field and $T_c(\varphi = 0) = T_c^0$ is unchanged, at small but finite flux, however, there is a correction,

$$T_c(\varphi) \approx T_c^0 \left(1 - \frac{\varphi^2}{r^2}\left(1 + \frac{w^2}{4R^2}\right)\right) \tag{16.31}$$

For small fluxes the T_c reduction is slightly stronger than for vanishing width.

The condition for a suppression of T_c^0 to zero close to $\varphi \sim 1/2$ can also be found. As for the case $w = 0$ the critical value for α, for which T_c vanishes, is $\alpha_{c0} = \pi T_c^0/2\gamma_E$. The critical flux, however, should now be found by equating α_{c0} to $\alpha_{0,0}(\varphi) = \frac{4}{\pi}\frac{\varphi^2}{r^2}T_c^0\left(1 + \frac{w^2}{4R^2}\right)$. The result for the critical flux is

$$\varphi_{c0} \approx 0.83\, r\left(1 - \frac{w^2}{8R^2}\right) \tag{16.32}$$

As expected, for a ring of finite width the critical flux is reduced.

In Fig. 16.8 we show the mean field transition line for two rings with the same radius $r = 0.66$, but different widths. The ring with vanishingly small width has a flux-periodic phase diagram and does not show a quantum phase transition, since $r > 0.6$. The other ring has a width of $w = R/3$. One observes three main changes. The maxima in T_c at finite flux are reduced compared to T_c^0. They are shifted towards smaller flux, i.e. they do no

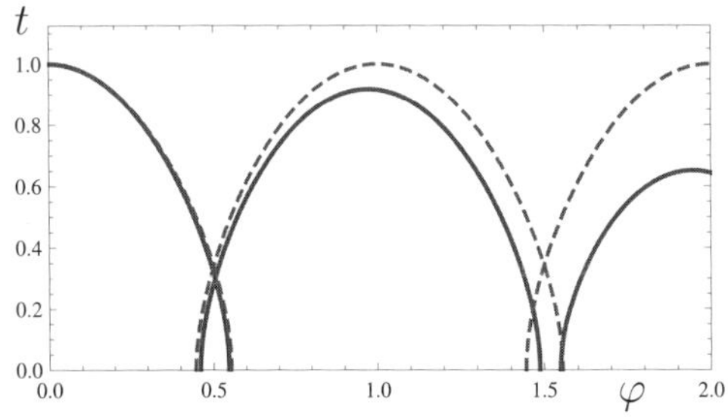

Fig. 16.8. The mean field transition line for a ring with finite width $w = R/3$ and $r = 2/3$ (solid line) and for a ring with vanishingly small width, but same radius (dashed line). It can be seen that for the ring with finite width a quantum phase transition occurs near $\varphi = 1.5$, while for the ideal one-dimensional ring there is no quantum critical point due to the periodicity of the phase diagram.

longer occur at integer values of φ. Finally, the ring exhibits a quantum phase transition close to $\varphi = 1.5$.

Magnetic impurities In this article we have so far discussed the role of an external magnetic field as the origin of the pair-breaking mechanisms. In principle, there are other effects that may cause pair-breaking. Among them are proximity effect, exchange field, magnetic impurities or interaction with the electromagnetic environment. Each pair-breaking mechanism will have its own pair-breaking parameter (see appendix A for examples), and to a good approximation the total pair breaking parameter $\alpha_{0;tot}$ is the sum of the individual ones. The effects of these pair-breaking mechanisms can be obtained formally by substituting α_0 by $\alpha_{0;tot}$ in the formulas discussed above. In particular, transition temperature $T_c(\varphi)$ can be obtained from Eq. (16.15) after the replacement $\alpha_0(\varphi) \to \alpha_{0,tot}(\varphi)$. The condition for the quantum critical point reads $\alpha_{0,tot}(\varphi) = \pi T_c^0 / 2\gamma_E$.

Of particular interest is the case of magnetic impurities, discussed by Bary-Soroker, Entin-Wohlman and Imry,[11,12] for which $\alpha_{0,mi} = 1/\tau_s$ is equal to the scattering rate caused by the magnetic impurities and independent of the flux. A sufficiently large concentration of magnetic impurities will reduce T_c to zero [even at vanishing flux] and a large persistent current is obtained due to the pair fluctuations. Ref. 11 suggest that the large persistent current observed in copper rings[22] may be attributed to such pairing fluctuations.

As emphasize earlier, in our case the addition of magnetic impurities may push the system to the quantum critical point at a smaller external magnetic field, since it provides an additional flux-independent pair-breaking mechanism. If the main experimental difficulty is to perform sensitive measurements at high magnetic fields, introducing magnetic impurities might therefore make possible the experimental observation of the flux tuned quantum critical point.

Acknowledgments

We thank E. Altman, H. Bary-Soroker, A. I. Buzdin, O. Entin-Wohlman, H. J. Fink, A. M. Finkel'stein, Y. Imry, Y. Liu, F. von Oppen for useful discussions, and K. Moler, N. Koshnick and H. Bluhm for stimulating discussions and for sharing their numerical results with us. We acknowledge financial support from the Minerva Foundation, DIP and ISF grants.

Appendix A

In this appendix we sketch the microscopic derivation of the Ginzburg-Landau functional for disordered superconductors in the presence of a pair-breaking mechanism. The purpose is to illustrate the relation between the phenomenological approach used in this article for the description of the thermal transition on the one hand and the microscopic approach used for the description of the quantum phase transition on the other hand. The material is well known[1] and included just for the convenience of the reader. As a starting point, we write the partition function $\mathcal{Z}$ of a disordered superconductor as a functional integral over fermionic fields χ, χ^* in the imaginary time formalism, $\mathcal{Z} = \int \mathcal{D}(\chi\chi^*)\ \exp(-\mathcal{S})$, where

$$\begin{aligned}\mathcal{S} = &\int_0^{1/T} d\tau d\mathbf{r}\ \chi_\sigma^*(\mathbf{r},\tau)\Big(\partial_\tau + \mathcal{H}_0 - \mu\Big)\chi_\sigma(\mathbf{r},\tau) \\ &-|\lambda|\int_0^{\beta} d\tau d\mathbf{r}\ \chi_\uparrow^*(\mathbf{r},\tau)\chi_\downarrow^*(\mathbf{r},\tau)\chi_\downarrow(\mathbf{r},\tau)\chi_\uparrow(\mathbf{r},\tau).\end{aligned} \tag{A.1}$$

In this formula, Here, $\mathcal{H}_0$ is the Hamiltonian of noninteracting electrons, σ denotes spin, μ is the chemical potential. In presence of a magnetic field and disorder potential $u(\mathbf{r})$ the Hamiltonian $\mathcal{H}_0$ of noninteracting electrons is $\mathcal{H}_0 = (-i\nabla - eA(\mathbf{r}))^2 + u(\mathbf{r})$. We consider the case of non-magnetic impurities and the disorder potential is assumed to be delta-correlated, $\langle u(\mathbf{r})u(\mathbf{r}')\rangle = \delta(\mathbf{r}-\mathbf{r}')/(2\pi\nu\tau)$. For the sake of brevity, we will sometimes use the notation $x = (\mathbf{r},\tau)$, $\int dx = \int_0^{1/T} d\tau d\mathbf{r}$. The above partition function is based on the standard BCS Hamiltonian. The singularity in the Cooper channel can be isolated by introducing the Hubbard-Stratonovich (HS) field Δ:

$$\exp\left(-|\lambda|\int dx\ \chi_\uparrow^*(x)\chi_\downarrow^*(x)\chi_\downarrow(x)\chi_\uparrow(x)\right) \tag{A.2}$$

$$= \int D(\Delta,\Delta^*)\ \mathrm{e}^{-\Delta^*(\mathbf{r},\tau)\chi_\downarrow(x)\chi_\uparrow(x)-\Delta(x)\chi_\uparrow^*(x)\chi_\downarrow^*(x)-|\lambda|^{-1}|\Delta(x)|^2} \tag{A.3}$$

where a trivial normalization constant has been suppressed. The main advantage of the HS decoupling is that the action $\mathcal{S}$ becomes quadratic in the fermion fields. The next step is to integrate them out and to organize the result in powers of the fields Δ, Δ^*. The resulting effective action up to fourth order in the expansion, which is sufficient for our purposes, can be written as $\mathcal{S}_{eff} = \mathcal{S}_{eff}^{(2)} + \mathcal{S}_{eff}^{(4)}$, where

$$\mathcal{S}_{eff}^{(2)} = T\sum_{\omega}\int d\mathbf{r}d\mathbf{r}'\Delta^*(\mathbf{r},\omega)\mathcal{L}^{-1}(\mathbf{r},\mathbf{r}',|\omega|)\Delta(\mathbf{r}',\omega) \tag{A.4}$$

$$\mathcal{S}_{eff}^{(4)} = \int dx_i\; \mathcal{B}(\{x_i\})\Delta^*(x_1)\Delta(x_2)\Delta^*(x_3)\Delta(x_4). \tag{A.5}$$

In this formula, the so-called fluctuation propagator $\mathcal{L}$ obeys

$$\mathcal{L}^{-1}(\mathbf{r},\mathbf{r}',\omega) = \delta(\mathbf{r}-\mathbf{r}')1/|\lambda| - \Pi(\mathbf{r},\mathbf{r}',\omega) \tag{A.6}$$

and for a disordered superconductor

$$\Pi(\mathbf{r},\mathbf{r}',\omega_k) = 2\pi\nu\tau T\sum_{n}\Theta(\varepsilon_n\varepsilon_{n+k})\,\mathcal{C}(\mathbf{r},\mathbf{r}',2\varepsilon_n+\omega_k) \tag{A.7}$$

Here, $\varepsilon_n = (2n+1)\pi T$ is a fermionic, and $\omega_n = 2\pi T$ a bosonic Matsubara frequency and the Cooperon $\mathcal{C}$ fulfills the equation

$$\Big\{D\big(-i\nabla - 2e\mathbf{A}(\mathbf{r}_1)\big)^2 + |\omega|\Big\}\mathcal{C}(\mathbf{r}_1,\mathbf{r}_2,\omega) = \frac{1}{\tau}\delta(\mathbf{r}_1-\mathbf{r}_2) \tag{A.8}$$

Explicit evaluation of the sum in ω gives

$$\hat{\Pi}(\omega_k) = \nu\psi\left(\frac{1}{2}+\frac{\omega_D}{2\pi T}+\frac{D\hat{p}^2+|\omega_k|}{4\pi T}\right) - \nu\psi\left(\frac{1}{2}+\frac{D\hat{p}^2+|\omega_k|}{4\pi T}\right) \tag{A.9}$$

The summation in frequencies has been restricted to the interval $|\omega| < \omega_D$, where ω_D is the Debey frequency. ψ is the digamma function,[29] and $\hat{p} = (-i\nabla - 2e\mathbf{A})$. In Fig. A.1 the Dyson equation for the fluctuation propagator is presented. The particle-particle bubble Π consists of two (disorder dressed) Green's functions. In a similar way, the vertex $\mathcal{B}$ entering $\mathcal{S}_{eff}^{(4)}$ can be represented with the help of four Green's functions.[1]

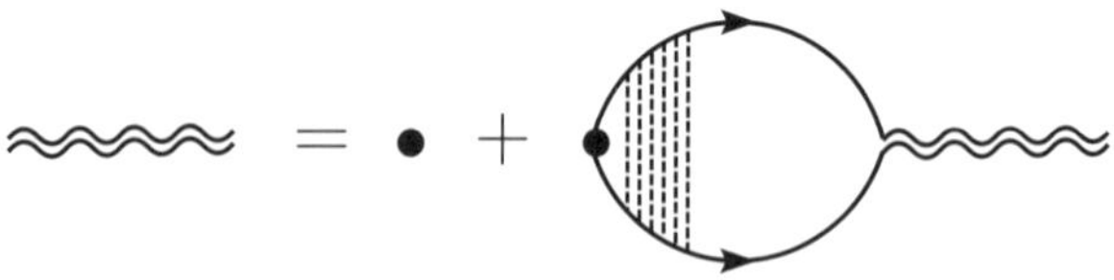

Fig. A.1. The Dyson equation for the fluctuation propagator (double wiggly line), compare to Eq. A.6. The dot symbolizes the bare interaction constant λ. The particle-particle bubble Π is dressed by impurity lines (dashed lines, see Ref. 36).

The mean field phase transition is determined by the instability of the *quadratic* action. When approaching the transition line from the normal phase the instability occurs for $\omega = 0$ as one can infer from the expression for $\mathcal{L}$. Therefore, the condition for the transition can generally be formulated as $\mathcal{L}^{-1}(\alpha_0, 0) = 0$.

For a bulk system (assuming translational invariance) in the absence of a magnetic field or magnetic impurities α_0 is equal to zero, the transition occurs for a spatially constant order parameter field Δ. Therefore, the condition $\mathcal{L}^{-1}(0,0) = 0$ determines the transition temperature T_c^0, for which one finds

$$T_c^0 = \frac{\omega_D}{2\pi} \exp\left(-\frac{1}{\nu|\lambda|} - \psi\left(\frac{1}{2}\right)\right) = \frac{2\gamma_E\omega_D}{\pi} \exp\left(-\frac{1}{\nu|\lambda|}\right) \qquad \text{(A.10)}$$

We used the following properties of the Digamma function: $\psi(x \gg 1) \sim \ln(x)$ for $x \gg 1$ and $\psi\left(\frac{1}{2}\right) = -\ln\gamma_E - \ln 4$, where $\gamma_E \approx 1.78$. One can use the relation between ω_D and T_c^0 to eliminate the ω_D dependence of $\mathcal{L}$ in favor of a dependence on T_c^0, which is easier to determine experimentally

$$(\nu\hat{\mathcal{L}})^{-1}(\omega) = \ln\left[\frac{T}{T_c^0}\right] + \psi\left[\frac{1}{2} + \frac{D\hat{p}^2 + |\omega|}{4\pi T}\right] - \psi\left[\frac{1}{2}\right]. \qquad \text{(A.11)}$$

The condition for a transition in the presence of a pair-breaking mechanism $\mathcal{L}^{-1}(\alpha_0, 0) = 0$ then reads

$$0 = \ln\left[\frac{T_c}{T_c^0}\right] + \psi\left[\frac{1}{2} + \frac{\alpha_0}{2\pi T_c}\right] - \psi\left[\frac{1}{2}\right]. \qquad \text{(A.12)}$$

This equation determines T_c for a given α_0 or the critical value $\alpha_c(T)$ for a given temperature. Let us give a few examples: For a thin disordered superconducting film with thickness t in a parallel magnetic field one finds $\alpha_0 = D(eBt)^2/6$, for a thin wire with radius R subject to a longitudinal magnetic field $\alpha_0 = D(eBR)^2/4$. In the case of superconducting rings or cylinders considered in this paper, $\alpha_0 = D\varphi^2/(2R^2)$ for $\varphi \in (0.0.5)$. Another important example, which demands a different derivation, is scattering on magnetic impurities. In this case, one can identify α_0 with $1/\tau_s$, the scattering rate. To conclude this brief discussion of pair-breaking mechanisms, let us note that to a first approximation different pair-breaking mechanisms are just additive, the effective α_0 is the sum of the individual pair-breaking parameters.

Now we can also make contact with the classical GL description. For a thermal transition, for which the transition temperature is not strongly suppressed ($T \sim T_c^0 \sim T_c(\varphi)$) one can expand the ψ-function for small

$\frac{D\hat{p}^2+|\Omega|}{4\pi T}$ in Eq. (A.11). Neglecting any magnetic field dependence for the fourth order term, focusing on the dominant $\omega = 0$ component of $\Delta(\mathbf{r}, \omega)$ by writing $\Delta(\mathbf{r}, \tau) \sim \Delta(\mathbf{r})$, and setting $T \to T_c^0$ wherever the dependence on temperature is smooth, one arrives at

$$\mathcal{S}_{eff} = \mathcal{S}_{eff}^{(2)} + \mathcal{S}_{eff}^{(4)} \tag{A.13}$$

$$\mathcal{S}_{eff}^{(2)} = \frac{1}{T_c^0} \int d\mathbf{r} \, \left(\nu\varepsilon |\Delta(\mathbf{r})|^2 + \nu\xi^2 |(-i\nabla - 2eA(\mathbf{r}))\Delta(\mathbf{r})|^2 \right) \tag{A.14}$$

$$\mathcal{S}_{eff}^{(4)} = \frac{1}{T_c^0} \int d\mathbf{r} \, \nu \frac{7\zeta(3)}{16\pi^2 (T_c^0)^2} |\Delta(\mathbf{r})|^4 \tag{A.15}$$

In this formula, the (zero temperature) coherence length ξ fulfills $\xi^2 = \pi D/(8T_c^0)$ and $\varepsilon = (T - T_c^0)/T_c^0$. The free energy given in Eq. (16.4) is recovered after introducing the order parameter field $\psi(\mathbf{r})$ as $\Delta(\mathbf{r}) = c\psi(\mathbf{r})$, where $c^2 = \alpha T_c^0/\nu$. The coefficient α is related to the mass m in Eq. (16.4) via $1 = 4m\alpha T_c^0 \xi^2$. A certain arbitrariness in the choice of parameters α and m remains, which is related to the arbitrary normalization of the order parameter field. While we do not make a particular choice here, because it does not affect any of the results, let us only remark that a popular choice is to identify m with the electron mass.

Appendix B

In this appendix we present the calculation for the partition function and the average $\langle |\psi_0|^2 \rangle$ in the zero-dimensional limit.[31] In this limit, the partition function can be presented as an ordinary integral over the complex variable ψ_0 in the form

$$\mathcal{Z} = \int D(\psi_0, \psi_0^*) \, \exp(-\mathcal{F}/T), \quad \mathcal{F} = a|\psi_0|^2 + \frac{b}{2V}|\psi_0|^4 \tag{B.1}$$

We first write the integration measure more explicitly $D(\psi_0, \psi_0^*) = d\Im(\psi_0) d\Re(\psi_0)$ up to an unimportant constant of proportionality. After introducing polar coordinates the trivial integration over the polar angle can be performed, since $\mathcal{F}$ depends on ψ_0 only via $|\psi_0|^2$. One may therefore replace $D(\psi_0, \psi_0^*) \to \pi d|\psi_0|^2$. When writing $y = |\psi_0|^2$ the expression for $\mathcal{Z}$ then reads

$$\mathcal{Z} = \pi \int_0^\infty dy \, \mathrm{e}^{-\frac{a}{T}y - \frac{b}{2TV}y^2} = \pi\sqrt{2TV/b} \, \mathrm{e}^{\frac{V}{2bT}a^2} \int_{a\sqrt{V/2bT}}^\infty dy \, \mathrm{e}^{-y^2} \tag{B.2}$$

In the last step a simple shift and rescaling of the integration variable was performed. Introducing the conjugated error function as $\mathrm{erfc}(x) =$

$\frac{2}{\sqrt{\pi}}\int_x^\infty \exp(-y^2)dy$, $\mathrm{erfc}(0)=1$, one arrives at the result

$$\mathcal{Z} = \frac{\pi\sqrt{\pi}}{\alpha Gi}\exp(x^2)\,\mathrm{erfc}(x) \tag{B.3}$$

where $x = \frac{\varepsilon}{Gi}$, $\varepsilon = (T - T_c^0)/T_c^0$ and $Gi = \sqrt{2b/\alpha^2 T_c V}$ is the $0d$ Ginzburg parameter. For the thermal average of $|\psi_0|^2$ one obtains

$$\langle|\psi_0|^2\rangle = \frac{2f(x)}{\alpha Gi}, \qquad f(x) = \frac{\exp(-x^2)}{\sqrt{\pi}\mathrm{erfc}(x)} - x \tag{B.4}$$

Appendix C

In this appendix we will show how to calculate the partition function when two modes ψ_0 and ψ_1 are taken into consideration. We start from Eq. (16.6) and introduce the notation $A_n = a_n/T_c^0$, $B = b/(2VT_c^0)$, $x = |\psi_0|^2$ and $y = |\psi_1|^2$. Then the expression for the partition function becomes

$$\mathcal{Z} = \pi^2\int_0^\infty dxdy\; \mathrm{e}^{-A_0x - A_1y - B(x^2+y^2+4xy)} \tag{C.1}$$

We want to perform one integration explicitly. To this end we first change integration variables to $w = x + y$ and $z = x - y$ and obtain $\mathcal{Z} = \frac{\pi^2}{2}\int_0^\infty dw\int_{-w}^{w} dz\; \exp(-wA_+ - zA_- - \frac{3}{2}Bw^2 + \frac{1}{2}Bz^2)$, where $A_\pm = (A_0 \pm A_1)/2$. After changing the order of integration we find

$$\mathcal{Z} = \frac{\pi^2\sqrt{\pi}}{\sqrt{3B}}\,\mathrm{e}^{A_+^2/6B}\int_0^\infty dz\,\mathrm{e}^{z^2}\cosh(A_-z\sqrt{2/B})\,\mathrm{erfc}(\sqrt{3}z + A_+/\sqrt{6B}) \tag{C.2}$$

Next we derive a formula for persistent current. Combining the formulas $I = (2mR^2\phi_0)^{-1}\sum_{n=0,1}(n-\varphi)\langle|\psi_n|^2\rangle$ and $\langle|\psi_n|^2\rangle = -\partial_{A_n}\ln\mathcal{Z}$ one finds

$$i = \frac{1}{4mR^2T_c^0}\left(\partial_{A_-} + 2\Delta\varphi\,\partial_{A_+}\right)\ln\mathcal{Z}, \tag{C.3}$$

where $\Delta\varphi = \varphi - 1/2 = 4mR^2T_c^0A_-$. Let us note that

$$\partial_{A_+}\ln\mathcal{Z} = \frac{A_+}{3B} - \frac{\pi^2\sqrt{\pi}}{12B\sqrt{B}}\frac{1}{\mathcal{Z}}\left(\mathrm{e}^{x_0^2}\mathrm{erfc}(x_0) + \mathrm{e}^{x_1^2}\mathrm{erfc}(x_1)\right) \tag{C.4}$$

$$\partial_{A_-}\ln\mathcal{Z} = -\frac{A_-}{B} - \frac{\pi^2\sqrt{\pi}}{4B\sqrt{B}}\frac{1}{\mathcal{Z}}\left(\mathrm{e}^{x_0^2}\mathrm{erfc}(x_0) - \mathrm{e}^{x_1^2}\mathrm{erfc}(x_1)\right)$$

where $x_n = A_n/(2\sqrt{B}) = \varepsilon_n/Gi$ coincides with the variable x_n used in the main text. After combining these results one finds

$$i_2 = -4\frac{r^2}{Gi}\mathcal{M}_- + \frac{8}{3}x_-\mathcal{M}_+, \tag{C.5}$$

where

$$\mathcal{M}_{\pm} = x_{\pm} \mp \frac{1}{\mathcal{P}}\left(\mathrm{e}^{x_0^2}\mathrm{erfc}(x_0) \pm \mathrm{e}^{x_1^2}\,\mathrm{erfc}(x_1)\right) \tag{C.6}$$

and

$$\mathcal{P} = 4\int_0^{\infty} dz\, \mathrm{e}^{3z^2+2(2x_1-x_0)z+x_1^2}\,\mathrm{erfc}(2z+x_1) \tag{C.7}$$

where $x_{\pm} = (x_0 \pm x_1)/2$. An analogous formula has been given in Ref. 15.

Most interesting for us is the quantity $\overline{\chi}(1/2) = -\,\partial i_2/\partial\varphi|_{\varphi=1/2}$. It is worth noting that for $\varphi = 1/2$ one finds $A_- = x_- = 0$ (i.e $x_0 = x_1$), and formulas simplify considerably. We can obtain the formula for $\overline{\chi}(1/2)$ directly by differentiating the result for i_2 or by first differentiating Eq. (C.3) and then using the relations in Eq. (C.5). In the latter case one obtains as an intermediate step the relation $\overline{\chi}(1/2) = -\frac{\phi_0^2}{2mR^2T_c^0\mathcal{Z}}(\partial_{A_+} + \frac{1}{8T_c^0mR^2}\partial_{A_-}^2)\,\mathcal{Z}|_{A_-=0}$. The result for $\overline{\chi}$ is stated in the main text, Eq. (16.10).

Appendix D

In this appendix we sketch the derivation of the expressions for the persistent current in the Gaussian approximation used in the main text. We also make contact with Ref. 12, where fluctuations at temperatures $T > T_c(\varphi = 0)$ have been examined [In the presence of magnetic impurities, $T_c(\varphi = 0)$ differs from T_c^0, the transition temperature in the absence of any pair-breaking mechanism, and may even vanish]. Our starting point is Eq. (16.16),

$$i_G = -t\sum_{n,\omega}\frac{\partial_{\varphi}\mathcal{L}^{-1}(n,\omega)}{\mathcal{L}^{-1}(n,\omega)} \tag{D.1}$$

Using standard methods for transforming the sum in n into an integral in the complex plane, we can write i as

$$i_G = -it\sum_{\omega}\oint_{C'} dz\, H(z,\varphi)\, R(z,\omega) \tag{D.2}$$

where

$$R(z,\omega) = \frac{\partial_z\tilde{\mathcal{L}}^{-1}(-iz,\omega)}{\tilde{\mathcal{L}}^{-1}(-iz,\omega)}, \quad H(z,\varphi) = \frac{\sin(2\pi\varphi)}{\cosh(2\pi z) - \cos(2\pi\varphi)} \tag{D.3}$$

and we use the notation

$$(\nu\tilde{\mathcal{L}})^{-1}(-iz,\omega) = \ln\left(\frac{T}{T_c^0}\right) + \psi\left(\frac{1}{2} + \frac{|\omega| - \varepsilon_T z^2}{4\pi T}\right) - \psi\left(\frac{1}{2}\right). \tag{D.4}$$

The contour $\mathcal{C}'$ includes all the poles of $H(z,\varphi)$ with $\Im z > 0$. These poles are located at $z = i\varphi, i(1-\varphi), i(1+\varphi), i(2-\varphi), \ldots$. The contour $\mathcal{C}'$ does not include poles of $R(z,\omega)$, see Fig. D.1. Our goal will be to deform the contour of integration in such a way that the integral can be evaluated at the poles of R with the help of the residue theorem. Poles of $R(z,\omega)$ occur either due to zeros or poles of $\tilde{\mathcal{L}}^{-1}(-iz,\omega)$. Due to the dependence of $\mathcal{L}^{-1}$ on z^2 the poles of $R(z,\omega)$ come in pairs, for each pole at a point z there is a pole at $-z$.

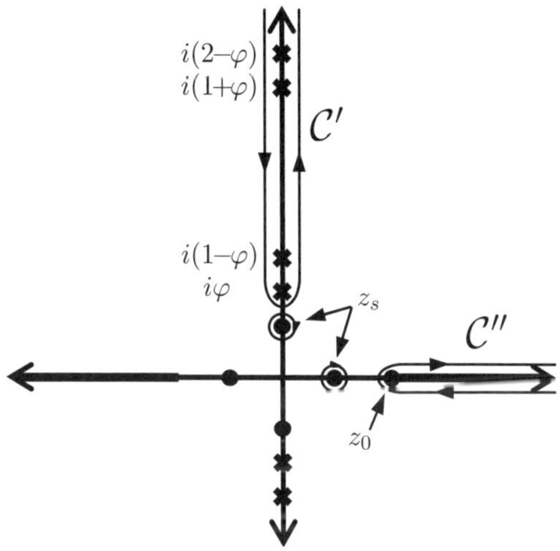

Fig. D.1. The complex z plane. The poles of $H(z,\varphi)$ (Eq. D.3) as a function of z are indicated by crosses together with the original integration contour $\mathcal{C}'$. Equivalently, the integral in Eq. D.2 can be calculated by integrating along the path $\mathcal{C}''$, and additionally taking into account the residue at the pole z_s. z_s can be either imaginary or real and is responsible for the dominant contribution i_s to the persistent current. The contour $\mathcal{C}''$ encloses z_n and $\tilde{z}_n$, the poles of $R(z,\omega)$, which are positioned on the real axis to the right of z_0. As the temperature decreases the distance between adjacent z_n and $\tilde{z}_n$ becomes small and the sum over the corresponding residues can be approximated by an integral along a branch cut.

Zeros of $\tilde{\mathcal{L}}^{-1}(-iz,\omega)$ can be classified according to the value of $A(z,\omega) = 1/2 + (|\omega| - \varepsilon_T z^2)/(4\pi T)$, the argument of the digamma function. There is a pair of zeros $\pm z_s$, for which A is positive, and there is a pair of zeros $\pm z_n$ for each interval $-n-1 < A < -n$, where $n = 0, 1, 2, \ldots$. The existence of such zeros is clear from the reflection formula $\psi(1-z) = \psi(z) + \pi\cot(\pi z)$. Poles of $\tilde{\mathcal{L}}^{-1}(-iz,\omega)$ originate from poles of the digamma function for $A = -n$, $n = 0, 1, 2, \ldots$ and we label the pairs accordingly,

$\pm\tilde{z}_n$, where $\tilde{z}_n = \sqrt{((2n+1)2\pi T + |\omega|)/\varepsilon_T}$. Note that all z_n and $\tilde{z}_n$ are real and $z_n < \tilde{z}_n < z_{n+1}$.

The zeros at $\pm z_s$ deserve special attention. For $T > T_c^0$, z_s is real. For $T < T_c^0$, however, z_s can be either real or imaginary. Indeed, due to the relation $\ln(T/T_c^0) = \psi(1/2) - \psi(1/2 + \alpha_c(T)/2\pi T)$, we can write $z_s = \varphi_c(T)\sqrt{|\omega|/2\alpha(T) - 1}$ for $T < T_c^0$. z_s is real for $\omega > 2\alpha(T)$ and purely imaginary $z_s = i\varphi_c(T)\sqrt{1 - |\omega|/2\alpha(T)}$ for $\omega < 2\alpha(T)$.

By deforming the contour $\mathcal{C}'$ and using the symmetry and convergence-properties of the integrand we can write i_s as an integral along a contour $\mathcal{C}''$ that encloses all poles of $R(z,\omega)$ on the positive imaginary axis for $0 < \Im z < \varphi$, and on the positive real axis in the complex z-plane, see Fig. D.1. The integral can then formally be calculated with the help of the residue theorem.

When ω is small and z_s imaginary, H can become large for $\varphi \sim \varphi_s$, reflecting the closeness to the phase transition. Evaluating the residue at z_s one finds the singular contribution to the persistent current

$$i_s = -2\pi t \sum_{\omega} H(z_s, \varphi) \tag{D.5}$$

The nonsingular contribution arises from all other poles of R, and can be written as

$$i_{ns} = -2\pi t \sum_{n} \sum_{\omega} [H(z_n, \varphi) - H(\tilde{z}_n, \varphi)]. \tag{D.6}$$

In general, the poles z_n need to be determined numerically. The derived representation is particularly useful in two limiting cases, either for very high or very low temperatures. For very high temperatures z_s is real the spacing between consecutive $z_s, z_n, \tilde{z}_n$ is large. Since $H(z,\varphi)$ decays fast as a function of z one can approximate the result well by considering just the smallest of the $z_s, z_n, \tilde{z}_n$ and in this way obtain relatively simple formulas. This approximation was utilized in Ref. 12, where a similar representation was derived for temperatures $T > T_c(\varphi = 0)$ (in the presence of magnetic impurities), when all poles of $R(z,\omega)$ are real. Another useful limit is the limit of very low temperatures, when z_s is possibly imaginary, but the other z_n, $\tilde{z}_n$ are closely placed on the real axis. Then, calculating the residues for i_{ns} is very similar to an integration along a branch cut as we will describe now. In this low temperature limit one can write

$$(\nu\mathcal{L})^{-1}_{n,\omega} = \psi\left[\frac{1}{2} + \frac{\alpha_n + \frac{|\omega|}{2}}{2\pi T}\right] - \psi\left[\frac{1}{2} + \frac{\alpha_c(T)}{2\pi T}\right] \overset{T \ll T_c^0}{\approx} \ln\left[\frac{\alpha_n + \frac{|\omega|}{2}}{\alpha_c(T)}\right] \tag{D.7}$$

Following the same analysis presented above with this approximate representation, i_s remains unchanged. For i_{ns} we have to perform an integration along a branch cut appearing for $z > z_0 \sim \sqrt{|\omega|/\varepsilon_T}$, which is related to the branch cut of the logarithm. It gives the result

$$i_{ns} = 2\pi t \sum_{\omega} \int_0^{\infty} \frac{dy}{y} \, \frac{H(\varphi_c(T)\sqrt{y + |\omega|/2\alpha_c(T)}, \varphi)}{\ln^2 y + \pi^2} \tag{D.8}$$

Simple substitution gives the formula Eq. (16.28) stated in the main text. In the limit $T \to 0$, it is more convenient to use the relation $[y(\ln^2 y + \pi^2)]^{-1} = \frac{1}{\pi}\partial_y \arctan(\ln(y)/\pi)$, perform a partial integration in y, subsequently use $\partial_y f(y + |\omega|/2\alpha_c(T)) = 2\alpha_c(T)\partial_{|\omega|} f(y + |\omega|/2\alpha_c(T))$ to perform the integral in ω and combine the result with the boundary term. The result is Eq. (16.29).

Finally, let us make contact with the analysis of Ref. 12. In this paper, fluctuations were analyzed for high temperatures $T > T_c(\varphi = 0)$ (in the presence of magnetic impurities) and the flux harmonics i_m of the persistent current $i = \sum_m i_m \sin(2\pi m\varphi)$ were calculated. In this situation one may use the Poisson summation formula to perform the sum over angular momentum modes, since the flux dependence of i is smooth for $T \gg T_c(\varphi = 0)$. [This is not so for $T < T_c(\varphi = 0)$ due to the phase transition to the superconducting state, or, in a mathematical language, due to the presence of the pole at purely imaginary z_s.] A comparable situation arises for $T > T_c^0$ in the absence of magnetic impurities: All poles of $R(z, \omega)$ (including z_s) are real, one can use Eqs. D.5 and D.6 and the relation

$$H(x, \varphi) = \frac{\sin(2\pi\varphi)}{\cosh(2\pi x) - \cos(2\pi\varphi)} = 2\sum_{m=1}^{\infty} \mathrm{e}^{-2\pi x m} \sin(2\pi m\varphi) \tag{D.9}$$

which is valid for $x > 0$, to write

$$i = -4\pi t \sum_{m=1}^{\infty} \sin(2\pi m\varphi) \sum_{\omega} \left[\mathrm{e}^{-2\pi z_s m} + \sum_n \left(\mathrm{e}^{-2\pi z_n m} - \mathrm{e}^{-2\pi \tilde{z}_n m} \right) \right], \tag{D.10}$$

from which one can read off the flux harmonics i_m.

Appendix E

In this appendix we consider the effects of a finite width w on the pair-breaking parameter α. We assume $w \ll \xi\,, \lambda$. We essentially follow Ref. 14.

The standard theory for pair-breaking transitions sketched in appendix A implies that one should find α from the equation

$$-\frac{D}{2}(\nabla - 2ie\mathbf{A})^2\Delta = \alpha\Delta \tag{E.1}$$

where $\mathbf{A}$ is the vector potential. To be more precise, α is the lowest eigenvalue of the operator on the left hand side. One should respect the boundary conditions $\mathbf{n}_i(\nabla - 2ie\mathbf{A})\Delta = 0$, $i = 1, 2$, where $\mathbf{n}_i$ are the unit vector perpendicular to the inner and outer surface. The boundary condition is chosen in analogy to the Ginzburg-Landau theory and means there is no electric current through the surface.

We make the simplifying assumption that Δ is constant in z direction. Then it is convenient to use cylindrical coordinates and the gauge $\mathbf{A} = A_\phi \mathbf{e}_\phi$, $A_\phi = Br/2$, where $\mathbf{e}_\phi$ is the unit vector in azimuthal direction. After expanding Δ in angular momentum modes, the differental equations for different components Δ_n and corresponding eigenvalues α_n decouple

$$-\frac{D}{2}\left[\frac{1}{r}\frac{\partial}{\partial r}\left(r\frac{\partial}{\partial r}\right) - \left(\frac{n}{r} - eBr\right)^2\right]\Delta_n(r) = \alpha_n\Delta_n(r) \tag{E.2}$$

and the boundary condition is $\partial_r\Delta_n = 0$. Let us repeat that this equation is already an approximation, since it assumes a constant magnetic field B. We denote the inner radius as $r_1 = R + w/2$ and the outer one as $r_2 = R + w/2$. In order to obtain the transition line, one needs to find the minimal α_{0n} for given B, R and t. We basically follow Ref. 14. It is immediately clear from Eq. (E.2) that a constant $\Delta_n(r)$ is not a solution. Nevertheless, spatial variations are energetically costly due to the condition $w \ll \xi$, and one can follow a kind of variational approach and determine α_{0n} for slowly varying Δ_n. To this end one may integrate over the volume of the ring. The term with radial derivatives does not give a contribution to the result due to the boundary condition. For the remaining integral can be calculated assuming Δ to be constant. For $r_1 = r_2 = R$ one directly obtains the width-independent part $\alpha_{0n}^{(0)} = \frac{D}{2R^2}(n-\varphi)^2$ used so far and by expansion the leading correction in w/R^{17} stated in Eq. (16.30).

References

1. A. I. Larkin and A. A. Varlamov, *Theory of fluctuations in superconductors.* (Oxford University Press, Oxford, 2005).
2. M. Tinkham, *Introduction to Superconductivity.* (McGraw-Hill, New York, London, 1996).

3. W. A. Little and R. D. Parks, Observation of quantum periodicity in the transition temperature of a superconducting cylinder, *Phys. Rev. Lett.* **9**, 9, (1962).
4. $\hbar = k_B = c = 1$ throughout.
5. G. Schwiete and Y. Oreg, Persistent current in small superconducting rings, *Phys. Rev. Lett.* **103**, 037001, (2009).
6. Y. Imry, *Introduction to mesoscopic physics.* (Oxford University Press, London, 2002).
7. Y. Liu, Y. Zadorozhny, M. M. Rosario, B. Y. Rock, P. T. Carrigan, and H. Wang, Destruction of the global phase coherence in ultrathin, doubly connected superconducting cylinders, *Science.* **294**, 2332, (2001).
8. N. C. Koshnick, H. Bluhm, M. E. Huber, and K. A. Moler, Fluctuation superconductivity in mesoscopic aluminum rings, *Science.* **318**, 1440, (2007).
9. P. G. de Gennes, *C. R. Acad. Sci. Ser. II.* **292**, 9, (1981).
10. P. G. de Gennes, Champ critique d'une boucle supraconductrice ramifie, *C. R. Acad. Sci. Ser. II.* **292**, 279, (1981).
11. H. Bary-Soroker, O. Entin-Wohlman, and Y. Imry, Effect of pair breaking on mesoscopic persistent currents well above the superconducting transition temperature, *Phys. Rev. Lett.* **101**, 057001, (2008).
12. H. Bary-Soroker, O. Entin-Wohlman, and Y. Imry, Pair-breaking effect on mesoscopic persistent currents, *Phys. Rev. B.* **80**, 024509, (2009).
13. A. V. Lopatin, N. Shah, and V. M. Vinokur, Fluctuation conductivity of thin films and nanowires near a parallel-field-tuned superconducting quantum phase transition, *Phys. Rev. Lett.* **94**, 037003, (2005).
14. N. Shah and A. V. Lopatin, Microscopic analysis of the superconducting quantum critical point: Finite temperature crossover in transport near a pair-breaking quantum phase transition, *Phys. Rev. B.* **76**, 094511, (2007).
15. M. Daumens, C. Meyers, and A. Buzdin, Little-parks effect for arbitrary geometry: fluctuations of the magnetic moment of mesoscopic loops, *Phys. Lett. A.* **248**, 445, (1998).
16. A. I. Buzdin and A. A. Varlamov, Critical fluctuations in a mesoscopic superconducting ring, *Phys. Rev. Lett.* **89**, 076601, (2002).
17. R. P. Groff and R. D. Parks, Fluxoid quantization and field-induced depairing in a hollow superconducting microcylinder, *Phys. Rev. B.* **176**, 567, (1968).
18. H. J. Fink and V. Grünfeld, Temperature dependence of fluxoid quantization in a superconducting hollow cylinder, *Phys. Rev. B.* **22**, 2289, (1980).
19. H. J. Fink and V. Grünfeld, Fluxoid periodicity in superconducting rings: Comparison to loops with josephson junctions, *Phys. Rev. B.* **33**, 6088, (1986).
20. $g = R_Q/R_\circ$, $R_Q = h/(2e^2) \approx 6.4k\Omega$, $R_\circ$: ring resistance.
21. Deep in the superconducting regime $I_S \sim \frac{T_c^0}{\phi_0} g$.
22. L. P. Levy, G. Dolan, J. Dunsmuir, and H. Bouchiat, Magnetization of mesoscopic copper rings: Evidence for persistent currents, *Phys. Rev. Lett.* **64**, 2074, (1990).

23. V. Ambegaokar and U. Eckern, Nonlinear diamagnetic response in mesoscopic rings of superconductors above t_c, *Europhys. Lett.* **13**, 733, (1990).
24. F. von Oppen and E. K. Riedel, Flux-periodic persistent current in mesoscopic superconducting rings close to t_c, *Phys. Rev. B.* **46**, 3203, (1992).
25. For rings with $\sqrt{g} \gg r$, below $T_c(\varphi \sim 1/2)$ the numerical procedure converges very slowly and therefore no comparison to the experimental data was performed in Ref. 8. In the analytical scheme this parameter regime does not pose a problem. In practice, non-equilibrium effects might become essential for the experiment.
26. Some important scales involved in the problem and notation we use: The zero temperature coherence length for disordered superconductors is $\xi^2 = \pi D/8T_c^0$, where T_c^0 is the transition temperature at vanishing flux. We measure the radius of the ring R in units of ξ: $r = R/\xi$. The reduced radius r is related to the critical flux φ_{c0} (at which the quantum phase transition occurs for sufficiently small rings) by $r^2 = 8\gamma_E\varphi_{c0}^2/\pi^2$, where $\gamma_E \approx 1.78$ $[\ln\gamma_E = -\psi(1/2) - \ln 4]$. The Thouless energy $\varepsilon_T = D/R^2$ and T_c^0 are related by $\varepsilon_T/T_c^0 = \pi/\gamma_E\varphi_{c0}^2 = 8/\pi r^2$. We often measure the persistent current I in units of T_c^0/ϕ_0, where $\phi_0 = \pi/e$ is the superconducting flux quantum,[4] $I = iT_c^0/\phi_0$, and in a similar way write for the susceptibility $\chi = \overline{\chi}T_c^0/\phi_0^2$. The zero-dimensional Ginzburg parameter Gi is given as $Gi = \sqrt{7\zeta(3)/4\pi^2\nu T_c^0 V}$, where ν is the density of states at the Fermi level and V is the volume of the ring. The parameter $\Lambda = 1/r^2 Gi$ determines whether a description in terms of a few modes only is a good approximation in the critical regime (for $\Lambda \gg 1$) or not. Defining the dimensionless conductance of the ring as $g = R_Q/R_\circ = 2e^2\nu D R_Q V/(2\pi R)^2$, $R_Q = \pi/e^2$, one can find an alternative expression, $\Lambda \approx 5\sqrt{g}/r$.
27. Y. Imry, D. J. Scalapino, and L. Gunther, Phase transitions in systems with coupled order parameters, *Phys. Rev. B.* **10**, 2900, (1974).
28. Ref. 8 uses parameters γ and $8(T - T_c)/(\pi E_c) \equiv \vartheta$, where $E_c = D/(4R^2)$. Note for comparison $\Lambda = 1/r^2 Gi = 2\pi^2/\sqrt{\gamma}$, $\vartheta = 4\varepsilon r^2$, $\varepsilon/Gi = \vartheta/Gi_\vartheta$, $Gi_\vartheta = 2\sqrt{\gamma}/\pi^2$. The rings with largest Λ displayed in 8 have $\Lambda \sim 70$, but rings with larger Λ have been measured.
29. M. Abramowitz and I. Stegun, *Handbook of Mathematical Functions.* (Dover Publ. NY, 1972).
30. Microscopic theory gives $\alpha^2/b = 8\pi^2\nu/(7\zeta(3))$ and $\xi^2 = \pi D/(8T_c)$ for disordered superconductors, D: diffusion constant, ν: density of states at the Fermi level.
31. B. Mühlschlegel, D. J. Scalapino, and R. Denton, Thermodynamic properties of small superconducting particles, *Phys. Rev. B.* **6**, 1767, (1972).
32. V. Ambegaokar and U. Eckern, Coherence and persistent current in mesoscopic ring, *Phys. Rev. Lett.* **65**, 381, (1990).
33. For values of $\sqrt{g}/r$ that are not very large the accuracy can be considerably improved by including the contribution of the other modes in a modified Gaussian approximation,[15] based on $\mathcal{F}_n = \alpha T_c \sum_{n\neq c}(\varepsilon_n + \alpha Gi^2 \sum_{k=c}\langle|\psi_k|^2\rangle)|\psi_n|^2$, where $c = 0$ or $c = (0,1)$. $\langle\ldots\rangle$ is defined with

respect to either $\mathcal{F}_0 = a_n|\psi_0|^2 + \frac{b}{2V}|\psi_0|^4$ or $\mathcal{F}$ of Eq.(16.6) for $\varphi \sim 0$ and $\varphi \sim 1/2$ respectively. Displayed i_2 and i_{20} include the (in this case very small) contribution of ψ_n with $n \neq 0, 1$.

34. The result is $i_2 = -4\frac{r^2}{Gi}\mathcal{M}_- + \frac{8}{3}x_-\mathcal{M}_+$, where $\mathcal{M}_\pm = x_\pm \mp \frac{1}{\mathcal{P}}\left(e^{x_0^2}\,\mathrm{erfc}(x_0) \pm e^{x_1^2}\,\mathrm{erfc}(x_1)\right)$ and $\mathcal{P} = 4\int_0^\infty dz\; e^{3z^2+2(2x_1-x_0)z+x_1^2}\,\mathrm{erfc}(2z+x_1)$, $x_\pm = (x_0 \pm x_1)/2$. An analogous formula has been given in Ref. 15.
35. A. C. Bleszynski-Jayich, W. E. Shanks, B. Peaudecerf, E. Ginossar, F. von Oppen, L. Glazman, and J. G. E. Harris. Persistent currents in normal metal rings: comparing high-precision experiment with theory, (2009). URL `http://www.citebase.org/abstract?id=oai:arXiv.org:0906.4780`.
36. A. A. Abrikosov, L. P. Gorkov, and I. E. Dzyaloshinski, *Methods of Quantum Field Theory in Statistical Physics.* (Prentice-Hall, Inc. Englewood Cliffs, New-Jersey, 1963).

Chapter 17

Phase-coherent Electron Transport through Double Dots

Martin Sigrist[1], Thomas Ihn[1], Klaus Ensslin[1], Vadim Puller[2] and Yigal Meir[2]

[1]*Laboratory for Solid State Physics, ETH Zurich, 8093 Zurich, Switzerland*
[2]*Dep. of Physics, Ben Gurion University, Beer Sheva 84105, Israel*
E-mail: ensslin@phys.ethz.ch

An Aharonov-Bohm interferometer with a quantum dot in each of the two arms of the ring is investigated. The phase coherence of charge carriers is probed for elastic and inelastic processes. Magnetic field-periodic Aharonov-Bohm oscillations display phase jumps of π at the onsets of inelastic transport as well as between onsets. The occurrence of these phase jumps is governed by the relevant inelastic processes between the dots and the leads.

17.1. Introduction

The phase coherence of electrons in mesoscopic conductors can be probed using the Aharonov-Bohm effect.[1] Such experiments have been performed for metallic systems[2] as well as for rings patterned in semiconductor heterostructures[3] made of GaAs/AlGaAs. Quantum dots,[4] on the other hand, resemble model systems, where electrons can be confined in all three dimensions. Electronic transport through such quantum dots is achieved via tunnel-coupled leads. The strength of the tunnel coupling as well as the charge occupancy of the dots can be controlled by external gate electrodes. Pioneering experiments [5] have demonstrated that transport through quantum dots in the low-bias regime is at least partially phase coherent. The phase of the measured signal changed by π each time an additional energy level was tuned into resonance with the Fermi levels in source and drain. The Weizmann group has furthermore shown that the investigation of the continuous phase evolution of electronic transport requires a four-terminal setup[6] in order to break the phase rigidity dictated by the Onsager symmetry.[7]

Here we investigate a two-terminal setup. The possible phases of the Aharonov-Bohm (AB) oscillations with respect to B=0 are either 0 or π because of time reversal invariance.[7] A quantum dot is embedded in each of the two arms of the AB ring. Coherent electron transport is investigated for different biases applied across the entire system. This gives access to elastic and inelastic cotunneling events. We find that the phase of the AB oscillations changes by π each time an additional excited state comes into the bias window. These phase jumps as well as the anti-symmetrized magnetic field-dependent conductance are analyzed by considering the processes that break phase rigidity.

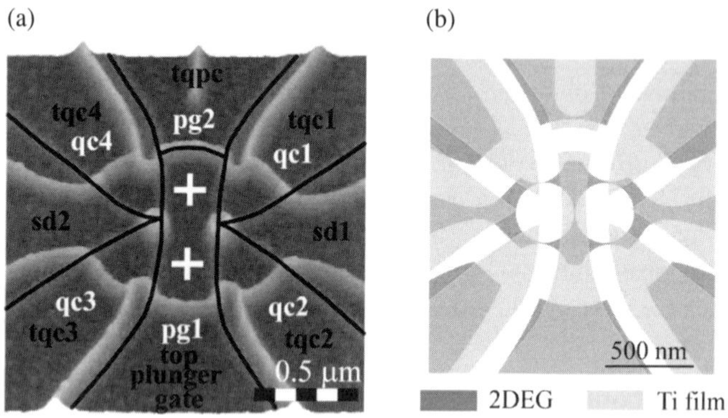

Fig. 17.1. (a) SFM-micrograph of the coupled DQD embedded in an AB interferometer fabricated by two-layer SFM-lithography. The GaAs oxide lines (bright lines) define the interferometer with in-plane gates (white letters). The thin Ti film is segmented by Ti oxide lines (indicated by black lines) into mutually isolated Ti top gates (black letters). The positions of the QDs are illustrated by white crosses. (b) The electronic structure of the sample is illustrated. The two oxide dots in the center of the structure form five internal QPCs in order to define the QDs in the interferometer. (adapted from Ref. 8)

17.2. Sample

The sample is fabricated by double-layer lithography with an atomic force microscope.[9] The details of the fabrication process and the sample operation are described in Refs. 10 and 11. An image of the sample topography taken with a scanning force microscope is shown in Fig. 17.1 (a). The schematic presented in Fig. 17.1 (b) shows the conducting areas for the 2DEG (dark grey), the insulating areas (white) and the top gates (bright grey). Electrons flow from the left through the ring geometry and leave the sample to the right. The coupling between the two quantum dots placed in the top and bottom arms of the ring can be tuned by the top gate finger reaching in from below. Electron transport at low temperatures in the regime where the two dots are tunnel coupled has been investigated before.[10] Here we focus on the regime where this tunnel coupling is negligibly small[11] and the dots are only capacitively coupled.

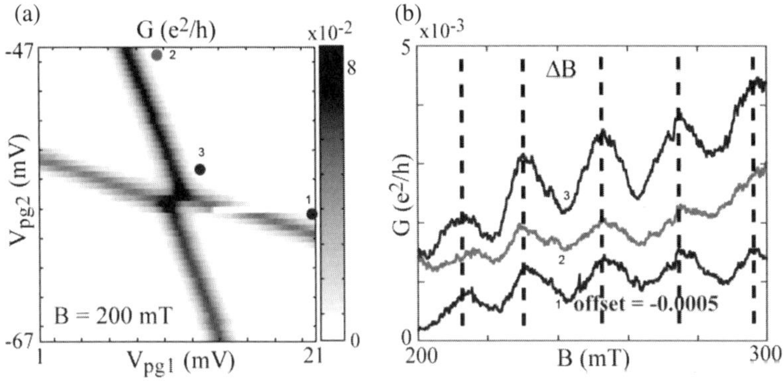

Fig. 17.2. (a) A pair of charge degeneracy points in conductance through capacitively coupled dots is mapped as a function of both in-plane gates. Each gate covers a range of about half the charging energy of a dot. (b) AB oscillations as a function of magnetic field are measured for fixed gate voltages (dots indicated and numbered in the left figure). We observe no change of phase. Vertical dashed lines mark the expected AB periodicity. The first AB period is smaller indicating a slight change in the interfering paths. (adapted from Ref. 8)

17.3. Linear Transport

Figure 17.2 (a) shows the conductance of the double dot system as a function of the two plunger gates which tune the charge occupancy of the two dots. The data is taken close to two degeneracy points (triple points). For a more general discussion of the charge stability diagram of a double dot system see Ref. 11. Here we focus on the linear transport regime, where transport occurs entirely through the ground states of the two dots. The black lines in Fig. 17.2 (a) which mark the transitions from the n to the n+1 ground state in each of the two dots are rather straight indicating that the coupling between the dots is predominately capacitive and tunneling between the dots can be neglected. Transport measurements as a function of magnetic field are presented in the low-bias regime in Fig. 17.2 (b). The three curves are taken at different points in the charge stability diagram (see color code in Fig. 17.2 (a) and (b)) and all display pronounced B-periodic oscillations. This means that transport through both quantum dots is at least partially coherent in the sequential tunneling regime investigated here. This is not unexpected considering the results on single-dot interferometers.[5] We would like to add that the amplitude of the AB oscillations is not only governed by the phase coherence of the relevant processes but also by the relative transmission through the two quantum dots. If transmission through one arm of the interferometer is much higher that through the other arm then direct transport through the more transmissive arm dominates the conductance and the amplitude of the AB oscillations can be below the experimental detection limit even if phase coherence is preserved. It is therefore crucial to tune the double dot into a regime as shown in Fig. 17.2 (a) where the conductances of the two dots are comparable, as shown by the grey levels of the lines with the two different slopes indicating transport through each of the dots, respectively.

17.4. From Linear to Non-linear Transport

In the linear transport regime the current through the interferometer is proportional to the applied bias voltage. Once the bias voltage exceeds internal energy scales of the system, inelastic excitations can become relevant. Figure 17.3 (a) shows again a part of the charge stability diagram of the capacitively coupled double dot system in the linear transport regime. In Fig. 17.3 (b) the differential conductance is plotted as a function of DC-bias voltage at a fixed gate setting and magnetic field (indicated by a black dot in Fig. 17.3 (a)). For (N,M) electrons in (dot 1, dot 2), we find a step-like conductance increase at a DC-bias voltage of +/- 75 μeV. For higher voltages, the differential conductance is reduced by about 0.2 x 10^{-3} e^2/h, then it increases again. At the onset, there is an overshoot in the differential conductance. For small bias in the elastic cotunneling regime, the differential conductance is about zero if an error of about 3 x 10^{-5} e^2/h in estimating the zero conductance offset of the I/V-converter is assumed.

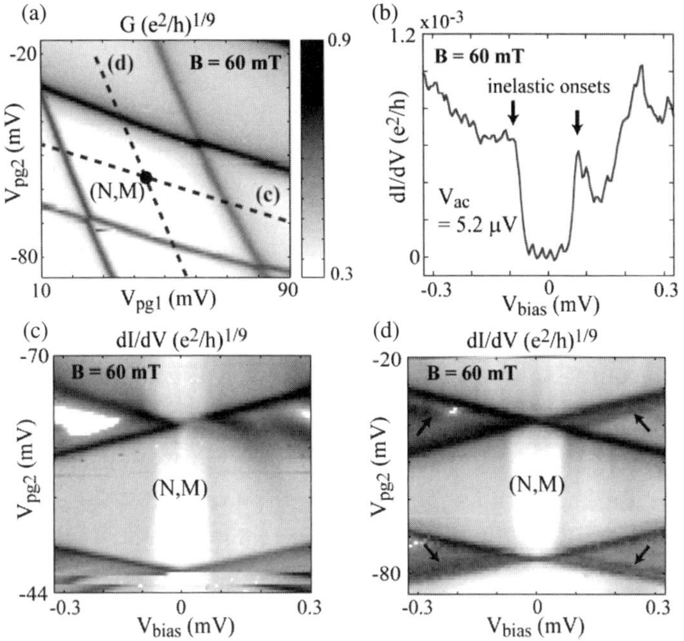

Fig. 17.3. (a) Conductance is plotted as a function of both in-plane gates at a magnetic field of 60 mT. Two traces for the lower figures are indicated. (b) Differential conductance as a function of DC-bias voltage. An onset is detected around 0.07 mV. (c) Differential conductance is measured along the indicated trace in the left upper figure. (d) Differential conductance is measured along the indicated trace in the left upper figure. An inelastic onset is detected in dot 2. (adapted from Ref. 8)

In order to find the origin of the inelastic onset, we measure Coulomb diamonds for both dots separately. The motivation for this experiment is that an inelastic onset in the Coulomb blockaded region can be linked to excited states in the conducting regions. In Fig. 17.3 (c), we keep the charge on dot 2 constant and observe the differential conductance as a function of DC bias voltage along the dashed line (c) in Fig. 17.3 (a). The inelastic onset, i.e. the vertical separation from bright to dark grey, is not influenced by charging electrons into dot 1 and it cannot be linked to any excited states in this dot. We observe only an increase in the elastic cotunneling conductance in the direction of adding more electrons to dot 1 due to tuning the tunnel barriers with the in-plane gates. We therefore expect the inelastic process to take place in dot 2. This is confirmed by measuring Coulomb diamonds for dot 2. The differential conductance was observed as a function of DC-bias voltage along the dashed line (d) in Fig. 17.3 (a). We find the inelastic onset only within the Coulomb diamond (N,M) and not for other charge states. The step-like conductance increase inside the Coulomb blockaded region can be linked to excited states (indicated by arrows) with the same slopes as the ground states of dot 2. Knowing in which dot the inelastic process takes place is crucial for the interpretation of a so-called which-path-experiment.

17.5. Aharonov-Bohm Oscillations in the Non-linear Regime

Next we establish phase coherent transport in the non-linear regime. In order to demonstrate that B-periodic AB oscillations occur in the non-linear regime a filter routine (for details see Ref. 8 and 10) has been developed to distinguish B-periodic features in the transport signal from other features modulating the background conductance. The extracted AB oscillations $I^{h/e}$ as a function of magnetic field and DC-bias voltage are shown in Fig. 17.4 (a). Minima and maxima in AB oscillations appear in black, vanishing amplitude is marked on white. We observe no AB-oscillations for small magnetic fields and small DC-bias voltage in the gap between the two inelastic onsets. We can therefore easily identify regions of clear AB oscillations in the 2d-plot of Fig. 17.4 (a). The grey scale is adapted to trace the AB oscillations which are small in amplitude. For low magnetic field and high DC-bias voltage of about ±0.3 mV, we observe strong AB oscillations. The differential conductance at this high DC-bias region is of the order of 10^{-3} e^2/h (see Fig. 17.4 (b)) compared to an average differential conductance peak amplitude on resonance of the order of 5 x 10^{-2} e^2/h. This indicates that we still measure in the cotunneling regime even at these high DC-bias voltages.

The visibility of the AB oscillations is shown as a function of magnetic field and DC bias voltage in Fig. 17.4 (b). It is calculated by dividing the extracted AB oscillations by the background current. The visibility tends to decrease for higher DC-bias voltage. It indicates relatively fewer phase-coherent contributions for higher DC-bias voltage. This can be partly attributed to phase shifts as observed for example in positive DC-bias voltage at low magnetic field. Another part may be because inelastic processes are less phase-coherent (if phase-coherent at all), however, contribute immensely to the background current.

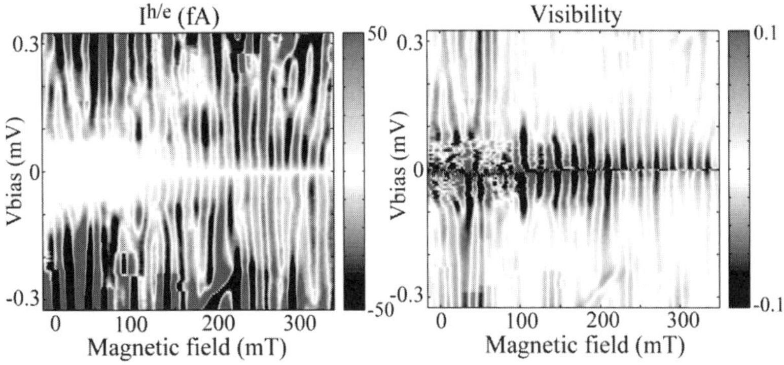

Fig. 17.4. (a) AB-oscillations with a period of 22 mT are extracted in the current measurement as a function of magnetic field and DC-bias voltage. (b) The visibility of the AB-oscillations shown in the left figure is plotted. (adapted from Ref. 8)

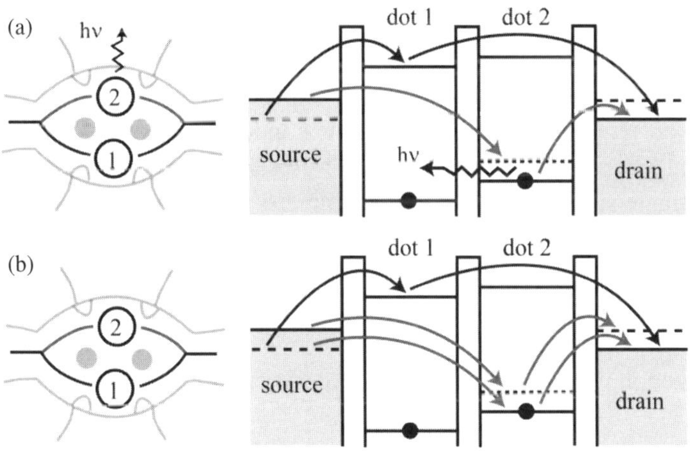

Fig. 17.5. (a) Interference scenario between an elastic cotunneling process through dot 1 and an inelastic process through dot 2 together with a phonon emission to relax dot 2 to the initial state. (b) Interference scenario between an elastic cotunneling process through dot 1 and a four-particle process through dot 2. (adapted from Ref. 8)

Interference can only take place if the initial and final state have the same energy. This is fulfilled for the interference scenario sketched in Fig. 17.5 (a). On the left, the interference paths are indicated in the schematic of the sample. On the right, the interfering processes are shown in the energy-level diagram of the double dot system. Elastic cotunneling through a virtual state of dot 1 interferes with inelastic cotunneling through dot 2. It relaxes to the ground state by emitting a phonon. The interference pattern vanishes if this phonon is "detected" by the environment. Further, we can think of interference scenarios which even do not allow "which-path" detection. An example is shown in Fig. 17.5 (b). Elastic cotunneling through dot 1 interferes with a four-particle process through dot 2. The relaxation of dot 2 back into the ground state is performed without emitting a phonon. The energy is dissipated in the leads, after closing the interference loop. This makes "which-path" detection impossible. The average time between successive electrons traversing the system needs to be shorter than the phonon relaxation time for this mechanism to be relevant. There are other conceivable scenarios which may contrinute to coherent transport in the inelastic tunneling regime.

17.6. Breaking of Phase Rigidity in the Non-linear Regime

For a two-terminal experiment one expects phase rigidity in the linear transport regime.[7] This means that the conductance is an even function of the magnetic field. The experiments presented here offer the chance to investigate such a magnetic field symmetry in the non-linear regime. The upper right panel of Fig. 17.6 shows experimental data in a larger bias regime, where several inelastic onsets are covered. The B-periodic signal displays π-shifts each time the bias voltage coincides with an inelastic onset. In between onsets the phase changes back by $-\pi$. This behavior is analyzed in more detail in Ref. 10 experimentally and theoretically.

The lower right panel of Fig. 17.6 shows the asymmetric (in magnetic field) component of the total conductance. The asymmetric signal is about an order of magnitude smaller than the full signal and is clearly odd in magnetic field.

In order to understand this behavior better the relevant higher order processes through the dot system have been investigated (for details see Ref. 13). The asymmetric contributions to the Aharonov-Bohm effect require 3rd order processes which are generally weaker than 2nd order processes. A special situation occurs when phase switching takes place, where 2nd order processes cancel out and the even-in-magnetic field Aharonov-Bohm signal vanishes in amplitude. In this particular situation 3rd order process dominate transport and give rise to an odd B-periodic signal. This has been calculated in detail and the corresponding data for the full conductance as well as its asymmetric part are displayed in the left column of Fig. 17.6.

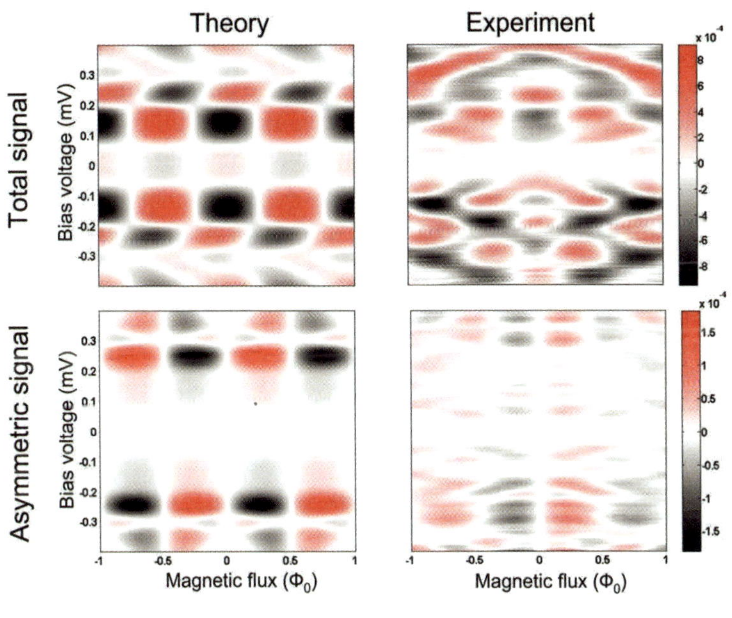

Fig. 17.6. Color plots of the differential conductance obtained from the theoretical model presented here (left panels) and from the experimental data (right panels). The upper and lower panels show, respectively, full and asymmetric components of the conductance. (adopted from Ref. 13)

The comparison of calculation (left column) and experimental data (right column) clearly demonstrates that such 3rd order transport processes can play an important role for the magnetic field symmetry. As a consequence the phase change of the Aharonov-Bohm signal at an inelastic onset is not completely abrupt, but may develop a weak magnetic field dependence.

17.7. Conclusions and Outlook

Coherence is an essential ingredient to understand transport processes in mesoscopic systems. We demonstrate that Aharonov-Bohm oscillations can be observed in a ring geometry with a quantum dot placed in each of the two halves of the ring. Coherent transport is observed in the sequential tunneling regime governed by linear transport as well as in the non-linear regime where the bias voltage is large enough to provide access to excited states. The phase of the Aharonov-Bohm oscillations jumps by π at the onsets of inelastic transport as well as in between these onsets. The anti-symmetrized conductance is odd in magnetic field. When the total conductance undergoes a phase change, the amplitude of the conductance oscillations becomes minimal and third-order transport processes between dots and leads become dominant and explain the effects in the anti-symmetrized conductance. Such situations are sensitive to external perturbations and can be used for detailed tests of decoherence.

Acknowledgments

This work was supported by the Swiss National Science Foundation (SNF) as well as by ISF and BSF.

References

1. Y. Aharonov and D. Bohm, Phys. Rev. 115, 485 (1959).
2. R. A. Webb, S. Washburn, C. P. Umbach, and R. B. Laibowitz, Phys. Rev. Lett. 54, 2696 (1985); for a review see S. Washburn and R. A. Webb, Adv. Phys. 35, 375 (1986).
3. G. Timp, A. M. Chang, J. E. Cunningham, T. Y. Chang, P. Mankiewich, R. Behringer, and R. E. Howard, Phys. Rev. Lett. 58, 2814 (1987).
4. For a review see L. P. Kouwenhoven, C. M. Marcus, P. McEuen, S. Tarucha, R. M. Westervelt, and N. Wingreen, in Nato ASI conference proceedings (Kluwer, Dordrecht, 1997).
5. A. Yacoby, M. Heiblum, D. Mahalu, and H. Shtrikman, Phys. Rev. Lett. 74, 4047 (1995).

6. R. Schuster, E. Buks, M. Heiblum, D. Mahalu, V. Umansky, and H. Shtrikman, Nature 385, 417 (1997).
7. L. Onsager, Phys. Rev. **38**, 2265 (1931); M. Büttiker, Phys. Rev. Lett. 57, 1761 (1986).
8. M. Sigrist, PhD thesis, ETH Zurich, 2006
9. M. Sigrist, A. Fuhrer, T. Ihn, K. Ensslin, D. D. Driscoll, and A. C. Gossard, Appl. Phys. Lett. **85**, 3558 (2004).
10. M. Sigrist, A. Fuhrer, T. Ihn, K. Ensslin, S. E. Ulloa, W. Wegscheider, and M. Bichler, Phys. Rev. Lett. 93, 66802 (2004), V. Puller and Y. Meir, Phys. Rev. B 77, 165421 (2008).
11. M. Sigrist, T. Ihn, K. Ensslin, D. Loss, M. Reinwald, and W. Wegscheider, Phys. Rev. Lett. **96**, 036804 (2006).
12. W. G. van der Wiel, S. De Franceschi, J. M. Elzerman, T. Fujisawa, S. Tarucha, and L. P. Kouwenhoven, Rev. Mod. Phys. **75**, 1 (2002).
13. V. Puller, Y. Meir, M. Sigrist, K. Ensslin, and T. Ihn, Phys. Rev. B **80**, 035416 (2009).

Chapter 18

Spin-Hall Effect in Chiral Electron Systems: From Semiconductor Heterostructures to Topological Insulators

P. G. Silvestrov[1] and E. G. Mishchenko[2]

[1] *Theoretische Physik III, Ruhr-Universität Bochum, 44780 Bochum, Germany*

[2] *Department of Physics, University of Utah, Salt Lake City, Utah 84112, USA*

The phenomenon of mesoscopic Spin-Hall effect reveals in a nonequilibrium spin accumulation (driven by electric current) at the edges of a ballistic conductor or, more generally, in the regions with varying electron density. In this paper we review our recent results on spin accumulation in ballistic two-dimensional semiconductor heterostructures with Rashba/Dresselhaus spin orbit interactions, and extend the method developed previously to predict the existince of spin-Hall effect on the surface of three-dimensional topological insulators. The major difference of the new Spin-Hall effect is its magnitude, which is predicted to be much stronger than in semiconductor heterostructures. This happens because in semiconductors the spin accumulation appears due to a small spin-orbit interaction, while the spin-orbit constitutes a leading term in the Hamiltonian of topological insulator.

18.1. Chiral Electron Systems

Chiral systems feature electron states whose quantum properties (e.g. spin) depend on the direction of propagation. One example of such chiral states arises from spin-orbit coupling that originates from relativistic corrections to the dispersion law of band electrons. In particular, for a typical two-dimensional electron gas (2DEG) the intrinsic asymmetry of a confining quantum well geometry is accompanied by a strong perpendicular (z) electric field that leads to spin-orbit coupling of the Rashba type,[1]

$$H_R = \lambda \mathbf{z} \cdot (\boldsymbol{\sigma} \times \mathbf{p}) = \lambda(\sigma_x p_y - \sigma_y p_x). \tag{18.1}$$

As a result, the spin degeneracy is lifted via *effective momentum-dependent* Zeeman field, $\mathbf{h_p} = \lambda(-p_y, p_x)$. The electron spin eigenstate is thus determined by its momentum, **p**.

Another important type of spin-orbit coupling occurs in 2DEG formed by semiconductors with broken inversion symmetry, e.g. GaAs, InAs. While only third order in momentum in the bulk GaAs, this coupling, known as Dresselhaus interaction,[2] is "upgraded" in two dimensions to the linear order by virtue of transverse momentum quantization and becomes,

$$H_D = \lambda_D(\sigma_x p_x - \sigma_y p_y). \tag{18.2}$$

Chiralities acquired from different spin-orbit couplings (18.1) and (18.2) are *opposite*, in a sense that the electron wave function acquires opposite Berry phases, $\pm\pi$, upon (adiabatic) completion of a loop in the momentum space (enclosing the degeneracy point $\vec{p} = 0$), depending on whether $\alpha > \beta$ or vice versa.

Dirac fermions in graphene represent another realization of the chiral system.[3] Due to hexagonal symmetry of underlying two-dimensional honeycomb atom arrangement the low-energy electron excitations are combined into two Dirac cones (valleys), K and $\widetilde{K}$, within the first Brillouin zone. The corresponding effective Hamiltonians are

$$H_K = v(\tau_x p_x + \tau_y p_y), \qquad H_{\widetilde{K}} = v(\tau_x p_x - \tau_y p_y). \tag{18.3}$$

Here $\boldsymbol{\tau}$ stands for the *pseudospin* operator that acts in the sublattice space.

It is interesting to note the analogy between *total* graphene Hamiltonian and a semiconductor with *equal* Rashba and Dresselhaus couplings ($\alpha = \beta$) in the case of a conventional 2DEG. The latter case features spin eigenstates that are momentum-independent[4] and, thus, somewhat trivial. However, since the two cones in graphene are *separated* in momentum space, the chiral physics can still be observed. (Yet some phenomena are trivially absent, e.g. pseudospin-Hall effect, cf. last section). It is important to emphasize that the chirality in graphene has nothing to do with spin-orbital coupling (which is rather weak in carbon allotropes) and is a consequence of the crystal geometry.

Another notable example of a chiral electron system is a topological insulator. The latter is different from the usual band insulator in that its valence and conduction bands originate not from different atomic orbitals but from *the same spin-orbit-split* orbital. In 2D HgTe quantum wells this gives rise to topologically protected edge states leading to a recently predicted[5] and discovered[6] quantum spin-Hall effect. Yet even more intriguing twist has recently been added by a discovery of two-dimensional states on the surface of 3D topological insulators[7,8] $Bi_{0.9}Sb_{0.1}$, Bi_2Sb_3, and Bi_2Te_3. The fundamental difference from graphene is that the number of pockets of the Fermi surface within the Brillouin zone is *odd*, with the complementary species residing on the opposite surface of a sample. Thus, these states are not simply separated in the momentum space (like in graphene), but also separated in *real* space. This removes the above mentioned "trivialization" that is present in graphene. In particular, spin-Hall effect can occur in topological insulators. As confirmed by the first-principle band structure calculations[9] spin structure of these states is indeed chiral. In particular, the low-energy Hamiltonian of Bi_2Te_3 can be deduced from symmetry considerations to be of the form,[10]

$$H_{TI} = v(\sigma_x p_y - \sigma_y p_x) + \frac{p^2}{2m^*} + \alpha\sigma_z p_x(p_x^2 + 3p_y^2). \tag{18.4}$$

For an ungated and undoped Bi_2Te_3 the last term is generally of the same order of magnitude as the first two. Still, as a starting point it is useful to neglect the effects of anisotropy.

18.2. Spin-Hall Effect

Spin-Hall effect[11] is the name given to a class of phenomena that exhibit boundary (surface, edge) spin polarization when electric current flows through a system with significant spin-orbit interaction. It has been observed in both 3D[12–14] and 2D systems.[15] It is customary to distinguish two mechanisms that could lead to this effect. The *extrinsic* mechanism is the dominant one in 3D semiconductors and originates from scattering off impurities.[16–19] Presence of impurities is unavoidable in high carrier density 3D semiconductors that rely on doping. Quite contrary, *intrinsic* mechanism[20,21] originates from spin-orbit-split band-structure, which induces spin precession when electric current is driven through the system. This mechanism can in principle be realized in coexistence with ballistic transport in 2D electron systems systems. Indeed, by placing dopant far enough from GaAs/AlGaAs interface one can reduce effects of disorder scattering.

In the present paper we concentrate on the intrinsic spin-Hall mechanism. It is known, however, that in two-dimensional electron systems with spin-orbit coupling linear in momentum (typical for n-doped heterostructures) *any* scattering that leads to a stationary electric current via deceleration of electrons by impurities, phonons, etc., will negate the precession due to external electric field and result in the exact cancellation[22–26] of the bulk spin-current in a dc case[a].

There are several ways to avoid such cancellation, in particular, to use ac currents with frequencies exceeding the inverse spin relaxation time.[22,27] The second possibility and the one of interest to us here, is make a system sufficiently small and clean (ballistic) so that the electron mean free time exceeds the time of flight across the systems. The corresponding scenario is known as the *mesoscopic spin-Hall effect*.[28] While initial theories of spin-Hall effect in infinite systems had addressed the auxiliary quantity of spin current (for a review see Refs.[29,30]), in a finite geometry it is both easier and more relevant to calculate spin polarization density, which is an experimentally measurable quantity.[31] Such edge polarization was considered by numerical methods in several earlier publications[28,32–35] as well as both analytically and numerically in our previous papers.[36,37]

A crucial note is due. The edge spin polarization in ballistic systems appears not as a result of electric field-driven acceleration of electrons and associated with it precession of spins. Indeed, electric field in a ballistic system is absent as the electric potential drop occurs over a contact region with the leads rather than over a bulk of a ballistic conductor. Nevertheless, spin precession does occur. It originates from accelerated electron motion in the boundary potential that provides lateral confinement. Bias applied to the contacts ensures that populations of left- and right-moving states are different and the net spin precession appears. Naturally, it is proportional to the applied bias V. The net spin accumulation near an edge of a 2DEG is *independent* of the shape of the boundary potential,

$$\int_{-\infty}^{\infty} s_z(x)dx = -\frac{\lambda^2 - \lambda_D^2}{12\pi^2 v_F^3} eV, \tag{18.5}$$

where v_F is the bulk value of the Fermi velocity[b]. Spin accumulation appears in the second order in spin-orbit interaction and vanishes for equal Rashba and Dresselhaus coupling strengths.

Position-resolved spin polarization $s_z(x)$ can be found analytically in two important situations. First is the case of a smooth confining potential $U(x)$ and the second is an infinite hard-wall boundary. We begin with

[a]This exact cancellation does not occur in 3D- or 2D hole-systems that feature non-linear spin-orbit couplings.

[b]Strictly speaking, the integration in Eq. (18.5) goes from a point far outside the 2DEG ($-\infty$) to some point deep inside 2DEG ($+\infty$), but still far away from the edges. Integration across the whole conductor would give zero, reflecting the fact that spin accumulation at the opposite edges has opposite signs.

analyzing semiclassical electron motion in smooth potentials in Sec. 18.3. In Sec. 18.4 we derive the result (18.5) for the net spin accumulation and illustrate it using the case of hard-wall boundary. In Sec. 18.5 we present a method of kinetic equation that allows to find local spin-polarization for smooth boundary potentials. In Sec. 18.6 singular dynamics near classical turning points is discussed. Finally, in Sec. 18.7 we use methods developed in the preceding chapters to describe spin-Hall effect in topological insulators.

Many of the results presented in this paper were published in the journal articles Refs. [36–38]. However, the derivation of the spin accumulation via the Kinetic equation in Sec. 18.5 and prediction of the nonequilibrium spin accumulation in topological insulators in Sec. 18.7 are presented here for the first time.

18.3. Semiclassical Electron Motion

Consider gated 2DEG with Rashba spin-orbit interaction[c], described by the Hamiltonian

$$H = \frac{p^2}{2m} + \lambda(p_y\sigma_x - p_x\sigma_y) + \frac{m\lambda^2}{2} + U(x,y). \tag{18.6}$$

Potential $U(x,y)$ is created by the external gates, or the edge potential ensuring the in-plane confinement of 2DEG. Possible effect of disorder on $U(x,y)$ are going to be neglected.

Classical electron dynamics described by the short wave length limit of the Hamiltonian (18.6) reveals a number of very unusual features. As a first step, in this section we demonstrate how the trajectories-based approach allows to describe propagation of fully (in-plane) polarized electric currents through mesoscopic constrictions.

Construction of semiclassical solutions of the Schrödinger equation with the Hamiltonian (18.6) follows the reasoning of the conventional WKB approach,[38–42] which is valid for a smooth potential, $\hbar|\nabla U| \ll \min(p^3/m, p^2\lambda)$.

Without the external potential U, the electron spectrum consists of the two subbands, $E_\pm(p_x,p_y) = (p \pm m\lambda)^2/2m$. The subbands meet at only one point, $p = 0$, and the spin in each subband is always aligned with one of the in-plane directions perpendicular to the momentum $\vec{p}$. The semiclassical electron dynamics[39] naturally captures the essential features

[c]Similar methods may be used to describe more complicated interactions, like Dresselhaus,[2] or combined Rashba and Dresselhaus interactions.

of this translationally invariant limit. The classical motion in each subband is determined by the equations of motion which follow from the effective Hamiltonian:

$$H_{\text{eff}} = \frac{(p \pm m\lambda)^2}{2m} + U(x, y). \tag{18.7}$$

Despite the fact that spin does not appear in this equation, one can easily construct semiclassical wave functions, which have spin pointed within the xy plane perpendicular to the momentum:

$$\psi = ue^{iS/\hbar}, \quad u = \sqrt{\frac{\rho}{2p}} \begin{pmatrix} \sqrt{p_y + ip_x} \\ \pm\sqrt{p_y - ip_x} \end{pmatrix}. \tag{18.8}$$

Here the action S is related to the momentum by $\vec{p} = \nabla S$, and $\rho = u^\dagger u$ is the classical density for a family of classical trajectories corresponding to a given energy E. The action S obeys the classical Hamilton-Jacobi equation,

$$\frac{(|\nabla S| \pm m\lambda)^2}{2m} + U(x, y) = E. \tag{18.9}$$

During its motion, an electron described by Eq. (18.8) changes the momentum p but always remains in the same spin-subband. To change the subband the electron trajectory should pass through the degeneracy point where both components of momentum vanish simultaneously, $\vec{p} = 0$, which is generically impossible. Moreover, with the proper use of potential barriers, one may realize a situation where electrons *of only one subband* are transmitted and the others are totally reflected. This leads to strong in-plane polarization of the transmitted electron flow.

To take into account the out-of-plain spin precession one has to go beyond the approximation Eq. (18.8), as was done in Ref. [38]. Instead of doing so we will describe in Sec. 18.4 a method allowing to find easily the expectation value of σ_z for potential depending only on one coordinate $U = U(x)$ (boundary potential).

18.3.1. *Sharvin conductance*

To give an example of a spin-polarized current let us consider transmission through a barrier, $U(x)$, varying along the direction of a current propagation. We assume periodic boundary conditions in the perpendicular direction ($y + L \equiv y$), thus p_y is an integral of motion[d]. For a smooth

[d]Spin polarized currents on a cylinder with x-dependent spin-orbit interaction were considered recently in Ref. [43].

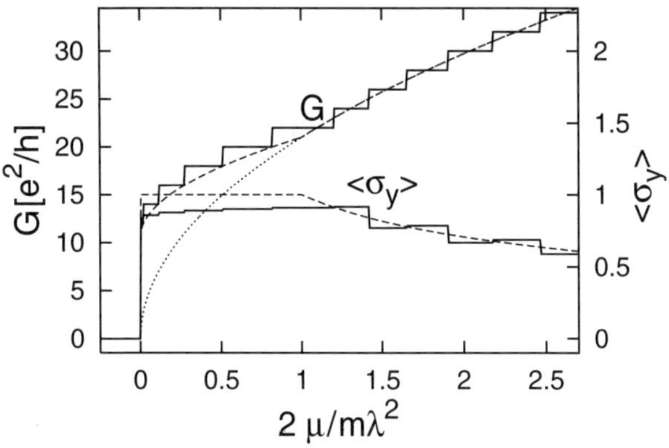

Fig. 18.1. Conductance (in units of e^2/h), and spin polarization of the current vs gate voltage (in units of $m\lambda^2/2$). Dashed lines show the smoothed curves (18.14),(18.15), solid lines show the quantized values for $m\lambda L/\hbar = 10.5\pi$. Dotted line shows the conductance without spin-orbit interaction.

potential $U(x)$ the conduction channels may either be perfectly transmitting or completely closed. The conserved transverse momentum takes the quantized values, $p_y^n = 2\pi\hbar n/L$. Consider the functions

$$E_\pm^n(p_x) = \frac{(p^n \pm m\lambda)^2}{2m}, \quad p^n = \sqrt{p_x^2 + p_y^{n\,2}} \tag{18.10}$$

For $n \neq 0$ the function $E_\pm^n(p_x)$ splits into two distinct branches. At any point x the equation

$$E_\pm^n(p_x) = E_F - U(x) \tag{18.11}$$

yields solutions p_x^L and p_x^R, corresponding to left- and right-moving electrons. Application of a small bias implies, e.g., the excess of right movers over left movers far to the left from the barrier. Particles are transmitted freely above the barrier if Eq. (18.11) has a solution, p_x^R, for any x. Let $\mu = E_F - U_{\max}$ be the difference between the Fermi energy and the maximum of the potential. The nth channel in the upper branch opens when

$$\mu = \left(2\pi\hbar|n| + m\lambda L\right)^2 / 2mL^2. \tag{18.12}$$

For the lower branch $E_-^n(p_x)$ Eq. (18.11) has four solutions (two for right and two for left movers) for $|n| < m\lambda L/2\pi\hbar$ and x close to the top of the barrier. However, far from the barrier (where the excess of right-movers is created) there are still only two crossings described by Eq. (18.11), one for right and one for left movers. As a result, all the extra electrons injected at $x = -\infty$ follow the evolution of a solution of Eq. (18.11) with the largest positive p_x. For all $|n| < m\lambda L/2\pi\hbar$ such a solution does exist for any positive μ. Thus, at $\mu = 0$ as many as $n_0 = m\lambda L/\pi\hbar$ channels *open up simultaneously*. The channels with higher values $|n| > m\lambda L/2\pi\hbar$ in the lower subband E_-^n open when

$$\mu = (2\pi\hbar|n| - m\lambda L)^2 / 2mL^2. \tag{18.13}$$

According to the Landauer formula, ballistic conductance is given by the total number of open channels multiplied by the conductance quantum $G_0 = e^2/h$

$$G = G_0 \frac{L}{\pi\hbar} \begin{cases} \sqrt{2\mu m} + m\lambda, & 0 < \mu < m\lambda^2/2 \\ 2\sqrt{2m\mu}, & \mu > m\lambda^2/2. \end{cases} \tag{18.14}$$

This dependence $G(\mu)$ is shown in Fig. 18.1. The striking evidence of the presence of spin-orbit interaction is the huge jump of the conductance at the pinch-off point, as opposed to the conventional square-root increase in the absence of spin-orbit coupling. This jump is a consequence of the "Mexican-hat" shape of the spectrum $E_-(p_x, p_y)$. Accuracy of Eqs. (18.12) and (18.13) is sufficient to resolve the steps in the conductance due to the discrete values of $|n| = 0, 1, 2, ...$, (conductance quantization), as shown in Fig. 18.1. The steps in $G(\mu)$ are abrupt in the limit $dU/dx \to 0$.

Close to the pinch-off, at $\mu \lesssim m\lambda^2$, the conserved p_y component of the electronic momentum varies for different transmitted channels within the range $|p_y| \lesssim m\lambda$. Therefore, far from the barrier, where the Fermi momentum is large $p_F \gg m\lambda$, we have $p_x \gg p_y$ and transmitted electrons propagate in a very narrow angle interval $|\theta| < \sqrt{m\lambda^2/2E_F} \ll 1$. Since the electron spin is perpendicular to its momentum, we conclude that the current due to electrons from each of the subbands is almost fully polarized. The total polarization of the transmitted current is given by the difference of two currents

$$\langle \sigma_y \rangle = \frac{\langle \psi^\dagger \sigma_y v_x \psi \rangle}{\langle \psi^\dagger v_x \psi \rangle} = \min(1, \sqrt{m\lambda^2/2\mu}), \tag{18.15}$$

which is also depicted in Fig. 18.1. This current polarization may also be viewed as a creation of in-plain nonequilibrium spin density, maximal on the barrier.

Derivation of Eqs. (18.14) and (18.15) was greatly simplified because of the periodic boundaries. Our next example demonstrates semiclassical treatment of realistic boundary conditions.

18.3.2. *Quantum Point Contact*

Let us consider probably the most experimentally relevant example of a quantum point contact, described by the potential

$$U(x,y) = -\frac{m\Omega^2 x^2}{2} + \frac{m\omega^2 y^2}{2}. \tag{18.16}$$

Even in this simple model the electron flow in the presence of spin-orbit interaction acquires a number of interesting and peculiar features. This become clear already from the figure 18.2, where we show three types of trajectories in such potential. Each kind of trajectories is necessary for calculation of conductance.

Classical equations of motion follow in the usual manner from the effective Hamiltonian (18.7): $\dot{\vec{r}} = \partial H_{\text{eff}}/\partial \vec{p}$, $\dot{\vec{p}} = -\partial H_{\text{eff}}/\partial \vec{r}$. We consider quantum point contact (QPC) close to the opening with only the lower E_- subband contributing to the conductance. A crucial property of the Hamiltonian H_{eff}, Eq. (18.7), is the existence of a circle of minima of the kinetic energy at $|p| = m\lambda$. Expanding around a point on this circle, $p_{x0} = m\lambda \cos\alpha$, $p_{y0} = m\lambda \sin\alpha$, one readily finds the equations of motion for $\mathcal{P} = p_x \cos\alpha + p_y \sin\alpha - m\lambda \ll m\lambda$,

$$\ddot{\mathcal{P}} + (-\Omega^2 \cos\alpha^2 + \omega^2 \sin\alpha^2)\mathcal{P} = 0 \ , \ \dot{\alpha} = 0. \tag{18.17}$$

The trajectory is found from the relations, $\dot{x} = \mathcal{P}\cos\alpha/m$, $\dot{y} = \mathcal{P}\sin\alpha/m$. We observe from Eq. (18.17) that only the trajectories within the angle

$$\tan|\alpha| < \tan\alpha_0 - \Omega/\omega \tag{18.18}$$

are transmitted through QPC. Trajectories with larger angles are trapped (oscillate) within the point contact. Examples of both types of trajectories are presented in Fig. 18.2. Quantization of trapped trajectories would give rise to a set of (extremely) narrow resonances in the conductance, specific for spin-orbit interaction. Below we consider only the smoothed conductance.

To calculate the current J through QPC one has to integrate over the phase space of the states which are transmitted from left to right,

$$J = \int dy \int e v_x \frac{d^2 p}{(2\pi\hbar)^2} = GV, \tag{18.19}$$

and have the energy within the interval $\mu - eV/2 < E_- < \mu + eV/2$, with V standing for the applied voltage. In this section we define μ as the difference between the Fermi energy and the value of the potential at the saddle point $\mu = E_F - U(0,0)$. The integral is most simply evaluated at $x = 0$ (with

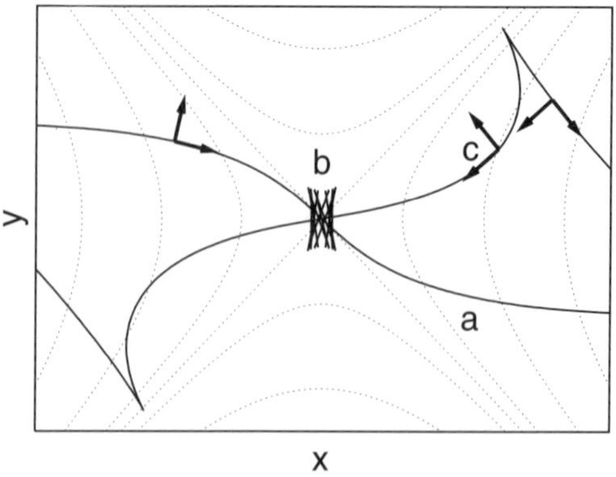

Fig. 18.2. Three kinds of trajectories in the point contact. a, transmitted trajectory whose momentum is always collinear with the velocity. b, trajectory bouncing inside the QPC. This trajectory is periodic in the linearized approximation described in the text, while the exact calculation for finite amplitude shows its slow drift. c, transmitted trajectory whose momentum inside the contact is opposite to the velocity. Electrons flow from left to right. Arrows show momentum and spin orientations. Few equipotential lines are also shown.

the velocity given by $v_x = \mathcal{P}\cos\alpha/m$). The allowed absolute values of the momentum are

$$2\mu - eV - m\omega^2 y^2 < \mathcal{P}^2/m < 2\mu + eV - m\omega^2 y^2. \tag{18.20}$$

The angle interval of transmitting trajectories consists of two domains: $|\alpha| < \alpha_0$, $\mathcal{P} > 0$, and $|\alpha - \pi| < \alpha_0$, $\mathcal{P} < 0$. The appearance of the latter range of integration is highly non-trivial. A simple reasoning shows that the particles with the velocity antiparallel to the momentum ($v_x > 0$, $p_x < 0$) should not contribute to the conduction in the case of a transition through a one-dimensional barrier $U = U(x)$, see Eq. (18.14). Despite corresponding to the right-moving electrons, these states *do not originate* in the left lead. Indeed, they exist only in the vicinity of $x = 0$, but disappear as $x \to -\infty$ and, thus, cannot be populated by the excess electrons (except due to the tunneling transitions which are irrelevant in the semiclassical regime). Such trajectories, however, *do exist* in QPC, Eq. (18.16), as demonstrated in Fig. 18.2. After passing through QPC the trajectory bounces at the wall reversing its velocity. This kind of classical turning points, where both components of the velocity vanish simultaneously, are specific for the effective Hamiltonian (18.7). The existence of transmitting trajectories with $|\alpha - \pi| < \alpha_0$, $\varrho < 0$ results in the doubling of the conductance. Simple calculation yields

$$G = G_0 \frac{4m\lambda \sin \alpha_0}{\pi \hbar \omega} \sqrt{\frac{2\mu}{m}}. \tag{18.21}$$

The presence of a threshold angle α_0, as well as the square-root dependence of $G(\mu)$, are in a sharp contrast to the well-known result $G = G_0\mu/\pi\hbar\omega$, in the absence of spin-orbit interaction.

Since Eq. (18.17) describes only the linearized electron dynamics, Eq. (18.21) is formally valid if $\mu \ll m\lambda^2$.[e] Nevertheless, the current remains totally polarized for $0 < \mu < m\lambda^2/2$ [similar to Eq. (18.15)]

$$\langle \sigma_y \rangle = \frac{\langle \psi^\dagger \sigma_y v_x \psi \rangle}{\langle \psi^\dagger v_x \psi \rangle} = 1. \tag{18.22}$$

With increasing the chemical potential, $\mu > m\lambda^2/2$, transmission via the upper subband E_+ kicks in and the degree of polarization gradually decreases, like it happened in Eq. (18.15).

In InAs-based heterostructures, typical value of spin-orbit coupling[44] is $\lambda\hbar = 2\times 10^{-11} eVm$. Characteristic spin-orbit length $l_R = \hbar/m^*\lambda = 100$ nm and energy $m^*\lambda^2/2 = 0.1$ meV. In order to have strongly spin-polarizing QPC, the latter should support many transmitting channels at chemical potential $\mu \sim m^*\lambda^2/2 \gg \hbar\omega$. This condition can, equivalently, be written in terms of the width of the point contact Δy, see Eq. (18.16), as $\Delta y \gg l_R$.

18.4. Edge Spin Accumulation. Hard Wall

Consider a semi-infinite ballistic 2DEG described by the Hamiltonian Eq. (18.6), where potential depending only on one coordinate, $U \equiv U(x)$, ensures boundary confinement (see Fig. 18.3). The system is attached to two ideal reflectionless leads injecting equilibrium electrons into 2DEG. The chemical potentials of the leads are shifted by the applied voltage, eV. The current flow along the y-direction in case of spin-orbit interaction Eq. (18.6) results in an edge spin accumulation $s_z(x)$. In this section we first derive an exact formula for the total amount of spin accumulated at the edge, $\int s_z dx$. Then we consider in more details the case of hard wall potential, $U(x<0) = \infty$, $U(x>0) = 0$, where exact results for $s_z(x)$ are available[f].

[e] Still the number of open channels should be large for semiclassics.

[f] Nonequilibrium edge spin accumulation at the hard wall boundary of disordered conductor was considered in Ref. [45]

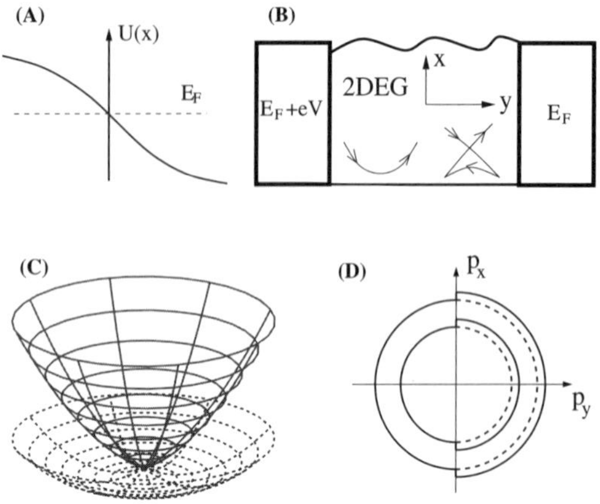

Fig. 18.3. Profile of a boundary potential $U(x)$. B) Geometry of the system: two-dimensional electron gas ($x > 0$) is attached to two ideal reflectionless metallic leads filled by equilibrium electrons up to different chemical potentials. C) Spin-orbit-split subbands structure. D) Difference in population of left- and right-moving electrons due to the applied bias eV.

18.4.1. *Net spin accumulation*

For the potential independent on y ($U = U(x)$) the corresponding momentum component, k_y, is an integral of motion. It is convenient to use the Fourier representation along the y-axis for the electron operators, $\hat{\psi}(\mathbf{r}) = \sum_{k_y} \hat{c}_{k_y}(x) e^{ik_y y}$. We employ here the second quantization formalism. One can derive the equation of motion for the expectation value of the electron spin operator, $\mathbf{s}(k_y, x) = \frac{1}{2}\langle \hat{c}^\dagger_{k_y}(x) \hat{\sigma} \hat{c}_{k_y}(x)\rangle$, which can be readily written in the form,

$$\partial_t s_y(k_y, x) = -\partial_x J^y_x(k_y, x) - 2\lambda k_y s_z(k_y, x). \tag{18.23}$$

Here J^y_x stands for the conventional operator of spin-current, i.e.,

$$J^y_x(k_y, x) = \frac{i}{4m}\langle \nabla_x \hat{c}^\dagger_{k_y} \hat{\sigma}_y \hat{c}_{k_y} - \hat{c}^\dagger_{k_y} \hat{\sigma}_y \nabla_x \hat{c}_{k_y}\rangle - \frac{\lambda}{2}\langle \hat{c}^\dagger_{k_y} \hat{c}_{k_y}\rangle.$$

In a steady state the lhs of Eq. (18.23) vanishes. Integrating Eq. (18.23) over the x-direction, we obtain for the net spin polarization,

$$\int_{-\infty}^{\infty} s_z(x)dx = -\frac{1}{2\lambda}\sum_{k_y} \frac{1}{k_y} J^y_x(k_y, \infty). \tag{18.24}$$

It is straightforward to calculate the value of the (k_y-resolved) spin current $J_x^y(k_y, \infty)$ inside the bulk of a 2D system:

$$J_x^y(k_y, \infty) = -\frac{1}{2} \sum_{\beta=\pm 1} \sum_{k_x} \left(\lambda + \frac{\beta k_x^2}{mk}\right) n_\beta(k_x, k_y), \tag{18.25}$$

where $n_\beta(k_x, k_y)$ stands for the population of different momentum states in the subband β. Only "uncompensated" states contribute to the non-equilibrium spin polarization given by Eqs. (18.24-18.25); these states describe electrons that originate in the left lead ($k_y > 0$) and belong to the energy interval near the Fermi energy, $E_F < (k + \beta\lambda)^2/2m < E_F + eV$. The integral (18.24) diverges logarithmically at $k_y \to 0$. Assuming the same infrared cut-off in both subbands, $\widetilde{k}$, we observe that the diverging $\ln \widetilde{k}$-contributions in the two subbands cancel each other, yielding in the linear (in V) response,

$$\int_{-\infty}^{\infty} s_z dx = \frac{eV}{2\lambda(2\pi)^2} \left(\frac{2\lambda}{v_F} - \ln \frac{v_F + \lambda}{v_F - \lambda}\right) \tag{18.26}$$

where $v_F = \sqrt{2E_F/m}$ is the Fermi velocity. Expanding this *general* result to the lowest non-vanishing order in λ/v_F we recover the net boundary polarization, Eq. (18.5).

18.4.2. *Evanescent modes*

We are now going to consider the edge spin density in the case of sharp (hard wall) edge potential. Since the sharp edge does not impose any finite length scale, the question arise, what would be the width of the edge spin distribution? Obvious candidate for that comes from the evanescent modes,[34] whose wave function do have an explicitly decaying component ($\sim \exp(-m\sqrt{v_F \lambda} x)$).

The reflection at the sharp boundary mixes the two bulk subbands. Evanescent contributions in the upper subband appear when the reflecting states from the lower subband belong to the domain, $m(v_F - \lambda) < k_y < m(v_F + \lambda)$. Repeating the calculations leading to Eq. (18.26) but now for the evanescent domain only, we obtain,

$$\int_{-\infty}^{\infty} s_z^{\text{ev}} dx = \frac{eV}{2\lambda(2\pi)^2} \left(2\sqrt{\frac{\lambda}{v_F}} - \ln \frac{1 + \sqrt{\lambda/v_F}}{1 - \sqrt{\lambda/v_F}}\right). \tag{18.27}$$

Remarkably, the net evanescent contribution turns out to be much larger than the full result Eq. (18.26). This means that this contribution is *largely* cancelled by the contribution from the normal domain $k_y < m(v_F - \lambda)$. Similar cancellation of all smooth long wavelength contributions takes place for the local spin density, as we discuss in the next section.

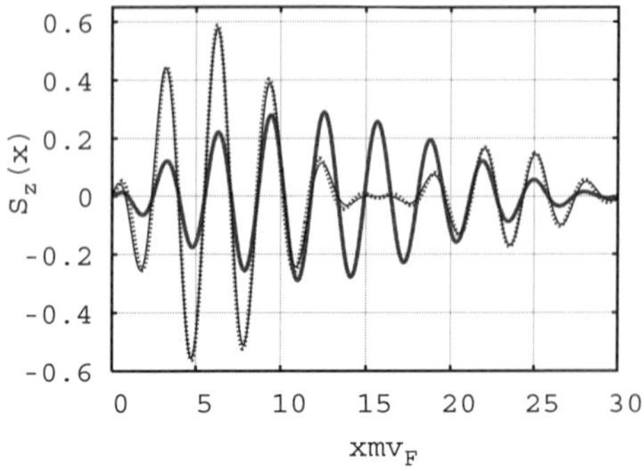

Fig. 18.4. Dependence of the local spin polarization, in units of $eVm/8\pi^2$, on the distance to the boundary for different values of spin-orbit coupling constant. Thick solid line: $\lambda/v_F = 0.1$, thick dotted line: $\lambda/v_F = 0.2$, thin black line utilizes the approximate formula (18.30) for $\lambda/v_F = 0.2$.

18.4.3. *Hard wall, results*

Exact lengthy explicit expression for $s_z(x)$ in the case of sharp boundary potential, Fig. 18.4, was given in Ref. [36]. Here we show only the spectral density

$$s_z(q) = 2\int_0^\infty dx s_z(x) \sin qx, \tag{18.28}$$

which is given by a simple piecewise continuous algebraic function defined in four domains. The surprising feature of the spectral distribution is its vanishing, $s_z(q) = 0$, in the whole long-wavelength domain, $0 < q < 2k^+$. In particular, this shows the exact cancelation between normal and evanescent modes. For larger values of q we obtain to the leading order in λ,

$$s_z(q) = \frac{eVq}{16\pi m v_F^2}\begin{cases} 0, & q < 2m(v_F - \lambda), \\ -1, & 2m(v_F - \lambda) < q < 2mv_F, \\ 1, & 2mv_F < q < 2m(v_F + \lambda), \\ -2/(q\xi)^4, & 2m(v_F + \lambda) < q. \end{cases} \tag{18.29}$$

The plot of the spectral distribution is illustrated in Fig. 18.5. Remarkably, the net spin polarization (given by $\pi^{-1}\int dq s_z(q)/q$) comes from the large-q tail ($\propto q^{-3}$) in the spectral density $s_z(q)$.

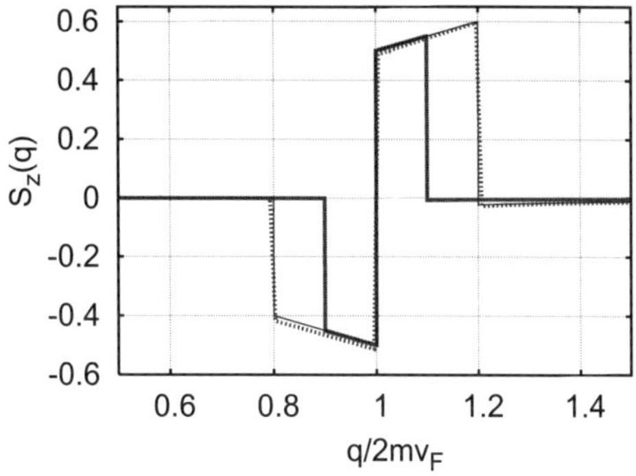

Fig. 18.5. Spectral distribution (18.29) of spin density in units of $eV/4\pi v_F$ for different values of spin-orbit coupling constant, $\lambda/v_F = 0.1$, and $\lambda/v_F = 0.2$.

The *approximate* spin density may be written in a simple form ($\hbar = 1$),

$$s_z(x) \approx \frac{eV}{2\pi^2 v_F x} \cos{(2mv_F x)} \sin^2{(m\lambda x)}. \tag{18.30}$$

Interestingly, the spin-orbit coupling constant enters via the period of beating only.

As is evident from Eq. (18.29) one should speak about the spin accumulation at the hard wall with certain caution. The spin density in this case comes from quickly oscillating functions and the notion of spin accumulation should be understood in the same mathematical sense as the finite value of the integral

$$\int_0^\infty dx \sin x = \lim_{\eta\to 0} \int_0^\infty dx\; e^{-\eta x} \sin x = 1. \tag{18.31}$$

18.5. Smooth Edge: Kinetic Equation

The second ballistic spin-Hall problem that allows analytical solution involves a smooth boundary potential, the exact condition to be presented below. Away from classical turning points (see the next section) spin accumulation can be obtained with the help of the matrix kinetic equation. In addition this method allows to find a complete electron distribution along the way. In the semiclassical approximation the 2×2 matrix $\hat{f}_{\mathbf{p}}(x)$ generalizes the usual distribution function. The equation for the matrix distribution function in the presence of a smooth confining potential $U(x)$

and in the absence of disorder takes the form,[46]

$$\frac{p_x}{m}\partial_x \hat{f}_{\mathbf{p}} - \lambda \left\{ \hat{s}_y, \partial_x \hat{f}_{\mathbf{p}} \right\} + 2i\lambda \left[p_y \hat{s}_x - p_x \hat{s}_y, \hat{f}_{\mathbf{p}} \right] = \frac{\partial \hat{f}_{\mathbf{p}}}{\partial p_x} \partial_x U(x). \quad (18.32)$$

To the zeroth order in the potential gradient $\partial_x U$ the solution is trivial and reduces to the equilibrium Fermi-Dirac distributions for the two spin-split subbands ($\beta = \pm 1$), $f_\beta(x) = n_F(\frac{1}{2m}(p+\beta\lambda)^2 + U(x) - \mu)$, with the local value of the potential $U(x)$ determining the elevation of the bottom of the subbands. To this order the electron spin remains adiabatically *within* the plane of 2DEG during the course of electron motion in the potential $U(x)$. The out-of-plane spin component arises from the non-adiabatic corrections that are first order in $\partial_x U$. The solution of kinetic equation (18.32) is rather straightforward and yields,

$$\hat{f}_{\mathbf{p}} = \frac{1}{2}\sum_\beta [1 + 2\beta(n_y \hat{s}_x - n_x \hat{s}_y)] f_\beta + \frac{\lambda n_y \hat{s}_z}{2p^2} \partial_x U \sum_\beta \frac{\partial(f_\beta/\lambda)}{\partial \lambda}, \quad (18.33)$$

where $\vec{n}$ is the direction of the electron momentum. The local value of spin-Hall density $s_z = \mathrm{Tr}\ \hat{s}_z \sum_{\mathbf{p}} \hat{f}_{\mathbf{p}}$ is obtained by integrating Eq. (18.33) over excess electron states that originate in the left lead, namely over those with $v_y > 0$ and the energy within the interval $[E_F, E_F + eV]$,

$$s_z(x) = \frac{dU}{dx}\frac{m\lambda}{2(2\pi)^2} \quad (18.34)$$
$$\times \frac{\partial}{\partial\lambda}\frac{1}{\lambda}\int \frac{dE}{\sqrt{2m[E-U(x)]}}\left(\frac{1}{p_+} - \frac{1}{p_-}\right)[n_F(E-eV) - n_F(E)].$$

Here $p_\pm$ are the momenta corresponding to a given energy: $p_\pm(x) = \sqrt{2m[E-U(x)]} \mp m\lambda$ when both subbands are occupied, $\sqrt{2m[E-U(x)]} < m\lambda$; and $p_\pm = m\lambda \mp \sqrt{2m[E-U(x)]}$ when the upper subband is empty, $\sqrt{2m[E-U(x)]} > m\lambda$. Calculation of this integral in the linear order in eV yields

$$\overline{s}_z(x) = -\frac{\lambda^2 eV}{(2\pi)^2 m(v_F^2 - \lambda^2)v_F^3}\frac{dU}{dx}. \quad (18.35)$$

The expression (18.35) has a number of interesting features. The local spin polarization is proportional to the force exerted on electrons by the boundary. As long as the Fermi energy is well above the bottom of the bands, $v_F \gg \lambda$, spin accumulation is small and only second order in the spin-orbit coupling constant. When the bottom of the band is elevated high enough, $v_F < \lambda$, the local spin polarization increases dramatically. The vicinities of the two singularities at $v_F(x) = 0$ (classical turning point) and $v_F(x) = \lambda$ (degeneracy point) have to be studied beyond kinetic equation (18.32).

In addition to a smooth classical spin distribution (18.35) quantum wiggles in $s_z(x)$ are present whose magnitude is not necessarily small compared to $\bar{s}_z(x)$. A convenient quantity (especially for numerical calculations) that averages out these wiggles is the "integrated spin density"

$$\mathcal{S}(x) = \sum_j \int_{-\infty}^{x} s_z(x')dx'. \tag{18.36}$$

Integrating Eq. (18.35) over x yields smooth part of the integrated spin density

$$\overline{\mathcal{S}}(x) = \frac{eV}{2\lambda(2\pi)^2}\left(\frac{2\lambda}{v_F(x)} - \ln\frac{v_F(x)+\lambda}{|v_F(x)-\lambda|}\right). \tag{18.37}$$

This formula presents a generalization of Eq. (18.26) for the case of a smooth variation of the confining potential.

Note that the net spin polarization across the edge is again independent of the shape of the boundary potential, $\mathcal{S}(\infty) = -\lambda^2 eV/12\pi^2 v_F^3$, and is expressed via the bulk value of the Fermi velocity $v_F(\infty)$. Here by $x = \infty$ we assume a point deep inside the 2DEG but yet far from its opposite edge. The latter has spin accumulation of the same absolute value and opposite sign.

18.6. Smooth Edge: Singular Spin Dynamics near Turning Points

Semiclassical Eqs. (18.35,18.37) do not offer the important information about the edge spin. First, these equations predict a singular spin density at $v_F(x) = 0$ and $v_F(x) = \lambda$. In addition, Eqs. (18.35,18.37) ignore any interference effects, which may be important for realistic boundary potentials. Both these problems may be addressed analytically, as it was done in Ref. [37]. In order to have more pedagogical discussion here we concentrate mostly on numerical results. Let us approximate the boundary by a linear potential

$$U(x) = -Fx, \tag{18.38}$$

with the constant force F.

Figure 18.6 shows the "raw" numerical data for $s_z(x)$ and $\mathcal{S}(x)$. The smoothness of the boundary implies that $F \ll m^2\lambda^3/\hbar$. We see from

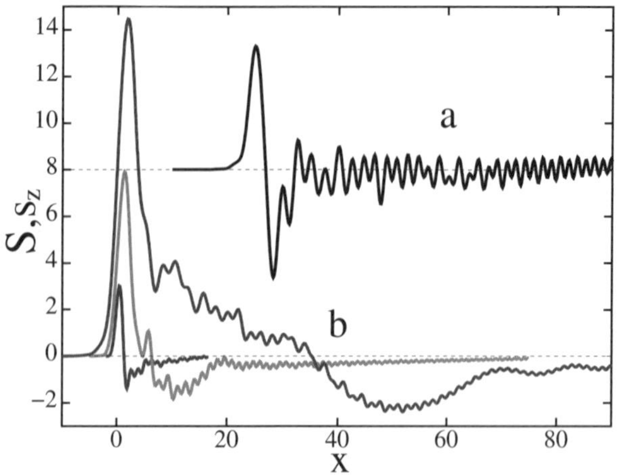

Fig. 18.6. a). Spin density for a force strength $F = 0.01m^2\lambda^3/\hbar\,(U = -Fx)$. [The curve offset both vertically and horizontally.] All coordinates are measured in units of $\hbar/m\lambda$. A smooth component of the density is hardly visible because of oscillating contributions. b). The smooth component is recovered in the integrated spin $\mathcal{S}(x) = \int_{-\infty}^{x} s_z dx'$. The three curves show the integrated spin for the force strength $F = \alpha m^2\lambda^3/\hbar$ with $\alpha = 0.25, 0.05, 0.01$, respectively. In all these cases we see three regions with different spin behavior. First, the spin density is the largest in narrow outer ($x \approx 0$) strip along the edge. This spin is compensated (and overcompensated) by the wide strip of negative smooth spin density. Finally, in the third strip (at $x > 50$ for $\alpha = 0.01$) the smooth component of the density changes sign to positive again. The width of all three strips and the amount of accumulated spin, which in each strip is much larger than in Eq. (18.5), increase [formally unlimited] with decreasing slope of the boundary potential. To obtain the values of the spin density and accumulated spin one need to multiply the numbers in the figure by $eVm/8\hbar\pi^2$ and $eV/8\lambda\pi^2$ respectively.

Fig. 18.6.a, how the rapid quantum oscillations make it hard to observe the mean value $\bar{s}_z(x)$ Eq. (18.35) even for $F\hbar/m^2\lambda^3 = 0.01$. The smooth component is recovered in the integrated spin on Fig. 18.6.b even for relatively steep boundary $F\hbar/m^2\lambda^3 = 0.25$.

Explicit comparison between numerics and analytical expression Eq. (18.37) is made on Fig. 18.7, where we plot the rescaleed numerical results for different values of the slope of the boundary potential.

According to Eq. (18.37) we find two regions of different smooth spin behavior. First, within the strip where $0 < v_F(x) < \lambda$ spin density is negative (which is seen in a downward slope of the integrated density $\mathcal{S}(x)$ in Fig. 18.7). Farther away, $s_z(x)$ changes sign for $v_F(x) > \lambda$, where both $s_z(x)$ and $\mathcal{S}(x)$ decrease gradually with increasing x.

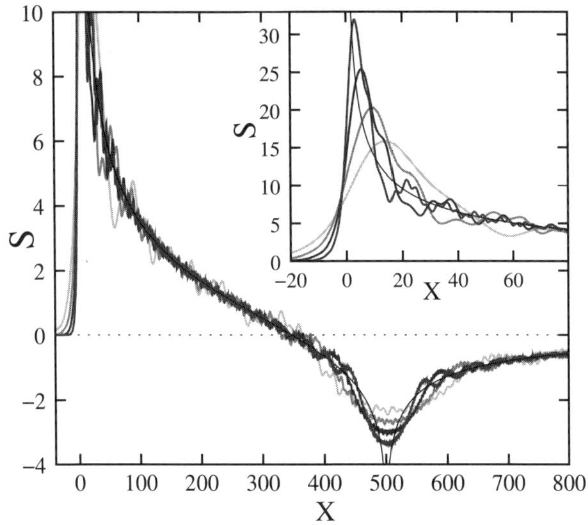

Fig. 18.7. Integrated spin density $\mathcal{S}(x) = \int_{-\infty}^{x} s_z dx$ for the potential $U(x) = -\alpha m^2\lambda^3 x/\hbar$ in units of $eV/8\lambda\pi^2$. The curves for $\alpha = 8, 4, 2, 1\times10^{-3}$ are shown. Both the peak at $x \approx 0$ and the dip at $x \approx 500$ became more pronounced with decreasing α. The horizontal coordinate is scaled differently for different curves, as x is measured in units of $10^3\alpha \times \hbar/m\lambda$. Narrow black lines stand for the semiclassical prediction, Eq. (18.26). The logarithimc behavior, $\sim \log\alpha$, of the dip at at $U(x) = -m\lambda^2/2$ ($x = 500$) is clearly seen. Inset magnifies the region near the edge of 2DEG ($x \approx 0$).

The most interesting is the behavior of spin at the borders of these regions, $v_F = 0$ and $v_F = \lambda$. At $v_F(x) = \lambda$ the accumulated spin $\overline{\mathcal{S}}(x)$ Eq. (18.37) diverges logarithmically. This singularity originates from the accumulation of classical turning points taking place when the conical crossing point in the spectrum of the Hamiltonian (18.6), see Fig. 18.3C, passes through the Fermi energy. This singularity is regularized as $\mathcal{S} \sim \log F$, according to Fig. 18.7.

Yet more peculiar is the behavior of both $s_z(x)$ and $\mathcal{S}(x)$ at the edge of 2DEG, near the point where $v_F(x) = 0$. The smooth part of the accumulated spin, Eq. (18.37), has an infinite jump here (from Eq. (18.37) it follows that $\overline{\mathcal{S}}(+0) = \infty$, while obviously $\overline{\mathcal{S}}(-0) = 0$). Development of such jump with decreasing slope of the potential is seen in the inset in Fig. 18.7. The jump in $\mathcal{S}(x)$ corresponds to the formation of a narrow strip with extremely large values of spin $s_z > 0$ along the border. This behavior will now be analyzed in more detail.

Classical dynamics of electrons with Rashba spin-orbit interaction is described by the effective Hamilton Eq. (18.7). with the boundary potential approximated by the linear function $U = -Fx$. The family of classical trajectories generated by this Hamiltonian, shown in Fig. 18.8, demonstrate a number of unusual features.

As seen from Fig. 18.8, those electrons from the lower subband that have $|p_y| < m\lambda$, pass *three* turning points in the course of their motion in the x direction, corresponding to three solutions of the equation $\partial H_{\text{eff}}/\partial p_x = 0$. Two of these turning points (those with $p = m\lambda$) correspond to simultaneous vanishing of *both* velocity components, $\vec{v}(x) = 0$, the behavior generically impossible in a 2DEG with the parabolic spectrum, $H = p^2/2m$.

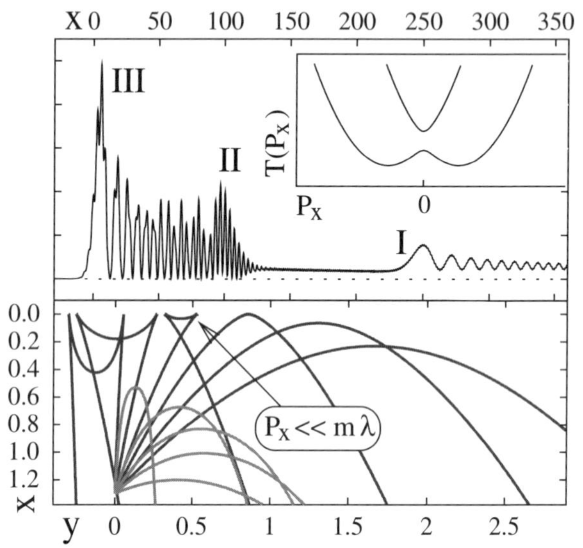

Fig. 18.8. *Bottom:* Family of classical trajectories at $E = E_F$ for different values of p_y, and $m = \lambda = F$. Trajectories for both lower (black) and upper (gray) spin-orbit split subbands are shown. An example of a trajectory contributing to the peak in spin density at $x \approx 0$, Eq. (18.40), is shown for "$p_x \ll m\lambda$" (see the text). *Top:* The electron density $\rho = \psi_1^\dagger\psi_1 + \psi_2^\dagger\psi_2$ for given longitudinal momentum and energy. Here $\psi_{1,2}$ are two eigenfunctions with $p_y = 0.2m\lambda, E_F = 0, F = 0.003m^2\lambda^3/\hbar$, ρ in arbitrary units, x in units of $\hbar/m\lambda$. Three classical turning points can be seen. The interference of incoming and reflected waves in the upper subband causes smooth oscillations to the right of the inner turning point I ($x > m\lambda^2/2F$). At the other turning points, II and III, the two kinds of oscillations are seen. Slow oscillations are caused by the interference of the incoming wave and the wave reflected at the turning point. Fast oscillations (wavelength $\sim \hbar/m\lambda$) are due to the interference of distant (in time) segments of the same trajectory. *Inset:* Kinetic energy (arbitrary units) $T_\pm(p_x) = (p \pm m\lambda)^2/2m$ for fixed $p_y = 0.2m\lambda$.

Analytical treatment[37] of the expectation value of the z-component of electron spin in the vicinity of turning point $v_F(x) = 0$ yields ($x \ll m\lambda^2/F$)

$$s_z = \frac{3meV}{4\pi\hbar}\frac{\partial}{\partial\widetilde{x}}\int_0^1 dz\ \mathrm{Ai}^2(-\widetilde{x}/z), \tag{18.39}$$

where $\widetilde{x} = x(2Fm/\hbar^2)^{1/3}$. In the asymptotic region $x \gg (2Fm/\hbar^2)^{-1/3}$ one can average over the oscillations of the Airy function. This allows us to recover the singular behavior of the smooth spin density (18.35): $\langle s_z\rangle \sim x^{-3/2}$. The integral in Eq. (18.39) features a logarithmic singularity at $x = 0$. With the logarithmic accuracy the (properly regularized) height of the peak of spin density is[37]

$$s_z(0) = \frac{meV}{10\sqrt{3}\pi^2\hbar}\ln\left(\frac{m^2\lambda^3}{\hbar F}\right). \tag{18.40}$$

Striking feature of this result is that this maximal value is virtually *independent* of the strength of spin-orbit coupling or the shape of the boundary potential (up to a weak logarithmic factor).

We thus conclude that the nonequilibrium spin-Hall spin accumulation near a smooth boundary of 2DEG ballistic conductor with spin-orbit interaction develops a narrow peak at the edge, with the width $\sim (\hbar^2/mF)^{1/3}$ and height given by Eq. (18.40). It is followed by a slow non-monotonic decay, as shown on Fig. 18.7. This smooth tail of spin density persists to much larger distances, $\gtrsim m\lambda^2/F$. The amount of spin accumulated in the peak (found as a maximum of the function $\mathcal{S}(x) = \int^x s_z dx$) equals $\mathcal{S}_{\max} \approx 0.04eV(m^2/\hbar F)^{1/3} > 0.04eV/\lambda$, where in the last inequality we utilize the fact that $F < m^2\lambda^3/\hbar$. We see that the spin accumulated at the edge described by a semiclassical boundary potential is inversely proportional to the strength of spin-orbit interaction and becomes progressively larger for smoother slopes. This prediction can be used for experimental observation of spin-Hall effect in realistic two-dimensional electron systems.

18.7. Spin-Hall Effect on Surfaces of Topological Insulators

The method developed in Secs. 18.4 and 18.5 is fully applicable to other chiral systems mentioned in the Introduction, as long as the corresponding terms in the Hamiltonians are *linear* in electron momentum. This is a good approximation in graphene and an acceptable one in Bi_2Te_3. In graphene, however, chiral structure arises from *pseudospin* (in sublattice space). There is nothing straightforward about experimental detection of pseudospin in graphene. In addition, as explained in the Introduction, the case of graphene is mapped onto "Rashba-like" chirality for one cone within the first Brillouin zone and "Dresselhaus-like" chirality for the other one. As a result, the net pseudospin accumulation *vanishes* when contributions from the two cones are added together, cf. Eq. (18.5).

The situation changes dramatically for topological insulators. First, the chirality originates from *true* spin, accessible by standard experimental techniques. Second, there is an *odd* number of Fermi valleys per (geometric) surface. This is in a sharp contrast to graphene which has two Dirac cones. Note that despite the equal number of "species" with opposite chiralities residing on the opposite surface, one can probe them *individually*, by virtue of their *spatial* separation.

Since the Fermi energy in topological insulators lies within the bandgap for 3D bulk electronic states, electron transport occurs only across the crystal surface. While not yet implemented in practice to our knowledge, it makes possible local control (via gate electrodes) of spatial distribution of electron density. We envisage the following experimental setup, see Fig. 18.9. Metallic contacts are attached to the surface of topological insu-

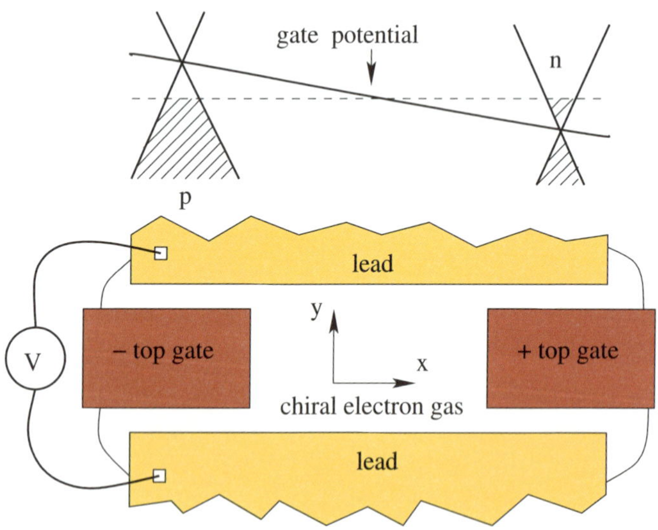

Fig. 18.9. Proposed experimental setup for creating a $n-p$ junction on the surface of topological insulator. Voltage V applied between the leads drives electric current in y-direction and creates a finite spin density along the junction (along the line $x=0$).

lator and drive dc electric current along y-direction. Gates are positioned (without direct contact) some distance above the surface. Application of electric potentials to the gates modulates position of electronic bands. Here, similarly to the rest of the paper, we assume that the topological insulator is disorder-free (ballistic). At the present time it is unclear how well this assumption is satisfied in contemporary samples, but undoubtedly sample quality is only going to improve in the near future.

Effective low-energy Hamiltonian for Bi_2Te_3[10] (neglecting cubic terms)

$$H = -iv(\sigma_x \partial_y - \sigma_y \partial_x) + U(x), \tag{18.41}$$

has a single Dirac point whose elevation is determined by the gate potential $U(x)$. There are two cases of interest here:
i) Potential $U(x)$ is not very strong, so that the Dirac point always lies below the Fermi-energy (which is also normal situation for ungated material).
ii) Potential is strong enough to lift the Dirac point *above* the Fermi level in some region on the surface, thus forming a p-n junction.

We now present our general method before analyzing these two cases separately. The method extends the approach of Sec. 18.4 for the Hamiltonian (18.41) and we present it here in detailed form for the reader's convenience. Utilizing the fact that in a ballistic system k_y is the integral of motion, we write the electron operators in the mixed representation,

$$\hat{\psi}(\vec{r}) = \sum_{k_y} \hat{c}_{k_y}(x) e^{ik_y y}. \tag{18.42}$$

Similarly, spin density in the mixed representation reads

$$\vec{s}(k_y, x) = \langle \hat{c}^\dagger_{k_y}(x) \vec{\sigma} \hat{c}_{k_y}(x) \rangle \tag{18.43}$$

and satisfies the equation

$$\partial_t s_y(k_y, x) = -\partial_x J^y_x(k_y, x) - 2vk_y s_z(k_y, x). \tag{18.44}$$

Here J^y_x is spin current operator (in general $J^i_j = \frac{1}{2}(\hat{\sigma}_i \hat{v}_j + \hat{v}_j \hat{\sigma}_i)$),

$$J^y_x(k_y, x) = -v \langle \hat{c}^\dagger_{k_y}(x) \hat{c}_{k_y}(x) \rangle. \tag{18.45}$$

In a steady state the left-hand side of Eq. (18.43) vanishes and we arrive at

$$s_z(k_y, x) = -\frac{1}{2vk_y} \partial_x J^y_x(k_y, x). \tag{18.46}$$

This yields the net spin polarization across the profile of the gate potential $U(x)$

$$\int_{-\infty}^{\infty} s_z(x) dx = -\frac{1}{2v} \sum_{k_y} \frac{1}{k_y} [J^y_x(k_y, \infty) - J^y_x(k_y, -\infty)], \tag{18.47}$$

expressing it via the values of spin current far away from the region where $U(x)$ is varied,

$$J_x^y(k_y, \pm\infty) = -v \sum_{k_x} n(k_x, k_y, \pm\infty), \tag{18.48}$$

with $n(k_x, k_y, \infty)$ denoting the corresponding distributions of electrons.

The summation in Eqs. (18.47)-(18.48) has to be performed over "uncompensated" states that originate from the left lead and belong to the energy interval $E_F < E < E_F + eV$, where V is the bias applied between electrical contacts. (The uncompensated states originate in the contact that has higher chemical potential and propagate towards the other contact.) The subsequent analysis will be performed separately for the two cases mentioned above.

i) n-n junction. When gate potential $U(x)$ is weaker than needed to elevate Dirac point above the Fermi-energy the electric current is carried by electrons only. Still, a smooth step-like potential $U(x)$ create a "junction" between half planes with different values of the Fermi momenta, k_L and k_R, respectively.

Since $dk_x = dEk/k_x v$ and $dE \to eV$ the summation over k_x is performed according to

$$\sum_{k_x} \to \frac{eVk_n}{\pi\sqrt{k_n^2 - k_y^2}}, \tag{18.49}$$

(note that both $k_x > 0$ and $k_x < 0$ contribute to this expression). From Eq. (18.47),

$$\int_{-\infty}^{\infty} s_z(x)dx = \frac{eV}{2\pi v}\left\{\int_0^{k_L} \frac{dk_y}{k_y}\frac{k_L}{\sqrt{k_L^2 - k_y^2}} - \int_0^{k_R} \frac{dk_y}{k_y}\frac{k_R}{\sqrt{k_R^2 - k_y^2}}\right\}. \tag{18.50}$$

Regularizing these formally divergent integrals (by assuming the same lower cutoff which subsequently drops out from the difference of the two terms) we find the amount of net spin accumulated at the junction,

$$\int_{-\infty}^{\infty} s_z(x)dx = \frac{eV}{2\pi v}\ln\left(\frac{k_L}{k_R}\right). \tag{18.51}$$

In addition, if the gate potential changes smoothly on the scale of $1/k_F$ the local spin polarization can be written as,

$$s_z(x) = \frac{eV\ \partial_x U(x)}{2\pi v^2 k_F(x)}. \tag{18.52}$$

ii) p-n junction. In the real experiment the strength of the gate potential $U(x)$ can be made significant enough to lift Dirac point above the Fermi energy over some region of the surface. For this setup we find

$$\int_{-\infty}^{\infty} s_z(x)dx = \frac{eV}{2\pi v}\left\{\int_0^{k_n}\frac{dk_y}{k_y}\frac{k_n}{\sqrt{k_n^2-k_y^2}} - \int_{-k_p}^{0}\frac{dk_y}{k_y}\frac{k_p}{\sqrt{k_p^2-k_y^2}}\right\}. \tag{18.53}$$

This result differs from Eq. (18.50), and from the spin accumulation around $v_F(x) = \lambda$ considered in Sec. 18.5, in that here the excess electrons to the left and to the right of the potential step have similar velocity, v_y, but opposite momentum, p_y, in the direction of current. In Eq. (18.53) the excess electrons originate from the lead $E_F + eV$ and thus have to have $v_y > 0$. On the contrary, in the case of edge spin accumulation (Sec. 18.5) the electrons arrive to the strip $0 < v_F(x) < \lambda$ from the bulk 2DEG, as is shown in Fig. 18.8, and thus must carry the bulk momentum $p_y > 0$, but may have the "wrong" sign of velocity. As a result the spin density in Eq. (18.53) has the same sign on both sides of the $n-p$ junction and two divergent logarithms add. Introducing a proper quantum mechanical cutoff to these integrals yields

$$\int_{-\infty}^{\infty} s_z(x)dx = \frac{eV}{2\pi v}\ln\left(\frac{4k_pk_n}{q_{\min}^2}\right), \quad q_{\min}^2 = \frac{1}{\hbar v}\left.\frac{dU}{dx}\right|_{x=0}. \tag{18.54}$$

Electrons with very small k_y start to tunnel thorough the $n-p$ junction, which would change our semiclassical predictions. The rude estimate of the quantum mechanical tunnelling exponent gives $\kappa = \int_0^{k_y} k_x(x)dx = k_y^2\hbar v/(2dU/dx)$. Requiring a small tunnelling probability, $e^{-\kappa} \ll 1$, gives the above cutoff $q_{\min}$.

We may now compare the strength of the Spin-Hall effect in semiconductor heterostructures Eq. (18.5) with the result for topological insulators Eqs. (18.51,18.54). The main difference is that in conventional semiconductors the chiral electron states appear because of the weak spin-orbit interaction. Consequently the result Eq. (18.5) acquires a small factor $(\lambda^2 - \lambda_D^2)/v_F^2$. There is now such suppression for the surface states in topological insulators[g].

---

[g]Spin accumulation Eq. (18.5) and Eqs. (18.51,18.54) depends also on the value of the Fermi velocity. The value of v_F in Eq. (18.5) depends on the bulk density of 2DEG. In real experiment however, the Fermi velocity in our two examples will probably be of the same order of magnitude.

18.8. Summary

In this paper we discussed several nonequilibrium spin-related phenomena occurring in 2-dimensional chiral electron systems.

We were mostly interested in the effects which may be explained in terms of semiclassical electron motion in smooth potentials. In this case it is easy (at least theoretically) to produce the strongly in-plane spin polarized currents (Sec. 18.3).

The main part of the paper was devoted to calculation of the out of plane spin density $\langle \sigma_z \rangle$. It is the out of plane spin component, that is usually investigated in the experiment via the measuring of the optical Kerr rotation.[12]

Finally we discussed the mesoscopic spin-Hall effect in novel materials such as graphene and topological insulators. Experimental investigation of pseudospin-Hall effect might prove not to be easy. On the contrary, observation of spin-Hall σ_z accumulation at the surface of 3D topological insulator looks most appealing. While the net spin accumulation in semiconductor heterostructures is of the second order in small spin-orbit interaction, $s \sim \lambda^2/v_F^3$ (18.5), there is no such restriction for topological insulator, $s \sim 1/v$ (18.51,18.54). As a result, the spin accumulation in the latter case should exceed that possible in traditional 2-dimensional electron gas by orders of magnitude.

Acknowledgments

We gratefully acknowledge our collaborator V. Zyuzin, who participated in obtaining a number of previously published results reviewed in this paper; we also have benefited from discussions with A. Andreev, G.E.W. Bauer, C.W.J. Beenakker, B.I. Halperin, M. Raikh and O. Starykh. P.G.S. was supported by the SFB TR 12; E.G.M. was supported by the DOE, Office of Basic Energy Sciences, Award No. DEFG02-06ER46313.

References

1. Yu.A. Bychkov and E.I. Rashba, J. Phys. C **17** 6039 (1984); F.T. Vas'ko, JETP Lett. **30**, 540 (1979).
2. G. Dresselhaus, Phys. Rev. **100**, 580 (1955).
3. A.H. Castro Neto, F. Guinea, N.M. Peres, K.S. Novoselov, and A.K. Geim, Rev. Mod. Phys. **81**, 109 (2009).
4. J. Schliemann, J.C. Egues, and D. Loss, Phys. Rev. Lett. **90**, 146801 (2004).

5. B.A. Bernevig, T.L. Hughes, S.-C. Zhang, Science **314**, 1757 (2006).
6. M. König, S. Wiedmann, Ch. Brune, A. Roth, H. Buhmann, L. W. Molenkamp, X-L. Qi, and S-C. Zhang, Science **318**, 766770 (2007).
7. D. Hsieh, D. Qian, L. Wray, Y. Xia, Y. S. Hor, R. J. Cava, and M.Z. Hasan, Nature **452**, 970-974 (2008).
8. Y. Xia, D. Qian, D. Hsieh, L. Wray, A. Pal, H. Lin, A. Bansil, D. Grauer, Y.S. Hor, R.J. Cava, M.Z. Hasan, Nature Physics **5**, 398 (2009).
9. H.-J. Zhang, C-X. Liu, X-L. Qi, X-Yu Deng, X. Dai, S-C. Zhang, Z. Fang, Phys. Rev. B **80**, 085307 (2009).
10. L. Fu, arXiv:0908.1418.
11. M.I. D'yakonov, V.I. Perel, Phys. Lett. A **35**, 459 (1971).
12. Y.K. Kato, R.C. Myer, A.C. Gossard, and D.D Awschalom, Science **306**, 1910 (2004).
13. V. Sih, R. C. Myers, Y. K. Kato, W. H. Lau, A. C. Gossardand D. D. Awschalom, Nature Physics **1**, 31 (2005).
14. S.O. Valenzuela and M. Tinkham, Nature **442**, 176 (2006).
15. J. Wunderlich, B. Kaestner, J. Sinova, and T. Jungwirth, Phys. Rev. Lett. **94**, 047204 (2005).
16. J.E. Hirsch, Phys. Rev. Lett. **83**, 1834 (1999).
17. Shufeng Zhang, Phys. Rev. Lett. **85**, 393 (2000).
18. H.A. Engel, B.I. Halperin, and E.I. Rashba, Phys. Rev. Lett. **95**, 166605 (2005).
19. W.-K. Tse, S. Das Sarma, Phys. Rev. Lett. **96**, 056601 (2006).
20. S. Murakami, N. Nagaosa, and S.-C. Zhang, Science **301**, 1348 (2003); Phys. Rev. B **69**, 235206 (2004).
21. J. Sinova, D. Culcer, Q. Niu, N.A. Sinitsyn, T. Jungwirth, and A.H. MacDonald, Phys. Rev. Lett. **92**, 126603 (2004).
22. E. G. Mishchenko, A. V. Shytov, and B. I. Halperin, Phys. Rev. Lett. **93**, 226602 (2004).
23. J.I. Inoue, G.E.W. Bauer, and L.W. Molenkamp, Phys. Rev. B **70**, 41303(R) (2004).
24. A. Khaetskii, Phys. Rev. Lett. **96**, 056602 (2006).
25. R. Raimondi and P. Schwab, Phys. Rev. B **71**, 33311 (2005).
26. O.V. Dimitrova, Phys. Rev. B **71**, 245327 (2005).
27. O. Entin-Wohlman, A. Aharony, Y. M. Galperin, V. I. Kozub, V. Vinokur, Phys. Rev. Lett. **95**, 086603 (2005).
28. B.K. Nikolic, L.P. Zarbo, and S.Souma, Phys. Rev. B **72**, 075361 (2005); B.K. Nikolic, S.Souma, L.P. Zarbo, and J. Sinova, Phys. Rev. Lett. **95**, 046601 (2005).
29. S. Murakami, Adv. in Solid State Phys. **45**, 197 (2005).
30. H.A. Engel, E.I. Rashba, and B.I. Halperin, *Theory of Spin Hall Effects in Semiconductors*, in Handbook of Magnetism and Advanced Magnetic Materials, H. Eds. Kronmüller and S. Parkin (John Wiley & Sons, 2007).
31. Non-local detection of spin Hall currents in multiterminal devices has recently been reported in K.C. Weng, *et al.*, arXiv:0804.0096; C. Bruene, *et al.*, arXiv:0812.3768.

32. K. Nomura, J. Wunderlich, J. Sinova, B. Kaestner, A.H. MacDonald, and T. Jungwirth, Phys. Rev. B **72**, 245330 (2005).
33. E.M. Hankiewicz, L.W. Molenkamp, T. Jungwirth, and J. Sinova, Phys. Rev. B **70**, 241301(R) (2004).
34. G. Usaj and C.A. Balseiro, Europhys. Lett. **72**, 631 (2005); A. Reynoso, G. Usaj and C.A. Balseiro, Phys. Rev. B **73**, 115342 (2006).
35. T.D. Stanescu and V. Galitski, Phys. Rev. B **74**, 205331 (2006).
36. V.A. Zyuzin, P.G. Silvestrov, and E.G. Mishchenko, Phys. Rev. Lett. **99**, 106601 (2007).
37. P.G. Silvestrov, V.A. Zyuzin, and E.G. Mishchenko, Phys. Rev. Lett. **102**, 196802 (2009).
38. P.G. Silvestrov and E.G. Mishchenko, Phys. Rev. B **74**, 165301 (2006).
39. R.G. Littlejohn and W.G.Flynn, Phys. Rev. A **44**, 5239 (1991); **45**, 7697 (1992).
40. J. Bolte and S. Keppeler, Phys. Rev. Lett. **81**, 1987 (1998).
41. M. Pletyukhov and O. Zaitsev, J. Phys. A: Math. Gen. **36**, 5181 (2003).
42. D. Culcer, J. Sinova, N. A. Sinitsyn, T. Jungwirth, A. H. MacDonald, and Q. Niu, Phys. Rev. Lett. **93**, 046602 (2004).
43. O. Entin-Wohlman, A. Aharony, Y. Tokura, Y. Avishai, arXiv:0911.1347.
44. D. Grundler, Phys. Rev. Lett. **84**, 6074 (2000).
45. E.B. Sonin, arXiv:0909.3156.
46. E.G. Mishchenko and B.I. Halperin, Phys. Rev. B **68**, 045317 (2003).

Chapter 19

The Aharonov-Bohm Effect and its Applications to Magnetic Field Observation

Akira Tonomura

Advanced Research Laboratory, Hitachi, Ltd., Hatoyama, Saitama, 350-0395, Japan
Advanced Science Institute, RIKEN, Wako, Saitama 351-0198, Japan
Okinawa Institute of Science and Technology, Uruma, Okinawa 394-2234, Japan

This chapter describes the Aharonov-Bohm (AB) effect of electron waves travelling in free space and its application to the observation of gauge fields (vector potentials). This article is dedicated to Professor Yoseph Imry on his 70th birthday for his great contributions to the exploration of the Aharonov-Bohm effect and to the founding of mesoscopic physics.

19.1. Introduction

The AB effect is inconceivable in classical physics since an observable effect is produced on electrons passing through field-free spaces. Electrons can be affected only by Lorentz forces due to electromagnetic fields. The situation is different in quantum mechanics, where electrons show wave properties: the concept of force is no longer relevant, so electric field $\boldsymbol{E}$ and magnetic field $\boldsymbol{B}$, defined as forces acting on a unit charge, take on secondary meanings. "Phase shifts" come into play, and the primary physical entities become neither $\boldsymbol{E}$ nor $\boldsymbol{B}$ but electrostatic potential V and vector potential $\boldsymbol{A}$. These potentials interact with electron waves and shift their phases.

In 1959, Aharonov and Bohm[1, 2] identified the physical significance of potentials V and $\boldsymbol{A}$ in quantum mechanics, which have no physical meaning in classical physics. This effect was later dubbed "the Aharonov-Bohm effect". They derived from the Schrödinger equation that a relative phase shift is produced between two electron waves even when they pass through a space free of $\boldsymbol{E}$ and $\boldsymbol{B}$. They thus predicted two versions of the AB effect, electric and magnetic.

The AB effect increased in significance in the 1970s in relation to the theories unifying all fundamental interactions in nature. The potentials were extended to "gauge fields" and became regarded as the most fundamental physical quantity.

In 1975, Wu and Yang[3] stressed the significance of the AB effect in relation to the physical reality of gauge fields as follows.

> "The concept of an SU_2 gauge field was first discussed in 1954. In recent years many theorists, perhaps a majority, believe that SU_2 gauge fields do exist. However, so far there is no experimental proof of this theoretical idea, since conservation of isotopic spin only suggests, and does not require, the existence of an isotopic spin gauge field. What kind of experiment would be a definitive test of the existence of an isotopic spin gauge field? A generalized Bohm-Aharonov experiment would be."

However, since potentials have long been regarded as mathematical auxiliaries, even the existence of the AB effect in electromagnetism was questioned, thus causing a controversy.[4, 5] Although the AB effect had been experimentally shown[6-9] to exist for the magnetic case soon after its prediction, these results were attributed to the effect of incident electron waves overlapping the magnetic field leaking from both ends of the finite solenoids or ferromagnets used in the experiments.[10, 11]

We made a series of experiments[12-15] exploring the AB effect by using tiny ferromagnetic toroidal samples. Such experiments had been proposed as a crucial test by Kuper.[16] We examined the effect of slight leaking field from the samples by quantitatively measuring the field using holographic interference microscopy.[17, 18] By the last experiment[14] using toroidal feromagnets covered with superconductors as proposed by Yang,[19] we conclusively confirmed the existence of the AB effect. Since the AB effect is fundamental to quantum mechanics, the effect began to be found in other fields. An important case is the AB effect for electrons in mesoscopic metals,[20] where Professor Imry has played an important role. This field is opening a new field in solid state physics.

We then used the principle underlying the AB effect to quantitatively observe magnetic lines of force[17,21] and quantized vortices in superconductors[22] as electron interference micrographs. We also used it to observe the movements of vortices[23,24] by Lorentz microscopy (out-of-focus transmission electron microscopy). In this chapter, I review the results of our experiments exploring the AB effect and the applications using coherent electron beams that we developed over the past 40 years.

19.2. Development of Coherent Electron Beams

Here I describe the historical development of electron holography together with advances in coherent beam technology, particularly the achievement of a "bright" electron beam field-emitted from a pointed tungsten tip.

We started our research on electron holography in 1968[25] as a way to break through the resolution limit of electron microscopes: the resolution was restricted not by the wavelength but by the aberrations in the electron lenses. As a means to compensate for lens aberration, electron holography was invented by Gabor[26]: an electron hologram of an object is formed as an interference pattern ("hologram") between an object wave and a reference wave, and then its image is optically reconstructed by illuminating a reference optical wave onto the hologram. The aberrations are optically compensated for in the optical reconstruction stage. However, we were convinced from our experiments[25] that bright electron beams, like laser beams in optics, would be needed to form high-precision holograms for practically applying electron holography.

We soon began developing bright and yet monochromatic electron beams, and we have continued to do so up to now. Crewe[27] *et al.* first developed such a beam as a practical electron gun for scanning electron microscopes (SEMs) and scanning transmission electron microscopes (STEMs), thus improving their resolutions greatly. This beam should be usable as a coherent beam for electron interferometery; its brightness can be much greater than that of a thermionic beam by several orders of magnitude. In addition, the energy spread of the beam can be as narrow as 0.3 eV.

However, since this electron source is extremely small, typically 50 Å in diameter, we had to overcome several technical difficulties. The tiny source should be neither blurred due to the aberrations accompanied by lens action when accelerating electrons, nor moved even a fraction of the source diameter: the slightest mechanical vibration of the tip, the accelerating tube, or the microscope column or the slightest deflection of the fine beam by a stray ac magnetic field should be avoided. Otherwise, the inherent high brightness of the electron beam will deteriorate.

Table 19.1. History of development of bright and coherent electron beams.

Year	Electron microscope	Brightness ($A/cm^2 \cdot ster$)	Application
1968	100-kV EM	1×10^6	Experimental feasibility of electron holography [25]
1979	80-kV FEEM	1×10^8	Direct observation of magnetic lines of force [17]
1982	250-kV FEEM	4×10^8	Conclusive proof of AB effect [14]
1989	350-kV FEEM	5×10^9	Dynamic observation of vortices In metal superconductors [23]
2000	1-MV FEEM	2×10^{10}	Observation of unusual behaviors of vortices in high-T_c superconductors [24]

※ FEEM: field-emission electron microscope

After ten years of work, we developed an 80-kV electron beam,[28] two orders of magnitude brighter than that of the then used thermal beams (see Table 19.1). Electron interference patterns became directly observable on a fluorescent screen, and as many as 3000 interference fringes were recorded on film, which had never been possible before. By using electron holography and bright electron beams, we obtained new information that could not be obtained by conventional electron microscopy. For example, magnetic lines of force inside and outside ferromagnetic samples were directly and quantitatively observed in *h*/*e* flux units[17, 21] using interference micrographs, which can be optically obtained from their electron holograms.

After that, we continued to develop even brighter electron beams. Every time we increased the brightness, new possibilities opened up, as shown in Table 19.1. A series of experiments on the AB effect[12-15] were carried out with a 250-kV electron microscope, magnetic vortices in metal superconductors[29] were observed with a 350-kV microscope,[30] and unusual behaviors of vortices peculiar to high-T_c superconductors[24, 31, 32] were observed with a 1-MV microscope.[33] These observations became possible by more precisely detecting the phase of an electron wave thanks to the increased brightness of the electron beams. In fact, the brightness of a 1-MV electron beam is 2×10^{10} A/cm^2·ster, which is four orders of magnitude higher than that of a 100-kV thermal electron beam, and the maximum number of biprism interference fringes obtained increased from 300 to 11,000.[34]

19.3. Phase Shift of Electron Wave due to Potentials — The Aharonov-Bohm Effect —

An electron wave transmitted through electromagnetic potentials, V and $\mathbf{A}$, is phase-shifted by them. The phase of an electron wave, S, when electromagnetic fields are weak enough for the WKB approximation to be valid, can simply be expressed as:

$$S = \frac{1}{\hbar}\int (m\boldsymbol{v} - e\mathbf{A})ds\,, \tag{19.1}$$

where the line integral is carried out along an electron path. The effect of electrostatic potential V is included in $\boldsymbol{v}$. It was theoretically predicted on the basis of the AB effect[1] that electrons passing through a space where there are no electromagnetic fields but potentials alone can produce observable effects as displacement of interferences fringes. In the magnetic AB effect, two electron waves one on each side of an infinitely long solenoid, are phase-shifted relative to each other, which is observed as fringe displacement (Fig. 19.1). This effect was attributed to the vector potentials on the two sides of the solenoid being directed in opposite

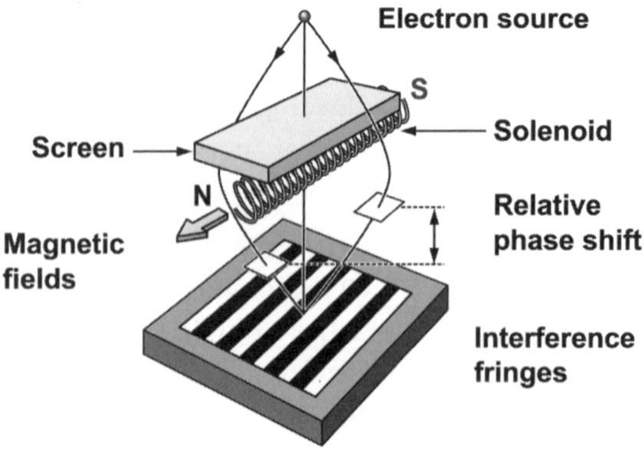

Fig. 19.1. The Aharonov-Bohm effect.

directions, causing the phases of the electron waves on the two sides of the solenoid to be in opposite directions.

Vector potential $\boldsymbol{A}$ and phase S are not uniquely determined for a given physical situation but have an arbitrariness due to gauge transformation. This does not mean, however, that phase S has no physical significance. A relative phase shift, ΔS, between two waves starting from a point source, passing through electromagnetic potentials and ending at another point, can be observed experimentally. This ΔS, which can be measured from the interference pattern, is given by

$$\Delta S = \frac{1}{\hbar}\oint (m\boldsymbol{v} - e\boldsymbol{A})ds\ . \qquad (19.2)$$

Here the integral is carried out along a closed path determined by connecting the two electron paths.

In a pure magnetic case,

$$\Delta S = -\frac{e}{\hbar}\oint \boldsymbol{A}ds = -\frac{e}{\hbar}\oint \boldsymbol{B}ds \qquad (19.3)$$

where the latter surface integral is performed over the surface bordered by the closed path. Since ΔS is proportional to the magnetic flux passing inside the closed circuit, it is uniquely determined.

The AB effect became regarded as fundamental in quantum mechanics and, in the 1970s, increased in significance especially in relation to the unfied theories of gauge fields. However, questions arose in 1978 about the existence of the AB effect[4)] and controversy ensued.[5)]

19.4. Controversy about the Existence of the AB Effect

Discussion of the physical reality of potentials dates back to the days of Faraday and Maxwell.[35)] Faraday look into consideration the law of electromagnetic induction he had discovered and surmised that electric and magnetic fields, $\boldsymbol{E}$ and $\boldsymbol{B}$, are not independent quantities but that there must exist a more fundamental quantity beyond them, which he called the "electrotonic state".

However, Faraday never found out what it was. Maxwell followed up this idea, and, in 1856, he found that Faraday's electrotonic state could be described using vector potentials as (in today's notation)[36)]

$$\boldsymbol{B} = \mathrm{rot}\,\boldsymbol{A}\,, \tag{19.4}$$

$$\boldsymbol{E} = -\frac{\mathrm{d}\boldsymbol{A}}{\mathrm{d}t}\,. \tag{19.5}$$

These equations tell us that both magnetic field $\boldsymbol{B}$ and electric field $\boldsymbol{E}$ can be obtained from vector potential $\boldsymbol{A}$. A magnetic field is produced when the spatial distribution of $\boldsymbol{A}$ has a rotation, or a vortex. An electric field is produced when $\boldsymbol{A}$ changes with time. Maxwell believed vector potential $\boldsymbol{A}$ to be the most fundamental quantity in electromagnetism and, in fact, called $\boldsymbol{A}$ "electromagnetic momentum".

However, Heaviside[37)] and Hertz,[38)] when they reformulated Maxwell's equations, discarded vector potential $\boldsymbol{A}$. Since then, $\boldsymbol{A}$ has been regarded as a mathematical quantity with no physical meaning and convenient only for calculation.

Vector potentials began to enter the central stage of physics again when a gauge theory was introduced by Weyl[39)] to unify gravity and electromagnetism although his theory was soon rejected by Einstein as producing an unrealistic result. Weyl assumed that his gauge fields, or vector potentials, changed the scale of space-time, but, in a new gauge theory established after the advent of quantum mechanics, it became evident that vector potentials change the phase of electrons, S. The unrealistic result noted by Einstein turned out to be physically possible in this new gauge theory and was in fact the AB effect.

After this gauge theory became the most probable candidate for all fundamental forces of nature, there arose a controversy[5)] over the existence of the AB effect.

We attempted to obtain conclusive evidence for it since the AB effect was also the fundamental principle behind our method of observing magnetic lines of force. We continued conducting a series of experiments on the AB effect until 1986[12-15)] to meet the repeated objections that arose about our results.

19.5. Experiments Confirming the AB Effect

We carried out a series of experiments to clarify the ambiguities raised during the controversy. Here we describe the last experiment,[14)] which is considered to be the most conclusive. We used a toroidal ferromagnet instead of a straight solenoid, which has inevitable leakage fluxes from both ends. An infinitely long solenoid is experimentally unattainable, but an ideal geometry with no flux leakage is achieved with the finite system of a toroidal magnetic field.[16)] Moreover, the toroidal ferromagnet was covered with a superconducting niobium layer to completely confine the magnetic field.

Electron waves were made incident to a tiny toroidal sample fabricated using the most advanced lithography techniques, and the relative phase shift, ΔS, between two waves passing through a hole and around the toroid was measured as an interferogram.

Although samples that had various magnetic flux values were measured, the measured ΔS was either 0 or π. The conclusion is now obvious. The photograph in Fig. 19.2 shows that a relative phase shift of π is produced, indicating the existence of the AB effect even when the magnetic field is completely confined within the superconductor and shielded from the electron wave. The waves must be physically affected by the vector potentials if we assume local interaction.

In this experiment, a quantization of the relative phase shift (0 or π) ensured that the niobium layer surrounding the magnet actually became superconductive. When a superconductor completely surrounds a magnetic flux, the flux is quantized to an integral multiple of $h/(2e)$. When an odd number of vortices are enclosed within the superconductor, the relative phase shift is π (mod. 2π). For an even number of vortices, the phase shift is 0 (mod. 2π). Therefore, the occurrence of flux quantization can be used to confirm that the niobium layer actually became superconductive, that the superconductor completely surrounded the magnetic flux, and that the Meissner effect prevented any flux leakage. Therefore, we can conclude that electron waves passing through the field-free regions inside and outside the toroidal magnet are phase-shifted by π although the waves never touch the magnetic fields.

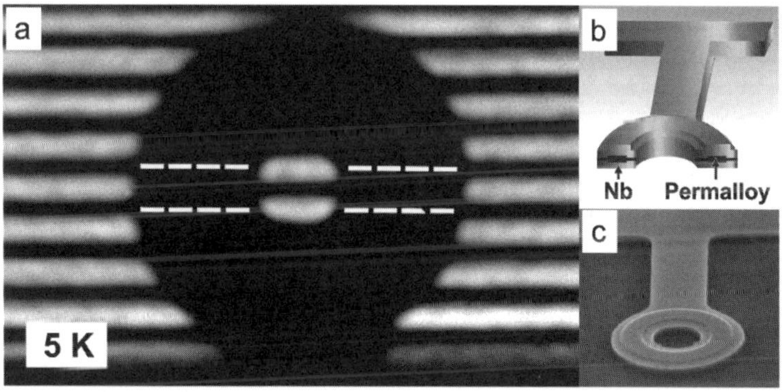

Fig. 19.2. Conclusive experiment for Aharonov-Bohm effect using a toroidal magnet covered with superconducting niobium layer. (a) Interference pattern indicating the existence of a non-zero phase difference. (b) Schematic of a toroidal sample. (c) Scanning electron micrograph of the sample.

Soon after the AB effect was conclusively proven by our experiments, electrons inside metals were also found to show the AB effect.[20] Webb of IBM used a tiny ring circuit to demonstrate that electrons inside metals also show interference and the AB effect.[40] If 100 electrons enter a ring circuit, it is a matter of course in classical physics that 100 electrons eventually exit. However, electrons now behave as waves, and even a single electron can split into two partial waves. Therefore, the number of electrons that exit the ring can become, for example, 90 or 110 because of constructive or destructive interference, depending on their relative phases. Therefore, when magnetic flux passes through the ring circuit and changes the relative electron phase between the two partial waves due to the AB effect, the electron current, or resistance, oscillates.

The AB effect was also detected in carbon nanotubes.[41)] Electrons in a cylinder can take many different paths to get from one point to another along the axis of the cylinder; a direct route or a right or left-handed path. If magnetic flux passes through a hole in the cylinder, the relative phases of the electrons change due to the AB effect, which changes the resistance. Ohm's law is no longer valid in this microscopic world, and the AB effect now plays an essential role in understanding the performance of ultra-microscopic devices. Metal carbon nanotubes can be changed into semiconductors due to the AB effect because a phase factor is added to the wavefunction, thus changing even the band structures.[42)]

As these examples show, the AB effect is becoming evident in the more macroscopic and more practical world.

19.6. Application of the AB Effect to Electromagnetic-field Observation

The AB effect principle has been used to observe microscopic distributions of electromagnetic fields by detecting the phase of a transmitted electron beam. More specifically, the thickness distribution of a specimen uniform in material can be observed as thickness contours in an interference micrograph obtained through an electron holography process[43)] because the phase of an electron wave is shifted by the inner potential of the specimen multiplied by the thickness when the wave passes through it.

Relative phase shifts can be detected from the conventional interference pattern with a $2\pi/4$ precision, but the precision increases to $2\pi/100$ if a phase-amplification technique peculiar to holography is used. In fact, this technique has been used to detect thickness changes due to monatomic steps[44)] and carbon nanotubes.[45)]

19.6.1. *Magnetic lines of force*

In the case a specimen is pure magnetic fields, the phase shift is produced by vector potentials. When the phase distribution is displayed as a contour map, the micrograph can be interpreted in the following straightforward way.[17)]

(1) Phase contours in the interference micrograph indicate magnetic lines of force, since no relative phase shift is produced between two beams passing through two points along a magnetic line.

(2) Phase contours show magnetic flux in units of *h/e*, since the relative phase shift between two beams enclosing a magnetic flux of *h*/*e* is 2π.

An example observation of magnetic lines of force inside a ferromagnetic fine particle is shown in Fig. 19.3. Only the triangular outline of this particle can be observed by electron microscopy. In its interference micrograph, two kinds of contour fringes are evident: narrow fringes parallel to the edges indicate the thickness contours in 200-Å units, and circular fringes in the inner region indicate in-plane magnetic lines of force in *h*/(2*e*) flux units since the micrograph is amplified two times and the specimen thickness is uniform there.

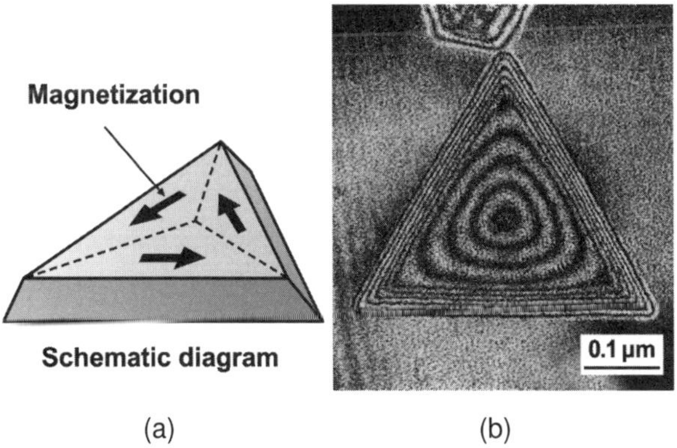

Fig. 19.3. Fine cobalt particle. (a) Schematic diagram. (b) Interference micrograph.

Another example is shown in Fig. 19.4. Magnetic lines of force radiating outward from a tiny magnetic head were observed as a holographic interference micrograph[46] (Fig. 19.4 (a)). This head is used for high-density perpendicular recording, so it must produce a very high and yet localized magnetic field. A strong magnetic field is rather simple to produce. However, this is not sufficient. We also need to be able to reproducibly control magnetic fields in any way we want. For example, in some cases we need to get a zero field outside the head, which poses some difficulty since the density of magnetic lines inside ferromagnets is a constant determined by the strength of the magnetization.

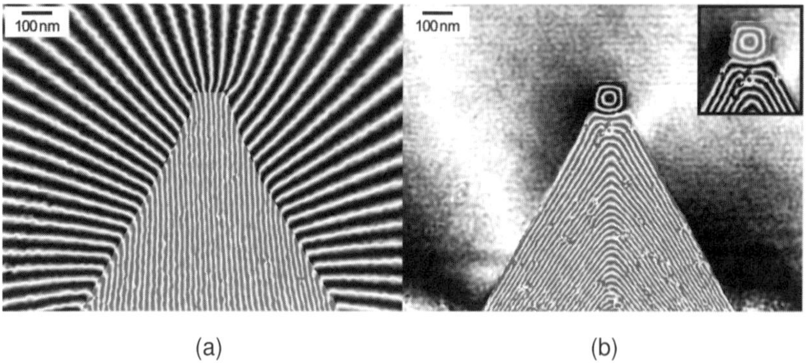

Fig. 19.4. Interference micrographs showing magnetic lines of force of a FeNi magnetic head for perpendicular recording: (a) ON; (b) OFF.
Magnetic lines extend radially from the magnetic head when it is magnetized in the upward direction by applying an external magnetic field (a). When the magnetic field is removed, magnetic lines outside the head almost completely disappear since magnetic lines inside it form a closed circuit (b).

For the FeNi head shown in Fig. 19.4, straight magnetic lines inside the head were changed to "U-turn" lines[46] (Fig. 19.4 (b)) by reducing the applied magnetic field to zero. At the pole tip, which is used to localize magnetic fields, the magnetic lines formed a tiny closed circuit.

19.6.2. *Vortices in superconductors*

Interference microscopy is not the only technique that can be used to visualize the phase distribution. For example, a phase object can be observed as an out-of-focus image because the phase change is transformed into an intensity change by defocusing the image. A quantized vortex in a superconductor, which acts as a pure weak phase object, has been visualized as a spot in a defocused image, or a Lorentz micrograph.[23)]

The experimental arrangement we used for observing vortices in a superconducting thin film is shown in Fig. 19.5. When a magnetic field is applied to the film, vortices are produced perpendicular to the film plane. Electrons passing through the tilted film are phase-shifted by the magnetic fluxes of the vortices due to the AB effect. The vortices can be observed by simply defocusing the electron microscopic image. That is, when the intensity of the electrons is observed in an out-of-focus plane, a vortex appears as a pair of bright and dark contrast features (see the simulated image in Fig. 19.5).

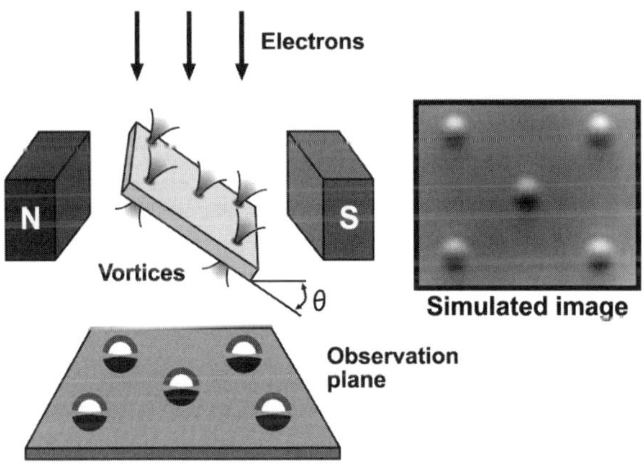

Fig. 19.5. Experimental arrangement for observing vortices by Lorentz microscopy.

Therefore, by applying Lorentz microscopy we can observe in real time the dynamics of vortices such as their behaviors at pinning centers and surface steps under various conditions of sample temperatures and applied magnetic fields. In fact, vortices move in interesting ways as if they were living organisms.

An interesting example[47)] is shown in Fig. 19.6, where two kinds of vortex images with opposite contrast appear in a single field of view. They are vortices and antivortices produced in niobium thin film when the

100-G magnetic field applied to the film was suddenly reversed and its magnitude was increased. The original vortices try to leave the film but cannot instantly do so since they are pinned by defects. While the antivortices whose magnetic fluxes are oppositely oriented begin to penetrate the film from its edges. Where a stream of vortices collides head-on with a stream of antivortices, the vortex-antivortex pairs of the heads of the two streams annihilate each other. The direct observation of this pair annihilation is analogous to that of particles and antiparticles since a vortex in a superconductor cannot be divided in two and can thus be regarded as an elementary particle.

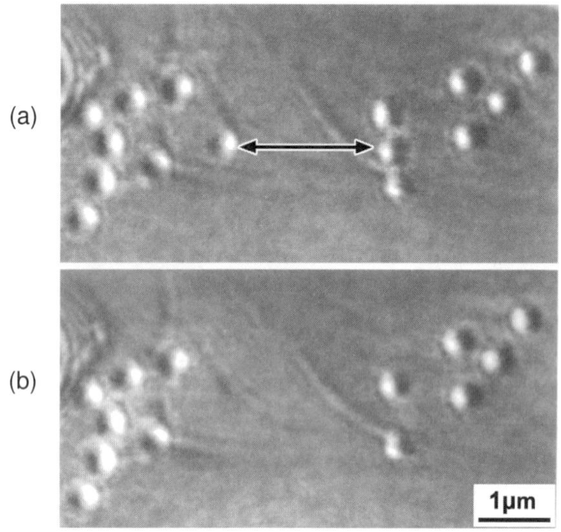

Fig. 19.6. Annihilation of vortices and antivortices in niobium thin film. (a) Before annihilation. (b) After annihilation.

Vortices show interesting behavior when artificial pinning centers are produced with a focused Ga-ion beam. When the centers form regular lattices, macroscopic measurement indicates that the critical current has peaks at specific values of the magnetic field. We microscopically investigated this matching effect by directly observing the behavior of individual vortices.[48)]

A regular array of artificial defects was produced in niobium thin film, and the vortex configuration was investigated by Lorentz microscopy under specific magnetic fields. Some of the micrographs obtained are shown in Fig. 19.7, where small black dots indicate the locations of defects, and the larger spots indicate vortex images. The vortices form regular and rigid lattices under the matching magnetic field ($H = H_1$), four times the matching field ($H = 4\ H_1$), and even 1/4th the matching field ($H = 1/4H_1$).

The reason the critical current has peaks at these specific fields is now evident; even if a vortex is depinned from a pinning site, it cannot find a stable "vacant site" to move to because the vortices form regular and rigid lattices. As a result, stronger forces are needed to move them, thus producing the matching effect of the critical current.

19.6.3. *Pinning of vortices by columnar defects*

High-T_c superconductors have been expected to become practical, but their critical current is, in general, very low because both the high temperature operation and the layered structure of the materials enable the vortices to easily move. To directly observe the unconventional behavior of vortices in high-T_c superconductors, we developed a 1-MV field-emission electron microscope[33] (Fig. 19.8). We need 1-MV electrons to observe the vortices because the electrons must penetrate a film thicker than the magnetic radius (penetration depth) of vortices in high-T_c superconductors. With this microscope, we first observed the internal behavior of vortices inside high-T_c Bi-2212 thin films.[24]

The columnar defects, which are produced by the irradiation of high-energy heavy ions and are considered optimal pinning traps for vortices in layered structure materials, are produced in Bi-2212 films in a tilted direction (see the electron micrograph shown in Fig. 19.9(a)). The tilted columns can be seen as tiny lines. When these images are defocused, the lines become blurred, spread out, and eventually disappear. However, when the images are defocused even further, vortex images appear since they are produced by the phase contrast. The resultant Lorentz micrograph of vortices is shown in Fig. 19.9(b). Some vortices are trapped at columnar defects and others are untrapped. The elongated images indicated by the

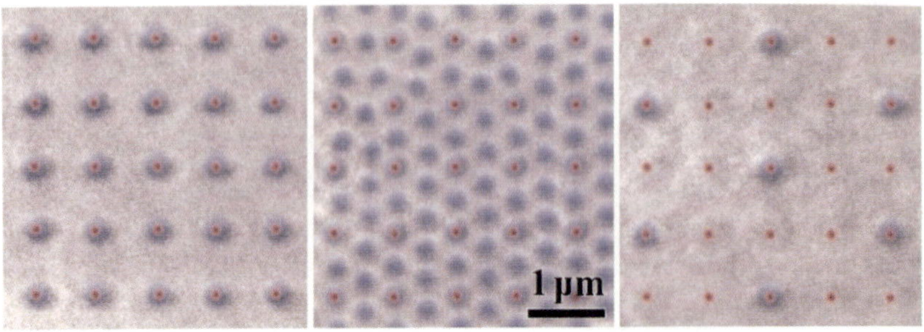

(a) (b) (c)

Fig. 19.7. Lorentz micrographs of vortices in niobium thin film with regular arrays of artificial defects. (a) $H = H_1$ (H_1: matching magnetic field). (b) $H = 4H_1$. (c) $H = 1/4H_1$. "Red" points indicate the locations of artificial defects, and "purple" dots indicate vortex images.

Fig. 19.8. 1-MV field-emission transmission electron microscope.

arrows in the micrograph (b) are produced at the locations of the columnar defects, and it can be seen by comparing (a) and (b) that they correspond to vortices trapped along the tilted columns.

We confirmed this phenomenon by simulation. The images of the untrapped vortex lines perpendicular to the film plane are circular spots having bright and dark regions. Vortex images located at the positions of the columnar defects are elongated spots with lower contrasts since these vortex lines are trapped at columnar defects tilted 70°. The circular images are produced in regions without defects (Fig. 19.9 (b)), so they must correspond to vortices perpendicularly penetrating the film.

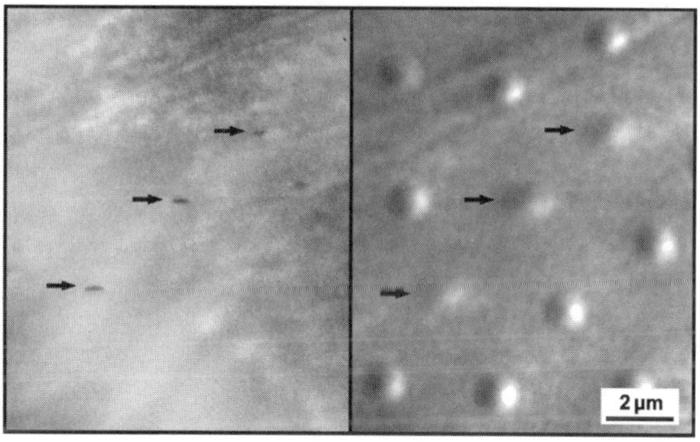

(a) (b)

Fig. 19.9. Comparison of columnar-defect images and vortex images in Bi-2212 thin film with tilted columnar defects. (a) Electron micrograph. (b) Lorentz micrograph. The vortex images indicated by the arrows correspond to vortices trapped at the defects.

An example of these vortices in a wider field of view is shown in Fig. 19.10. The vortex images indicated by arrows correspond to vortices trapped by the columnar defects. When a driving force is applied, the difference in their pinning forces becomes evident. Untrapped vortices soon begin to move, but trapped vortices do not. We can investigate the behavior of trapped and untrapped vortices by using these different vortex images under various conditions.[23]

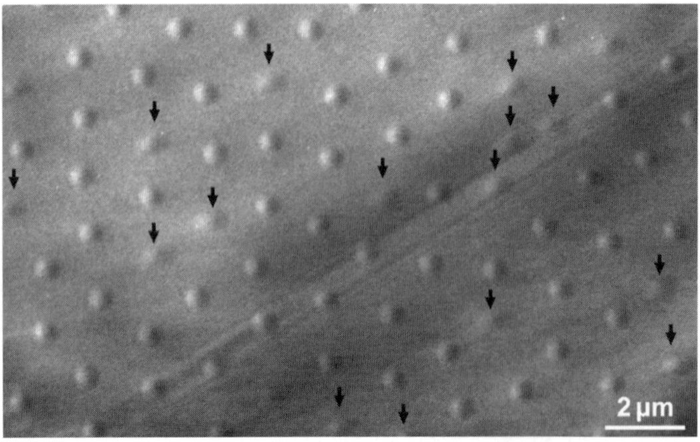

Fig. 19.10. Lorentz micrograph of vortices trapped and untrapped along tilted columnar defects in Bi-2212. Images indicated by the arrows are trapped.

19.6.4. *Unusual arrangements of vortices in high-T_c superconductors*

Vortices usually form a closely packed triangular lattice. This is the case even for anisotropic high-T_c superconductors, as long as the magnetic field is directed along the anisotropy c-axis.

When the magnetic field is strongly tilted away from the *c*-axis, however, Bitter images show that the vortices no longer form a triangular lattice. Instead, in the case of $YBaCu_3O_{7.8}$ (YBCO), they form arrays of linear chains in the direction of the tilted field[49] (Fig. 19.11(a)). For Bi-2212,[50] they form alternating domains of chains and triangular lattices (Fig. 19.11(b)). While the chain state for YBCO was explained by the tilting of the vortex lines within the framework of the anisotropic London theory,[51] the chain-lattice state for Bi-2212 has long been a topic of discussion. For example, it has been attributed to two sets of vortex lines with different orientations,[52] one set forming chains and the other forming triangular lattices.

Koshelev[53] proposed an interesting model for the chain-lattice state: elliptical Josephson vortices penetrate the layer planes in the Bi-2212, and the vortices that perpendicularly intersect the Josephson vortices form chains with the remaining vortices, forming triangular lattices.

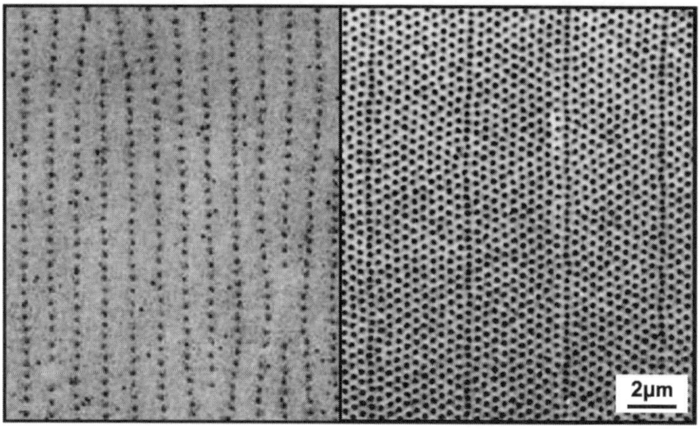

(a) (b)

Fig. 19.11. Bitter patterns of vortices in tilted magnetic fields. (a) YBCO. (b) Bi-2212.

However, no direct evidence for such mechanisms was experimentally produced because of the lack of methods for observing the arrangements of vortex lines inside superconductors. In addition, this model used to be thought to be difficult to accept because no interaction takes place between the two perpendicular magnetic fields. However, Koshelev considered the second-order approximation and determined an energy reduction in this vortex arrangement. As a result of the interaction between the two kinds of vortices, a vertical vortex line winds slightly in the opposite directions above and below the crossing Josephson vortex since the circulating supercurrent around the Josephson vortex exerts Lorentz forces onto the vertical vortex line.

We used Lorentz microscopy with our 1-MV electron microscope to determine whether or not vortex lines in the chain states inside high-T_c superconductors are tilted. For YBCO, we found that vortices tilted together in the direction the applied magnetic field tilted.[31)] This is evident from the obtained Lorentz micrographs (Fig. 19.12), in which the vortex images become more elongated and form linear chains as the tilting angle of the magnetic field increases. When the tilt angle exceeds 75°, the vortex images begin to elongate and, at the same time, form arrays of linear chains. This implies that vortices in YBCO are produced by an attractive force between the tilted vortex lines and form chains.

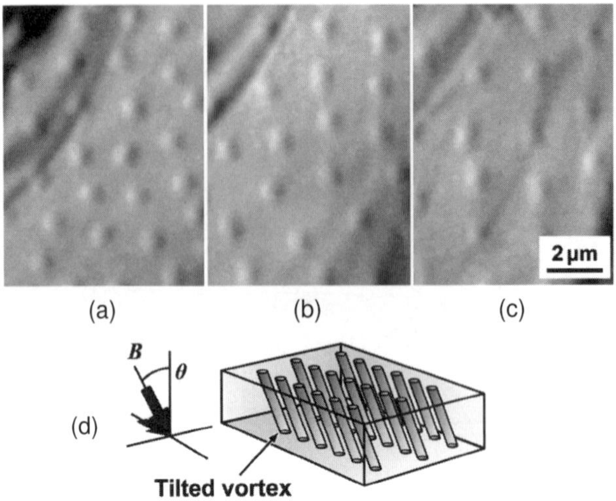

Fig. 19.12. Lorentz micrographs of vortices in YBCO film in tilted magnetic fields. (temperature T = 30 K; perpendicular magnetic field B_p = 3 G). (a) θ = 75°. (b) θ = 82°. (c) θ = 83°. (d) Schematic of tilted vortex lines.
When tilt angle θ is larger than 75°, the vortex images begin to elongate and, at the same time, to form arrays of linear chains. This implies that chain vortices in YBCO are produced by attractive forces between tilted vortex lines.

For Bi-2212, Lorentz microscopy observation under various defocusing conditions showed that neither chain nor lattice vortices tilted but both stood perpendicular to the layer plane,[32)] as shown in Fig. 19.13(a). If vortex lines are strongly tilted at an angle comparable to that of the applied magnetic field, the vortex images should be elongated, as shown in the inset of Fig. 19.13(a).

Our finding that both chain and lattice vortices stand perpendicular and do not tilt is clear evidence of the Koshelev mechanism although the clearest evidence for this model would of course be if the Josephson vortices were directly observed along with the chain vortices. However, the magnetic field of a Josephson vortex extends widely between the layers, making it difficult to detect with our method. In fact, we could not directly observe Josephson vortices, but we found evidence for their existence: we observed that vertical vortices always penetrate the sample along straight lines. These straight lines must be determined by the Josephson vortices.

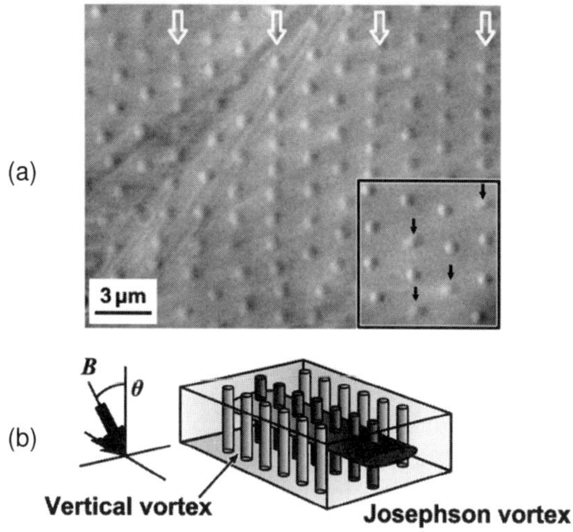

Fig. 19.13. Chain-lattice state of vortices in tilted magnetic field in Bi-2212 thin film. (a) Lorentz micrograph ($T = 50$ K, $B = 50$ G, $\theta = 80°$). (b) Schematic.
The chains are indicated by white arrows in (a). If vortex lines are tilted, the vortex images must be elongated and weak in contrast. Since the images of chain vortices, as well as those of lattices vortices, are not elongated but circular, all the vortex lines are not tilted but perpendicular to the layer plane. If tilted, the vortex line images become elongated, as shown in the inset in (a). A Josephson vortex, illustrated in (b), is not evident in (a).

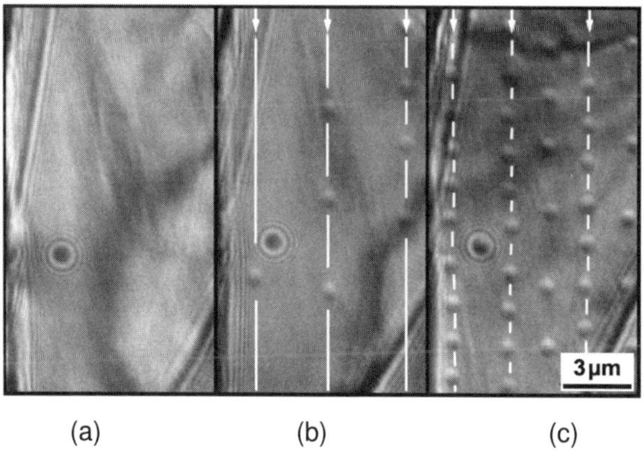

Fig. 19.14. Series of Lorentz micrographs of vortices in Bi-2212 thin film when magnetic field B_p perpendicular to layer plane begins to be applied and increases with fixed in-plane magnetic field of 50 G at $T = 50$ K. (a) $B_p = 0$ G. (b) $B_p = 0.2$ G. (c) $B_p = 1$ G.
Vortices begin to penetrate the film along straight lines (white lines in the micrographs), which are determined by Josephson vortices.

An example of this is shown in Fig. 19.14. When we apply an in-plane magnetic field, no vortex images can be seen in the Lorentz micrograph (Fig. 19.14(a)). However, when perpendicular magnetic field B_p is additionally applied and slowly increased, images of vertical vortices begin to appear in this field of view. They do not appear in the triangular-lattice form but only along straight chains, indicated by white lines in Fig. 19.14(b), which are considered to be determined by Josephson vortices. Vortices are always located along straight chains, even in case of large intervals between the vortices. Therefore, we can find no other reason for the production of chain vortices than the assumption that vertical vortices crossing Josephson vortices form chains, as illustrated in Fig. 19.13(b). When $B_p > 1$ G, vertical vortices also appear between the chain vortices, as shown in Fig. 19.14(c).

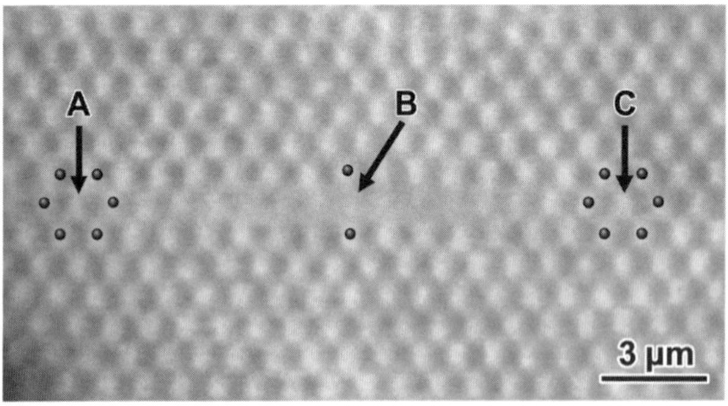

Fig. 19.15. Lorentz micrograph of disappearing chain vortices in Bi-2212 thin film ($T = 57$ K, $B_p = 8$ G, $\theta = 80°$).
The partial disappearance of the vortex images along the chain is attributed to the oscillation of vortices since chain vortices are incommensurate with the surrounding triangular lattice vortices. Vortex image B disappears but vortex images A and C can be observed since they are surrounded by six vortices.

We also found that only the images of chain vortices in Bi-2212 begin to disappear at temperatures much lower than critical temperature T_c.[31] An example is shown in Fig. 19.15. The chain of vortices in this case began to disappear at 50 K, *i.e.*, well below T_c (=83 K). This does not mean that the vortices themselves disappear. This is because the vortex images gradually fade away with increasing temperature. Moreover, the vortex images, even along a single chain, partially disappear depending on their positions relative to the surrounding lattice vortices. That is, in Fig. 19.15, vortices A and C can be clearly seen in the chain, but vortices that are far from them begin to disappear. Vortices A and C are located "stably" in the midst of six surrounding vortices, indicated by the points in the figure, while vortex B is sandwiched "unstably" between two vortices above and below.

This vortex arrangement may be stable at low temperatures, but, at high temperatures where vortices vibrate thermally, vortex B begins to oscillate back and forth, like a pinball connected by springs to incommensurate periodic potentials, as in the Frenkel-Kontorova model. We attribute the disappearance of chain vortices to such longitudinal oscillations of vortices along chains.

19.7. Conclusion

Thanks to recent advances in technology such as the development of coherent electron beams, highly sensitive electron detectors, and photolithography, experiments once regarded as "Gedanken" experiments can now be carried out. In addition, the wave nature of electrons can be utilized to observe microscopic objects previously unobservable. Examples are the quantitative observation of both the microscopic distribution of magnetic lines of force in *h*/*e* units by interference microscopy and the dynamics of quantized vortices in superconductors by Lorentz microscopy. This measurement and observation technique is expected to play an even more important role in future research and development in nano-science and technology.

References

1. Y. Aharonov and D. Bohm, *Phys. Rev.* **115**, 485 (1959).
2. W. Ehrenberg and R. W. Siday, *Proc. Phys. Soc. London* **B62**, 8 (1949).
3. T. T. Wu and C. N. Yang, *Phys. Rev.* **D12**, 3845 (1975).
4. P. Bocchieri and A. Loinger, *Nuovo Cimento* **47A**, 475 (1978).
5. M. Peshkin and A. Tonomura, *Lecture Notes in Physics* **340**, Springer, Heidelberg (1989).
6. R. G. Chambers, *Phys. Rev. Lett.* **5**, 3 (1960).
7. H. A Fowler, L. Marton, J. A. Simpson and J. A. Suddeth, *J. Appl. Phys.* **32**, 1153 (1961).

8. G. Möllenstedt and W. Bayh, *Phys. Bl.* **18**, 299 (1962).
9. H. Boersch, H. Hamisch and K. Grohmann, *Z. Phys.* **169**, 263 (1962).
10. S. M. Roy, *Phys. Rev. Lett.* **44**, 111 (1980).
11. P. Bocchieri and A. Loinger, *Lett. Nuovo. Cimento* **30**, 449 (1981).
12. A. Tonomura, T. Matsuda, R. Suzuki, A. Fukuhara, N. Osakabe, H. Umezaki, J. Endo, K. Shinagawa, Y. Sugita and H. Fujiwara, *Phys. Rev. Lett.* **48**, 1443 (1982).
13. A. Tonomura, H. Umezaki, T. Matsuda, N. Osakabe, J. Endo and Y. Sugita, *Phys. Rev. Lett.* **51**, 331 (1983).
14. A. Tonomura, N. Osakabe, T. Matsuda, T. Kawasaki, J. Endo, S. Yano and H. Yamada, *Phys. Rev. Lett.* **56**, 792 (1986).
15. N. Osakabe, T. Matsuda, T. Kawasaki, J. Endo, A. Tonomura, S. Yano and H. Yamada, *Phys. Rev.* **A34**, 815 (1986).
16. C. G. Kuper, *Phys. Lett.* **794**, 413 (1980).
17. A. Tonomura, T. Matsuda, J. Endo, T. Arii and K. Mihama, *Phys. Rev. Lett.* **44**, 1430 (1980).
18. A. Tonomura, *Electron Holography -2nd Edition*, Springer, Heidelberg (1999).
19. C. N. Yang, In Proc. Int. Symp. on Foundations of Quantum Mechanics, Tokyo, 1983, (ed. S. Kamefuchi *et al.*) Phys. Society of Japan, Tokyo, 27 (1984).
20. Y. Imry and R. A. Webb, *Scientific American* **260**, 56 (1989)
21. A. Tonomura, T. Matsuda, J. Endo, T. Arii and K. Mihama, *Phys. Rev.* **B34**, 3397 (1986).
22. J. E. Bonevich, K. Harada, T. Matsuda, H. Kasai, T. Yoshida, G. Pozzi and A. Tonomura, *Phys. Rev. Lett.* **70**, 2952 (1993).
23. K. Harada, T. Matsuda, J. Bonevich, M. Igarashi, S. Kondo, G. Pozzi, U. Kawabe and A. Tonomura, *Nature* **360**, 51 (1992).
24. A. Tonomura, H. Kasai, O. Kamimura, T. Matsuda, K. Harada, Y. Nakayama, J. Shimoyama, K. Kishio, T. Hanaguri, K. Kitazawa, M. Sasase and S. Okayasu, *Nature* **412**, 620 (2001).
25. A. Tonomura, A. Fukuhara, H. Watanabe and T. Komoda, *Jpn. J. Appl. Phys.* **7**, 295 (1968).
26. D. Gabor, *Proc. R. Soc. London, Ser.* A 197454 (1949).
27. A. V. Crewe, J. Wall and J. Langmore, *Science* **168**, 1338 (1970).
28. A. Tonomura, T. Matsuda, J. Endo, H. Todokoro and T. Komoda, *J. Electron Microsc.* **28**, 1 (1979).
29. K. Harada, T. Matsuda, J. Bonevich, M. Igarashi, S. Kondo, G. Pozzi, U. Kawabe and A. Tonomura, *Nature* **360**, 51 (5 November 1992).
30. T. Kawasaki, T. Matsuda, J. Endo and A. Tonomura, *Jpn. J. Appl. Phys.* **29**, L508 (1990).
31. T. Matsuda, O. Kamimura, H. Kasai, K. Harada, T. Yoshida, T. Akashi, A. Tonomura, Y. Nakayama, J. Shimoyama, K. Kishio, T. Hanaguri and K. Kitazawa, *Science* **294**, 2136 (7 December 2001).
32. A. Tonomura, K. Kasai, O. Kamimura, T. Matsuda, K. Harada, T. Yoshida, T. Akashi, J. Shimoyama, K. Kishio, T. Hanaguri, K. Kitazawa, T. Masui, S. Tajima, N. Koshizuka, P. L. Gammel, D. Bishop, M. Sasase and S. Okayasu, *Phys. Rev. Lett.* **88**, 237001 (2002).

33. T. Kawasaki, T. Yoshida, T. Matsuda, N. Osakabe, A. Tonomura, I. Matsui and K. Kitazawa, *Appl. Phys. Lett.* **76**, 1342 (2000).
34. T. Akashi, K. Harada, T. Matsuda, H. Kasai, A. Tonomura, T. Furutsu, N. Moriya, T. Yoshida, T. Kawasaki, K. Kitazawa and H. Koinuma, *Appl. Phys. Lett.* **81**, 1922 (2002).
35. C. N. Yang, In Quantum Coherence and Decoherence (ed. Fujikawa K., and Ono Y. A.) North-Holland Delta Series, ELSEVIER, Amsterdam, 307 (1996).
36. J. C. Maxwell, *Trans. Camb. Phil.* Soc. **10**, 27(1856).
37. For example, see O. Heaviside (1893-1912) Electromagnetic Theory. 3 vols. London: Electrician Co. Repr. New York: Chelsea, 1971.
38. H. R. Hertz, *Wiedemann's Annalen* **23**, 84 (1884).
39. H. Weyl, *Ann. der Physik* **59**, 101 (1919).
40. R. A. Webb, S. Washburn, C. P. Umbach and R. B. Laibowitz, *Phys. Rev. Lett.* **54**, 2696 (1985).
41. A. Bachtold, C. Strunk, J. P. Salvetat, J. M. Bonard, L. Forro, T. Nussbaumer and C. Schönenberger, *Nature* **397**, 673 (1999).
42. E. D. Minot, Y. Yalsh, V. Sazonova and P. L. McEuen, *Nature* **428**, 536 (2004).
43. A. Tonomura, J. Endo and T. Matsuda, *Optik* **53**, 143 (1979).
44. A. Tonomura, T. Matsuda, T. Kawasaki, J. Endo and N. Osakabe, *Phys. Rev. Lett.* **54**, 60 (1985).
45. Q. Ru, J. Endo, T. Tanji and A. Tonomura, *Appl. Phys. Lett.* **59**, 2372 (1991).
46. J. J. Kim, K. Hirata, Y. Ishida, D. Shindo, M. Takahashi and A. Tonomura, *Appl. Phys. Lett.* **92**, 162501 (2008).
47. K. Harada, H. Kasai, T. Matsuda, M. Yamasaki and A. Tonomura, *J. Electron Microsc.* **46**, 3227 (1997).
48. K. Harada, O. Kamimura, H. Kasai, T. Mastuda and A. Tonomura, *Science* **274**, 1167 (1996).
49. C. A. Bolle, P. L. Gammel, D. G. Grier, C. A. Murray, D. J. Bishop, D. B. Mitzi and A. Kapitulnik, *Phys. Rev. Lett.* **66**, 112 (1991).
50. P. L. Gammel, D.J. Bishop, J. P. Rice and D.M. Ginsberg, *Phys. Rev. Lett.* **68**, 3343 (1992).
51. A. I. Buzdin and A. Y. Simonov, *JETP Lett.* **51**, 191 (1990).
52. I. V. Grigorieva and J. W. Steeds, *Phys. Rev.* **B51**, 3765 (1995).
53. A. E. Koshelev, *Phys. Rev. Lett.* **83**, 187 (1999).

Chapter 20

Dimensional Crossover of the Dephasing Time in Disordered Mesoscopic Rings: From Diffusive through Ergodic to $0D$ Behavior

M. Treiber, O. M. Yevtushenko, F. Marquardt and J. von Delft

Arnold Sommerfeld Center and Center for Nano-Science, Ludwig Maximilians University, Munich, D-80333, Germany

I. V. Lerner

School of Physics and Astronomy, University of Birmingham, Birmingham, B15 2TT, UK

We analyze dephasing by electron interactions in a small disordered quasi-one dimensional ($1D$) ring weakly coupled to leads, where we recently predicted a crossover for the dephasing time $\tau_\varphi(T)$ from diffusive or ergodic $1D$ ($\tau_\varphi^{-1} \propto T^{2/3}, T^1$) to $0D$ behavior ($\tau_\varphi^{-1} \propto T^2$) as T drops below the Thouless energy E_{Th}.[1] We provide a detailed derivation of our results, based on an influence functional for quantum Nyquist noise, and calculate all leading and subleading terms of the dephasing time in the three regimes. Explicitly taking into account the Pauli blocking of the Fermi sea in the metal allows us to describe the $0D$ regime on equal footing as the others. The crossover to $0D$, predicted by Sivan, Imry and Aronov for $3D$ systems,[2] has so far eluded experimental observation. We will show that for $T \ll E_{\mathrm{Th}}$, $0D$ dephasing governs not only the T-dependence for the smooth part of the magnetoconductivity but also for the amplitude of the Altshuler-Aronov-Spivak oscillations, which result only from electron paths winding around the ring. This observation can be exploited to filter out and eliminate contributions to dephasing from trajectories which do not wind around the ring, which may tend to mask the T^2 behavior. Thus, the ring geometry holds promise of finally observing the crossover to $0D$ experimentally.

20.1. Introduction

Over the last twenty-five years many theoretical and experimental works addressed quantum phenomena in mesoscopic disordered metallic rings.[3]

This subject was launched in part by several seminal papers by Joe Imry and his collaborators [4–11], and continues to be of great current interest. One intensively-studied topic involves persistent currents, which can flow without dissipation due to quantum interference in rings prepared from normal metals.[4,12–15] Attention was also paid to Aharonov-Bohm oscillations in the conductance through a mesoscopic ring attached to two leads,[5–7,10,16] and the closely related oscillations of the negative weak localization (WL) correction to the magnetoconductivity.[17,18] These oscillations result from the interference of closed trajectories which have a non-zero winding number acquiring the Aharonov-Bohm phase. Both persistent currents and magnetooscillations require the ring to be phase coherent, since any uncertainty of the quantum phase due to the environment or interactions immediately suppresses all interference phenomena.[19]

The mechanism of dephasing in electronic transport and its dependence on temperature T in disordered conductors was studied in numerous theoretical[2,19–27] and experimental[28–35] works. The characteristic time scale of dephasing is called the dephasing time τ_φ. At low temperatures phonons are frozen out and dephasing is mainly due to electron interactions, with the dephasing time $\tau_\varphi(T)$ increasing as T^{-a} when $T \to 0\,,\ a > 0$.

The scaling of the dephasing time with temperature depends on the dimensionality of the sample.[20] It was predicted in a pioneering paper by Sivan, Imry and Aronov[2] that the dephasing time in a disordered quantum dot shows a dimensional crossover from $\tau_\varphi \propto T^{-1}$, typical for a $2D$ electron gas,[20] to $\tau_\varphi \propto T^{-2}$ when the temperature is lowered into the $0D$ regime:

$$\hbar/\tau_\varphi \ll T \ll \hbar/\tau_{\mathrm{Th}}\,, \tag{20.1}$$

where $\tau_{\mathrm{Th}} = \hbar/E_{\mathrm{Th}}$ is the Thouless time, i.e. the time required for an electron to cross (diffusively or ballistically) the mesoscopic sample; E_{Th} is the Thouless energy. In this low-T, 0D regime, the coherence length and the thermal length are both larger than the system size, independent of geometry and real dimensionality of the sample. In this regime WL is practically the only tool to measure the T-dependence of dephasing in mesoscopic wires or quantum dots (the mesoscopic conductance fluctuations go over to a universal value of order e^2/h for $T \ll E_{\mathrm{Th}}$[3]).

Although the $\tau_\varphi \propto T^{-2}$ behavior is quite generic, arising from the fermionic statistics of conduction electrons, experimental efforts[31–33] to observe it have so far been unsuccessful. The reasons for this are today still unclear. Conceivably dephasing mechanisms other than electron interactions were dominant, or the regime of validity of the $0D$ description had not been reached. In any case, other ways of testing the dimensional crossover for τ_φ are desirable.

In a recent paper,[1] we described the crossover of the dephasing time to the $0D$ regime in a mesoscopic ring weakly coupled to leads. We considered a ring of the type shown in Fig. 20.1 with dimensionless $1D$ conductance

$$g_1 = \frac{h\sigma_0}{e^2}\frac{A}{L}, \tag{20.2}$$

where A and L are the ring's cross section and circumference and σ_0 is its classical Drude conductivity. In the present paper we give a detailed derivation of our results based on an influence functional approach for *quantum noise*. This approach explicitly takes into account the Pauli blocking of the electrons in the metal, which will allow us to describe quantitatively all regimes of the dephasing time in a quasi-$1D$ ring on an equal footing and to calculate first order correction terms to the dephasing time. In particular, we will see that Pauli blocking dominates the regime of $0D$ dephasing. We find that in the $0D$ regime, T^{-2} behavior also emerges for the amplitude of the Altshuler-Aronov-Spivak (AAS) oscillations of the conductivity[17] in a magnetic field, which arise from pairs of time-reversed paths encircling the ring at least once. A necessary requirement to reach this regime is that electron trajectories are effectively confined in the system. Thus the conductance through the contact, $g_{\rm cont}$, is assumed to be much smaller than g_1, such that the time an electron spends inside the ring, the dwelling time $\tau_{\rm dw}$, is much larger than the time an electron needs to explore the whole ring, i.e. the Thouless time $\tau_{\rm Th}$.

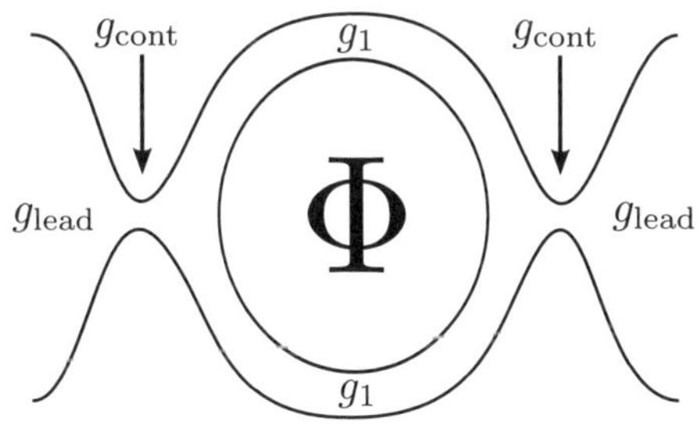

Fig. 20.1. A ring weakly coupled to leads: We assume a metallic system, where the conductance at the contacts $g_{\rm cont}$ is much smaller than the conductance of the ring g_1 and of the lead $g_{\rm lead}$, i.e. $(g_{\rm lead}, g_1) \gg g_{\rm cont} \gg 1$. This assures (a) that the average time electrons spend in the ring ($\tau_{\rm dw}$) is much larger than the average time they need to explore the whole ring ($\tau_{\rm Th}$) and (b) that the probability for electrons which escaped from the ring to return back to it is small.

We will show below that after subtracting from the amplitude of the AAS-oscillations the non-oscillating background, only contributions to dephasing from paths encircling the ring will contribute. However, some of these paths may involve loops which not only encircle the ring, but along the way also enter the lead and reenter the ring (see Fig. 20.6(b) below). Such lead-ring cross-contributions to dephasing will contribute a non-$0D$ T-dependence to the conductance and hence tend to mask the $0D$ behavior. We shall argue that by additionally choosing the conductance of the connected leads, $g_{\rm lead}$, to be larger than $g_{\rm cont}$, dephasing due to lead-ring cross-contributions, can be neglected, and the remaining contributions will be characterized by $0D$ dephasing.

20.2. Dephasing and Weak Localization

In a disordered metal, the conductivity is reduced by coherent backscattering of the electrons from impurities, an effect known as weak localization (WL). In a semi-classical picture it can be understood as the constructive interference of closed, time-reversed random-walks through the metal's impurity landscape. It is most pronounced in systems of low dimensionality d where the integrated return probability becomes large for long times. For an infinite system characterized by the diffusion constant $D = v_F l/d$ (v_F is the Fermi velocity and l is the mean free path), the probability of a random walk of duration t to return back to its origin is given by

$$\mathcal{C}_0(t) = (4\pi D t)^{-d/2} . \tag{20.3}$$

To leading order in $1/g_1$, the relative correction to the conductance (20.2) can be written as

$$\Delta g = \frac{\Delta\sigma}{\sigma_0} = -\frac{1}{\pi\nu}\int_0^\infty \mathrm{d}t\, \mathcal{C}(t) , \tag{20.4}$$

where ν is the density of states per volume in the ring and we have set $\hbar = 1$ henceforth. The function $\mathcal{C}(t)$ is the so called Cooperon propagator corresponding to the interference amplitude of the time-reversed random walks. $\mathcal{C}(t)$ reduces to Eq. (20.3) if time-reversal symmetry is fully preserved. Processes which destroy this symmetry lead to a suppression of this contribution at long times, since the random walks and their time-reversed counterparts acquire a different phase. The model we are considering assumes a suppression of the Cooperon of the following form

$$\mathcal{C}(t) \equiv \mathcal{C}_0(t) \exp\left[-t/\tau_H - t/\tau_{\rm dw} - \mathcal{F}(t)\right] . \tag{20.5}$$

In Eq. (20.5), we consider dephasing due to the effect of an external magnetic field leading to the cutoff $\tau_H \sim 1/H$ of the integral in Eq. (20.4).[43] Furthermore, our model of an almost isolated ring assumes an average dwelling time, τ_{dw}, of the electrons in the ring.[45]

Our primary interest is the effect of electron interactions, which we describe in terms of the *Cooperon decay function* $\mathcal{F}(t)$, which grows with time and may be used to define a dephasing time via

$$\mathcal{F}(\tau_\varphi) = 1\,. \tag{20.6}$$

Dephasing due to electron interactions can be understood roughly as follows: At finite temperatures the interactions lead to thermal fluctuations (noise) of the electron's potential energy $V(\mathbf{x}, t)$. Then the closed paths contributing to WL and their time-reversed counterparts effectively "see" a different local potential, leading to a phase difference. This is most clearly seen in a path integral representation of the Cooperon in a time-dependent potential,[20] which is given by

$$\mathcal{C}(t) \propto \int_{\mathbf{x}(0)=\mathbf{x}_0}^{\mathbf{x}(t)=\mathbf{x}_0} \mathcal{D}\mathbf{x}\; e^{i\varphi(t)} e^{-\int_0^t \mathrm{d}t_1\, \mathcal{L}(t_1)}\,. \tag{20.7}$$

Here the Lagrangian $\mathcal{L}(t_1) = \dot{\mathbf{x}}^2(t_1)/4D$ describes diffusive propagation, and $\varphi(t)$ is a phase corresponding to the time-reversed structure of the Cooperon:

$$\varphi(t) = \int_0^t \mathrm{d}t_1\, \left[V(\mathbf{x}(t_1), t_1) - V(\mathbf{x}(t_1), t - t_1)\right]\,. \tag{20.8}$$

Assuming that the noise induced by electron interactions is Gaussian, the decay function $\mathcal{F}(t)$ in Eq. (20.5) can be estimated from $\mathcal{F}(t) = \frac{1}{2}\langle\overline{\varphi^2}\rangle_{\mathrm{crw}}$, where $\overline{\cdots}$ denotes averaging over realizations of the noise and $\langle\ldots\rangle_{\mathrm{crw}}$ over closed random walks of duration t from $\mathbf{x}_0$ back to $\mathbf{x}_0$. $\mathcal{F}(t)$ is then given in terms of a difference of the noise correlation functions, taken at reversed instances of time:

$$\mathcal{F}(t) = \int_0^t \mathrm{d}^2 t_{1,2} \left\langle \overline{VV}(\mathbf{x}_{12}, t_{12}) - \overline{VV}(\mathbf{x}_{12}, \bar{t}_{12}) \right\rangle_{\mathrm{crw}}\,. \tag{20.9}$$

Here $t_{12} = t_1 - t_2$ and $\bar{t}_{12} = t_1 + t_2 - t$, while $\mathbf{x}_{12} = \mathbf{x}(t_1) - \mathbf{x}(t_2)$ is the distance of two points of the closed random walk taken at times t_1 and t_2. For an infinite wire and the case of classical Nyquist noise (defined in Eq. (20.11) below), Eq. (20.9) has been shown[26,27] to give results practically equivalent to the exact results obtained in Ref. [20].

20.3. Thermal Noise due to Electron Interactions

Electron interactions in the metal lead to thermal fluctuations of the electric field $\mathbf{E}$, producing so-called Nyquist noise. In the high temperature limit, it can be obtained from the classical Fluctuation-Dissipation Theorem leading to a field-field correlation function in 3D of the form

$$\overline{\mathbf{EE}}(\mathbf{q},\omega) \xrightarrow{|\omega|\ll T} \frac{2T}{\sigma_0}. \tag{20.10}$$

Note that the fluctuations of the fields do not depend on $\mathbf{q}$ or ω, i.e. they correspond to white noise in space and time. To describe dephasing in a quasi-1D wire, we need the correlation function of the corresponding scalar potentials V in a quasi-1D wire. Since $E = \frac{1}{e}\nabla V$, the noise correlator that corresponds to the classical limit (20.10) has the form

$$\overline{VV}_{\text{class}}(\mathbf{q},\omega) = \frac{2Te^2}{\sigma_0}\frac{1}{\mathbf{q}^2}. \tag{20.11}$$

This so-called classical Nyquist noise is frequency independent, i.e. corresponds to "white noise". For present purposes, however, we need its generalization to the case of quantum noise, valid for arbitrary ratios of $|\omega|/T$. In particular, $\overline{VV}$ is expected to become frequency-dependent: it should go to zero for $|\omega| \gg T$, since the Pauli principle prevents scattering processes into final states occupied by other electrons in the Fermi sea.[38] A careful analysis of quantum noise has been given recently in Ref. [26] and Ref. [27]. The authors derived an effective correlation function for the quantum noise potentials that properly accounts for the Pauli principle. It is given by

$$\overline{VV}(\mathbf{q},\omega) = \text{Im}\mathcal{L}^R(\mathbf{q},\omega)\,\frac{\omega/2T}{\sinh\left(\omega/2T\right)^2} \tag{20.12}$$

with

$$\mathcal{L}^R(\mathbf{q},\omega) = -\frac{D\mathbf{q}^2 - i\omega}{2\nu D\mathbf{q}^2 + (D\mathbf{q}^2 - i\omega)/V(\mathbf{q})}; \tag{20.13}$$

$V(q)$ is the Fourier-transformed bare Coulomb potential (not renormalized due to diffusion) in the given effective dimensionality.

If the momentum and energy transfer which dominates dephasing is small then the second term of the denominator of Eq. (20.13) can be neglected so that Eq. (20.13) reduces to

$$\text{Im}\mathcal{L}^R(\mathbf{q},\omega) \simeq \frac{\omega}{2\nu D\mathbf{q}^2}. \tag{20.14}$$

This simplification holds true, in particular, in the high temperature (diffusive) regime where $\omega \ll T$.[20] We will argue below (see Eqs. (20.54) to (20.57)) that the same simplification can be used in the low temperature regime where $|\omega| \sim T \lesssim E_{\mathrm{Th}}$.[1]

Inserting Eq. (20.14) in Eq. (20.12) with $\sigma_0 = 2e^2\nu D/A$, where A is the cross-section perpendicular to the current direction, we obtain

$$\overline{VV}(\mathbf{q},\omega) = \frac{2e^2\,T}{\sigma_0 A}\,\frac{1}{\mathbf{q}^2}\left(\frac{\omega/2T}{\sinh(\omega/2T)}\right)^2 . \tag{20.15}$$

In the time and space domain, this correlator factorizes into a product of time- and space-dependent parts:

$$\overline{VV}(\mathbf{x},t) = \frac{2e^2\,T}{\sigma_0 A}\,Q(\mathbf{x})\,\delta_T(t)\,, \tag{20.16}$$

where $\delta_T(t)$ is a broadened delta function of width $1/T$ and height T:

$$\delta_T(t) = \pi T w(\pi T t)\,, \qquad w(y) = \frac{y\coth(y) - 1}{\sinh^2(y)}\,. \tag{20.17}$$

The fact that the noise correlator (20.16) is proportional to a *broadened* peak $\delta_T(t)$ is a direct consequence of the effects of Pauli blocking. Previous approaches often used a sharp Dirac-delta peak instead. In the frequency domain this corresponds to white noise and leads to (20.11), instead of our frequency-dependent form (20.12). Such a "classical" treatment reproduces correct results for the dephasing time when processes with small energy transfers $|\omega| \ll T$ dominate. However, it has been shown in Ref. [2] that this is in fact not the case in the $0D$ limit $T \ll E_{\mathrm{Th}}$, where the main contribution to dephasing is due to processes with $|\omega| \simeq T$. Thus, the results become dependent on the form of the cutoff that eliminates modes with $|\omega| > T$ to account for the Pauli principle. For such purposes, previous treatments typically introduced a sharp cutoff, $\theta(T - |\omega|)$, by hand. However, the precise form of the cutoff becomes important in an analysis interested not only in qualitative features, but quantitative details. The virtue of (20.11) is that it encodes the cutoff in a quantitatively reliable fashion. (For example, it was shown[26] to reproduce a result first obtained in Ref. [23], namely the subleading term in an expansion of the large-field magnetoconductance (for quasi-1D wires) in powers of the small parameter $1/\sqrt{T\tau_H}$.)

The position-dependent part of Eq. (20.16), the so-called diffuson at zero frequency $Q(\mathbf{x})$, is the time-integrated solution of the diffusion equation. In the isolated system, it satisfies

$$-\Delta Q(\mathbf{x}) = \delta(\mathbf{x})\,, \tag{20.18}$$

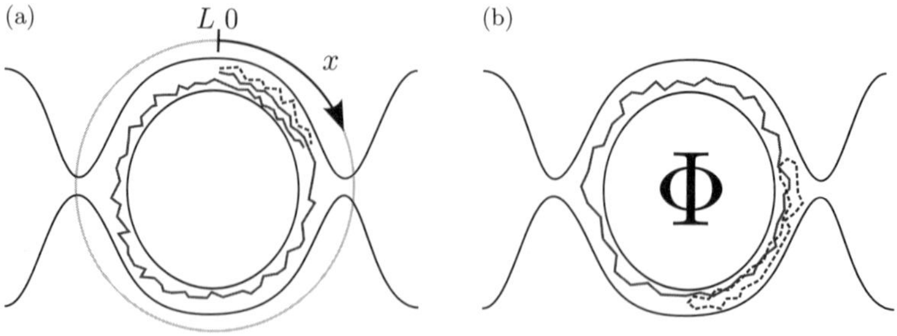

Fig. 20.2. (a) Illustration of our choice of the coordinate system: Both paths have the same start and end point ($0 \mapsto x$), but the dashed path has winding number $n = 0$ and the solid path $n = 1$. (b) Two *closed* paths in the ring contributing to the Cooperon. The contribution of the solid path (with winding number $n = 1$) is affected by the flux Φ, since the path (and it's time-reversed counterpart) acquire an Aharonov-Bohm phase when interfering with itself at their origin. This gives rise to the Altshuler-Aharonov-Spivak oscillations. The dashed path with $n = 0$ is unaffected by the flux, since the acquired phase at the origin is zero.

with given boundary conditions, which govern the distribution of the eigenmodes of Q. In an isolated system, where a $\mathbf{q} = 0$ mode is present, $Q(\mathbf{x})$ diverges. However, the decay function is still regular, since terms in Q which do not depend on $\mathbf{x}$ simply cancel out in Eq. (20.9) and cannot contribute to dephasing.

To evaluate the decay function Eq. (20.9), we note that only the factor $Q(\mathbf{x})$ in Eq. (20.16) depends on $\mathbf{x}$, thus, the average $\langle Q(\mathbf{x}) \rangle_{\mathrm{crw}}$ has to be calculated. This will be done in the next section for an almost isolated ring. Then, after a qualitative discussion in 20.5, we proceed by evaluating $\mathcal{F}(t)$ in section 20.6.

20.4. Diffusion in the almost Isolated Ring

The probability density of a random walk in a $1D$, infinite, isotropic medium to travel the distance x in time t is given by

$$P_0(x,t) = \frac{1}{\sqrt{4\pi Dt}}\, e^{-x^2/4Dt} \,. \tag{20.19}$$

In an isolated ring, electrons can reach each point without or after winding around the ring n times, where n is called *winding number*. Denoting the probability density for the latter type of path by $P_n(x,t)$, the diffusion probability density can be expanded in n as

$$P(x,t) = \sum_{n=-\infty}^{+\infty} P_n(x,t)\,, \quad P_n(x,t) = \frac{1}{\sqrt{4\pi Dt}}\, e^{-(x+nL)^2/4Dt}\,, \tag{20.20}$$

where L is the circumference of the ring and $x \in [0, L]$ is the cyclic coordinate along the ring, see Fig. 20.2(a). To model the effect of the two contacts of the ring, we assume that an electron, on average, stays inside the ring only for the duration of the dwelling time $\tau_{\rm dw}$, introduced in Eq. (20.5), and then escapes with a vanishing return probability. This simplified model of homogeneous dissipation, strictly applicable only in the limit $\tau_{\rm Th} \ll \tau_{\rm dw}$ and for a very large lead conductance, captures all the essential physics of the $0D$ crossover we are interested in. Our present assumptions lead to the following replacement of the diffusion probability density:

$$P(x,t) \to P(x,t)\; e^{-t/\tau_{\rm dw}}\,. \tag{20.21}$$

Furthermore, the spatial dependence of the noise correlation function (20.13) acquires an additional dissipation term in the denominator. Thus, in contrast to the isolated case, $Q(x)$ now satisfies the Laplace transform of the diffusion equation, given by

$$\left[\frac{1}{L_{\rm dw}^2} - \Delta\right] Q(x) = \delta(x)\,, \tag{20.22}$$

where $L_{\rm dw} = \sqrt{\tau_{\rm dw} D}$. For a ring with circumference L we obtain

$$Q(x) = \frac{L_{\rm dw}}{2}\, \frac{\cosh\Big([L - 2|x|]/2L_{\rm dw}\Big)}{\sinh(L/2L_{\rm dw})}\,. \tag{20.23}$$

We can expand Eq. (20.23) for the almost isolated ring in powers of $\tau_{\rm Th}/\tau_{\rm dw} \ll 1$:

$$Q(x) \approx C - \frac{|x|}{2}\left(1 - \frac{|x|}{L}\right) + \mathcal{O}\left(\frac{\tau_{\rm Th}}{\tau_{\rm dw}}\right)\,, \tag{20.24}$$

where the x-independent first term, $C = L\tau_{\rm dw}/\tau_{\rm Th}$, describes the contribution of the zero mode. As expected, see the discussion after Eq. (20.18), it diverges in the limit $\tau_{\rm Th}/\tau_{\rm dw} \to 0$.

Having described the diffuson in our model of the almost isolated ring, we can proceed by calculating the closed random walk average (crw) of Eq. (20.24). We will see below that we need to consider the random walk average with respect to closed paths with a specific winding number n only. For an isolated ring, using Eq. (20.20), it can be written as

$$\langle Q\rangle_{\rm crw}(t_{12},n) = \int_0^L {\rm d}^2 x_{1,2}\, Q(x_{12}) P_{\rm crw}(x_{12},t_{12},n)\,, \tag{20.25}$$

with

$$P_{\rm crw}(x_{12},t_{12},n) = \sum_{i+j+k=n} \frac{P_i(x_{01},t_1)P_j(x_{12},t_{21})P_k(x_{20},t-t_2)}{P_n(0,t)}\,, \tag{20.26}$$

where we used the notation $x_{\alpha\beta} = x_\alpha - x_\beta$ and $t_{\alpha\beta} = t_\alpha - t_\beta$. Obviously, the replacement (20.21) does not affect this averaging procedure, so that it remains valid in our model of homogeneous dissipation. Note that the expression (20.26) depends in fact only on x_{12} and not on x_0, as can be verified by integrating both sides of the equation over x_0 using the following semi-group property in the ring:

$$\int_0^L {\rm d}x_2\, P_l(x_{12},t_1)P_m(x_{23},t_2) = P_{l+m}(x_{13},t_1+t_2)\,. \tag{20.27}$$

Doing the average of Eq. (20.23) according to Eq. (20.25), we finally obtain

$$\langle Q\rangle_{\rm crw}(t_{12},n) = C - \frac{L}{2}\sum_{k=1}^{\infty} \frac{\cos(2\pi k n\, t_{12}/t)}{(\pi k)^2} e^{-(2\pi k)^2\, E_{\rm Th}\, t_{12}(1-t_{12}/t)}\,. \tag{20.28}$$

It follows that a finite dissipation rate does not affect the decay function to leading order in $\tau_{\rm Th}/\tau_{\rm dw}$.

For the Cooperon, an expansion similar to Eq. (20.20) can be done. In addition to that, the dependence of the Cooperon on an external magnetic field changes due to the ring geometry. It not only leads to the suppression of the Cooperon at long times, but also, due to the Aharonov-Bohm effect, to Altshuler-Aronov-Spivak oscillations[17] of the WL-correction, see Fig. 20.2(b). Combining these remarks with Eq. (20.5) and inserting Eq. (20.3) with $d = 1$, we write the Cooperon in our model as

$$\mathcal{C}(t) = \sum_{n=-\infty}^{+\infty} \frac{e^{-(nL)^2/4Dt}}{\sqrt{4\pi Dt}} e^{-t/\tau_H - t/\tau_{\rm dw} - \mathcal{F}_n(t)} e^{in\theta}\,, \tag{20.29}$$

where $\theta = 4\pi\phi/\phi_0$ and $\phi = \pi(L/2\pi)^2 H$ is the flux through the ring ($\phi_0 = 2\pi c/e$ is the flux quantum). Note that the decay function $\mathcal{F}$ is now a function of n: Since we used an expansion in winding numbers n, we should consider the phase (20.8) acquired by paths with the winding number n only. Thus, the crw-average in Eq. (20.9) has to be performed with respect to paths with given winding number n only, as anticipated in Eq. (20.25).

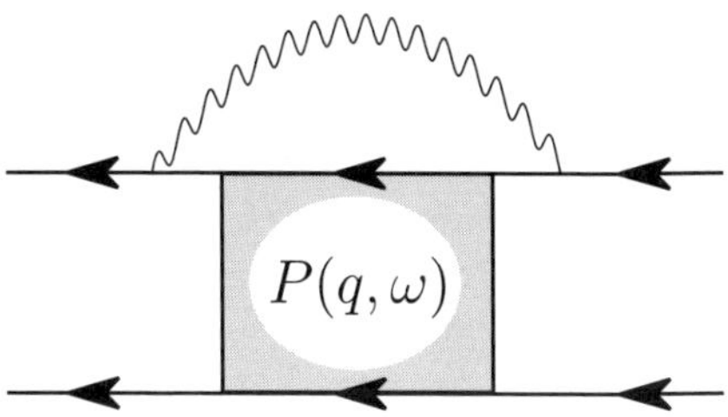

Fig. 20.3. Typical diagram from the expansion of the Cooperon self energy. The shaded area denotes impurity lines, described by the diffusion propagator Eq. (20.31). The upper solid lines correspond to a retarded electron Green's function and the lower to an advanced (or vice versa). Wiggly lines denote electron interactions, described by Eq. (20.16).

20.5. Qualitative Picture from the Perturbative Expansion of the Cooperon

In our previous paper[1] we showed how all the regimes of the dephasing time in an isolated ring can be understood qualitatively from the influence functional picture. In particular, we demonstrated how $0D$ dephasing emerges from the assumption of a noise field that is effectively "frozen" on the time scale τ_{Th} (since $|\omega| \simeq T \ll E_{\mathrm{Th}}$), leading to a drastically reduced dephasing rate. The qualitative behavior of τ_φ also follows from the standard perturbative expression for the Cooperon self-energy. Such self-energy diagrams are of the type shown in Fig. 20.3 and were first evaluated in Ref. [21]. This diagram and its complex conjugate give contributions to the dephasing time of the form

$$\frac{1}{\tau_\varphi} \propto \int \mathrm{d}\omega \int \mathrm{d}q \, \overline{VV}(q,\omega) \, \mathrm{Re}[P(q,\omega)] \,, \tag{20.30}$$

where the diffuson $P(q,\omega)$ is given by the Fourier transform of Eq. (20.19):

$$P(q,\omega) = \frac{1}{Dq^2 - i\omega} \,. \tag{20.31}$$

We have already mentioned that large energy transfers are suppressed according to Eq. (20.15) leading to an upper cutoff at T of the frequency integration. Furthermore, it was shown in Refs. [26,27] that vertex contributions to these self-energy diagrams cure the infrared divergences in the frequency integration, leading to a cutoff at $1/\tau_\varphi$. Such fluctuations are simply too slow to influence the relevant paths. Note that in contrast to the perturbative treatment presented in this section, the path integral calculation leading to the expression Eq. (20.9) for the decay function is free of these IR divergences. In fact, it was shown that the first term of Eq. (20.9) corresponds, when compared to a diagrammatic evaluation of the Cooperon self-energy, to the so-called self-energy contributions (shown in Fig. 20.3), while the second term corresponds to the so-called vertex contributions.

In the ring geometry, the diffuson has quantized momenta and the $q = 0$ mode can not contribute. For a qualitative discussion we may take this into account by inserting a lower cutoff $1/L$ of the momentum integration.

Taking into account the above remarks, we can estimate the dephasing time as

$$\frac{1}{\tau_\varphi} \propto \frac{T}{g_1 L} \int_{1/\tau_\varphi}^{T} \mathrm{d}\omega \int_{1/L}^{\infty} \mathrm{d}q \, \frac{D}{(Dq^2)^2 + \omega^2} \,. \tag{20.32}$$

Eq. (20.32) illustrates succinctly that the modes dominating dephasing lie near the infrared cutoff ($\omega \simeq \tau_\varphi^{-1}$ or E_{Th}) for the diffusive or ergodic regimes, but near the ultraviolet cutoff $\omega \simeq T$ for the $0D$ regime, which is why, in the latter, the broadening of $\delta_T(t)$ becomes important. Performing the integrals in Eq. (1.28) and solving for τ_φ self-consistently, we find three regimes:

(1) The *diffusive regime*, for $\tau_T \ll \tau_\varphi \ll \tau_{\mathrm{Th}}$, with

$$\tau_\varphi \propto (g_1/\sqrt{E_{\mathrm{Th}}T})^{2/3} \,; \tag{20.33}$$

(2) the *ergodic regime*, for $\tau_T \ll \tau_{\mathrm{Th}} \ll \tau_\varphi$, with

$$\tau_\varphi \propto g_1/T \,; \tag{20.34}$$

(3) and the $0D$ *regime*, reached at $\tau_{\mathrm{Th}} \ll \tau_T \ll \tau_\varphi$, with

$$\tau_\varphi \propto g_1 E_{\mathrm{Th}}/T^2 \,. \tag{20.35}$$

Here, $\tau_T = \sqrt{D/T}$ is the thermal time. Expressing (20.35) in terms of the level spacing $\delta = E_{\mathrm{Th}}/g_1$ we find $\tau_\varphi \delta \propto E_{\mathrm{Th}}^2/T^2$. This ratio is $\gg 1$ in the 0D regime, implying that dephasing is so weak that the dephasing rate $1/\tau_\varphi$ is smaller than the level spacing.

20.6. Results for the Cooperon Decay Function

For a systematic analysis of the Cooperon decay function, we rewrite Eq. (20.9) in terms of an integral over the dimensionless variable $u = t_{12}/t$:

$$\mathcal{F}_n(t) = \frac{4\pi T t}{g_1} \int_0^1 \mathrm{d}u \, z(u) \, q_n(u) \,, \tag{20.36}$$

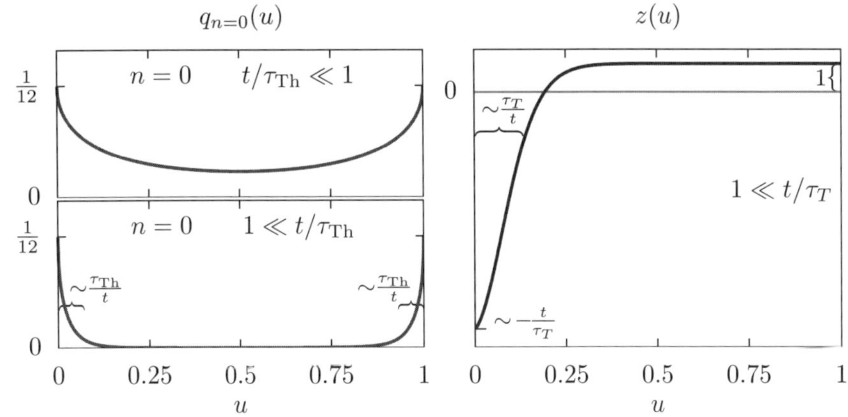

Fig. 20.4. Functions $q_{n=0}(u)$ (left panel) defined in Eq. (20.38) and $z(u)$ (right panel) defined in Eq. (20.37).

where the kernel

$$z(u) = -2\pi Tt\,(1-u)\,w(\pi Ttu) + \int_{\pi Ttu}^{\pi Ttu} \mathrm{d}v\,w(v) \tag{20.37}$$

depends on the parameter $tT = t/\tau_T$, and the dimensionless crw-averaged diffuson

$$q_n(u) = \frac{\langle Q\rangle_{\rm crw}(ut) - C}{L} = -\frac{1}{2}\sum_{k=1}^{\infty}\frac{\cos(2\pi knu)}{(\pi k)^2}e^{-(2\pi k)^2\,(t/\tau_{\rm Th})\,u(1-u)}\,, \tag{20.38}$$

depends on $t/\tau_{\rm Th}$, see Eq. (20.28). Note that we can add or subtract an arbitrary number from $q_n(u)$ without changing the result, since constant terms in $q_n(u)$, describing the zero mode, do not contribute to dephasing, because of the following property of $z(u)$:

$$\int_0^1 \mathrm{d}u\,z(u) = 0\,. \tag{20.39}$$

Both functions, Eq. (20.37) and Eq. (20.38), are illustrated in Fig 20.4 in all relevant limiting cases. Note that in the regime of WL we always have $\tau_T \ll t$. In the opposite regime the interaction correction to the conductivity (Altshuler-Aronov correction) originating from the Friedel oscillations dominate,[48] which we do not consider here.

We proceed with an asymptotic evaluation of Eq. (20.36). For large $t/\tau_{\rm Th}$, $q_n(u)$ is non-zero ($\simeq \frac{1}{12}$) only in the intervals $0 < u < \tau_{\rm Th}/t$ and

$1-\tau_{\rm Th}/t<u<1$, see Fig. 20.4. For small $t/\tau_{\rm Th}$ and $n=0$ we can use the expansion

$$q_{n=0}(u) \approx \frac{1}{12} - \frac{1}{\sqrt{\pi}}\sqrt{\frac{t}{\tau_{\rm Th}}(u(1-u))} + \frac{t}{\tau_{\rm Th}}(u(1-u))\,. \tag{20.40}$$

For larger n the exponential function in Eq. (20.38) can be expanded since the sum converges at $k \simeq 1$.

For $\tau_T \ll t$, $z(u)$ is large ($\sim -t/\tau_T$) in the interval $0<u<\tau_T/t$ and $z(u)\approx 1$ otherwise. Thus, it will be convenient to decompose $z(u) = \overline{z} + \tilde{z}(u)$ into a constant part $\overline{z}=+1$ and a peaked part $\tilde{z}(u) = z(u)-1$. For contributions of the peaked type one observes that

$$\int_0^1 \mathrm{d}u\, \tilde{z}(u)\, u^s = \begin{cases} -1, & s=0\,; \\ -\sqrt{\frac{\tau_T}{t}}\frac{\sqrt{2\pi}}{4}|\zeta\left(\frac{1}{2}\right)|, & s=1/2\,; \\ -\frac{\tau_T}{t}, & s=1\,. \end{cases} \tag{20.41}$$

We identify the following 3 regimes:

Diffusive regime $\tau_T \ll t \ll \tau_{\rm Th}$ and $n=0$: Here we can use the expansion Eq. (20.40). The constant term does not contribute, due to Eq. (20.39). The main contribution to the integral comes from values of u where $z(u)\approx 1$. Thus, we decompose $z(u)=\overline{z}+\tilde{z}(u)$ as suggested above. The leading result and corrections $\propto \sqrt{t/\tau_{\rm Th}}$ due to the second and third term in Eq. (20.40) stem from $\overline{z}$. Corrections $\propto \sqrt{t/\tau_T}$ can be calculated with the help of Eq. (20.41) with $s=1/2$ from the $\tilde{z}(u)$ part. In total we obtain for $n=0$:

$$\mathcal{F}_{n=0}(t) = \frac{\pi^{3/2}\sqrt{E_{\rm Th}}}{2g_1}\,Tt^{3/2}\left(1+\frac{2^{3/2}\zeta\left(\frac{1}{2}\right)}{\pi}\frac{1}{\sqrt{tT}} - \frac{4}{3\sqrt{\pi}}\sqrt{\frac{t}{\tau_{\rm Th}}}\right)\,. \tag{20.42}$$

Diffusive regime $\tau_T \ll t \ll \tau_{\rm Th}$ and $|n|>0$: For winding numbers larger than zero, we expand the exponential function in (20.38). In contrast to the case of $n=0$, the leading result comes here from the peaked part $\tilde{z}(u)$. After expanding the exponential function and doing the sum over k, we can apply Eq. (20.41) with $s=0$ and $s=1$ to find the leading result and a correction $\sim \tau_T/t$. For $\overline{z}$, we observe that the first term vanishes since the integral is over n full periods of cos. The second term of the expansion gives a correction $\sim t/\tau_{\rm Th}$ and in total for $0<|n|\ll t/\tau_T$:

$$\mathcal{F}_n(t) = \frac{\pi}{3g_1}\,Tt\left(1-\frac{2}{n^2}\frac{t}{\tau_{\rm Th}} - \frac{6}{\pi n}\frac{\tau_T}{t}\right)\,. \tag{20.43}$$

Note that in the diffusive regime, winding numbers $|n| > 0$ only contribute weakly to the conductivity, see Eq. (20.29).

Ergodic regime $\tau_T \ll \tau_{\mathrm{Th}} \ll t$: In this regime, the main contribution to the conductivity will not depend on n, since we may neglect the cos-term of $q_n(u)$ as long as $|n| \ll t/\tau_{\mathrm{Th}}$. This restriction on n is justified by the fact that large values of $|n|$ give contributions smaller by a factor of $\sim \exp(-n^2 t/\tau_{\mathrm{Th}})$, see Eq. (20.29).

Again, we decompose $z(u) = \overline{z} + \tilde{z}(u)$. For the $\tilde{z}(u)$ part, we use the expansion of $q_{n=0}(u)$, Eq. (20.40), where the constant term 1/12 will yield the main result. Corrections due to the second term of Eq. (20.40) are $\sim \sqrt{\tau_T/\tau_{\mathrm{Th}}}$, because of Eq. (20.41) with $s = 1/2$. For $\overline{z}$, we do the integral over u directly using $\int_0^1 \mathrm{d}u \, \exp(-xu(1-u)) \xrightarrow{x\to\infty} 2/x$. From this we obtain a correction $\sim \tau_{\mathrm{Th}}/t$ and in total

$$\mathcal{F}_n(t) = \frac{\pi}{3g_1} \, Tt \left(1 - \frac{6}{\sqrt{2\pi}}\sqrt{\frac{\tau_T}{\tau_{\mathrm{Th}}}} - \frac{1}{30}\frac{\tau_{\mathrm{Th}}}{t}\right) . \tag{20.44}$$

It is not surprising that the case $|n| > 0$ in the diffusive regime gives, to leading order, the same results as all n of the ergodic regime (compare Eq. (20.44) to (20.43)), since higher winding numbers are by definition always ergodic: The electron paths explore the system completely.

0*D* regime $\tau_{\mathrm{Th}} \ll \tau_T \ll t$: In this regime, $q_n(u)$ is more sharply peaked than $z(u)$, since $\tau_T/t \gg \tau_{\mathrm{Th}}/t$. This means that the electron reaches the fully ergodic limit (where $q(u) = \mathrm{const}$ and no dephasing can occur) before the fluctuating potential changes significantly. Thus, the potential is effectively frozen and only small statistical deviations from the completely ergodic limit yield a phase difference between the two time-reversed trajectories. The width of the peak of $z(u)$ becomes unimportant, instead, we can expand $z(u)$ around $u = 0$ and $u = 1$. Furthermore, we can expand the argument of the exponential function in $q_n(u)$ and then extend the integral to $+\infty$ and scale u by $k\pi$:

$$\mathcal{F}_n(t) = \frac{4\pi Tt}{g_1} \int_0^{\infty} \mathrm{d}u \left[\frac{2\pi Tt}{3} - 1 - \frac{4\pi^3}{15}(tT)^3 u^2\right] \sum_{k=1}^{\infty} \frac{\cos(2nu)}{(k\pi)^3} e^{-4k\pi E_{\mathrm{Th}} tu} \tag{20.45}$$

(the -1 in the integrand stems from the region $u \approx 1$). Now, assuming $|n| \ll t/\tau_{\mathrm{Th}}$, the integral over u can be done and then the sum over k

evaluated. The result is

$$\mathcal{F}_n(t) = \frac{\pi^2 \tau_{\mathrm{Th}}}{270\, g_1}\, T^2 t \left(1 - \frac{3}{2\pi}\frac{1}{Tt} - \frac{\pi^2}{210}\frac{T^2}{E_{\mathrm{Th}}^2}\right). \tag{20.46}$$

Note that, as mentioned before, the precise form of the shape of $\tilde{z}(u)$, corresponding to the broadened delta function Eq. (20.17), matters only in $0D$ regime.

To summarize, we found the following regimes:

$$\mathcal{F}_n(t) \simeq \begin{cases} \dfrac{\pi^{3/2}}{2g_1}\sqrt{E_{\mathrm{Th}}}Tt^{3/2} & , \tau_T \ll t \ll \tau_{\mathrm{Th}}, n = 0; \quad (20.47\mathrm{a}) \\ \dfrac{\pi T t}{3g_1}, & \tau_T \ll t \ll \tau_{\mathrm{Th}}, |n| > 0; \quad (20.47\mathrm{b}) \\ \dfrac{\pi T t}{3g_1}, & \tau_T \ll \tau_{\mathrm{Th}} \ll t, \text{all } n; \quad (20.47\mathrm{c}) \\ \dfrac{\pi^2}{270\, g_1}\dfrac{T^2 t}{E_{\mathrm{Th}}}, & \tau_{\mathrm{Th}} \ll \tau_T \ll t, \text{all } n. \quad (20.47\mathrm{d}) \end{cases}$$

Note that the crossover temperatures where $\tau_\varphi^{\mathrm{diff}} \simeq \tau_\varphi^{\mathrm{erg}}$ or $\tau_\varphi^{\mathrm{erg}} \simeq \tau_\varphi^{0D}$, namely $c_1 g_1 E_{\mathrm{Th}}$ or $c_2 E_{\mathrm{Th}}$, respectively, involve large prefactors, $c_1 = 27/4 \simeq 7$ and $c_2 = 90/\pi \simeq 30$. This can be seen in a numerical evaluation of Eq. (20.36), which is presented in Fig. 20.6. In particular, one observes that the onset of the $0D$ regime is already at temperatures smaller than $30E_{\mathrm{Th}}$, i.e. well above E_{Th}. This should significantly simplify experimental efforts to reach this regime.

20.7. Correction to the Conductance

Inserting Eq. (20.29) into Eq. (20.4), we obtain the temperature dependent correction to the conductance

$$\Delta g(T,\phi) = -\frac{4L}{g_1 \tau_{\mathrm{Th}}} \int_0^\infty \mathrm{d}t \sum_{n=-\infty}^{+\infty} \frac{e^{-(n/2)^2 \tau_{\mathrm{Th}}/t}}{\sqrt{4\pi D t}} e^{-t/\tau_H - t/\tau_{\mathrm{dw}} - \mathcal{F}_n(t)} \cos(4\pi n\, \phi/\phi_0)\,. \tag{20.48}$$

The resulting value of $|\Delta g(T,\phi)|$ increases with decreasing T in a manner governed by τ_φ. We recall that in the high temperature regime dephasing can be relatively strong, so that one can neglect effects of dissipation (i.e.

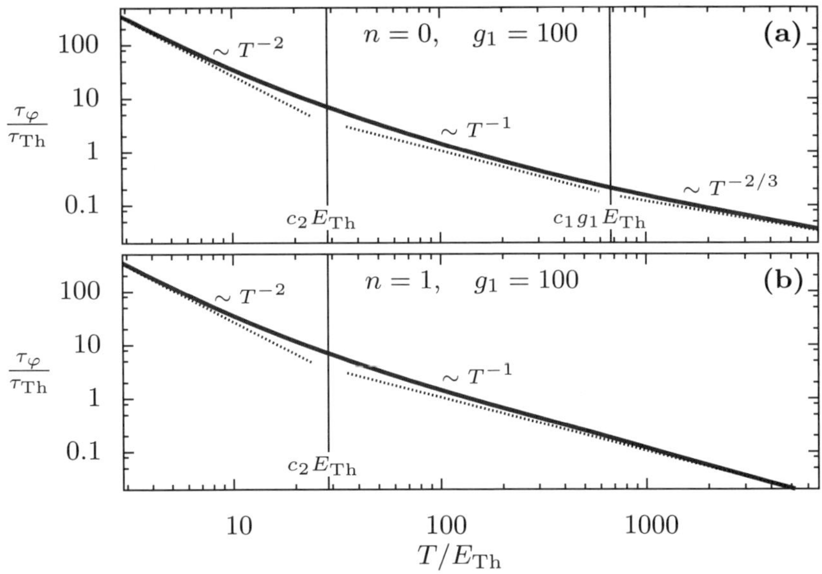

Fig. 20.5. Dephasing time τ_φ extracted from Eq. (20.36) and $\mathcal{F}_n(\tau_\varphi) = 1$ for $g_1 = 100$. (a) Shows the result for zero winding number $n = 0$ and (b) for $n = 1$. For winding numbers $|n| > 0$ the diffusive regime, $\tau_\varphi \sim T^{-2/3}$, is absent.

particle escape out of the ring) and of the external magnetic field on the Cooperon if $\tau_\varphi(T) \ll \min[\tau_H, \tau_{\rm dw}]$. In the diffusive regime, $\tau_\varphi \ll \tau_{\rm Th}$, $\Delta g(T, \phi)$ is dominated by the trajectories with $n = 0$ since the contribution of the trajectories with $|n| \geq 2\sqrt{t/\tau_{\rm Th}} \sim \sqrt{\tau_\varphi/\tau_{\rm Th}}$ is exponentially small. Thus we arrive at:[20]

$$|\Delta g| \simeq \frac{2}{g_1}\sqrt{\frac{\tau_\varphi}{\tau_{\rm Th}}} \propto \left(\frac{E_{\rm Th}}{g_1^2 T}\right)^{1/3} . \tag{20.49}$$

In contrast, the trajectories with large winding number contribute in the ergodic regime, $\tau_T \ll \tau_{\rm Th} \ll \tau_\varphi$, therefore, converting the sum to the integral $\sum_n \exp(-(n/2)^2 \tau_{\rm Th}/t) \simeq \int {\rm d}n \exp(-(n/2)^2 \tau_{\rm Th}/t) \sim \sqrt{t/\tau_{\rm Th}}$, Eq. (20.48) yields[24,25]

$$|\Delta g| \simeq \frac{4}{g_1}\frac{\tau_\varphi}{\tau_{\rm Th}} \propto \frac{E_{\rm Th}}{T} . \tag{20.50}$$

Dephasing due to electron interactions becomes *weak* in the $0D$ regime and, therefore, the situation drastically changes at the crossover from the ergodic regime to the $0D$ one. In particular, we find $\tau_\varphi \gg g_1 \tau_{\rm Th}$, see Eq. (20.47c) and as far as g_1 is large, one may enter a low temperature regime where $\tau_\varphi \geq \tau_{\rm dw}$. In this case, the temperature independent parts of the Cooperon decay must be taken into account. In our model, with decreasing T, the growth of $|\Delta g(T,\phi)|$ saturates towards $|\Delta g(0,\phi)|$ once τ_φ increases past $\min[\tau_H, \tau_{\rm dw}]$ (a more quantitative consideration is given in the next section). Nevertheless, the temperature dependence of Δg is still governed by $\tau_\varphi(T)$ and we can single it out by subtracting the conductance from its limiting value at $T = 0$. Then the difference

$$|\Delta g(0,\phi)| - |\Delta g(T,\phi)| \simeq \frac{4}{g_1}\frac{\tau_{\rm dw}^2}{\tau_{\rm Th}\tau_\varphi} \propto \left(\frac{\tau_{\rm dw}T}{g_1}\right)^2 \tag{20.51}$$

shows T^2-behavior in the $0D$ regime.

20.8. Suggested Experiments

Our theoretical predictions should be observable in real experiments, provided that several requirements are met. We list these conditions in accordance with their physical causes, focusing below on the example of a ring prepared from a quasi-$1D$ wire of width L_W on a $2D$ surface.

20.8.1. *Validity of theoretical predictions*

$1D$ diffusion: We have used the theory of $1D$ diffusion which calls for the following inequalities

$$L \gg (\ell, L_W) \gg \lambda_{\rm F}\,; \tag{20.52}$$

$\lambda_{\rm F}$ is the Fermi wavelength.

Weak localization regime: Eq. (20.48) describes the leading weak localization correction to the conductance. Subleading corrections can be neglected if (a) the classical conductance of the ring is large

$$g_1 \propto (\ell/L)(L_W/\lambda_{\rm F}) \gg 1\,; \tag{20.53}$$

and (b) the leading correction to the conductivity is smaller than its classical value, $|\Delta g| < 1$. The former condition can be assured by a proper choice of the ring geometry and of the material while, in the low temperature regime, the latter is provided by finite dissipation.

Time-/spatial-dependence of the noise correlation function: We have used the noise correlation function (20.16) where dependencies on time- and space-coordinates are factorized. This simplified form requires the following condition (see Section 20.3):

$$2\nu D\mathbf{q}^2 \gg (D\mathbf{q}^2 - i\omega)/V(\mathbf{q})\,. \tag{20.54}$$

In the $0D$ regime we can roughly estimate typical values, $D\mathbf{q}^2 \sim \omega \sim E_{\mathrm{Th}}$, arriving at the inequality

$$\nu V(\mathbf{q}) \gg 1\,. \tag{20.55}$$

For a quasi-$1D$ wire on a $2D$ structure, ν and V can be written as (restoring $\hbar$)

$$\nu_{\mathrm{2D}} = \frac{m_e}{2\pi\hbar^2}\,, \tag{20.56}$$

where m_e is the electron mass, and

$$V_{\mathrm{1D}}(q) = \frac{e^2 L_W}{4\pi\epsilon_0}|\ln(L_W^2 q^2)| \tag{20.57}$$

(in SI units). Thus, (20.55) implies that L_W cannot be taken to be overly small. Inserting material parameters, however, this condition turns out not to be very restrictive, as long as ν_{2D} is reasonably large.

Contacts (dissipation and absence of the Coulomb blockade): The presence of contacts, through which electrons can escape into leads, is mimicked in our model through the homogeneous dissipation rate $1/\tau_{\mathrm{dw}}$. We have assumed weak dissipation:

$$\tau_{\mathrm{Th}} \lesssim \tau_{\mathrm{dw}}\,. \tag{20.58}$$

This ensures that the winding trajectories with $|n| \geq 1$, responsible for AAS oscillations, are relevant. On the other hand, τ_{dw} cannot be taken to be arbitrarily large, since the growth of the WL correction to the conductance with decreasing temperature is cut off mainly due to this temperature independent dissipation, and this cutoff has to occur sufficiently soon that the relative correction remains small, else we would leave the WL regime. Choosing the zero temperature limit, somewhat arbitrary, as $|\Delta g(0,\phi)| = 1/2$, we find from Eq. (20.50)

$$\tau_{\mathrm{dw}}/\tau_{\mathrm{Th}} \lesssim g_1/8\,. \tag{20.59}$$

We note that our assumptions imply $\tau_\varphi \gg g_1\tau_{\rm Th} > \tau_{\rm dw}$ in the $0D$ regime, i.e., dephasing due to electron interactions is weak (each electron contributing to transport through the ring is dephased only a little bit during the course of its stay in the ring). Nevertheless, we demonstrate below (see Fig. 20.8) that the T^2-dependence of the conductance should be visible in real experiments.

To choose a suitable value for the conductance at the contacts, we estimate $\tau_{\rm dw}/\tau_{\rm Th} \simeq g_1/g_{\rm cont}$, which results in

$$8 \lesssim g_{\rm cont}\,. \tag{20.60}$$

We suppose that the contacts are open and have a maximal transmission per channel at the contact

$$T_{\rm cont} = 1 \;\Rightarrow\; g_{\rm cont} = T_{\rm cont} N = N\,, \tag{20.61}$$

(N is the number of transmitting reflectionless channels at the contact). This choice allows one to maximize the WL effect and, simultaneously, to minimize any Coulomb blockade effects, which we have neglected.

20.8.2. *Possible experimental setup*

Temperature range: The relevant temperature range, $[T_{\rm dil}, T_{\rm ph}]$, is limited from below by dilution refrigeration ($T_{\rm dil} \simeq 10$mK) and from above by our neglect of phonons ($T_{\rm ph} \simeq 5$K). Furthermore, the ring should be small enough that $c_2 E_{\rm Th} \gtrsim T_{\rm dil}$; $c_2 E_{\rm Th}$ is the upper estimate for the temperature of the crossover to the $0D$ regime, see the discussion after Eq. (20.47).

Contributions from the leads: We have considered an ideal situation and calculated the Cooperon decay function for the isolated ring, where the finite dissipation rate $1/\tau_{\rm dw}$ does not affect the decay function up to leading order in $\tau_{\rm Th}/\tau_{\rm dw}$. This means that the Cooperons are assumed to live completely inside the ring and not influenced by dephasing in the leads, i.e. it corresponds to the situation shown in Fig. 20.6(a).

In real experiments, the correction to the conductance, Δg, is sensitive to dephasing in the leads because Cooperons exist which either belong to the lead (e.g. the situation shown in Fig. 20.6(c)) or extend over both the ring and the lead (Fig. 20.6(b))[49]. (Note that in contrast to Ref. [51] or Ref. [49], we do not consider Cooperons with a Hikami-box directly at the contact, since we chose $T_{\rm cont} = 1$.)

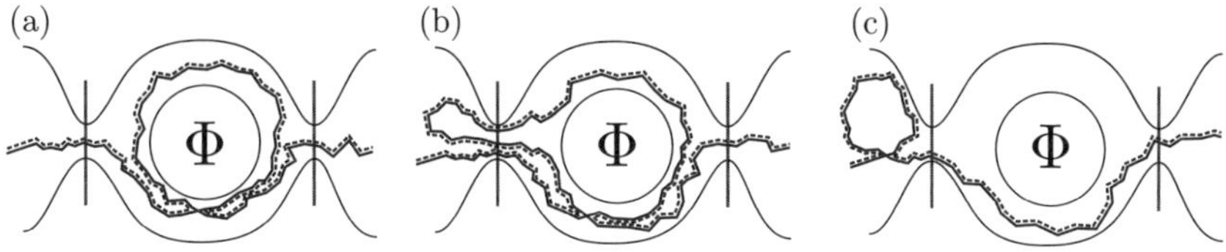

Fig. 20.6. (a) a "ring"-Cooperon, confined entirely to the ring; (b) a "cross"-Cooperon, extending from the ring to the lead and back; (c) a "lead"-Cooperon, confined entirely to the leads.

Contributions of such trajectories might mask the signatures of dephasing in the confined region (the ring). This concern also applies to quantum dots connected to leads (cf. the $\tau_\varphi \propto T^{-1}$-behavior observed in Refs. [31–33]), or finite-size effects in a network of disordered wires,[35] where paths encircling a given unit cell might spend significant time in neighboring unit cells as well (cf. $T^{-1/3}$-behavior observed in Ref. [35] at $\tau_\varphi/\tau_{\rm Th} \geq 1$).

We will now argue that if the lead dimensionless conductance is larger than the contact conductance[50]

$$g_{\rm lead} \gg g_{\rm cont} = N\,, \tag{20.62}$$

then the ring-Cooperon yields the dominating contribution to the WL corrections. Let us focus on the ergodic and 0D regimes, for which $\tau_\varphi \gg \tau_{\rm Th}$, so that that the Cooperon ergodically explores the entire ring. Then the probability to find a closed loop in the ring is proportional to the dwell time, $p_{\rm ring} \propto \tau_{\rm dw}/\nu$, which is $\propto 1/g_{\rm cont}$. Thus we can estimate:

- the probability to enter the ring as $p_{\rm in} \sim g_{\rm cont}/g_{\rm lead}$;
- the probability to find a closed loop in the ring as $p_{\rm ring} \sim (\tau_{\rm dw}/\nu) \sim 1/g_{\rm cont}$;
- the probability to exit the ring as $p_{\rm out} \sim (\tau_{\rm dw}/\nu) g_{\rm cont} \sim 1$;
- the probability to find a closed loop in the diffusive lead as $p_{\rm lead} \sim 1/g_{\rm lead}$

Using these estimates, the probabilities to find a ring-, cross-, or lead-Cooperon are

$$P_{\rm C-ring} \sim p_{\rm in} \times p_{\rm ring} \times p_{\rm out} \sim 1/g_{\rm lead}\,; \tag{20.63}$$

$$P_{\rm C-cross} \sim p_{\rm in} \times p_{\rm ring} \times p_{\rm out} \times p_{\rm in} \times p_{\rm out} \sim g_{\rm cont}/g^2_{\rm lead}\,; \tag{20.64}$$

$$P_{\rm C-lead} \sim p_{\rm lead} \times p_{\rm in} \times p_{\rm out} \sim g_{\rm cont}/g^2_{\rm lead}\,, \tag{20.65}$$

respectively. Thus we arrive at:

$$P_{\text{C-lead}} \sim P_{\text{C-cont}} \sim P_{\text{C-ring}} \times g_{\text{cont}}/g_{\text{lead}} \ll P_{\text{C-ring}}\,, \qquad (20.66)$$

which proves that the ring-Cooperon dominates the WL correction for our choice of parameters if $g_{\text{lead}} \gg g_{\text{cont}}$.

Since the $0D$ regime implies weak dephasing it is highly desirable to improve the "signal-to-noise" ratio by filtering out contributions which do not show $0D$ dephasing. This can be done[35] by constructing from $|\Delta g(T,\phi)|$ its non-oscillatory envelope $|\Delta g_{\text{en}}(T,\phi)|$, obtained by setting $\theta = 0$ in Eq. (20.29) while retaining $\tau_H \neq 0$, and studying the difference

$$\Delta\overline{g}(T,\phi) = |\Delta g_{\text{en}}(T,\phi)| - |\Delta g(T,\phi)|\,. \qquad (20.67)$$

This procedure is illustrated in Fig. 20.7. The lead-Cooperons do not have the Aharonov-Bohm phase and are eliminated by this filtering procedure. Unfortunately, cross-Cooperons cannot be filtered in this manner, since they do experience the Aharonov-Bohm phase. Nevertheless, if the condition $g_{\text{cont}} \ll g_{\text{lead}}$ holds, $\Delta\overline{g}$ is completely dominated by paths residing only in the ring in accordance with the estimate Eq. (20.66).

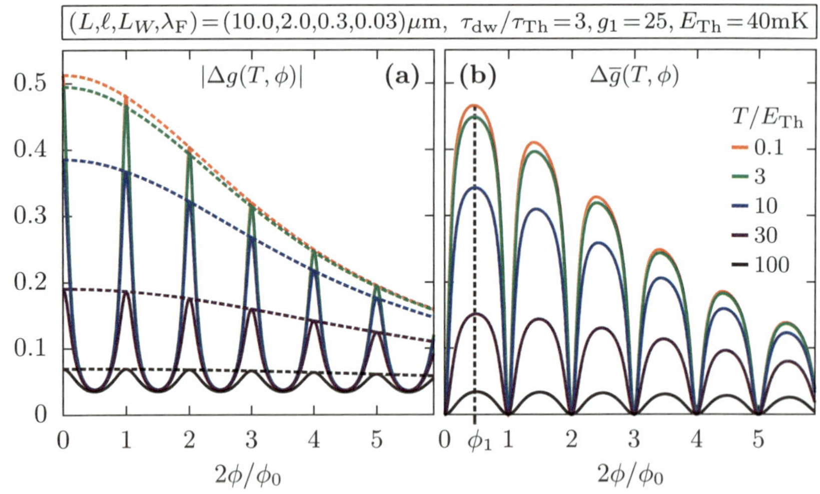

Fig. 20.7. (a) The WL correction $|\Delta g(T,\phi)|$ (solid lines), its envelope $|\Delta g_{\text{en}}(T,\phi)|$ (dashed lines) and (b) their difference $\Delta\overline{g} = |\Delta g_{\text{en}}| - |\Delta g|$, plotted as function of magnetic flux $2\phi/\phi_0$, for five different temperatures between $0.1E_{\text{Th}}$ and $100E_{\text{Th}}$ (increasing from top to bottom).

20.8.3. *Numerical results for 2D GaAs/AlGaAs heterostructures*

All above-mentioned constraints can be met, e.g., with rings prepared from a $2D$ GaAs/AlGaAs heterostructure. In such systems, diffusive behavior emerges from specular boundary scattering of the electrons, see Ref. [46], leading to the following dephasing time due to the external magnetic field:

$$\tau_H = 9.5(c/eH)^2 \times (l/DL_W^3)\,. \tag{20.68}$$

Furthermore, inserting Eq. (20.56) into the $2D$ conductivity $\sigma_{2\mathrm{D}} = 2e^2\nu_{2\mathrm{D}}D$ with $D = v_{\mathrm{F}}\ell$, we obtain the corresponding dimensionless conductance:

$$g_1 = \frac{h}{e^2}\frac{\sigma_{2\mathrm{D}}L_W}{L} = 4\pi\frac{L_W\ell}{\lambda_F L}\,. \tag{20.69}$$

A typical Fermi wavelength in a GaAs/AlGaAs heterostructure is $\lambda_{\mathrm{F}} \approx$ 30nm ($v_F \approx 2.5 \cdot 10^5$m/s).[28,35,39,47] Thus, by suitably choosing L, L_W and ℓ we can adjust g_1 and E_{Th} to make all regimes of the dephasing time accessible.

Numerical results for $|\Delta g|$ and $\Delta\overline{g}$, obtained from Eq. (20.4) using experimentally realizable parameters, are shown in Figs. 20.7 and 20.8 for several combinations of these parameters. The regime where Δg exhibits

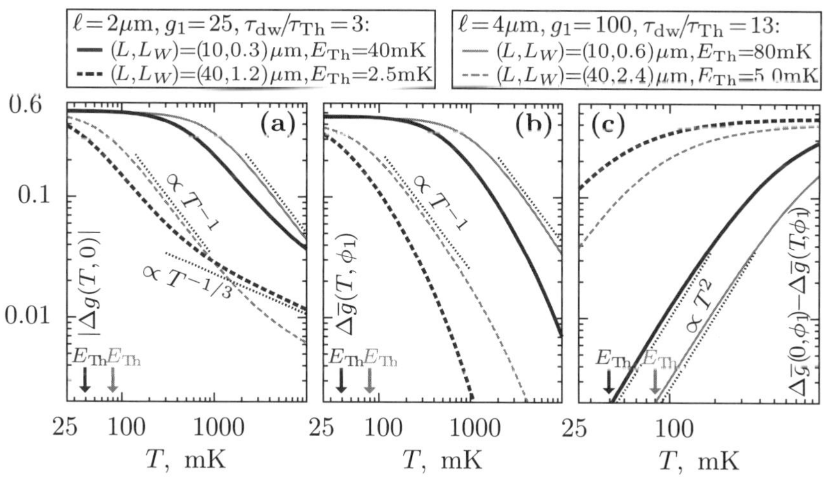

Fig. 20.8. T-dependence of (a) the WL correction at zero field, $|\Delta g(T, 0)|$ and (b) at finite field with envelope subtracted, $\Delta\overline{g}(T, \phi_1)$; (c) the difference $\Delta\overline{g}(0, \phi_1) - \Delta\overline{g}(T, \phi_1)$, which reveals a crossover to T^2-behavior for $T \ll 30E_{\mathrm{Th}}$. The flux ϕ_1, which weakly depends on T, marks the first maximum of $\Delta\overline{g}(T, \phi)$, see Fig. 20.7(b). [This figure is reproduced from Ref. [1]]

diffusive $T^{-1/3}$ behavior ($7g_1E_{\rm Th} \ll T \ll T_{\rm ph}$) is visible only for our smallest choices of both g_1 and $E_{\rm Th}$ (Fig. 20.8(a), heavy dashed line). AAS oscillations in $|\Delta g|$ and $\Delta\overline{g}$ (Fig. 20.7), which require $\tau_{\rm Th} \ll \tau_\varphi$, first emerge at the crossover from the diffusive to the ergodic regime. They increase in magnitude with decreasing T, showing ergodic T^{-1} behavior for $30E_{\rm Th} \ll T \ll 7g_1E_{\rm Th}$ (Figs. 20.8(a),(b)), and eventually saturate towards their $T = 0$ values, with $\Delta\overline{g}(0,\phi) - \Delta\overline{g}(T,\phi)$ showing the predicted $0D$ behavior, $\propto T^2$, for $T \lesssim 5E_{\rm Th}$, see Fig. 20.8(c).

20.9. Conclusions

For an almost isolated disordered quasi-$1D$ ring with $T \gg E_{\rm Th}$, the T-dependence of the dephasing time has been known to behave as $\tau_\varphi \propto T^{-2/3}$ (Ref. [20]) or $\propto T^{-1}$ (Refs. [24,25]) in the diffusive or ergodic regimes, respectively. Here we showed how it crosses over, for $T \ll 30E_{\rm Th}$, to $\tau_\varphi \propto T^{-2}$, in agreement with the theory of dephasing in $0D$ systems (Ref. [2]). This crossover manifests itself in both the smooth part of the magnetoconductivity and the amplitude of the AAS oscillations. Importantly, the latter fact can be exploited to decrease the effects of dephasing in the leads, by subtracting from the magnetoconductivity its smooth envelope. While we did not give an exhaustive study of all contributions to dephasing in the connected ring, we were able to show that the leading contribution results only from trajectories confined to the ring. Thus, an analysis of the T-dependence of the AAS oscillation amplitude may offer a way to finally observe, for $T \lesssim 5E_{\rm Th}$, the elusive but fundamental $0D$ behavior $\tau_\varphi \sim T^{-2}$. Its observation, moreover, would allow *quantitative* experimental tests of the role of temperature as ultraviolet frequency cutoff in the theory of dephasing. An interesting challenge for future works consists in a more realistic model of the connection to the leads. Work on the model of an N-channel ring attached via two arms with fewer channels to absorbing boundaries is currently in progress.[52]

Acknowledgments

We acknowledge illuminating discussions with B. L. Altshuler, C. Bäuerle, N. O. Birge, Ya. M. Blanter, P. W. Brouwer, L. I. Glazman, Y. Imry, V. E. Kravtsov, J. Kupferschmidt, A. D. Mirlin, Y. V. Nazarov, A. Rosch, D. Weiss and V. I. Yudson. We acknowledge support from the DFG through SFB TR-12, the Emmy-Noether program and the Nanosystems Initiative

Munich Cluster of Excellence; from the NSF, Grant No. PHY05-51164; and from the EPSRC, Grant No. T23725/01.

References

1. M. Treiber, O. M. Yevtushenko, F. Marquardt, J. von Delft, and I. V. Lerner, Phys. Rev. B **80**, 201305(R) (2009).
2. U. Sivan, Y. Imry, and A. G. Aronov, Europhys. Lett. **28**, 115 (1994).
3. Y. Imry, in *Introduction to Mesoscopic Physics*, Oxford University Press (1997)
4. M. Büttiker, Y. Imry, and R. Landauer, Phys. Rev. A **96**, 365 (1983).
5. M. Büttiker, Y. Imry, and M. Ya. Azbel, Phys. Rev. A **30**, 1982 (1984).
6. Y. Gefen, Y. Imry, and M. Ya. Azbel, Phys. Rev. Lett. **52**, 129 (1984).
7. M. Büttiker, Y. Imry, R. Landauer, and S. Pinhas, Phys. Rev. B **31**, 6207 (1984).
8. M. Murat, Y. Gefen, and Y. Imry, Phys. Rev. B **34**, 659 (1985).
9. Y. Imry, and N. S. Shiren, Phys. Rev. B **33**, 7992 (1986).
10. A. D. Stone, and Y. Imry, Phys. Ref. Lett. **56**, 189 (1986).
11. B. L. Altshuler, Y. Gefen, and Y. Imry, Phys. Rev. Lett. **66**, 88 (1990).
12. L. P. Levy, G. Dolan, J. Dunsmuir, and H. Bouchiat, Phys. Rev. Lett. **64**, 2074 (1990).
13. H. Bluhm, N. C. Koshnick, J. A. Bert, M. E. Huber, and K. A. Moler, Phys. Rev. Lett. **102**, 136802 (2009).
14. A. C. Bleszynski-Jayich, W. E. Shanks, B. Peaudecerf, E. Ginossar, F. von Oppen, L. I. Glazman, and J. G. E. Harris, arXiv:0906.4780 (2009).
15. Y. Imry, Physics **2**, 24 (2009).
16. R. A. Webb, S. Washburn, C. P. Umbach, and R. B. Laibowitz, Phys. Rev. Lett. **54**, 2696 (1985).
17. B. L. Altshuler, A. G. Aronov, and B. Z. Spivak, JETP Lett. **33**, 94 (1981).
18. D. Y. Sharvin, and Y. V. Sharvin, Pis'ma Zh. Teor. Eksp. Fiz. **34**, 285 (1981).
19. A. Stern, Y. Aharonov, and Y. Imry, Phys. Rev. A **41**, 3436 (1990).
20. B. L. Altshuler, A. G. Aronov, and D. E. Khmelnitsky, J. Phys. C **15**, 7367 (1982).
21. H. Fukuyama and E. Abrahams, Phys. Rev. B, **27**, 5976 (1983).
22. B. L. Altshuler and A. G. Aronov, in *Electron-Electron Interactions in Disordered Systems*, edited by A. L. Efros and M. Pollak, North-Holland, Amsterdam, 1 (1985).
23. I. L. Aleiner, , B. L. Altshuler, and M. E. Gershenson, Waves in Random Media **9**, 201 (1999).
24. T. Ludwig and A. D. Mirlin, Phys. Rev. B **69**, 193306 (2004).
25. C. Texier and G. Montambaux, Phys. Rev. B **72**, 115327 (2005).
26. F. Marquardt, J. von Delft, R. A. Smith, and V. Ambegaokar, Phys. Rev. B **76**, 195331 (2007).
27. J. von Delft, F. Marquardt, R. A. Smith, and V. Ambegaokar, Phys. Rev. B **76**, 195332 (2007).

28. C. M. Marcus, R. M. Westervelt, P. F. Hopkins, and A. C. Gossard, Phys. Rev. B **48**, 2460 (1993).
29. A. Yacoby, M. Heiblum, H. Shtrikman, V. Umansky, and D. Mahalu, Semicond. Sci. Tech. **9**, 907 (1994).
30. B. Reulet, H. Bouchiat, and D. Mailly, Europhys. Lett. **31**, 305 (1995).
31. A. G. Huibers, M. Switkes, C. M. Marcus, K. Campman, and A. C. Gossard, Phys. Rev. Lett. **81**, 200 (1998);
32. A. G. Huibers, S. R. Patel, C. M. Marcus, P. W. Brouwer, C. I. Duruöz, and J. S. Harris, Jr., Phys. Rev. Lett. **81**, 1917 (1998);
33. A. G. Huibers, J. A. Folk, S. R. Patel, C. M. Marcus, C. I. Duruöz and J. S. Harris, Jr., Phys. Rev. Lett. **83**, 5090 (1999).
34. A. B. Gougam, F. Pierre, H. Pothier, D. Esteve, and N. O. Birge, J. Low Temp. Phys. **118**, 447 (2000).
35. M. Ferrier, A. C. H. Rowe, S. Gueron, H. Bouchiat, C. Texier, and G. Montambaux, Phys. Rev. Lett. **100**, 146802 (2008).
36. I. L. Aleiner, B. L. Altshuler, and Y. M. Galperin, Phys. Rev. B **63**, 201401 (2001).
37. I. L. Aleiner and Y. M. Blanter, Phys. Rev. B **65**, 115317 (2002).
38. Y. Imry, arXiv: cond-mat/0202044 (2002)
39. Y. Niimi, Y. Baines, T. Capron, D. Mailly, F.-Y. Lo, A. D. Wieck, T. Meunier, L. Saminadayar, and C. Bäuerle, Phys. Rev. Lett. **102**, 226801 (2009).
40. L. P. Gor'kov, A. I. Larkin, and D. E. Khmel'nitskii, JETP Lett. **30**, 228 (1979).
41. E. Akkermans and G. Montambaux, *Mesoscopic physics of electrons and photons* (Cambridge University Press, Cambridge, 2007).
42. D. E. Khmelnitskii, Physica B&C **126**, 235 (1984).
43. B. L. Altshuler and A. G. Aronov, JETP Lett. **33**, 499 (1981).
44. P. W. Brouwer and C. W. J. Beenakker, Phys. Rev. B **55**, 4695 (1997);
45. E. McCann and I. V. Lerner, Phys. Rev. B **57**, 7219 (1998).
46. C. W. J. Beenakker and H. van Houten, Phys. Rev. B **38**, 3232 (1988).
47. O. Yevtushenko, G. Lütjering, D. Weiss, and K. Richter, Phys. Rev. Lett. **84**, 542 (2000).
48. B. L. Altshuler, A. G. Aronov, and P. A. Lee, Phys. Rev. Lett. **44**, 1288 (1980).
49. Ya. M. Blanter, V. M. Vinokur, and L. I. Glazman, Phys. Rev. B **73**, 165322 (2006).
50. J. N. Kupferschmidt, and P. W. Brouwer, (private communication).
51. R. S. Whitney, Phys. Rev. B **75**, 235404 (2007).
52. M. Treiber, O. M. Yevtushenko, F. Marquardt, J. von Delft, and I. V. Lerner, to be published.

Chapter 21

Single-Electron Tunneling and Fluctuation Theorem

Y. Utsumi[1], D. S. Golubev[2], M. Marthaler[3], T. Fujisawa[4,5], and Gerd Schön[2,3]

[1]*Institute for Solid State Physics, University of Tokyo, Kashiwa, Chiba 277-8581, Japan*
[2]*Institut für Nanotechnologie, Karlsruhe Institute of Technology, 76021 Karlsruhe, Germany*
[3]*Institut für Theoretische Festkörperphysik and DFG Center for Functional Nanostructures (CFN), Karlsruhe Institute of Technology, 76128 Karlsruhe, Germany*
[4]*NTT Basic Research Laboratories, NTT Corporation, Morinosato-Wakamiya, Atsugi 243-0198, Japan*
[5]*Research Center for Low Temperature Physics, Tokyo Institute of Technology, Ookayama, Meguro, Tokyo 152-8551, Japan*

Experiments on the direction-resolved full-counting statistics of single-electron tunneling allow testing the fundamentally important Fluctuation Theorem (FT). At the same time, the FT provides a frame for analyzing such data. Here we consider tunneling through a double quantum dot system which is coupled capacitively to a quantum point contact (QPC) detector. Fluctuations of the environment, including the shot noise of the QPC, lead to an enhancement of the effective temperature in the FT. We provide a quantitative explanation of this effect; in addition we discuss the influence of the finite detector bandwidth on the measurements.

21.1. Introduction

The second law of thermodynamics states that the entropy of a macroscopic system driven out of equilibrium grows with time, and the dynamics of such a system is irreversible. The entropy of a mesoscopic system also grows in the long-time limit, but it may decrease over sufficiently short periods of time. Hence, the entropy production ΔS during a time interval τ

is a random variable, characterized by a distribution $P_\tau(\Delta S)$. The 'Fluctuation Theorem' (FT)[1] states that the probabilities of positive and negative entropy changes at sufficiently long τ are related by

$$\frac{P_\tau(\Delta S)}{P_\tau(-\Delta S)} = \exp(\Delta S). \tag{21.1}$$

Remarkably, this simple and universal relation remains valid even far from equilibrium. It has been proven for thermostated Hamiltonian systems,[1] Markovian stochastic processes,[2–4] quantum systems,[5] and mesoscopic conductors.[6–11] The FT is fundamentally important for transport theory. One of its consequences is the Jarzynski equality,[12,13] which in turn leads to the second law of thermodynamics. It also leads to the fluctuation-dissipation theorem and Onsager symmetry relations,[14] as well as to their extensions to nonlinear transport.[6–11]

The FT was first verified in an experiment measuring the distribution of the work done on a colloidal particle placed in a water flow and trapped by an optical tweezer.[15] By monitoring the position fluctuations it is possible to estimate the work done on the particle. For this classical experiment, as well as for other related ones performed at room temperature,[16] the thermal fluctuations are quite large and the FT has been confirmed. In contrast, experiments for mesoscopic quantum systems[17] were lacking until very recently.[18,19]

Let us now discuss the implications of FT for mesoscopic systems. The first experimental test of the FT applied to single-electron transport has been performed recently in Refs.[18,20] A system of two coupled quantum dots in a 2DEG at the GaAs/AlGaAs interface was operated in the Coulomb blockade regime (The single-electron charging energy of a single dot was of the order of 100 μeV, and the sample was cooled to 100 milli-Kelvin). The single-electron tunneling through the double dot system was detected via the current through a nearby quantum point contact (QPC). The time resolution of the readout was better than 0.1 ms. By using an asymmetric setup the direction of tunneling could be resolved, and the probability distribution of forward and backward tunneling processes could be determined.

The entropy production in this experiment is related to Joule heating and reads $\Delta S = qeV_S/k_BT$, where q is the number of electrons (with charge e) transfered through the conductor during time τ and V_S is the bias voltage. Hence the FT can be formulated in terms of the distribution of transfered charge $P_\tau(q)$ at sufficiently long times, $\tau \gtrsim e/I$, where I is the current, as follows

$$\frac{P_\tau(q)}{P_\tau(-q)} = \exp\left(\frac{qeV_{\rm S}}{k_{\rm B}T}\right). \quad (21.2)$$

Unlike in classical systems, the fluctuations of the charge transfered through the quantum dots are strongly affected by the environment (phonons and electromagnetic environment) and by the measurement back-action. In what follows we discuss various environmental effects and demonstrate their importance for the interpretation of the experiment. We show that the non-equilibrium electromagnetic fluctuations caused by the non-equilibrium shot noise of the QPC detector lead to an apparent violation of the FT. However, we find the FT to be satisfied if we replace the temperature T in Eq. (21.2) by an enhanced effective temperature T^*, which we relate to the tunneling rates for the various relevant processes. We also study the effect of finite bandwidth of the detector[21–23] and show that in the parameter regime of our experiment, this effect may also be accounted for by an effective temperature.

The paper is organized as follows. In Sec. 21.2, we briefly discuss the experimental results. In Sec. 21.3, we discuss the theory relevant for our experiment assuming a perfect detector. In Sec. 21.4, we discuss the effect of the environment. The realistic case, where the time resolution of the detector is limited by its bandwidth, will be discussed in Sec. 21.5. There we will show that it is still possible to recover the FT with properly corrected tunneling rates. In Sec. 21.6, we show how the tunneling rates are determined experimentally and how the finite bandwidth of the detector affects the experimentally obtained value. Section 21.7 summarizes our discussion.

21.2. Experimental Test of the FT in Single-electron Counting

Most of the experiments on single-electron counting are performed with a single quantum dot capacitively coupled to a QPC detector.[24] In this case the QPC current switches between the two values corresponding to an occupied and an empty quantum dot. Such a detector cannot resolve the direction of the electron tunneling and, therefore, is not suitable for testing the FT. The simplest system which does resolve the direction of the tunneling consists of two serially coupled quantum dots which are asymmetrically coupled to a QPC detector[20] (Fig. 21.1 a). The left and right gate voltages, $V_{\rm GL}$ and $V_{\rm GR}$, applied to the quantum dots are tuned in such a way that only three charge states of the DQD need to be considered. In the experiment[20] those states are $|D\rangle$ (both dots are occupied), $|L\rangle$ (left dot is occupied by one electron) and $|R\rangle$ (right dot is occupied). Accordingly, the current through the QPC, which is coupled asymmetrically to the DQD, switches between three different values (Fig. 21.1 b). This setup allows distinguishing electron tunneling in different directions and between the dots and leads.

From the time trace of the current taken during time τ one obtains the distribution of transfered charges between the two dots, $P_\tau(q)$, an example of which is shown in the inset of Fig. 21.2. In Fig. 21.2 we perform a test of the FT (21.2). The combination $\ln[P_\tau(q)/P_\tau(-q)]$ depends indeed linearly on the transfered charge q with a slope $eV_S/k_B T^*$. Here $V_S = 300\ \mu$V is the applied DQD bias voltage, but the effective temperature $T^* = 1.37$ K fitting the data (dashed line) strongly exceeds the bath temperature of the leads of $T = 130$ mK (dot-dashed line).

In order to understand this apparent violation, we have to consider the total system. The FT for the system composed of the DQD and QPC should be formulated in terms of the joint probability distribution $P_\tau(q, q')$, where q and q' charges are transmitted through the DQD and the QPC,

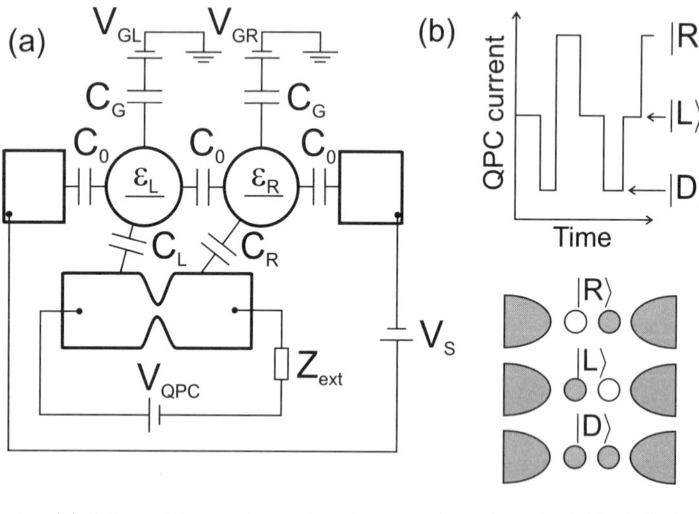

Fig. 21.1. (a) Setup of the system with two quantum dots (DQD) with single-level energies ε_L and ε_R coupled to a quantum point contact (QPC). (b) The QPC current switches between three values corresponding to the three charge states of the DQD.

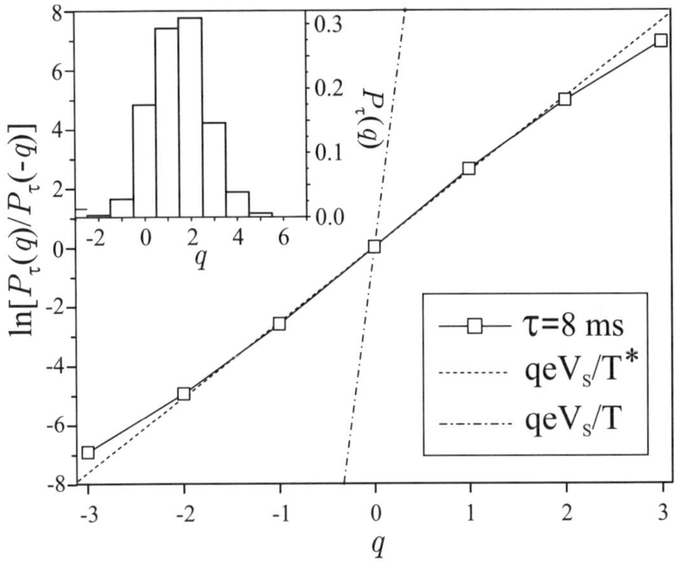

Fig. 21.2. Test of FT (21.2) with the measurement time $\tau = 8$ ms. Lines with squires: logarithm of lhs of Eq. (21.2); dashed line: $q\,eV_S/k_\mathrm{B}T^*$ with $T^* = 1.37$ K; dot-dashed line: $q\,eV_S/k_\mathrm{B}T$. Inset: the distribution $P_\tau(q)$ at $\tau = 4$ ms.

respectively (Fig. 21.3).[9] It satisfies

$$P_\tau(q, q') = \exp\left(\frac{qeV_\mathrm{S} + q'eV_\mathrm{QPC}}{k_\mathrm{B}T}\right) P_\tau(-q, -q'). \qquad (21.3)$$

Since only the number of charges q is measured, Eq. (21.3) should be summed over q'. As a result, the right hand side deviates from $\exp[eV_\mathrm{S}/k_\mathrm{B}T]\,P_\tau(-q)$ (except for $V_\mathrm{QPC} = 0$), which appears to violate the FT. In the following section we will provide more detailed discussions and further the reasoning why the FT is recovered when we introduce the effective temperature.

21.3. FT in the Single-electron Transport

In the experiment, electrons tunnel through the dots sequentially. The system is then fully characterized by the vector of the occupation probabilities of the DQD charge states, $\mathbf{p}^T = (p_L, p_R, p_D)$, which satisfies the following master equation

$$\partial_t \mathbf{p} = \mathbf{\Gamma}(\lambda)\mathbf{p}. \qquad (21.4)$$

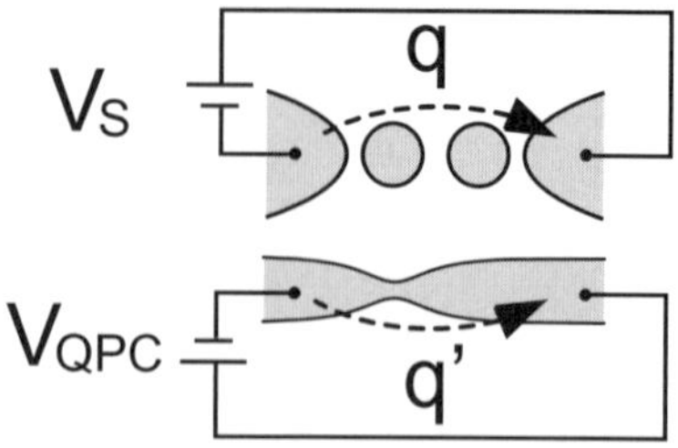

Fig. 21.3. Schematic picture of the DQD capacitively coupled to the QPC illustrating the apparent violation of the FT. In addition to the source-drain bias V_S applied to the DQD the voltage $V_{\rm QPC}$ is applied to the QPC. Thus, the total system is a 4-terminal setup, and the FT should be formulated for the joint probability distribution of transmitted charges through the QPC and DQD, $P(q,q')$. The FT for $P(q)$ is valid only for $V_{\rm QPC}=0$.

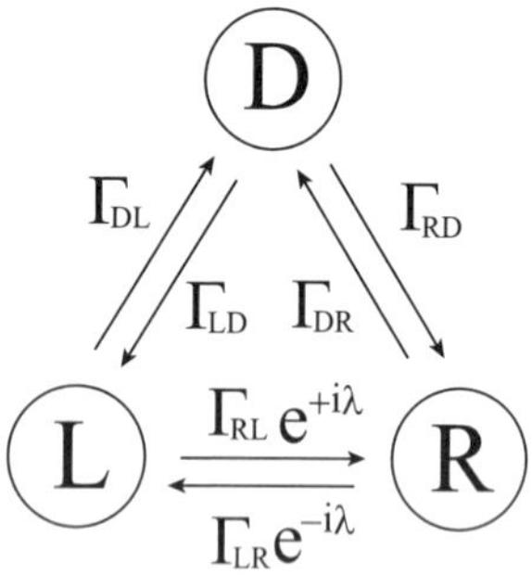

Fig. 21.4. The relevant transition processes. Circles represent the double-dot states and arrows represent the directions of the transitions. The factor $\mathrm{e}^{\pm i\lambda}$, needed for the full-counting statistics, indicates that the electron number is 'counted' at the center barrier.

Here the transition matrix is given by,

$$\mathbf{\Gamma}(\lambda)=\begin{pmatrix} -\Gamma_{RL}-\Gamma_{DL} & \Gamma_{LR}e^{-i\lambda} & \Gamma_{LD} \\ \Gamma_{RL}e^{i\lambda} & -\Gamma_{LR}-\Gamma_{DR} & \Gamma_{RD} \\ \Gamma_{DL} & \Gamma_{DR} & -\Gamma_{LD}-\Gamma_{RD} \end{pmatrix}. \tag{21.5}$$

Figure 21.4 indicates six transitions with Γ_{ij} between three charge states. Following the recipe of the full-counting statistics (FCS) of Bagrets and Nazarov,[25] we introduced the counting field λ, which keeps track of the electrons transfered through the tunnel barrier between the two quantum dots (Fig. 21.4). Then the probability distribution of the charge transfered through this barrier during the time τ is given by the Fourier transform

$$P_\tau(q) = \int_{-\pi}^{\pi} \frac{d\lambda}{2\pi}\, e^{-i\lambda q} \mathcal{Z}_\tau(\lambda), \tag{21.6}$$

where

$$\mathcal{Z}_\tau(\lambda) = p_L(\tau) + p_R(\tau) + p_D(\tau), \tag{21.7}$$

is the characteristic function. In the long time limit $\tau \gg I/e$, the characteristic function takes the form $\mathcal{Z}(\lambda) \sim e^{\tau \mathcal{F}(\lambda)}$, where $\mathcal{F}(\lambda)$ is the eigenvalue of the matrix $\mathbf{\Gamma}(\lambda)$ with the largest real part. The function $\mathcal{F}(\lambda)$ has to be found from the characteristic equation

$$\begin{aligned} 0 = \det[\mathbf{\Gamma}(\lambda) - \mathcal{F}\mathbf{I}] &= \mathcal{F}^3 + K\mathcal{F}^2 + K'\mathcal{F} \\ &+ \Gamma_{DR}\Gamma_{RL}\Gamma_{LD}(\mathrm{e}^{i\lambda} - 1) + \Gamma_{DL}\Gamma_{LR}\Gamma_{RD}(\mathrm{e}^{-i\lambda} - 1), \end{aligned} \tag{21.8}$$

where K and K' are parameters independent of the counting field,

$$K = \sum_{i \neq j} \Gamma_{ij}, \quad K' = \sum_{i \neq j} \sum_{k \neq i,j} \Gamma_{ik}\Gamma_{kj} + \sum_{i \neq j} \sum_{k \neq i,j} \Gamma_{ki}\Gamma_{kj}/2\,. \tag{21.9}$$

Without solving the characteristic equation, we observe that $\mathcal{F}(\lambda)$ and hence $\mathcal{Z}(\lambda)$ in the long time limit satisfy the identity

$$\mathcal{Z}(\lambda) = \mathcal{Z}\left(-\lambda + i\frac{eV_S}{k_{\mathrm{B}}T^*}\right), \tag{21.10}$$

where the effective temperature T^* is

$$T^* = \frac{eV_S}{k_{\mathrm{B}} \ln w}, \tag{21.11}$$

$$w = \frac{\Gamma_{DR}\Gamma_{RL}\Gamma_{LD}}{\Gamma_{DL}\Gamma_{LR}\Gamma_{RD}}\,. \tag{21.12}$$

Performing the inverse Fourier transformation of Eq. (21.10), we arrive at the relation (21.2) for the distribution $P_\tau(q)$ with T being replaced by T^*.

One can demonstrate[18] that the effective temperature (21.11) is equal to the base temperature, $T^* = T$, when the tunneling rates Γ_{ij} satisfy the local detailed balance relation[2]

$$\frac{\Gamma_{ij}}{\Gamma_{ji}} = \exp\left(\frac{\Delta_j - \Delta_i}{k_{\mathrm{B}}T}\right), \tag{21.13}$$

where Δ_j are the electrochemical potentials of the charge states of the DQD. In real experiments the QPC is biased and generates a non-equilibrium shot noise. Under these conditions the local detailed balance is violated and, as we will show later in Sec. 21.4, we have $T^* > T$. Although the canonical form of the FT (21.2) is violated in this case, the more general form (21.1) still holds. The reason is the underlying simplicity of the considered system, in which a forward transfer of a single electron occurs only through the following cycle of transitions $|D\rangle \to |L\rangle \to |R\rangle \to |D\rangle$ (see Fig. 21.4). It enables us to introduce the macroscopic affinity $\ln w$ uniquely.[2–4,26] For a general system with more cycles in its state transition diagram, the FT holds only when the affinities for all cycles associated with current flow into/out of a particular lead coincide.[4,26] Later in Sec. 21.5, we will show that a detector with finite bandwidth leads to a violation of this condition.

21.4. Effects of Backaction and Environments

Mesoscopic electron transport suffers from environmental effects. In GaAs nanostructures, acoustic phonons strongly couple to electrons via the deformation potential and the piezoelectric coupling.[27] For a QPC measurement, one cannot avoid the Coulomb interaction between the dots and QPC leads, which is marked by C_L and C_R in Fig. 21.1 a.[22,28,29] This interaction is unwanted and causes a measurement backaction, since the nonequilibrium QPC shot noise leads to QD level fluctuations. In addition, the external circuit acts as the electromagnetic environment, which further affects both of the DQD and QPC.

Such environmental effects can be accounted for by the so-called $P(E)$-theory[30] and more systematically by using the real-time diagrammatic technique.[31] The phonon and electromagnetic environments, as well as the nonequilibrium QPC current noise generate QD level fluctuations, δV_L and δV_R. Then the tunnel rates connecting the three charge states are modified as,

$$\Gamma_{LR} = \pi \left| t_{dd} \right|^2 P_{dd}(E_L - E_R) \,, \tag{21.14}$$

$$\Gamma_{DL} = \Gamma_R \int d\omega f(E_D - E_L - V_S/2 - \omega) \, P_R(\omega) \,, \tag{21.15}$$

$$\Gamma_{RD} = \Gamma_L \int d\omega f(\omega - E_R + V_S/2 + E_D) \, P_L(\omega) \,. \tag{21.16}$$

(Γ_{RL}, Γ_{LD} and Γ_{RD} are given in a similar manner). Here t_{dd} is a tunnel matrix element describing the central barrier. The total energies of the charge states E_j include the electrostatic energy. The tunneling rates Γ_{DL} and Γ_{RD} are also affected by the thermal broadening of the reservoir levels through the Fermi distribution, $f(\omega) = 1/(1 + e^{\omega/k_B T})$.

The Fourier transform of the correlation function, $P_j(\omega) = \int dt\, e^{i\omega t} P_j(t)/(2\pi)$ induces additional broadening. It is determined by the fluctuating dot potentials, $\varphi_{L/R}(t) = \int^t dt' \delta V_{L/R}(t')$ and $\varphi_{dd} = \varphi_R - \varphi_L$, as

$$P_j = \left\langle e^{i\hat{\varphi}_j(t)} e^{-i\hat{\varphi}_j(0)} \right\rangle = \exp\left[\int d\omega \frac{S_{\delta V j}(\omega)(e^{-i\omega t} - 1)}{\omega^2} \right], \tag{21.17}$$

where the correlation function for the level fluctuations is determined by the properties of the environment.

For acoustic phonons the correlation function takes the following form

$$S^{\mathrm{ph}}_{\delta V j}(\omega) = \frac{A^{\mathrm{ph}}_j(\omega)}{1 - e^{-\omega/k_B T}} \tag{21.18}$$

with super-ohmic, $A^{\mathrm{ph}}_{dd} \propto \omega^3$, or ohmic, $A^{\mathrm{ph}}_{L/R} \propto \omega$, phonon spectral functions. Though phonons cause some additional broadening, for the experiment considered with low measurement current, $\langle I_{\mathrm{QPC}} \rangle \approx 12\mathrm{nA}$, the heating effect is negligible.[27] As long as the lattice and electronic systems are isothermal the local detailed balance holds, and as a consequence the FT, Eq. (21.2), is satisfied.

The situation changes when the environment itself is out of equilibrium. Such an environment is generated by the QPC current fluctuations S_I^{QPC}, which give rise to the voltage fluctuation spectrum

$$S^{\mathrm{QPC}}_{\delta V j}(\omega) = \kappa_j |Z_t(\omega)|^2 S_I^{\mathrm{QPC}}(\omega). \tag{21.19}$$

($\kappa_{L/R} = 1$, $\kappa_{dd} = 4$). The non-symmetrized current noise of the QPC is given by

$$S_I^{\mathrm{QPC}} = \frac{2}{R_K} \left[\sum_{\pm} \frac{\mathcal{T}_{\mathrm{QPC}}(1 - \mathcal{T}_{\mathrm{QPC}})(\omega \pm V_{\mathrm{QPC}})}{1 - e^{-(\omega \pm V_{\mathrm{QPC}})/k_B T}} + \frac{2 \mathcal{T}_{\mathrm{QPC}}^2 \omega}{1 - e^{-\omega/k_B T}} \right]. \tag{21.20}$$

The impedance $Z_t(\omega) = 1/(i\omega \bar{C} + 1/\bar{R})$ characterizes the capacitive coupling between the QPC and the dot-level fluctuations δV_r. Here $\bar{R}$ is written with the QPC resistance R_{QPC} as $\bar{R} = R_{\mathrm{QPC}}/[1 + \bar{C}(C_L^{-1} + C_R^{-1})]$. The capacitance is $\bar{C} = (3C_0 + C_G)/2$, where the capacitances C_0 and C_G characterizes the coupling between dots, leads and gate electrodes (Fig. 21.1 a). For the experiment,[20] the QPC transparency for each spin is estimated as $\mathcal{T}_{\mathrm{QPC}} = R_K/(2R_{\mathrm{QPC}}) \approx 0.19$ and $V_{\mathrm{QPC}} = 0.8\mathrm{mV}$. Equation (21.19) is reduced to the equilibrium form Eq. (21.18), when $V_{\mathrm{QPC}} = 0$ and the local detailed balance is satisfied. However, for $V_{\mathrm{QPC}} \neq 0$, the local detailed balance and thus the FT, Eq. (21.2), are violated. Note that the violation is not contradict to the FT for the total DQD and QPC system, Eq. (21.3).

Generally the electromagnetic environment suppresses the phase correlations. In the long-time limit, it decays exponentially $P_j(t) \approx \exp(-\Gamma_j^{\rm QPC} t/2)$. For realistic parameters, $\bar{C}=5{\rm fF}$ and $\bar{C}(C_L^{-1}+C_R^{-1})=0.02$, the decay rate is rather big, $\Gamma_C^{\rm QPC}/2 = \pi |Z_t(0)|^2 S_I(0) \approx 50\mu{\rm eV}$. Even for the ideal case, i.e. there is no capacitive coupling between the QPC and the DQD, $C_{L/R}=0$, the correlation function P_j decays exponentially. It is because of an *intrinsic* backaction often discussed in the context of the weak measurement.[32] For the ideal case, the decay rate is,

$$\frac{\Gamma_C^{\rm QPC}}{2} \approx -\frac{V_{\rm PC}\, t}{\pi} \ln\Big(\sqrt{\mathcal{T}_{|L\rangle}}\sqrt{\mathcal{T}_{|R\rangle}}+\sqrt{1-\mathcal{T}_{|L\rangle}}\sqrt{1-\mathcal{T}_{|R\rangle}}\Big)\,, \tag{21.21}$$

in the limit of $t \to \infty$ for $T \ll V_{\rm QPC}$. Here $\mathcal{T}_{|s\rangle}$ means the transmission probability through the QPC when the DQD is in the state $|s\rangle$. However, for the experiment,[20] $I_{|L/R\rangle} \sim 12 \pm 0.1{\rm nA}$, and thus we estimate $\Gamma_C^{\rm QPC}/2 \sim 2.1{\rm nV}$, which is negligible.

21.5. Effect of Finite Detector Bandwidth

In this section we discuss the effect of the finite bandwidth of the measurement device on the FT. If the detector bandwidth is finite, the QPC current does not follow the switching between the DQD charge states immediately. Naaman and Aumentado[21] proposed to describe such a system by doubling the number of states. In our system the DQD switches between three states $|D\rangle$, $|L\rangle$, $|R\rangle$, and the QPC current takes three values corresponding to those states $|D\rangle_{\rm QPC}$, $|L\rangle_{\rm QPC}$, $|R\rangle_{\rm QPC}$. Then, we should describe the system by 9 states $|r\rangle|r'\rangle_{\rm QPC}$ $(r, r' = L, R, D)$, and we have to consider a vector of 9 occupation probabilities

$$\mathbf{p}^T = (p_{LL}, p_{RL}, p_{DL}, p_{LR}, p_{RR}, p_{DR}, p_{LD}, p_{RD}, p_{DD}), \tag{21.22}$$

where the first index refers to the state of the DQD and the second one to the value of the QPC current. As shown in Fig. 21.5, the detector will always change to the state corresponding to the dot-state. We model the fact that the detector needs a finite time for this switching, by introducing the detector rate Γ_D. For an ideal detector, $\Gamma_D \to \infty$, only the states are p_{LL}, p_{RR} and p_{DD}.

Because of the one-way transitions between detector states the considered system is outside the class discussed in Ref.[4] In order to describe the experiment and the consequences for the FT we calculate the cumulant

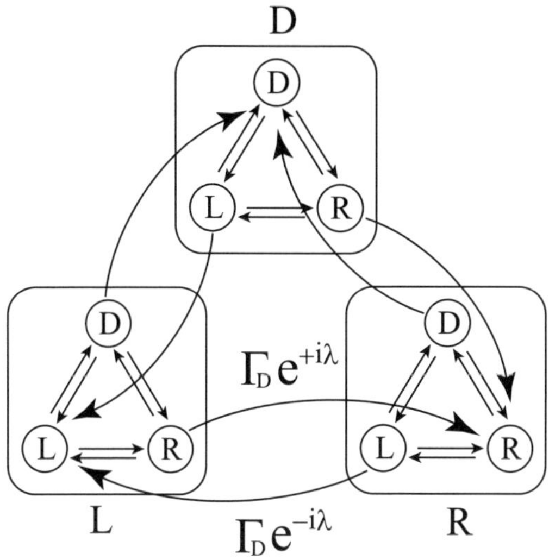

Fig. 21.5. State transition diagram for the double dot and the QPC detector with finite bandwidth. Circles represent the double-dot states and arrows show the directions of the transitions. The additional states representing the detector states (squares) are introduced. Within each square, transitions between all the dot states are possible. Between the 3 detector states, one-way transition, the relaxation from the 'false' detector state to the 'true' detector state, occurs.

generating function. We introduce the master equation of the total system,

$$\partial_t \mathbf{p} = \boldsymbol{M}(\lambda)\,\mathbf{p}, \tag{21.23}$$

where the transition matrix is a 9×9 matrix,

$$\boldsymbol{M}(\lambda) = \begin{pmatrix} \boldsymbol{\Gamma}(0) - \boldsymbol{\Gamma}_{D2} - \boldsymbol{\Gamma}_{D3} & \boldsymbol{\Gamma}_{D1} e^{-i\lambda} & \boldsymbol{\Gamma}_{D1} \\ \boldsymbol{\Gamma}_{D2} e^{i\lambda} & \boldsymbol{\Gamma}(0) - \boldsymbol{\Gamma}_{D1} - \boldsymbol{\Gamma}_{D3} & \boldsymbol{\Gamma}_{D2} \\ \boldsymbol{\Gamma}_{D3} & \boldsymbol{\Gamma}_{D3} & \boldsymbol{\Gamma}(0) - \boldsymbol{\Gamma}_{D1} - \boldsymbol{\Gamma}_{D2} \end{pmatrix}, \tag{21.24}$$

with sub-matrices given by

$$\boldsymbol{\Gamma}_{Di} = \Gamma_D \begin{pmatrix} \delta_{i1} & 0 & 0 \\ 0 & \delta_{i2} & 0 \\ 0 & 0 & \delta_{i3} \end{pmatrix}. \tag{21.25}$$

Note that the counting field is associated with the rate Γ_D. I.e., in the present model the switching of the QPC current are counted and not the transitions in the DQD system. Thus the model provides information about the experimentally accessible statistics of the detector rather than that of the DQD, which is not directly measurable. The model outlined above fits the measured higher cumulants for the single-dot case quite accurately.[22,23]

The FCS cumulant generating function $\mathcal{F}$ is obtained by solving the characteristic equation for the eigenvalue

$$\begin{aligned} 0 &= \det|\boldsymbol{M}(\lambda) - \mathcal{F}\mathbf{I}| \\ &= \det|\boldsymbol{M}(0) - \mathcal{F}\mathbf{I}| + \Gamma_+(\mathcal{F})\left(e^{i\lambda} - 1\right) + \Gamma_-(\mathcal{F})\left(e^{-i\lambda} - 1\right), \end{aligned} \tag{21.26}$$

where $\Gamma_\pm$ are factorized as,

$$\Gamma_+(\mathcal{F}) = \Gamma_D^6(x\,\Gamma_{DL} + \Gamma^*_{DL})(x\,\Gamma_{LR} + \Gamma^*_{LR})(x\,\Gamma_{RD} + \Gamma^*_{RD}), \tag{21.27}$$

$$\Gamma_-(\mathcal{F}) = \Gamma_D^6(x\,\Gamma_{LD} + \Gamma^*_{LD})(x\,\Gamma_{RL} + \Gamma^*_{RL})(x\,\Gamma_{DR} + \Gamma^*_{DR}). \tag{21.28}$$

We introduced $x = \mathcal{F}/\Gamma_D$ and the corrected tunnel rates

$$\Gamma^*_{ij} = \Gamma_{ij}\left(1 + \frac{\Gamma_{ik} + \Gamma_{jk}}{\Gamma_D}\right) + \frac{\Gamma_{ik}\Gamma_{kj}}{\Gamma_D}, \quad (k \neq i, j). \tag{21.29}$$

In order to check whether the FT is satisfied, we consider the following ratio, generalizing w of the ideal case Eq. (21.12),

$$w^* = \frac{\Gamma_+(\mathcal{F})}{\Gamma_-(\mathcal{F})}. \tag{21.30}$$

The FT has to be exact if this ratio does not depend on $\mathcal{F}$. It is obvious that generally this is not the case and therefore the FT is violated. However, in two limits, for a fast detector $\Gamma_D \gg \Gamma_{ij}$ and for a slow detector $\Gamma_D \ll \Gamma_{ij}$, we are able to show analytically that the FT holds. In the former case, $\Gamma_D \gg \Gamma_{ij}$, we expand w^* in powers of $1/\Gamma_D$. Since $\mathcal{F}_D \sim \Gamma_{ij}$, we arrive at the following result

$$w^* = w + \frac{1-w}{\Gamma_D}\left(\frac{\Gamma_{LD}\Gamma_{DR}}{\Gamma_{LR}} + \frac{\Gamma_{RL}\Gamma_{LD}}{\Gamma_{RD}} + \frac{\Gamma_{DR}\Gamma_{RL}}{\Gamma_{DL}}\right) + \mathcal{O}\left(\frac{\Gamma_{ij}^2}{\Gamma_D^2}\right). \tag{21.31}$$

We observe that in the lowest and next to lowest orders in the parameter Γ_{ij}/Γ_D the ratio w^* does not depend on $\mathcal{F}$ and thus the FT holds. Comparing Eqs. (21.29) and (21.31) we note that within the accuracy of our approximation, i.e. up to the terms $\sim 1/\Gamma_D$, the ratio w^* may be written in the same form as for an ideal detector but with modified tunnel rates,

$$w^* \approx \frac{\Gamma_+(0)}{\Gamma_-(0)} = \frac{\Gamma^*_{DR}\Gamma^*_{RL}\Gamma^*_{LD}}{\Gamma^*_{LR}\Gamma^*_{RD}\Gamma^*_{DL}}, \tag{21.32}$$

and effective temperature

$$T^* = \frac{eV_S}{k_B \ln w^*}. \tag{21.33}$$

In the opposite limit of a slow detector, $\Gamma_D \ll \Gamma_{ij}$, one can show that $\mathcal{F} \sim \Gamma_D \ll \Gamma_{ij}$ and therefore one can put $\mathcal{F} = 0$ in Eq. (21.30). Surprisingly, this means that Eqs. (21.32-21.33) also become valid in the opposite limit, that of a very slow detector. Numerical calculations of $\mathcal{F}$ confirm these results, but as expected $\mathcal{F}$ does not have the right symmetries for intermediate values of Γ_D.

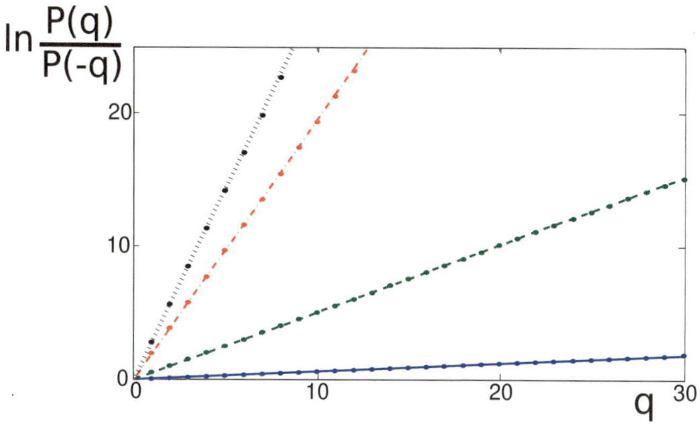

Fig. 21.6. The ratio of the probability of transfered charge as a function of the number of charges in the long-time limit. The solid lines are the results given by the analytical expression for the effective temperature [see Eq. (21.33)], the dots are from fully numerical calculations. The different colors correspond to different rates for the detector bandwidth: (black dotted-line) $\Gamma_D = 100\,\text{kHz}$, (red dash-dotted line) $\Gamma_D = 10\,\text{kHz}$, (green dashed line) $\Gamma_D = 1\,\text{kHz}$, (blue solid line) $\Gamma_D = 0.1\,\text{kHz}$. The rates for the transitions on the dot are the same as used in Ref.[18] and are of the order of 1kHz: $\Gamma_{DR}=4$kHz, $\Gamma_{RD}=0.3$kHz, $\Gamma_{DL}=1$kHz, $\Gamma_{LD}=1.5$kHz, $\Gamma_{LR}=1.7$kHz, and $\Gamma_{RL}=1.8$kHz.

In Fig. 21.6 we compare our result for the effective temperature with numerical results. Varying the detector bandwidth over four orders of magnitude we see that our expression for the effective temperature Eq. (21.33) provides good fits for rather wide range of parameters of q and Γ_D. This means that although formally the detector model introduced by Naaman and Aumentado violates the FT, practically, the finite bandwidth effect can be accounted for simply by the effective temperature. Thus Eqs. (21.32-21.33) should describe the experiment reasonably well, regardless of the value of Γ_D.

To conclude, in this section we have demonstrated that a finite detector bandwidth in general distorts the measured statistics of the charge transfer and leads to the formal violation of FT. However the effect of the finite bandwidth can be accounted for by using the expression (21.33) for the effective temperature.

21.6. Tunneling Rates in the Experiment

Let us now discuss how to determine the values of Γ_{ij} and Γ_D from experimental data. The standard way of doing this is to generate dwell time histograms for every current state. Let us consider the QPC current state corresponding to the DQD state $|L\rangle$. One should count how many times during sufficiently long observation time the QPC current has switched from the state $|L\rangle$ to either state $|R\rangle$ or $|D\rangle$ within the time interval from τ and $\tau + \Delta\tau$, where $\Delta\tau$ is sufficiently short. Denoting the corresponding numbers $\Delta N_{L\to R}$ and $\Delta N_{L\to D}$, one gets the histograms plotting the values $\Delta N_{L\to R,\, L\to R}(\tau)$ as a function of time τ. At sufficiently long time τ the numbers $\Delta N_{L\to R,\, L\to R}(\tau)$ decay in time exponentially,

$$\Delta N_{L\to R}(\tau) = K_{RL} e^{-\Gamma_L^* \tau}, \quad \Delta N_{L\to D}(\tau) = K_{DL} e^{-\Gamma_L^* \tau}. \qquad (21.34)$$

The parameters Γ_L^*, K_{RL} and K_{DL} are extracted by fitting the histograms.

In the case of a fast detector, $\Gamma_D \gg \Gamma_{ij}$ one can easily express the DQD tunneling rates Γ_{RL} and Γ_{DL} in terms of the parameters Γ_L^*, K_{RL}, K_{DL}. Indeed, the theory in this case predicts $\Gamma_L^* = \Gamma_{RL} + \Gamma_{DL}$ and $K_{RL}/K_{DL} = \Gamma_{RL}/\Gamma_{DL}$ and, therefore

$$\Gamma_{RL} = \frac{K_{RL}}{K_{RL} + K_{DL}} \Gamma_L^*, \quad \Gamma_{DL} = \frac{K_{DL}}{K_{RL} + K_{DL}} \Gamma_L^*. \qquad (21.35)$$

The remaining four rates are determined analogously.

In the following, we will present our analysis in detail. Suppose at time $\tau = 0$ the QPC current has switched to the state L. Since the state of the DQD remains unknown, the total occupation probability of such a state is given by the sum $p_L = p_{LL} + p_{RL} + p_{DL}$. To find out how this probability decays in time we have to solve the equation for the probabilities $\mathbf{p}^T = (p_{LL}, p_{RL}, p_{DL})$,

$$\partial_t \mathbf{p} = (\mathbf{\Gamma}(0) - \mathbf{\Gamma}_{D2} - \mathbf{\Gamma}_{D3})\, \mathbf{p}\,. \qquad (21.36)$$

This equation is a sub-block of a more general equation (21.23) where only the outgoing processes from the QPC state L are kept. For a large number of events the histograms should converge to the following expressions

$$\Delta N_{L\to R}(\tau) = N\Gamma_D p_{RL}, \quad \Delta N_{L\to D}(\tau) = N\Gamma_D p_{DL}, \qquad (21.37)$$

where N is the normalization factor. Solving the differential equation and considering the long-time limit we arrive at Eqs. (21.34) with

$$\Gamma_L^* = \Gamma_{RL} + \Gamma_{DL} - \frac{\Gamma_{RL}\Gamma_{LR} + \Gamma_{DL}\Gamma_{LD}}{\Gamma_D} + \mathcal{O}\left(\frac{1}{\Gamma_D^2}\right) \tag{21.38}$$

and

$$\frac{K_{RL}}{K_{DL}} = \frac{\Gamma_{RL}^*}{\Gamma_{DL}^*} + \mathcal{O}\left(\frac{1}{\Gamma_D^2}\right). \tag{21.39}$$

Thus the ratio of the prefactor is sufficient to estimate the effective temperature (21.33).

One can also obtain the full time-dependence of $\Delta N_{L\to R}(\tau)$ and $\Delta N_{L\to D}(\tau)$. To this end we put $p_{LL}(0) = 1, p_{RL}(0) = p_{DL}(0) = 0$ and in the limit $\Gamma_D \gg \Gamma_{ij}$ obtain the result

$$\Delta N_{L\to R} = \frac{N\Gamma_D[\Gamma_{RL}(\Gamma_D + \Gamma_{LD} + \Gamma_{RD} - \Gamma_L^*) + \Gamma_{RD}\Gamma_{DL}]\left[e^{-\Gamma_L^*\tau} - e^{-\Gamma_D\tau}\right]}{(\Gamma_D + \Gamma_{LR} + \Gamma_{D2} - \Gamma_L^*)(\Gamma_D + \Gamma_{LD} + \Gamma_{RD} - \Gamma_L^*) - \Gamma_{DR}\Gamma_{RD}},$$

$$\Delta N_{L\to D} = \frac{N\Gamma_D[\Gamma_{DL}(\Gamma_D + \Gamma_{LR} + \Gamma_{DR} - \Gamma_L^*) + \Gamma_{DR}\Gamma_{RL}]\left[e^{-\Gamma_L^*\tau} - e^{-\Gamma_D\tau}\right]}{(\Gamma_D + \Gamma_{LR} + \Gamma_{D2} - \Gamma_L^*)(\Gamma_D + \Gamma_{LD} + \Gamma_{RD} - \Gamma_L^*) - \Gamma_{DR}\Gamma_{RD}}.$$

With the aid of these expressions one can extract Γ_D from the experimental data.

21.7. Summary

We have discussed the possibility of experimental verification of the FT (21.2) in single-electron counting experiments. Although the experiments allow testing the FT, the properties of the detector turn out to be very important. We have shown that in a generic system of a double quantum dot the FT (21.2) is robust against the backaction of a QPC detector in the sense that the effect can be absorbed in an effective temperature (21.11). We also investigated the influence of the finite bandwidth of the detector. We found that finite bandwidth results in the distortion of the signal properties and formally cannot be reduced to the renormalization of the effective temperature in Eq. (21.2). However, in practice these deviations turn out to be small, and Eq. (21.2) with modified temperature (21.33) should describe the data quite well, even if they are obtained with a slow detector. We hope the experimental test of the FT in the single-electron transport,[18] as well as that in the Aharonov-Bohm ring,[19] would stimulate the development in the nonequilibrium statistical physics and the mesoscopic quantum physics.

Acknowledgment

We thank M. Hettler, K. Kobayashi, and K. Saito for valuable discussions. This work has been supported by Strategic International Cooperative Program of the Japan Science and Technology Agency (JST) and by the German Science Foundation (DFG).

References

1. D. J. Evans, E. G. D. Cohen, and G. P. Morriss, Phys. Rev. Lett **71**, 2401 (1993).
2. J. L. Lebowitz and H. Spohn, J. Stat. Phys. **95**, 333 (1999).
3. D. Andrieux and P. Gaspard, J. Stat. Mech. P01011 (2006).
4. D. Andrieux and P. Gaspard, J. Stat. Phys. **127**, 107 (2007).
5. J. Kurchan, cond-mat/0007360.
6. J. Tobiska and Yu. V. Nazarov, Phys. Rev. B **72**, 235328 (2005).
7. M. Esposito, U. Harbola, and S. Mukamel, Phys. Rev. B **75**, 155316 (2007).
8. H. Förster and M. Büttiker, Phys. Rev. Lett. **101**, 136805 (2008).
9. K. Saito and Y. Utsumi, Phys. Rev. B **78**, 115429 (2008).
10. Y. Utsumi and K. Saito, Phys. Rev. B **79**, 235311 (2009).
11. D. Andrieux, P. Gaspard, T. Monnai, and S. Tasaki, New J. Phys. **11**, 043014 (2009).
12. C. Jarzynski, Phys. Rev. Lett. **78**, 2690 (1997).
13. M. Campisi, P. Talkner, and P. Hänggi, Phys. Rev. Lett. **102**, 210401 (2009).
14. G. Gallavotti, Phys. Rev. Lett. **77**, 4334 (1996).
15. G. M. Wang, E. M. Sevick, E. Mittag, D. J. Searles, and D. J. Evans, Phys. Rev. Lett. **89**, 050601 (2002).
16. J. Liphardt, S. Dumont, S. B. Smith, I. Tinoco Jr., C. Bustamante, Nature **296**, 1832 (2002).
17. Y. Imry, *Introduction to Mesoscopic Physics* (Oxford University Press, Oxford, 1997) 2nd ed.
18. Y. Utsumi, D. S. Golubev, M. Marthaler, K. Saito, T. Fujisawa, and G. Schön, Phys. Rev. B **81**, 125331 (2010).
19. S. Nakamura, Y. Yamauchi, M. Hashisaka, K. Chida, K. Kobayashi, T. Ono, R. Leturcq, K. Ensslin, K. Saito, Y. Utsumi, A. C. Gossard, Phys. Rev. Lett. **104**, 080602 (2010).
20. T. Fujisawa, T. Hayashi, R. Tomita, and Y. Hirayama, Science **312**, 1634 (2006).
21. O. Naaman and J. Aumentado, Phys. Rev. Lett. **96**, 100201 (2006).
22. S. Gustavsson, R. Leturq, T. Ihn, K. Ensslin, R. Reinwald, and W. Wegescheider, Phys. Rev. B **75**, 075314 (2007).
23. C. Flindt, C. Fricke, F. Hohls, T. Novotny, K. Netocny, T. Brandes, and R.J. Haug, Proc. Natl. Acad. Sci. USA **106**, 10116 (2009).
24. S. Gustavsson, R. Leturcq, M. Studer, I. Shorubalko, T. Ihn, K. Ensslin, D.C. Driscoll, A.C. Gossard, Surface Science Reports **64**, 191 (2009).

25. D. A. Bagrets and Yu. V. Nazarov, Phys. Rev. B **67**, 085316 (2003); D.A. Bagrets, Y. Utsumi, D.S. Golubev, and G. Schön, Fortschritte der Physik **54**, 917 (2006).
26. J. Schnakenberg, Rev. Mod. Phys. **48**, 571 (1976).
27. U. Gasser, S. Gustavsson, B. Küng, K. Ensslin, T. Ihn, D. C. Driscoll, and A. C. Gossard, Phys. Rev. B **79**, 035303 (2009).
28. M. Hashisaka, Y. Yamauchi, S. Nakamura, S. Kasai, T. Ono, and K. Kobayashi, Phys. Rev. B **78**, 241303(R) (2008).
29. R. Aguado and L. P. Kouwenhoven, Phys. Rev. Lett. **84**, 1986 (2000).
30. G.-L. Ingold and Y. V. Nazarov, in *Single Charge Tunneling*, eds. H. Grabert and M. Devoret, NATO ASI, Series B: Physics (Plenum, N.Y., 1992), Vol. 294, pp. 21-107.
31. J. König, H. Schoeller, and G. Schön, Phys. Rev. Lett. 76, 1715 (1996); J. König, J. Schmid, H. Schoeller, and G. Schön, Phys. Rev. B 54, 16820 (1996).
32. D. V. Averin and E. V. Sukhorukov, Phys. Rev. Lett. **95**, 126803 (2005).

Chapter 22

Phase Relaxation of Electrons in Disordered Thin Ferromagnetic Films

P. Wölfle and K. A. Muttalib

Institut für Theorie der Kondensierten Materie and DFG-Center for Functional Nanostructures, Karlsruhe Institute of Technology (KIT), D-76128 Karlsruhe, Germany
E-mail: peter.woelfle@kit.edu
Department of Physics, University of Florida, P.O. Box 118440, Gainesville, FL 32611-8440
E-mail: muttalib@phys.ufl.edu

The phase relaxation caused by scattering off spin wave excitations in disordered ferromagnetic films is considered. In the case that the spin wave spectrum is gapped, it is shown that the corresponding phase relaxation rate at temperatures above the gap follows a linear dependence in both two and three dimensions, similar to what has been found previously for clean systems. In the absence of the gap the behavior is controlled by the spin wave damping characteristics, typically leading to fractional power laws in temperature. The effect of finite external magnetic field is considered and shown to affect the prefactor of the linear T law.

22.1. Introduction

The transport properties of thin metallic disordered ferromagnetic films show quantum corrections at low temperatures, studied in detail in recent experiments.[1,2] In particular a weak localization contribution to both the longitudinal and the Hall conductivity has been identified. This may seem surprising at first sight, since a magnetic field (as applied in these experiments), spin flip scattering or spin-orbit scattering may be expected to limit the phase coherence in these systems to short length scales. While it is certainly true that the above mechanisms severely limit the phase relaxation time (the corresponding rate may be estimated to be of order $1K$), at higher temperatures $T > 5K$ a temperature dependent contribution to the phase relaxation rate appears to be visible in the data.

It has been proposed some time ago that scattering of electrons off spin waves plays an important role in transport in thin ferromagnetic films.[3,4] In a clean system (without disorder) at low temperatures T , the inelastic scattering rate due to spin-waves $(\tau_\varphi)^{-1}$ has been obtained,[3] within a simple self-energy approximation, to be proportional to T/Δ , where Δ is the gap in the spin-wave excitation spectrum. This result has been used to argue[1] that while the phase relaxation rate from the electron-electron interaction is too small for weak localization effects to be observable in thin ferromagnetic films, the much larger contribution obtained from the scattering off spin waves provides a window of temperature where weak localization effects are experimentally accessible. In particular, it is the phase relaxation length which determines whether a film is effectively in two or three dimensions at a given disorder and temperature. Since the spin-wave contribution dominates in a ferromagnetic film, it is important to know the disorder as well as temperature dependence of (τ_φ). However, a calculation of $(\tau_\varphi)^{-1}$ in the diffusive limit does not exist yet. In this work we will calculate $(\tau_\varphi)^{-1}$ for both two and three dimensions, in the clean as well as the diffusive limits. We will show that in all different cases, $(\tau_\varphi)^{-1}$ has a linear dependence on temperature T, except in the clean 3D limit where the dependence is $T \ln T$.

Note that in case of thin films, the internal magnetic field is equal to the external field since the magnetization inside is cancelled by surface polarizations.[3] Thus in the absence of an external field, the Zeeman splitting $B = 0$. In considering the anomalous Hall effect which is measured in the presence of a large magnetic field, one needs to keep a finite B.

22.2. The Model

The Hamiltonian we will consider is given as

$$\begin{aligned} H = &\sum_{\mathbf{k}\sigma}(\epsilon_{\mathbf{k}} - \frac{1}{2}\sigma B)c^+_{\mathbf{k}\sigma}c_{\mathbf{k}\sigma} \\ &+ \sum_{\mathbf{k},\mathbf{k}'\sigma,\mathbf{j}} V(\mathbf{k}-\mathbf{k}')e^{i(\mathbf{k}-\mathbf{k}')\cdot\mathbf{R}_j}c^+_{\mathbf{k}'\sigma}c_{\mathbf{k}\sigma} \\ &+ \sum_q \omega_q a^+_q a_q + J\sum_{q,k}[a^+_q c^+_{k+q\downarrow}c_{k\uparrow} + h.c.] \end{aligned}$$

where the annihilation and creation operators c_k, c^+_k refer to electrons and a_q, a^+_q refer to spin-waves, J is the effective spin exchange interaction and B is the Zeeman splitting. The impurity scattering is modelled by the potential V. The spin wave has a dispersion relation $\omega_q = \Delta + Aq^2$, where Δ is the spin-wave gap and A is the spin stiffness. The spin-wave propagator in Matsubara representation is

$$S_{\uparrow\downarrow}(q,\omega_l) = \frac{1}{i\omega_l - \omega_q + i\gamma(q)sign(\omega_l)} = [S_{\downarrow\uparrow}]^*$$

where $\omega_l = 2\pi lT$ is the bosonic Matsubara frequency, and $\gamma(q)$ is a phenomenological damping parameter that depends on q as some power q^α with $\alpha \geq 2$. The spin-wave mediated effective electron interaction is then given by

$$v(q,\omega_l) = nJ^2[S_{\uparrow\downarrow}(q,\omega_l) + S_{\downarrow\uparrow}(q,\omega_l)] \tag{22.1}$$

$$= -\frac{2nJ^2\omega_q}{(|\omega_l| + \gamma)^2 + \omega_q^2} \tag{22.2}$$

which is attractive. Here n is the density of conduction electrons.

22.3. Inelastic Scattering Time from Self Energy: Known Result

We will first reproduce the results of Ref. 3 where the inelastic scattering is obtained simply from the imaginary part of the self energy. Fig. 22.1 shows the diagram for the self energy, corresponding to the expression

$$\begin{aligned}
\Sigma(k,\epsilon_n) &= nJ^2\sum_q T\sum_{\omega_l} G_{k-q}(i\epsilon_n - i\omega_l)v(q,i\omega_l) \\
&\simeq nJ^2\sum_q T\sum_{\omega_l} \frac{1}{i\epsilon_n - i\omega_l - \epsilon_{k-q}}\frac{2\Delta}{(i\omega_l)^2 - \Delta^2} \\
&= -nJ^2\sum_q \left[\frac{2\Delta b(i\epsilon_n - \epsilon_{k-q})}{(i\epsilon_n - \epsilon_{k-q})^2 - \Delta^2} + \frac{b(\Delta)}{i\epsilon_n - \epsilon_{k-q} - \Delta} + \frac{b(-\Delta)}{i\epsilon_n - \epsilon_{k-q} + \Delta}\right]
\end{aligned}$$

where we approximated the spin wave spectrum by $\omega_q \simeq \Delta$ and the summation on ω_l has been done by contour integration.

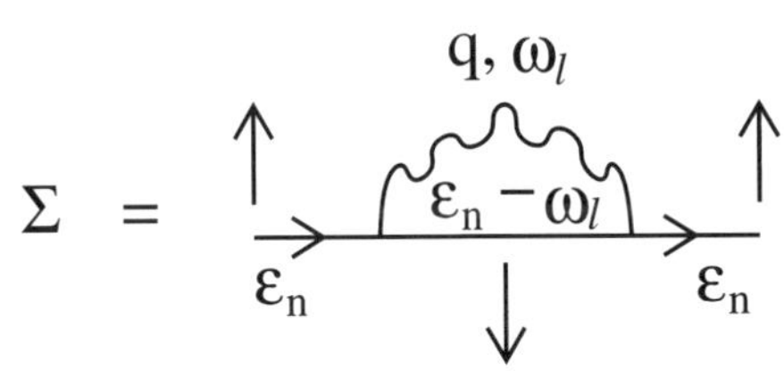

Fig. 22.1. Self energy diagram for spin wave interaction.

We now use the properties of the Bose function, $b(i\epsilon_n - \epsilon_{k-q}) = -[1 - f(\epsilon_{k-q})]$, and $b(-\Delta) = -[b(\Delta) + 1]$ to get

$$\Sigma(k, \epsilon_n) = -nJ^2 \sum_q \left[\frac{f(\epsilon_{k-q}) + b(\Delta)}{i\epsilon_n - \epsilon_{k-q} + \Delta} + \frac{1 - f(\epsilon_{k-q}) + b(\Delta)}{i\epsilon_n - \epsilon_{k-q} - \Delta} \right].$$

Converting the integration on $\mathbf{q}$ into an integration over energy, which may be performed by contour integration, this becomes

$$\Sigma(k, \epsilon_n) = 2\pi n J^2 N_0 [f(i\epsilon_n + \Delta) + b(\Delta) + (1 + b(\Delta)) - f(i\epsilon_n - \Delta)].$$

In the limit $\epsilon_n \to 0$, and using $f(\Delta) + b(\Delta) = 1/\sinh(\beta\Delta)$, where $\beta = 1/T$ (we will use units such that the Boltzmann constant $k_B = 1$), we get finally

$$\frac{1}{2\tau_\epsilon^0} = Im[\Sigma] = \frac{4\pi N_0 n J^2}{\sinh(\beta\Delta)} \tag{22.3}$$

where N_0 is the density of states at the Fermi level. At high temperatures $T >> \Delta$ this gives a linear T dependence as found in Ref. 3. The result for the energy relaxation rate $1/\tau_\epsilon^0$ found here might not be appropriate for use as the cut off energy for weak localization. Rather, it is necessary to calculate the dephasing rate appearing in the Cooperon propagator. This will be done in the next section.

22.4. Phase Relaxation Time of Weak Localization: $B = 0$

The phase relaxation time τ_φ limiting phase coherence in a weak localization situation appears in the particle-particle diffusion propagator $C(q, \omega)$ (Cooperon) in the form

$$C(q, \omega_l) = \frac{1}{2\pi N_0 \tau^2} \frac{1}{Dq^2 + |\omega_l| + 1/\tau_\varphi}.$$

Here N_0 is the density of states at the Fermi level, τ is the elastic scattering time and $D = \frac{1}{d} v_F^2 \tau$ is the diffusion coefficient, where d is the dimensionality and v_F is the Fermi velocity. We have used $\omega_l = 2\pi l T$ as the Matsubara frequency. If we label the Cooperon propagator in the absence of interactions as C_0, then we can write

$$\frac{1}{\tau_\varphi} = \frac{1}{2\pi N_0 \tau^2} [C^{-1} - C_0^{-1}].$$

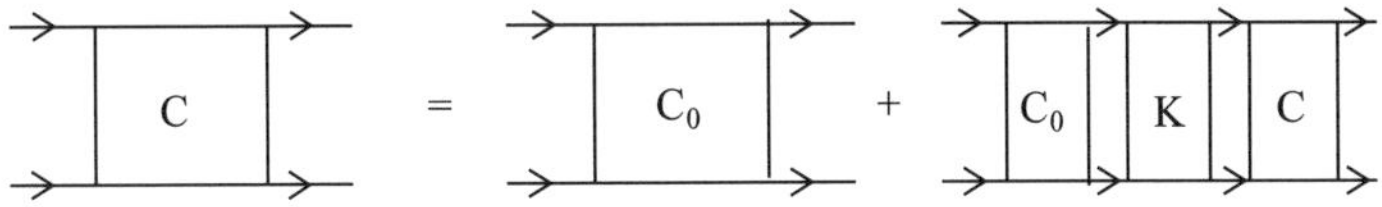

Fig. 22.2. Full particle-particle diffusion propagator C with spin wave interaction, in terms of C_0 in the absence of interaction and the iinteraction vertex K.

In general, $C(q,\omega)$ can be evaluated diagrammatically in the presence of interactions and disorder in a ladder approximation[5] that can be symbolically written as

$$C = C_0 + C_0 K C$$

where the interaction vertex K contains self energy as well as vertex corrections due to both interactions and disorder. It then follows that $1/\tau_\varphi$ is given by

$$\frac{1}{\tau_\varphi} = -\frac{1}{2\pi N_0 \tau^2} K. \tag{22.4}$$

We will evaluate the diffusion propagator diagrammatically, in the presence of spin-wave mediated electron-electron interactions, both in the clean and in the diffusive limits, in two and three dimensions, in the absence or presence of Zeeman splitting.

An approximate diagrammatic calculation of the inelastic scattering time for electron-electron (e-e) interactions was performed by Fukuyama and Abrahams (FA)[5] who obtained $1/\tau_\varphi \propto T \ln T$. A later exact calculation by Altshuler and Aronov (AA)[6] gave $1/\tau_\varphi \propto T$. However, the AA method can not be easily adapted for spin-wave interactions. On the other hand, as we will see, in the diffusive limit including the spin-wave interaction, $1/\tau_\varphi$ does not acquire a $\ln T$ term that rendered the perturbative result in the e-e- case less reliable.[7] We will therefore follow the method of FA to calculate diagrammatically the leading temperature and disorder dependence of the inelastic scattering time.

Fig. 22.3 shows the vertex K to be calculated, in the absence of disorder. Note that compared to the e-e interaction, there are only self energy corrections and no vertex correction because the spin wave operator creates a down spin from an up spin on one vertex but an up spin from a down spin on the other. This will exist only if the two initial spins are not of the same type. Since vertex corrections were argued to be negligible in the e-e case, we can essentially use the FA results with e-e interactions replaced

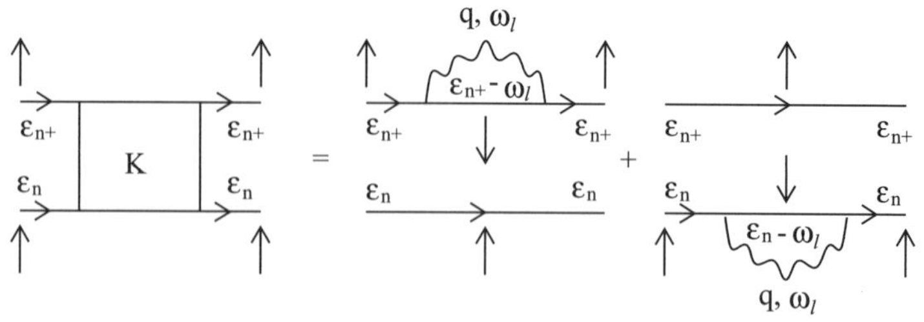

Fig. 22.3. Diagrams for inelastic lifetime due to scattering off spin waves.

by the spin-wave interactions. For definiteness, we will consider only two initial up spins.

22.4.1. *Clean limit: finite spin-wave gap*

The presence of a spin-wave gap provides a natural low-frequency cut-off, which makes the calculation of τ_φ simpler. In the absence of Zeeman splitting and in the clean limit, FA obtain the following result:

$$K^0 = 8\pi\tau^2 \sum_q \int \frac{dx}{\sinh(\beta x)} A_q(x) Im\{v(q,x)\}$$

where

$$A_q(x) = \sum_k \delta(\epsilon - \epsilon_k)\delta(\epsilon + x - \epsilon_{k+q}).$$

In $d = 2, 3$ dimensions we get

$$A_q(x) = \frac{A_d}{q}; \quad A_2 \equiv \frac{N_0}{\pi v_F}; \quad A_3 \equiv \frac{N_0}{2v_F}.$$

The inelastic scattering time is then given by

$$\frac{1}{\tau_\varphi} = \frac{8A_d}{(2\pi)^d N_0} nJ^2 \int \frac{d^d q}{q} \frac{1}{\sinh(\beta\omega_q)}.$$

The integral over q is from 0 to some upper limit q_0 determined essentially by the lattice spacing. Using the spin-wave dispersion relation $\omega_q = \Delta + Aq^2$, we can approximate the integral in the regime where $T \gg \Delta$ as

$$I_d \equiv \int_0^{q_0} \frac{d^d q}{q} \frac{1}{\sinh \frac{\Delta + Aq^2}{T}} \approx \frac{T}{A} \int_0^{\sqrt{T/A\Delta}} \frac{d^d q}{q} \frac{1}{q^2 + \Delta/A}.$$

This gives us

$$I_2 \approx \frac{\pi}{2}\frac{T}{\Delta}; \quad d = 2 ;$$
$$I_3 \approx \frac{T}{2A}\ln\frac{T}{\Delta}; \quad d = 3.$$

Thus, in the clean limit, the leading temperature dependence of the inelastic scattering rate is given by

$$\frac{1}{\tau_\varphi} = \frac{1}{4\pi^2}\frac{nJ^2}{v_F}\frac{T}{\Delta}; \quad d = 2 ; \tag{22.5}$$
$$\frac{1}{\tau_\varphi} = \frac{1}{4\pi^3}\frac{nJ^2}{v_F}\frac{T}{2A}\ln(\frac{T}{\Delta}); \quad d = 3. \tag{22.6}$$

The 2D result agrees in its temperature dependence with Ref. 3.

22.4.2. *Diffusive limit: finite spin-wave gap*

In the diffusive limit the quantum corrections are given by the diagrams depicted in Fig. 22.2. Again we can copy the results from FA with spin-wave interaction replacing the screened Coulomb interaction. The result is that the q-integral has an extra particle-particle propagator and the inelastic scattering rate is given by

$$\frac{1}{\tau_\varphi} = \frac{4}{\pi}nJ^2\int_0^{q_0}\frac{d^dq}{\sinh(\beta\omega_q)}\frac{Dq^2 + 1/\tau_\varphi}{(Dq^2 + 1/\tau_\varphi)^2 + \omega_q^2}.$$

In the presence of a spin-wave gap the integrals are convergent and if $1/\tau_\varphi <$ Δ we can neglect the $1/\tau_\varphi$ inside the integral. The q-integral then becomes

$$J_d \approx \int_{\sqrt{\Delta/A}}^{q_0}\frac{d^dq}{q}\frac{1}{\sinh(\beta Aq^2})\frac{Dq^2}{(D^2 + A^2)q^4}.$$

For small $\Delta \ll T$, we then get,

$$J_d \approx \frac{T}{A}\frac{D}{(D^2 + A^2)}\int_{\sqrt{\Delta/A}}^{\infty}\frac{dqq^{d-1}}{q^4}$$

where we have extended the upper limit to infinity since the main contribution to the integral comes from small q. The inelastic scattering rate is given, in this limit, by

$$\frac{1}{\tau_\varphi} \approx \frac{2}{\pi}nJ^2\frac{D}{(D^2 + A^2)}\frac{T}{\Delta}; \quad d = 2 \tag{22.7}$$
$$\frac{1}{\tau_\varphi} \approx \frac{4}{\pi}nJ^2\frac{D}{(D^2 + A^2)}\frac{T}{\sqrt{A\Delta}}; \quad d = 3. \tag{22.8}$$

Thus the temperature dependence remains linear in both two and three dimensions. The disorder dependence for large D is $1/D$.

In the opposite limit $1/\tau_\varphi > \Delta$ the cutoff at low q is provided by $1/\tau_\varphi$ rather than Δ, which leads to a self consistency relation for $1/\tau_\varphi$, with the solution

$$\frac{1}{\tau_\varphi} \approx [\frac{2}{\pi} nJ^2 \frac{D}{(D^2+A^2)}]^{1/2} T^{1/2}; \quad d=2\,; \tag{22.9}$$

$$\frac{1}{\tau_\varphi} \approx [\frac{4}{\pi} nJ^2 \frac{D}{(D^2+A^2)\sqrt{A}}]^{2/3} T^{2/3}; \quad d=3. \tag{22.10}$$

22.4.3. *Diffusive limit: zero spin-wave gap*

The above results for $1/\tau_\varphi$ break down when the spin-wave gap goes to zero. In that case we can no longer approximate the imaginary part of the interaction function as sum of two delta functions, but must use the actual spin-wave interaction with a finite damping term. Using the interaction given in Eq. (22.2), we have

$$Im[v(q,\omega)] = \frac{4\gamma\omega\omega_q}{(\omega^2-\omega_q^2-\gamma^2)^2+(2\gamma\omega)^2}.$$

Note that γ is a function of q, which goes to zero for small q as q^α, the precise exponent depending on the model. Because of the presence of the $1/\sinh(\beta\omega)$ term, the major contribution to the frequency integral comes from the small ω region and we can put $\omega = 0$ everywhere in the denominator of $Im[v]$. Thus we use the approximate form

$$Im[v(q,\omega)] \approx \frac{4\gamma\omega\omega_q}{(\omega_q^2+\gamma^2)^2}.$$

Using this form, the inelastic scattering rate becomes

$$\frac{1}{\tau_\varphi} \approx \frac{2}{\pi} nJ^2 \int \frac{d^dq}{(2\pi)^d} \frac{4\gamma\omega_q}{(\omega_q^2+\gamma^2)^2} \int_{-\infty}^{\infty} \frac{xdx}{\sinh(\beta x)} \frac{b}{b^2+x^2}; \quad b \equiv Dq^2 + 1/\tau_\varphi.$$

The integral may be estimated as follows:

$$\begin{aligned} I &= \int_{-\infty}^{\infty} \frac{xdx}{\sinh(\beta x)} \frac{b}{b^2+x^2} \\ &\approx \frac{2}{b} \int_0^b \frac{xdx}{\sinh(\beta x)} + b \int_b^{\infty} \frac{dx}{x\sinh(\beta x)}. \end{aligned}$$

For $b \ll T$, the first term gives $2T$, while the second term is approximately T. For $b \gg T$, the first term is estimated as $\propto T^2/b$ and the second term is negligible. Inserting these results we get

$$\frac{1}{\tau_\varphi} \approx \frac{2}{\pi} n J^2 T \int_0^{\sqrt{T/D}} \frac{d^d q}{(2\pi)^d} \frac{4\gamma\omega_q}{(\omega_q^2+\gamma^2)^2} + \frac{2}{\pi} n J^2 \frac{T^2}{D} \int_{\sqrt{T/D}}^{\infty} \frac{d^d q}{(2\pi)^d} \frac{4\gamma\omega_q}{q^2(\omega_q^2+\gamma^2)^2}.$$

The result now depends sensitively on the exponent α in $\gamma \propto q^\alpha$. Assuming $\alpha = 4$, in three dimensions one finds

$$\frac{1}{\tau_\varphi} \approx \frac{2}{\pi} n J^2 \frac{T^{3/2}}{D^{1/2}}. \tag{22.11}$$

In two dimensions a larger exponent, $\alpha > 4$, is required to give a finite result for $1/\tau_\varphi$.

22.5. Phase Relaxation Time at Finite Magnetic Field

For finite Zeeman splitting B, the Green's functions G_σ for $\sigma = \uparrow, \downarrow$ have different energies. The particle-hole and particle-particle propagators $\Gamma^{ph}_{\sigma,\sigma}$, $\Gamma^{pp}_{\sigma,\sigma}$ with incoming electrons of the same spin type remain the same, but for propagators with different incoming spins the diffusion pole will be cut off by the Zeeman energy. We choose the energies $\epsilon_{k\uparrow} = \epsilon_{k\downarrow} + B$. Explicitly, the equation for the particle-hole propagators is given by

$$\Gamma^{ph}_{\uparrow\downarrow}(q,\omega_l,B) = \Gamma_0 + \Gamma_0 \sum_k G^+_{k+q,\uparrow} G^-_{k,\downarrow} \Gamma^{ph}_{\uparrow\downarrow}(q,\omega_l,B).$$

Here Γ_0 is the impurity potential correlator with $2\pi N_0 \tau \Gamma_0 = 1$. To solve the equation for the particle-hole propagator we use the expansion of G for small q and ω_l ($\omega\tau \ll 1$, $q v_F \tau \ll 1$):

$$G^+_{k+q}(\epsilon_n+\omega_l) = G^+_k(\epsilon_n) - i|\omega_l|(G^+_k)^2(\epsilon_n) + (q\cdot v)^2 (G^+_k)^3(\epsilon_n) + \cdots. \tag{22.12}$$

We use the results

$$\sum_k G^+_{k,\uparrow} G^-_{k,\downarrow} = 2\pi N_0 \tilde{\tau}^*$$
$$\sum_k (G^+_{k,\uparrow})^2 G^-_{k,\downarrow} = -2\pi N_0 i (\tilde{\tau}^*)^2$$
$$\sum_k (G^+_{k,\uparrow})^3 G^-_{k,\downarrow} = -2\pi N_0 (\tilde{\tau}^*)^3$$

where we have defined (the superscript star denotes complex conjugation)

$$\frac{1}{\tilde{\tau}} \equiv \frac{1}{\tau} - iB,$$

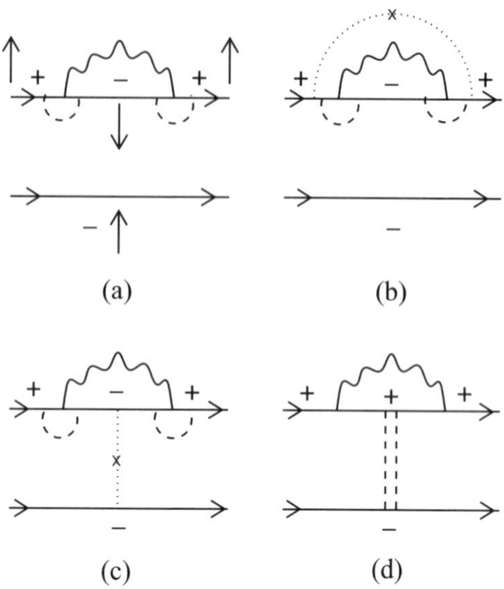

Fig. 22.4. Quantum corrections to K. Here broken and double broken lines are particle hole and particle-particle propagators, respectively. The dotted line with a cross is an impurity scattering and the wavy line is the spin-wave propagator. There are similar diagrams with the spin-wave propagator in the lower line.

to get

$$\sum_k G^+_{k+q,\uparrow} G^-_{k,\downarrow} = 2\pi N_0 \tilde{\tau}^* \left[1 - |\omega_l|\tilde{\tau}^* - \tilde{D}^* q^2 \tilde{\tau}^*\right].$$

The expression for the p-h propagator then becomes

$$\Gamma^{ph}_{\uparrow\downarrow}(q,\omega_l,B) = \frac{1}{2\pi N_0 \tau} \frac{\tau}{\tilde{\tau}^{*2}} \frac{1}{|\omega_l| + \tilde{D}^* q^2 + \tilde{\delta}^*} \tag{22.13}$$

It is easy to see that switching the spin indices $\uparrow \to \downarrow$ in effect switches $B \to -B$, and $\tilde{\tau} \to \tilde{\tau}^*$. The p-p propagator is obtained in a similar way as

$$\Gamma^{pp}_{\uparrow\downarrow}(q,\omega_l,B) = \frac{1}{2\pi N_0 \tau} \frac{\tau}{\tilde{\tau}^2} \frac{1}{|\omega_l| + \tilde{D} q^2 + \tilde{\delta}} = \Gamma^{ph*}_{\uparrow\downarrow}. \tag{22.14}$$

The diagrams contributing to K are shown in Fig. 22.4, with energy labels similar to Fig. 22.3. We will choose for definiteness $\epsilon_{n+} > 0$. Then dominant contributions to K come from the regions with $\epsilon_n < 0$ for all four diagrams and with $\epsilon_{n+} - \omega_l < 0$ for figures (a) - (c) and $\epsilon_{n+} - \omega_l > 0$ for figure (d). The contribution of diagrams (a)-(c) can be written in the form

$$N^{(a+b+c)} = T\sum_{\omega_l}\sum_{q} L^{(a+b+c)}(q,\omega_l)\left[\Gamma^{ph}_{\uparrow\downarrow}(q,\omega_l,B)\right]^2 v(q,\omega_l)\Theta(\omega_l-\epsilon_{n+}), \tag{22.15}$$

where $L^{(a+b+c)} \equiv L^{(a)} + L^{(b)} + L^{(c)}$ and

$$L^{(a)} = \sum_{k'}(G^{+}_{k',\uparrow})^2 G^{-}_{k',\uparrow}G^{-}_{k'-q,\downarrow}\sum_{k''}G^{+}_{k'',\uparrow}G^{-}_{k''-q,\downarrow}\sum_{k'''}G^{+}_{k''',\uparrow}G^{-}_{k'''-q,\downarrow}$$

$$\begin{aligned} L^{(b)} &= \Gamma_0\sum_{k'}(G^{+}_{k',\uparrow})^2 G^{-}_{k',\uparrow}\sum_{k''}(G^{+}_{k'',\uparrow})^2 G^{-}_{k''-q,\downarrow} \\ &\times\sum_{k'''}G^{+}_{k''',\uparrow}G^{-}_{k'''-q,\downarrow}\sum_{k''''}G^{+}_{k'''',\uparrow}G^{-}_{k''''-q,\downarrow}\end{aligned}$$

$$\begin{aligned} L^{(c)} &= \Gamma_0\sum_{k'}G^{+}_{k',\uparrow}G^{-}_{k',\uparrow}G^{-}_{k'-q,\downarrow}\sum_{k''}G^{+}_{k'',\uparrow}G^{-}_{k''-q,\downarrow} \\ &\times\sum_{k'''}G^{+}_{k''',\uparrow}G^{-}_{k''',\uparrow}G^{-}_{k'''-q,\downarrow}\sum_{k''''}G^{+}_{k'''',\uparrow}G^{-}_{k''''-q,\downarrow}\end{aligned}$$

Diagram (d) is of a different form, given by

$$N^{(d)} = T\sum_{\omega_l}\sum_{q} L^{d}(q,\omega_l)\left[\Gamma^{pp}_{\downarrow\uparrow}(q,\omega_l,B)\right] v(q,\omega_l)\Theta(\epsilon_{n+}+\omega_l)$$

with

$$L^{(d)} = \sum_{k'}G^{+}_{k',\uparrow}G^{-}_{k',\uparrow}G^{+}_{k'-q,\downarrow}\sum_{k''}G^{+}_{k'',\uparrow}G^{-}_{k'',\uparrow}G^{+}_{k''-q,\downarrow}.$$

To these, we need to add the diagrams $\bar{a},\bar{b},\bar{c},\bar{d}$, where the interaction lines are at the lower Greens functions. Details of the evaluation may be found in Appendix A.1. The total contribution to N from all diagrams can be written as

$$N \equiv N^{(a+b+c+d)} + N^{(\bar{a}+\bar{b}+\bar{c}+\bar{d})} = (2\pi N_0\tau)^2\sum_{q}[\Gamma_1(q)-\Gamma_2(q)], \tag{22.16}$$

where

$$\Gamma_1 = \left[\frac{(\tilde{\tau}^*)^3}{\tau}T\sum_{\omega_l>\epsilon_{n+}}\Gamma^{ph}_{\uparrow\downarrow}(q,\omega_l) + \frac{(\tilde{\tau})^3}{\tau}T\sum_{\omega_l>-\epsilon_n}\Gamma^{ph}_{\downarrow\uparrow}(q,\omega_l)\right]v(q,\omega_l),$$

$$\Gamma_2 = \left[(\tilde{\tau})^2 T\sum_{\omega_l>-\epsilon_{n+}}\Gamma^{pp}_{\downarrow\uparrow}(q,\omega_l+\omega_\lambda) + (\tilde{\tau})^{*2}T\sum_{\omega_l>\epsilon_n}\Gamma^{pp}_{\uparrow\downarrow}(q,\omega_l+\omega_\lambda)\right]v(q,\omega_l).$$

In the limit ω_λ very small, we use

$$\sum_{\omega_l > -\epsilon_n} \approx \sum_{\omega_l > \epsilon_{n+}} \, ; \quad \sum_{\omega_l > \epsilon_n} \approx \sum_{\omega_l > -\epsilon_{n+}}$$

and rewrite

$$\Gamma_1 \approx T \sum_{\omega_l > \epsilon_{n+}} \left[\frac{(\tilde{\tau}^*)^3}{\tau} \Gamma^{ph}_{\uparrow\downarrow}(q, \omega_l) + \frac{(\tilde{\tau})^3}{\tau} \Gamma^{ph}_{\downarrow\uparrow}(q, \omega_l) \right] v(q, \omega_l), \quad (22.17)$$

$$\begin{aligned} \Gamma_2 &\approx T \sum_{\omega_l > -\epsilon_{n+}} \left[(\tilde{\tau})^2 \Gamma^{pp}_{\downarrow\uparrow}(q, \omega_l + \omega_\lambda) + (\tilde{\tau})^{*2} \Gamma^{pp}_{\uparrow\downarrow}(q, \omega_l + \omega_\lambda) \right] v(q, \omega_l) \\ &= T \sum_{\omega_l > -\epsilon_{n+}} \Gamma_p(q, \omega_l + \omega_\lambda) v(q, \omega_l) \end{aligned} \quad (22.18)$$

where we defined

$$\Gamma_p(q, \omega_l, \omega_\lambda) = 2Re \left[\tilde{\tau}^2 \Gamma^{pp}_{\downarrow\uparrow}(q, \omega_l + \omega_\lambda) \right]. \quad (22.19)$$

Note that using Eqs. (22.13) and (22.14), we have

$$\frac{(\tilde{\tau}^*)^3}{\tau} \Gamma^{ph}_{\uparrow\downarrow}(q, \omega_l) = \frac{1}{2\pi N_0 \tau} \frac{\tilde{\tau}^*}{|\omega_l| + \tilde{D}^* q^2 + \tilde{\delta}^*},$$

$$\tilde{\tau}^2 \Gamma^{pp}_{\downarrow\uparrow}(q, \omega_l + \omega_\lambda) = \frac{1}{2\pi N_0 \tau} \frac{\tau}{|\omega_l + \omega_\lambda| + \tilde{D} q^2 + \tilde{\delta}}.$$

The spin-wave mediated interaction $v(q, \omega_l)$ is given in Eq. (22.2).

We now consider the sum over the frequency in Γ_1 and Γ_2. Note that while Γ_2 includes summation over $\omega_l = 0$, Γ_1 excludes it. Since the bosonic frequency ω_l will include the Bose function which is singular at $\omega_l = 0$, the dominant contribution will come from Γ_2 which includes the sum over zero frequency. Thus we will only consider Γ_2. The summation can be done by analytic continuation to the complex plane, as shown in Appendix A.2. We obtain

$$\Gamma_2 \approx -\frac{4nJ^2 \gamma \omega_q}{\pi} \int_{-\infty}^{\infty} \Gamma_p(q, -i\omega) \frac{\omega}{(-\omega^2 + \omega_q^2 + \gamma^2)^2 + 4\gamma^2 \omega^2} n(\omega) d\omega, \quad (22.20)$$

where now we have taken the limit $\omega_\lambda = 0$.

22.5.1. *Limit of large B*

In order to do the integral in Eq. (22.20) for Γ_2 in the large B limit we will consider a spin-wave spectrum with a gap smaller than T, i.e. $\omega_q = \Delta + Aq^2$ with $\Delta < T$. Note that we also have an integral over q to do. The major contributions to the ω and q integrals come from the small ω regime because of the factor $n(\omega)$, and from the small q regime because of the damping parameter $\gamma(q)$. In this limit of small ω and q, and for $B\tau \gg 1$, we can neglect the ω dependence of $\Gamma_p(q)$ defined in Eq. (22.19) and write

$$\Gamma_2 \approx -\frac{4nJ^2\gamma\omega_q\Gamma_p(q)}{\pi}\int_{-\infty}^{\infty}\frac{x}{\sinh(\beta x)}\frac{1}{x^4+bx^2+a},$$

where we defined $a \equiv (\omega_q^2+\gamma^2)$ and $b \equiv 2(\gamma^2-\omega_q^2)$. Since the dominant contribution comes from small x, we use $\sinh(\beta x) \approx \beta x$ and cut off the integral at an upper limit $1/\beta$. We then get

$$\Gamma_2 \approx -\frac{8nJ^2\gamma\omega_q\Gamma_p(q)}{\pi}I_0.$$

Here we defined

$$I_0 \equiv T\int_0^T \frac{1}{x^4+bx^2+a}$$
$$= \frac{T}{4Q^3\sin\alpha}\left[\sin\frac{\alpha}{2}\ln\frac{T^2+2QT\cos\frac{\alpha}{2}+Q^2}{T^2-2QT\cos\frac{\alpha}{2}+Q^2}+2\cos\frac{\alpha}{2}\left(\arctan\frac{T^2-Q^2}{2QT\sin\frac{\alpha}{2}}+\frac{\pi}{2}\right)\right],$$

where we have used Ref. 8 with $Q \equiv \sqrt{\omega_q^2+\gamma^2}$ and $\sin\alpha \equiv \frac{2\gamma\omega_q}{\omega_q^2+\gamma^2}$. In the limit $\omega_q \ll T$, this gives

$$I_0 \approx \frac{\pi T}{4\gamma(\omega_q^2+\gamma^2)}.$$

Note that in this limit the logarithm gives a small correction term ω_q/T. In the opposite limit $\omega_q \gg T$, we get

$$I_0 \approx \frac{T^2}{2\omega_q^2(\omega_q^2+\gamma^2)}.$$

We are left with doing the q-integral. In the limit $T \gg \omega_q$, we have small q such that $q \ll (T-\Delta)/A$. Then for $B\tau \gg 1$, using Eq. (22.19), we obtain

$$\Gamma_p \approx \frac{1}{2\pi N_0}\left[\frac{1}{\tilde{\delta}}+\frac{1}{\tilde{\delta}^*}\right] = -\frac{\tau}{\pi N_0}\frac{1}{B^2\tau^2}$$

which is independent of q. Thus we get

$$\Gamma_2 \approx -\frac{2nJ^2}{\pi B}\frac{1}{\pi N_0 \tau}\frac{T}{B}\frac{\omega_q}{\omega_q^2+\gamma^2}.$$

This finally leads to a linear T dependence for the relaxation rate,

$$\frac{1}{\tau_\varphi} \approx \frac{4nJ^2}{B\tau}\left(\frac{T}{B}\right)\int_0^{q_0}(dq)\frac{\omega_q}{\omega_q^2+\gamma^2}. \tag{22.21}$$

Note that for finite spin wave gap, the q-integral is finite. Since $\gamma(q) \sim q^\alpha$, the integral contributes most for small q.

22.5.2. *Limit of small B*

More generally, for $\omega \lesssim T$ and $B > T$, we can still neglect the ω dependence of Γ_p defined in Eq. (22.19), with

$$\begin{aligned}\Gamma_p(q) &\approx \frac{1}{2\pi N_0}\left[\frac{1}{\tilde{D}q^2+\tilde{\delta}}+\frac{1}{\tilde{D}^*q^2+\tilde{\delta}^*}\right]\\ &= \frac{1}{2\pi N_0}\frac{2Dq^2-2B^2\tau(1+B^2\tau^2}{D^2q^4-4B^2\tau^2q^2+B^2(1+B^2\tau^2)^2}.\end{aligned}$$

The T dependence is still the same, but the disorder and B dependence would be different. For $B \to 0$, the frequency integral has to be done with the ω dependent diffuson.

22.6. Summary and Conclusion

In ferromagnetic metals phase relaxation, in the sense of weak localization, may take place by excitation of spin waves. In the above we have calculated the phase relaxation rate $1/\tau_\varphi$ in leading order in the interaction of electrons with spin wave excitations. For a clean system the inelastic scattering rate (imaginary part of the self energy) had been determined before, in the case when the spin wave spectrum is gapped. Here we extend this result to finite disorder and a finite magnetic field and to vanishing spin wave gap, including vertex corrections. We employed a diagrammatic approach for calculating the effects of interaction on the Cooperon propagator. As a result we find that the phase relaxation rate is a linear function of temperature in both two and three dimensions, provided the spin wave spectrum is gapped. In the absence of a spin wave gap the result depends on the spin wave damping in the limit of zero wave vector. The dependence on disorder is given by a factor $D/(D^2+A^2)$, where D is the diffusion constant and A is the spin wave stiffness. In the limit of high magnetic field, Zeeman splitting $B >> 1/\tau$ where τ is the elastic relaxation rate, we find $1/\tau_\varphi \sim 1/B^2$.

Acknowledgements

We dedicate this paper to Prof. Y. Imry, on the occasion of his 70th birthday. This work has been stimulated by experimental results obtained in the group of Art Hebard at the University of Florida. We acknowledge support by the DFG-Center for Functional Nanostructures at the Karlsruhe Institute of Technology (KIT).

Appendix A.1: Evaluation of N

Consider the expression for $N^{(a+b+c)}$ given in Eq. (22.15). We can write

$$L^{(a+b+c)} = \left(\sum_{k'} G^+_{k'',\uparrow} G^-_{k''-q,\downarrow} \right)^2 F$$

where

$$F = \sum_{k'} (G^+_{k',\uparrow})^2 G^-_{k',\uparrow} G^-_{k'-q,\downarrow} + \Gamma_0 \sum_{k'} (G^+_{k',\uparrow})^2 G_{k',\uparrow} \sum_{k''} (G^+_{k'',\uparrow})^2 G^-_{k''-q,\downarrow}$$
$$+ \Gamma_0 \left(\sum_{k'} G^+_{k',\uparrow} G^-_{k',\uparrow} G^-_{k'-q,\downarrow} \right)^2 .$$

Using the small (ω_l, q) expansion for $G_{k-q}(\epsilon_n - \omega_l)$ similar to Eq. (22.12), and the results

$$\sum_{k'} (G^+_{k',\uparrow})^2 G^-_{k',\uparrow} (G^-_{k',\downarrow})^n = -i2\pi N_0 \tau (i\tilde{\tau}^*)^n (\tau + n\tilde{\tau}^*), \quad n = 1, 2, 3$$

for (a) and similar results for (b) and (c) we obtain, for small q and ω_l,

$$F = 2\pi N_0 \tilde{\tau}^* \tau \left[(\tau - \tilde{\tau}^*) + (2\tilde{\tau}^* - \tau)(|\omega_l|\tilde{\tau}^* + \tilde{D}^* q^2 \tilde{\tau}^*) \right].$$

Then

$$L^{(a+b+c)} = (2\pi N_0 \tilde{\tau}^*)^2 \left[1 - |\omega_l|\tilde{\tau}^* - \tilde{D}^* q^2 \tilde{\tau}^* \right]^2 \times F.$$

Note that for $B \to 0$, the leading term in F, $\propto (\tau - \tilde{\tau}^*) \to 0$, and the rest cancel one of the diffuson poles in $N^{(a+b+c)}$. This leads to the result in FA.[5] In the present case,$(\tau - \tilde{\tau}^*) = iB\tau\tilde{\tau}^*$ and the first term dominates if $B\tau \gg 1$. Let us rewrite F as

$$F = 2\pi N_0 \tilde{\tau}^* \tau \left[(\tau - \tilde{\tau}^*) + (\tilde{\tau}^*)^2(|\omega_l| + \tilde{D}^* q^2) - (\tau - \tilde{\tau}^*)\tilde{\tau}^*(|\omega_l| + \tilde{D}^* q^2) \right].$$

The first two terms equal to $1/2\pi N_0\Gamma^{ph}_{\uparrow\downarrow}$ and cancel one of the diffuson propagators in N. In other words,

$$L^{(a+b+c)}\left(\Gamma^{ph}_{\uparrow\downarrow}\right)^2 = (2\pi N_0\tilde{\tau}^*)^2\tau\tilde{\tau}^*\Gamma^{ph}_{\uparrow\downarrow}\left[1 - 2\pi N_0\tilde{\tau}^*(\tau-\tilde{\tau}^*)(|\omega_l| + \tilde{D}^*q^2)\Gamma^{+-,ph}_{\uparrow\downarrow}\right].$$

A similar evaluation of diagram (d) gives

$$L^{(d)} = -(2\pi N_0\tau\tilde{\tau})^2\left[1 - |\omega_l|\tilde{\tau} - \tilde{D}q^2\tilde{\tau}\right]^2.$$

If we neglect the last term in F, then

$$\begin{aligned} N^{(a+b+c+d)} &= T\sum_{\omega_l}\sum_q (2\pi N_0\tau)^2 v(q,\omega_l) \\ &\times\left[\frac{(\tilde{\tau}^*)^3}{\tau}\Gamma^{ph}_{\uparrow\downarrow}(q,\omega_l)\Theta(\omega_l - \epsilon_{n+}) - (\tilde{\tau})^2\Gamma^{pp}_{\downarrow\uparrow}(q,\omega_l+\omega_\lambda)\Theta(\epsilon_{n+}+\omega_l)\right]. \end{aligned}$$

Note that while $\Gamma^{ph}_{\uparrow\downarrow}(q,\omega_l)$ is evaluated at the same frequency as $v(q,\omega_l)$, $\Gamma^{pp}_{\downarrow\uparrow}(q,\omega_l+\omega_\lambda)$ is evaluated at a shifted frequency where $\epsilon_{n+} = \epsilon + \omega_\lambda$ defines the shift.

To this, we add the diagrams $\bar{a}, \bar{b}, \bar{c}, \bar{d}$, where the interaction lines are at the lower Greens functions. Since switching $+ \to -$ makes $G \to G^*$, we find that

$$L^{(\bar{a}+\bar{b}+\bar{c}+\bar{d})} = [L^{(a+b+c+d)}]^*,$$

and

$$\begin{aligned} N^{(\bar{a}+\bar{b}+\bar{c}+\bar{d})} &= T\sum_{\omega_l}\sum_q (2\pi N_0\tau)^2 v(q,\omega_l) \\ &\times\left[\frac{(\tilde{\tau})^3}{\tau}\Gamma^{ph}_{\downarrow\uparrow}(q,\omega_l)\Theta(\omega_l + \epsilon_n) - (\tilde{\tau})^{*2}\Gamma^{pp}_{\uparrow\downarrow}(q,\omega_\lambda+\omega_l)\Theta(\omega_l - \epsilon_n)\right]. \end{aligned}$$

Thus finally, the total contribution to N from all diagrams can be written as sum $N = N^{(a+b+c+d)} + N^{(\bar{a}+\bar{b}+\bar{c}+\bar{d})}$, giving Eq. (22.16).

Appendix A.2: Evaluation of Γ_2

We now consider the sum over the frequency in Γ_2 given by Eq. (22.18). The summation can be done by analytic continuation to the complex plane,

$$\Gamma_2 = \int_C \Gamma_p(q,-i\omega,\omega_\lambda)v(q,-i\omega)$$

where the contour C includes all the bosonic poles ω_l on the imaginary axis such that $\omega_l > -\epsilon_n - \omega_\lambda$. In this case, $v(q,\omega_l)$ has a branch cut at $\omega_l = 0$

while $\Gamma_{\downarrow\uparrow}(q, \omega_l + \omega_\lambda)$ has a branch cut at $\omega_l = -\omega_\lambda$. Since the lower limit is above the branch cut at $\omega_l = -\omega_\lambda$, the contour can be deformed around the real axis The frequency integral becomes

$$\begin{aligned}\Gamma_2 &= \frac{1}{2\pi i} \int_{-\infty - i(\epsilon_n + \omega_\lambda)}^{\infty - i(\epsilon_n + \omega_\lambda)} \Gamma_p(q, -i\omega, \omega_\lambda) v(q, -i\omega) n(\omega) \\ &+ \frac{1}{2\pi i} \left[\int_{-\infty + i\eta}^{\infty + i\eta} - \int_{-\infty - i\eta}^{\infty - i\eta} \right] \Gamma_p(q, -i\omega, \omega_\lambda) v(q, -i\omega) n(\omega).\end{aligned}$$

Changing the variables $\omega \to \omega + i(\epsilon_n + \omega_\lambda)$ in the first integral and using the fact that $n(\omega + i(\epsilon_n + \omega_\lambda)) = -f(\omega)$, where $f(\omega)$ is the Fermi function, we get

$$\begin{aligned}\Gamma_2 = &-\frac{1}{2\pi i} \int_{-\infty}^{\infty} \Gamma_p(q, -i\omega - (\epsilon_n + \omega_\lambda), \omega_\lambda) v(q, -i\omega - (\epsilon_n + \omega_\lambda)) f(\omega) \\ &+ \frac{1}{2\pi i} \int_{-\infty}^{\infty} \Gamma_p(q, -i\omega, \omega_\lambda) \tilde{v}(q, -i\omega) n(\omega)\end{aligned}$$

where we have defined $\tilde{v}(q, -i\omega)$ as the discontinuity in $v(q, -i\omega)$ across the real axis given by

$$\begin{aligned}\tilde{v}(q, -i\omega) &= -2nJ^2\omega_q \left[\frac{1}{(-i\omega + \gamma)^2 + \omega_q^2} - \frac{1}{(-i\omega - \gamma)^2 + \omega_q^2} \right] \\ &= -8nJ^2 \frac{i\omega\gamma\omega_q}{(-\omega^2 + \omega_q^2 + \gamma^2)^2 + 4\gamma^2\omega^2}.\end{aligned}$$

Clearly, since the Fermi function is non-singular at $\omega = 0$, the integral containing the Bose function will contribute dominantly. This, in the limit $\omega_\lambda = 0$, leads to Eq. (22.20).

References

1. P. Mitra, R. Misra, A.F. Hebard, K.A. Muttalib and P. Wölfle, Phys. Rev. Lett. 99, 046804 (2007).
2. R. Misra, A.F. Hebard, K.A. Muttalib and P. Wölfle, Phys. Rev. B 79, 140408(R) 2009.
3. G. Tatara, H. Kohno, E. Bonet and B. Barbara, Phys. Rev. B 69, 054420 (2004).
4. M. Plihal, D.L. Mills and J. Kirschner, Phys. Rev. Lett. **82**, 2579 (1999).
5. H. Fukuyama and E. Abrahams, Phys. Rev. B 27, 5976 (1983).
6. B.I. Altshuler and A.G. Aronov in *Electron-electron interactions in disordered systems*, eds. A.L. Efros and M. Pollak (Elsevier, Amsterdam 1985).

7. A linear T contribution without the logarithm was later obtained within the diagrammatic approach by H. Fukuyama, J. Phys. Soc. Japan, 53, 3299 (1984).
8. I.S. Gradshtein and I. M. Ryzhik, Table of Integrals, series and Products, Elsevier (1966).

Subject Index